Lecture Notes in Chemistry

Edited by G. Berthier M. J. S. Dewar H. Fischer
K. Fukui G. G. Hall H. Hartmann H. H. Jaffé J. Jortner
W. Kutzelnigg K. Ruedenberg E. Scrocco

34

Nicolaos Demetrios Epiotis

Unified Valence Bond Theory of Electronic Structure
Applications

Springer-Verlag
Berlin Heidelberg New York Tokyo 1983

Author

Nicolaos Demetrios Epiotis
Department of Chemistry
University of Washington
Seattle, WA 98195, USA

ISBN-13: 978-3-540-12000-1   e-ISBN-13: 978-3-642-93239-7
DOI: 10.1007/978-3-642-93239-7

Library of Congress Cataloging in Publication Data
Epiotis, N. D., 1944- Unified valence bond theory of electronic structure applications.
(Lecture notes in chemistry; 34)
1. Molecular orbitals. 2. Valence (Theoretical chemistry) I. Title. II. Series.
QD462.E653 1983 541.2'24 83-2482 ISBN-13:978-3-540-12000-1 (U.S.)

2152/3140-543210

PROLOGUE

A senior undergraduate and a doctoral candidate, a high school teacher and
a university professor, an industrial researcher and an academician, in short,
every chemist can, more or less, provide a believable explanation of an isolated
chemical phenomenon. With the same facility, chemists can use different models
to rationalize different facts in different areas of chemistry. Indeed, it is
often said that a "good man" is the one who can judiciously choose the right model
for the right problem. In recent years, the increased sophistication of laboratory
instrumentation and the advent of the computer in the everyday life of the
scientist have generated multitudes of new facts and have caused a great prolifer-
ation of conceptual models in chemistry. The present status can be summarized as
follows: <u>There are many theories for many problems but not one theory for all
problems</u>, where by "theory" we mean a conceptual framework for the comprehension,
rationalization, and prediction of chemical and physical phenomena whether
occuring in the laboratory or reproduced (or simulated) by explicit computations.

Some years ago, I sought a solution to the problem stated above. That is to
say, I searched for a quantum mechanical formalism which can provide the basis
for a <u>general</u> theory of chemistry. The result was the development of the MOVB
theory which has been described in a previous monograph along with the reasons why
I believe that this is the best way to approach chemistry in a self-consistent
manner. The original work was entitled "Unified Valence Bond Theory of Electronic
Structure". In it, I stated that "our present 'understanding' of chemistry has
been often illusory". This work <u>is an "across chemistry" application of the
qualitative VB and MOVB theory presented in the original monograph and it constitutes a defense of the utilization of the term "Unified" and a justification of</u>

<u>the assertion just mentioned</u>. For, I now show that one and the same set of
concepts can be applied to organic and inorganic problems, ground and excited
states, normal and "hypervalent" molecules, static stereochemistry and reaction
stereoselection, the design of "strange" molecules, etc., in a way that has never
been accomplished before by any single (<u>non-numerical</u>) theoretical model. In
doing so, I reveal common denominators of apparently unrelated problems, I show
that certain problems long thought to be related are not so, I compare the
predictions of VB theory with those of crude monodeterminantal MO models, and, in
general, I demonstrate that the way in which electrons "behave" is very different
from what I myself used to think a decade ago. Since chemists think and argue
by analogy, MOVB theory, by virtue of annihilating conceptual interdisciplinary
barriers, opens new vistas, it suggests new experiments and new computations, and
it defines new problems for investigation. Thus, in a certain sense, the original
monograph can be re-titled "Qualitative VB Theory - How to Derive It" and this one
"Qualitative VB Theory - How to Apply It". Of course, from the standpoint of the
practicing chemist, this second volume will be much more entertaining than the
first one as it contains chemical applications in its entirety. Indeed, the
reader who thinks that all important concepts of chemistry have been discovered
and their implications have been understood will be surprised to find in this
work a multitude of new ideas: The principle of configuration aromaticity, the
notion of weak and strong overlap binding, the concept of electronic anticooperativity and sigma-pi hybridization, the idea of coulomb polarization, etc. In
short, this book is good reading for anyone who believes that "small molecule"
chemistry is a saturated field wherein everything has already been discovered.

Qualitative FO-PMO theory, as commonly practiced today, is an approximate form of HMO theory. The latter is a crude approximation of SCF-MO theory which is only an approximation of SCF-MO-CI theory. The conceptual superstructure of chemistry is presently based largely on FO-PMO and HMO theory. Even high level SCF-MO-CI computations are nowadays analyzed by using HMO concepts! This work asks the reader to make the intellectual commitment to retrain his thinking at the level of SCF-MO-CI theory, i.e., at the level of VB and MOVB theory. This is not an easy task! The author hopes, however, that even a casual reading of this work, based on the original monograph published a year ago, will provide sufficient incentive for every chemist to take this bold step. For after all, at some point in the future, this will become inevitable, assuming that human intellectual curiosity will continue to exist even in an age when computers may start doing the thinking for us humans.

A word about the organization of the material. The book is divided into two parts for the sole purpose of underscoring the two fundamental aspects of qualitative MOVB theory: its conceptual power and its formal correctness. The various chapters are arranged in such away so that the previous concept leads to the development of the next one. In addition, apparently unrelated topics are discussed in consecutive chapters so that the common denominator is exposed. For, after all, it is not only the individual applications but the over all cohesion of this work which qualifies it as a new theory of chemistry as a whole, where the term "new" means that, aside from the fact that this treatise is based on the Schrödinger equation, the MO and VB recipes of constructing the electronic wavefunction, and the variation and perturbation methods of solving the Schrödinger equation, the theoretical formalism, the ensuing concepts, and the resulting applications contained in this work have, to a very large extent, no precedent.

Because of this, I have an unorthodox suggestion for the "busy" and/or "impatient" reader who wants to find out immediately whether there are good reasons for learning a new chemical language as the author suggests: First read this opus in reverse, starting with the Epilogue of the second volume and ending with the Prologue of the first volume. Thus, having seen why I have claimed that our understanding of chemistry has often been illusory, read it in the proper order always trying to compare what we espouse to the standard MO theory practices of today.

This work has been made possible by the contributions of many people to whom I am grateful. First, I would like to mention the many researchers whose work provided the necessary checks of the approach espoused here. For it is true that I frequently found myself wavering in the application of the same MOVB concepts I developed only to find reassurance and guidance by the experimental and theoretical literature. Mr. Hugh Eaton, Ms. Angela Diamond, Ms. Barbara Lau, and Dr. James Larson carried out several test computations, some of which are included in this work. The Department of Chemistry provided the needed logistical support and Mrs. Martha Kady was the miraculous "secretary-artist-editor" without whom this work could never be produced in the relatively short period of one year. Ms. Linda Daniel was the precious companion who supported me during this under- taking. Finally, Dr. F. L. Boschke and the Editors of this series are the ones responsible for making this work available to the international community of chemists. In closing, I add that the theory described in this and the companion monograph was conceived, tested, refined, and applied without any grant support from private or federal U.S. agencies.

Nicolaos D. Epiotis

TABLE OF CONTENTS

TABLE OF CONTENTS (CONTINUED)

PART ONE

THE CONCEPTUAL POWER

OF

MOLECULAR ORBITAL - VALENCE BOND (MOVB)

THEORY

The bond diagrammatic representation of molecules is the foundation of MOVB
theory. To a certain extent, this kind of representation is analogous to the one
on which "resonance theory" is based and this fact can be projected by a comparison
of the various ways in which MOVB theory depicts a species made up of three core
and two ligand MO's which define two subsystems containing a total of six electrons
and the ways in which "resonance theory" (i.e., qualitative VB theory) depicts a
six-electron-six-AO species such as the pi system of $CH_2=CH-CH=CH-CH=O$. The
different pictorial representations are shown in Scheme 1 so that the analogies
are made evident. First of all, the total MOVB diagrammatic representation of the
6/5 species is obtained by a linear combination of three complete bond diagrams, as
in A1, which describe the optimal linear combination of all MOVB Configuration
Wavefunctions (CW's). By the same token, a total VB diagrammatic representation of
the 6/6 species can be obtained by writing a "dot structure", as in B1, and taking
this to mean the optimal linear combination of all VB CW's. Next, we can approximate the MOVB wavefunction of the 6/5 species by one complete (or detailed) bond diagram (A2). No simple VB representation analogy can be given in this case. Alternatively, we can approximate the MOVB wavefunction by a linear combination of compact
bond diagrams, as in A3, in the way described before. These compact bond diagrams
have common CW's and they also exclude a set of extrinsic CW's which must necessarily
be added in quantitative as well as in some qualitative applications of MOVB theory.
The VB analogue in this case is the hybrid representation of the 6/6 species by a
set of "important" VB CW's, as in B2. Finally, we can obtain a compact representation
of the 6/5 species by simply eliminating all compact bond diagrams except the one
"containing" the dominant CW or CW's, as in A4. Correspondingly, we can obtain a
compact representation of the 6/6 species by eliminating all but the CW which makes

the major contribution to the VB resonance hybrid, as in B3. Henceforth, we shall use the letter Θ to symbolize a total MOVB wavefunction, Ψ_i to symbolize a complete bond diagram, Ξ_i to denote a compact bond diagram, and Ω_i to denote a sybsystem and the associated wavefunction. In this work, we shall make frequent use of the approximate MOVB representation, as this is adequate for the treatment of most problems of interest, and we shall always differentiate between complete and compact MOVB bond diagrams, explicitly or implicitly. Furthermore, we shall use the convéntion of assigning charges to the core and the ligand by reference to the perfect pairing CW of the total system. We shall revert to the total representation only when the problem demands so.

Some notable aspects of the MOVB representations are the following:

a. The MOVB differ from the conventional VB representations to the extent that MOVB diagrams represent <u>a collection of MOVB CW's</u> while the familiar VB resonance structures represent individual VB CW's.

b. A species having only one subsystem can be fully represented by a single detailed bond diagram. A species having more than one subsystem can be fully represented only by a linear combination of detailed bond diagrams. The reason is that a single detailed bond diagram fully accounts for interfragmental bond "correlation" while a detailed set of complete bond diagrams is required for the reproduction of intrafragmental correlation. At the limit of tight core-ligand binding, higher energy detailed bond diagrams play a small role and can be neglected.

c. The Ψ_i's as well as the Ξ_i's are resonance bond diagrams.

d. Neither total nor approximate MOVB wavefunctions can be reproduced by monodeterminantal SCF-MO theory. The total MOVB wavefunction can be obtained by complete SCF-MO-CI theory and the approximate MOVB wavefunction by SCF-MO-CI

4

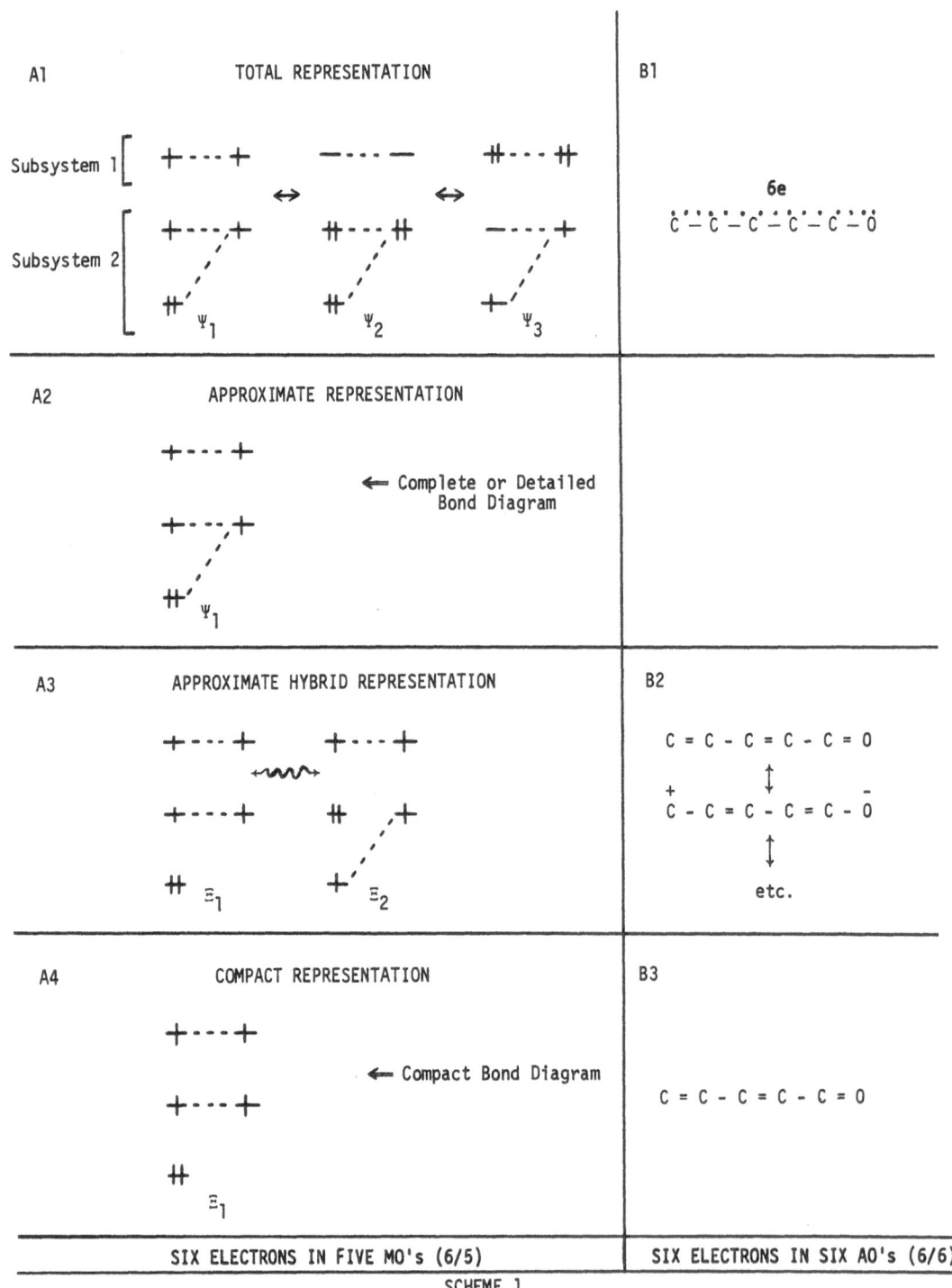

A1   TOTAL REPRESENTATION   B1

Subsystem 1

Subsystem 2

Ψ_1 Ψ_2 Ψ_3

6e

$$C - C - C - C - C - O$$

A2   APPROXIMATE REPRESENTATION

← Complete or Detailed Bond Diagram

Ψ_1

A3   APPROXIMATE HYBRID REPRESENTATION   B2

Ξ_1 Ξ_2

$$C = C - C = C - C = O$$

$$\updownarrow$$

$$\overset{+}{C} - C = C - C = C - \overset{-}{O}$$

$$\updownarrow$$

etc.

A4   COMPACT REPRESENTATION   B3

← Compact Bond Diagram

$$C = C - C = C - C = O$$

Ξ_1

SIX ELECTRONS IN FIVE MO's (6/5) | SIX ELECTRONS IN SIX AO's (6/6)

SCHEME 1

theory with a truncated configuration basis which does not permit intrafragmental "correlation". These aspects will be better projected by actual examples as this series of articles unfolds.

e. A version of approximate MOVB theory which is ideally suitable for analyzing (ground or excited) chemical stereoselection is the Independent Bond Model, described in the original monograph, according to which a given species can be viewed as the product of MOVB subsystem wavefunctions with its total energy being a simple sum of subsystem energies.

With this background, we can now outline a general MOVB theory of chemical bonding which can be routinely implemented by following these steps:

a. A molecule or complex is divided into a core fragment (C) and a fragment which contains all ligands (L).

b. A bond diagram is constructed for every assumed geometry of the molecule or complex by following these steps:

1) The core and ligand symmetry adapted orbitals are generated either from first principles or by explicit computation.

2) The electrons are arranged in the core and ligand orbitals in a way which which generates the reference, "perfect pairing" (R) CW subject to the symmetry constraints imposed by the geometry in question. This is the open shell CW which places core and ligand electron pairs in the lowest energy orbitals of core and ligand subject to the requirement that it generates the maximum number of core-ligand bonds through spin pairing. For example, the R CW of a six-electron-five-orbital system where the lowest two core (ω_1 and ω_2) and the lowest ligand (σ_1) orbitals are of one symmetry type and the highest core (ω_3) and ligand (σ_2) orbitals are of a different symmetry type is written as follows:

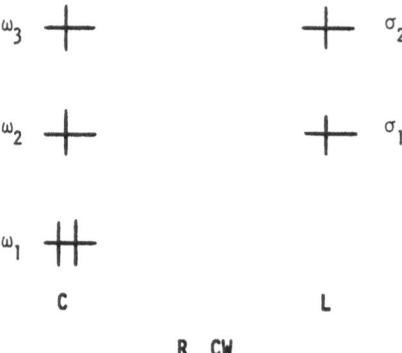

C   L

R   CW

3) The detailed bond diagram for the geometry in question is constructed by adding dashed lines to the drawing of the R CW in order to denote all possible CW's which can be generated by the implied electron shifts under the imposed symmetry constraints. For example, the detailed bond diagram corresponding to the previous case becomes:

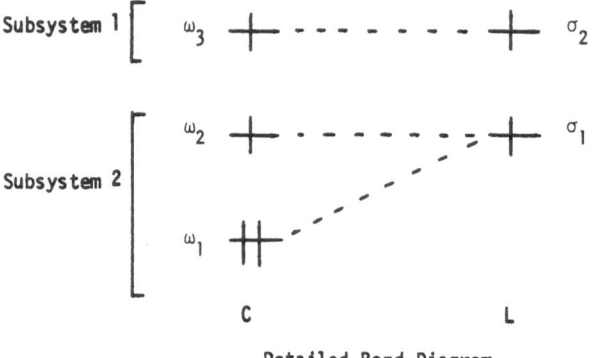

C   L

Detailed Bond Diagram

Each detailed bond diagram is a pictorial approximate representation of the optimal MOVB wavefunction of the system in question. In addition, it shows explicitly the number and types of independent "many electron-many center" bonds which join the core and the ligand fragments. For the reader's convenience, the symmetry of each orbital is specified either by the formal point group label or by the letters S (symmetric) and A (antisymmetric) which define the behavior of the orbital upon performance of an obvious symmetry operation (e.g., rotation about an axis, reflection through a plane, etc.).

In dealing with detailed bond diagrams, it must be kept in mind that, because of the way in which the R CW is defined and the way in which the detailed bond diagram is constructed, neither the "parent" R CW is <u>necessarily</u> the lowest energy CW nor the principal compact bond diagram ($\equiv_1$ in Scheme 1) is <u>necessarily</u> the one directly reflected by the detailed bond diagram <u>as written</u>, although in most problems of interest this is indeed the case. However, neither of these semantic difficulties is an obstacle to the qualitative application of MOVB theory. **Consistent with the above stated** philosophy, we assign <u>formal charges</u> to C and L <u>by reference to the perfect pairing</u> <u>CW projected by the bond diagram as written</u>. Thus, for example, we have written C and L underneath the bond diagram shown above because we assumed that neutral C has four electrons and neutral L two electrons. Had neutral C had three and neutral L also three electrons, we would have written C^- and L^+ underneath the bond diagram, keeping always in mind that these are formal charges with the real ones being determined by all CW's which make up the MOVB wavefunction. Obviously, the formal charge distribution will tend to be a good approximation of the real charge distribution as the perfect pairing CW tends to make a dominant contribution to the total MOVB wave function. Finally, we assume that the lowest and highest energy orbitals of a diatomic core play no significant role insofar as electron delocalization is concerned. This assumption is valid in most problems tackled in this monograph.

c. Each independent "many electron-many center" bond is alternatively called a subsystem. Different subsystems do not interact in a one-electron sense and, in dealing with the stereochemistry of ground state closed shell systems, their bielectronic interaction can often be neglected. In such an event, the energy change of a composite system is equal to the sum of the energy changes of the individual subsystems.

d. Core and ligand orbital symmetry dictates three different types of subsystem bonding: D bonding which permits electrons to descend to low lying orbitals, U bonding which confines some of the electrons to high lying orbitals, and H bonding which represents a hybrid of D and U bonding sometimes involving conservation of spatial orbital overlap (H' bonding) and other times reduction of spatial orbital overlap (H" bonding). Finally, N and N' bonding represent two different types of simple two electron-two orbital bonding with the prime denoting impairment of spatial orbital overlap.

e. The result of a structural change on each subsystem is spelled out in the form of an equation. For example, if the structural modification changes an $(\omega, \sigma/2)$ subsystem bonding from N- to N'-type, we write: $N(\omega, \sigma/2) \rightarrow N'(\omega, \sigma/2)$.

f. The critical subsystem conversion(s) is(are) singled out. This becomes necessary because the energetics of some subsystems may or may not change as a result of a structural modification.

g. A conclusion regarding the effect of a modification, i.e., stabilizing or destabilizing, is reached from appraisal of the critical sybsystem conversions according to the following delocalization rules:

1) Hybridization Rule: H bonding becomes increasingly favorable relative to D bonding as primary CT occurs in a direction which prevents overlap repulsion and fosters secondary delocalization.

2) Deexcitation Rule: H bonding becomes increasingly favorable relative to U bonding as primary CT occurs in a direction which turns off secondary delocalization.

These rules can be restated in an alternative language:

1) H bonding becomes increasingly more favorable relative to D bonding as V-type CW's attain low energy.

2) H bonding becomes increasingly favorable relative to U bonding as I-type CW's attain low energy.

It ought to be emphasized that these general rules are applicable to actual chemical problems only when the structural variation under scrutiny affects the promotional energies of the interacting CW's but leaves the overlap interaction terms relatively unchanged. When the latter condition is not adequately met, exceptions are expected which, in toto, may define a new chemical concept.

In tailoring VB theory to an operationally useful tool for the practicing chemist, we are less interested in detail and more interested in immediate application which can inspire new experiments or computations. To this extent, one may alternatively use the compact MOVB theory of chemical bonding which can be implemented by following these steps:

a. In comparing two isomers, one constructs the compact MOVB diagram (Ξ_1 in Scheme 1) for each one of them and determines the type of bonding imposed by core and ligand orbital symmetry in each case recalling that there are three bonding "flavors": D bonding which permits electrons to descend to low lying orbitals, U bonding which confines some of the electrons to high lying orbitals, and H bonding which represents a hybrid of D and U bonding accompanied by impairment of core-ligand spatial overlap, at least in most cases of interest. Loss of spatial overlap is indicated by affixing a dagger superscript to the appropriate letter, most often H. The letter (U, $H^\dagger$, and D) assignment is always made in a _relative_ _sense_ and the selection rules are: D is always better than $H^\dagger$ and U bonding but $H^\dagger$ may be superior or inferior to U bonding depending on whether deexcitation is more important than loss of spatial overlap or _vice_ _versa_. This latter principle constitutes the quantum mechanical rationalization of stereochemical diversity in nature.

b. If the compact bond diagrams of two different isomers belong to the same bond type (e.g., both involve D-bonding, to a first approximation), one proceeds to examine the consequences of the principal direction of interfragmental Charge Transfer (CT) in each of the core-ligand bonds of the two isomers. The way in which primary and secondary CT differentially affect two isomers has been extensively discussed in the original work.[15] This last procedure effectively amounts to a one-to-one energetic comparison of the compact bond diagrams (Ξ_i's) the linear combinations of which constitute the approximate hybrid bond diagrammatic representations of the two species.

c. When there are core and/or ligand orbital degeneracies, the two compact bond diagrams may differ in that one of them describes stronger core-ligand spatial overlap while core and ligand excitation remains constant. In such an event, we say that the conversion of one isomer to the other involves rebonding symbolized in any one of the following three ways: $U \rightarrow U^{\ddagger}$, $D \rightarrow D^{\ddagger}$, or, $N \rightarrow N'$. The selection rule is now that the isomer with the impaired core-ligand spatial overlap ($U^{\ddagger}$, $D^{\ddagger}$, or, N' bonding) is the more unstable one.

d. In (a)-(c), it has been assumed that the two bond diagrams described the same number of core-ligand bonds. If this is not the case, the isomer having the maximum number of core-ligand bonds is the more stable one, at least in the vast majority of cases.

In Part One of this work, we will use the MOVB bond diagrammatic method in order to tackle problems, most of which lie "within" monodeterminantal MO theory, in order to demonstrate the conceptual power of MOVB theory.

Chapter 1. The Induced Deexcitation Model.

In complex scientific disciplines, such as chemistry, a command of the literature and an ability to recognize the common denominator of many apparently unrelated experimental observations often leads to the formulation of concepts and rules of broad applicability. One demonstration of how such a marriage of knowledge and intuition can bear offspring was given by Walsh, who, many years ago, recognized that many experimental facts, which were known at that time, could be explained in a self-consistent manner by assuming that the hybridization of a central atom (or core) depends on the electronic nature of the ligands attached to it and proposed the following rule: "If a Group X attached to Carbon is replaced by a more Electro-negative Group Y, then the Carbon Valency towards Y has more p Character than it had towards X".[1] We shall refer to this as <u>Walsh's rehybridization rule</u>. In recent times, this rule, in one form or another, has been applied to a variety of interesting chemical problems by many, most notably by Bent.[2]

An interesting application of Walsh's rehybridization rule, which is probably familiar to most chemists, is shown below:[3,4]

$$H_3C \overset{\displaystyle O}{\diagup \diagdown} CH_3 \qquad\qquad F \overset{\displaystyle O}{\diagup \diagdown} F$$

$$112° \qquad\qquad\qquad 103°$$

The angle decrease is ascribed to the increase of ligand electronegativity, which, allegedly, causes an increased utilization of the oxygen 2p AO's for bonding. The data shown below can also be rationalized in a similar fashion.[4,5]

$$Cl \overset{\displaystyle O}{\diagup \diagdown} Cl \qquad\qquad F \overset{\displaystyle O}{\diagup \diagdown} F$$

$$111° \qquad\qquad\qquad 103°$$

We now reexamine these concepts using MOVB theory.[6]

In a previous work , we applied MOVB theory to the problem of the stereochemistry of H_2O and its derivatives. The salient features of our approach can be illustrated by reference to the conversion of Linear (L) to Bent (B) H_2O. The appropriate bond diagrams are shown in Figure 1 and the subsystem convention is:

s, p_y, $\sigma/4$	p_x, $\sigma*/2$	$p_z/2$

The energetic consequences of the L $\rightarrow$ B transformation are as follows:

$$U\ (s,p_y,\sigma/4) \rightarrow H'\ (s,p_y,\sigma/4)$$

$$N\ (p_x,\sigma*/2) \rightarrow N'\ (p_x,\sigma*/2)$$

$$(p_z/2) \rightarrow (p_z/2)$$

Clearly, the critical subsystem conversions are the first two. Specifically, the N $\rightarrow$ N' conversion acts in a manner which favors the maximum overlap L form. On the other hand, the U $\rightarrow$ H' conversion acts in an opposite sense favoring the B form which allows an electron pair to occupy the low lying 2s AO of the core atom. Since the deexcitation accompanying the U $\rightarrow$ H' conversion occurs over a large energy gap, $\Delta\varepsilon$, equal, to a first approximation, to the energy difference between the 2s and 2p AO's of oxygen, the U $\rightarrow$ H' dominates the N $\rightarrow$ N' conversion and water ends up being bent.

We can now design systematically derivatives of H_2O where the bending tendency is either diminished or enhanced. This can be done by tuning the U $\rightarrow$ H' conversion. The four electron-three orbital wavefunctions for the U and H' subsystems of H_2O are shown in Figure 2. We recapitulate the conditions for the selective modification of the energetic advantage of H' over U:

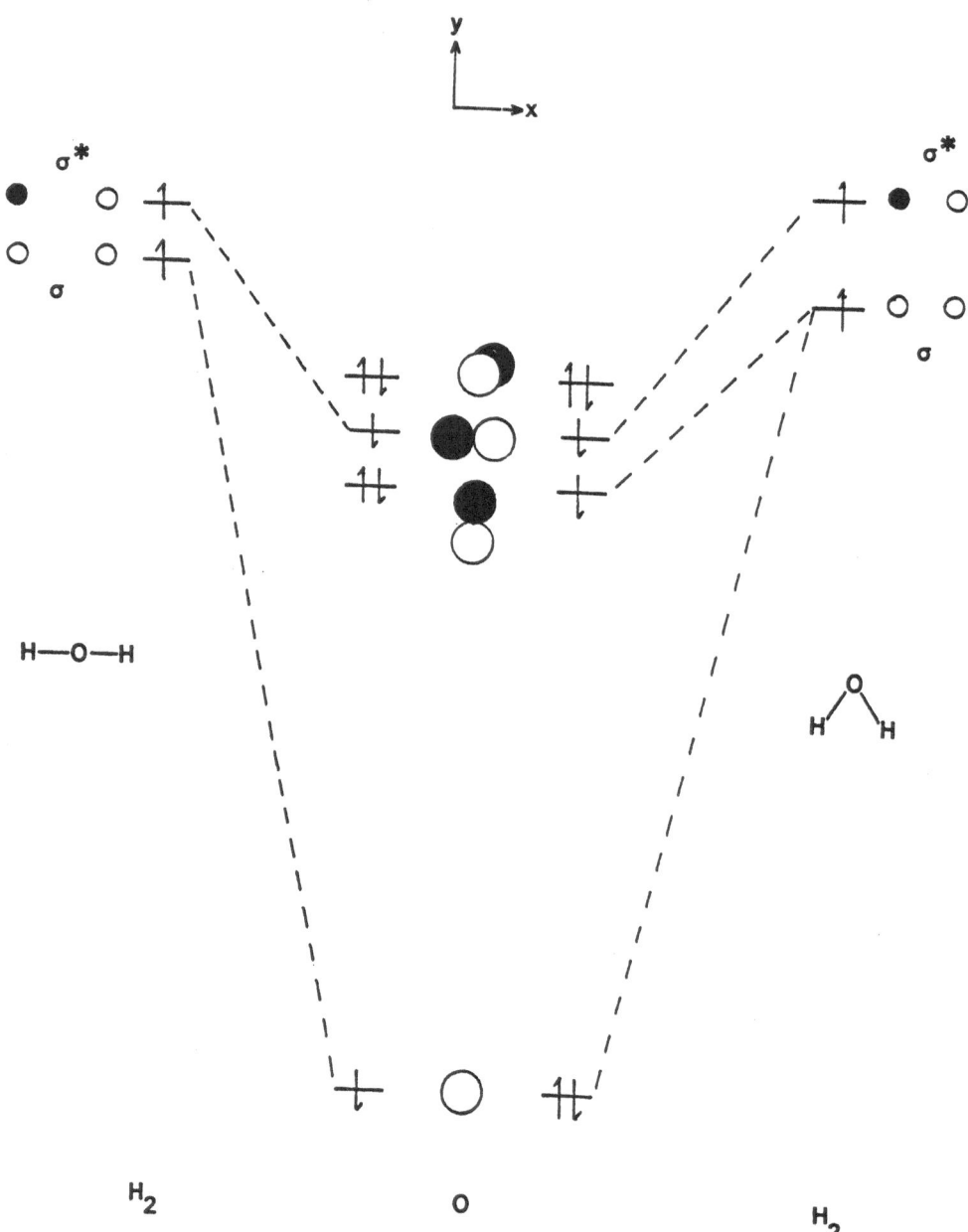

Figure 1. Detailed MOVB bond diagrams for linear and bent H_2O.

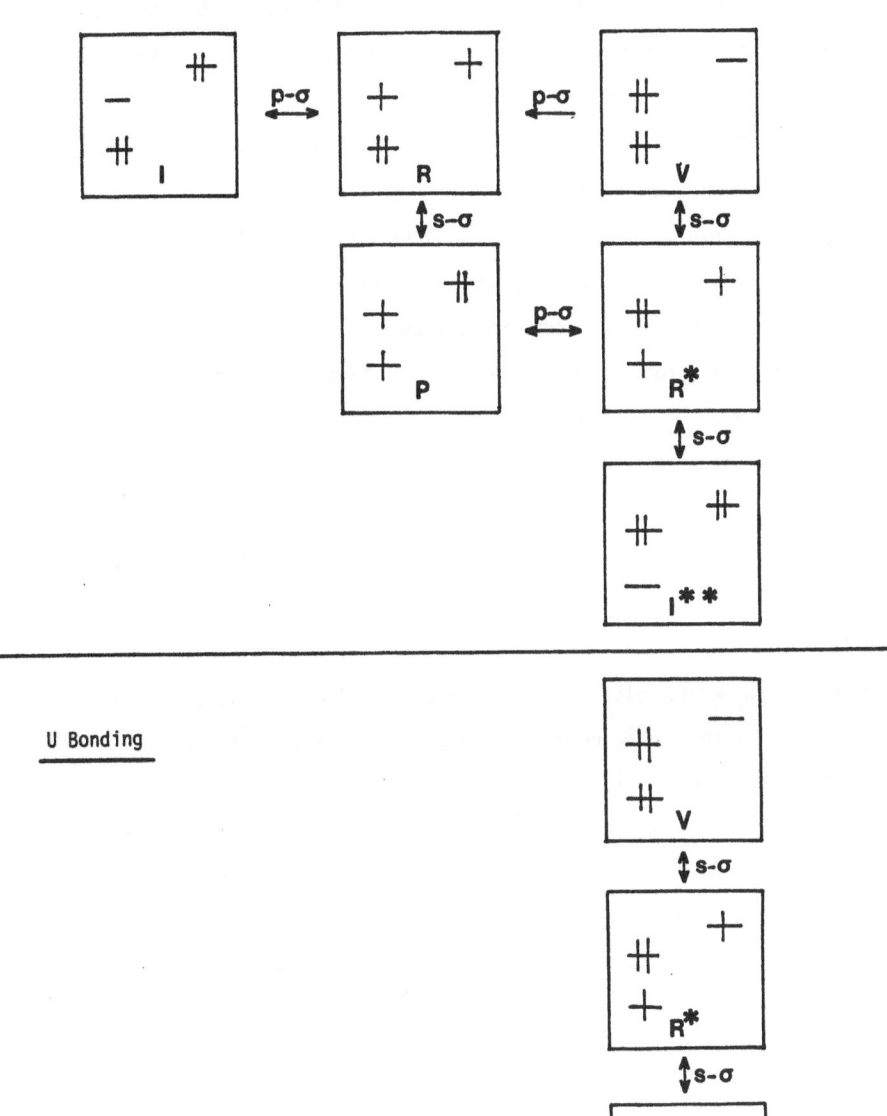

Figure 2. (2s, 2p$_y$, σ/4) wavefunctions for linear (U) and bent (H') H$_2$O. Double arrows indicate charge transfer CW interactions. The orbital pair denoted atop the double arrow defines the origin and terminus of one electron transfer.

a. An increase of the ligand electronegativity will render the CW I dominant and, thus, increase the energetic advantage of H'. This occurs because I is a contributor to the H' but not to the U wavefunction. As a result, the $\widehat{XOX}$ angle will decrease.

b. An increase of the ligand electropositivity will have the opposite effect because it will render V dominant and both H' and U wavefunctions contain V. As a result, the $\widehat{XOX}$ angle will increase.

An alternative way of predicting the effect of ligand replacement on the structure of H_2O is to start with the wavefunction of H_2O in its equilibrium geometry, focus attention on the H' subsystem, and determine the manner in which rehybridization will occur as a result of the substitution. Replacement of hydrogens by electronegative ligands will render the I CW relatively more important and the original H' will tend to become an HD' subsystem, i.e., an H' subsystem with major D character. On the other hand, replacement of hydrogens by electropositive ligands will render the V CW relatively more important and the original H' will tend to become an HU' subsystem. As a result, electronegative substituents will cause the central oxygen to bind to them using primarily the O2p AO's while electropositive substituents will enforce utilization of the O2p and O2s AO's for bonding.

The sense of the above analysis can be conveyed in a pictorial sense in the way indicated below. The H_2O molecule is pictured as a hybrid of two resonance bond diagrams, D and U, with the relative weights of them (λ_D and λ_U) varying according to the nature of the ligands in such a way that λ_D/λ_U tends to be greater than one when the ligands are electronegative and less than one when they are electropositive.

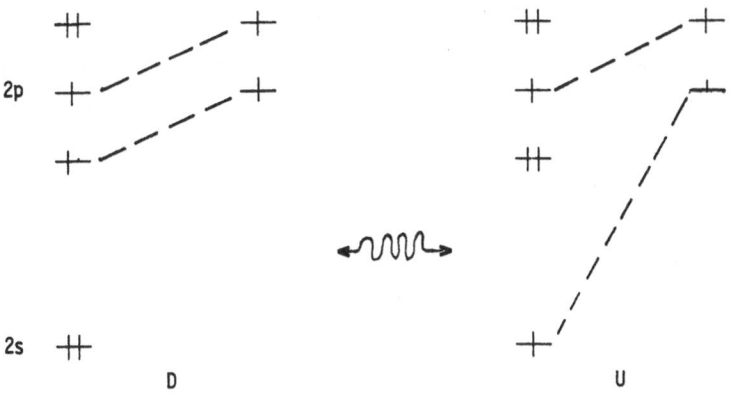

$$\Psi_{H_2O} = \lambda_D \, D + \lambda_U \, U$$

Clearly, this very analysis, based on the stated assumptions, leads to the same predictions as Walsh's rehybridization rule.

We can now formulate the following important question: Do the experimental facts cited before imply that the delocalization rules, derived on the assumption that a core or ligand replacement, in general, readjusts the relative promotional energies of the interacting CW's while leaving all overlap interaction terms relatively unaffected, represent a satisfactory solution of the problem of substituent effects, or, do the same experimental facts define accidental correlations which conceal the operation of yet another important factor which remains to be discovered? In other words, how are the delocalization rules obtained before on the basis of the stated assumptions become modified if we now recognize that a chemical modification, in the most general sense, affects

<u>all</u> energy matrix elements, $<\Phi_i|\hat{H}|\Phi_j>$, where Φ_i represents a Configuration Wave-function (CW)? This problem will be discussed after we examine how AO resonance integrals, which largely determine the overlap interaction of the core and ligand fragments, depend upon the nature of the atoms which sustain the AO overlap interaction.

I. The Relative Overbinding Abilities of Atoms

According to VB theory, the simple two electron-two orbital bond of H-H is formed in the way indicated in Figure 3. The functional dependence of each of the critical quantities ΔE and δE_n (n=1-4), graphically defined in Figure 3, is the following:

a. ΔE is a measure of the promotional energy difference between the H··H and H^+H^- CW's and it is equal to the difference between the ionization potential, I, of one hydrogen atom and the electron affinity, A, of the other.

b. δE_2 is approximately equal to the overlap term $2\beta s + K$ where β is the AO resonance integral, s the AO overlap integral, and K the bielectronic exchange integral:

$$\beta = <1s_H|\hat{H}|1s_H'> \tag{1}$$

$$s = <1s_H|1s_H'> \tag{2}$$

$$K = <1s_H1s_H'|1s_H1s_H'> \tag{3}$$

c. δE_4, much like δE_2, is proportional to the overlap term $2\beta s + K$.

d. δE_1 depends on the strength of the interaction of Φ and Φ^+. This, in turn, depends on the corresponding interaction matrix element, H^+, which is a function of β.

$$H^+ = <\Phi^+|\hat{H}|\Phi> \alpha <H^+H^-|\hat{H}|H··H> \alpha \beta \tag{4}$$

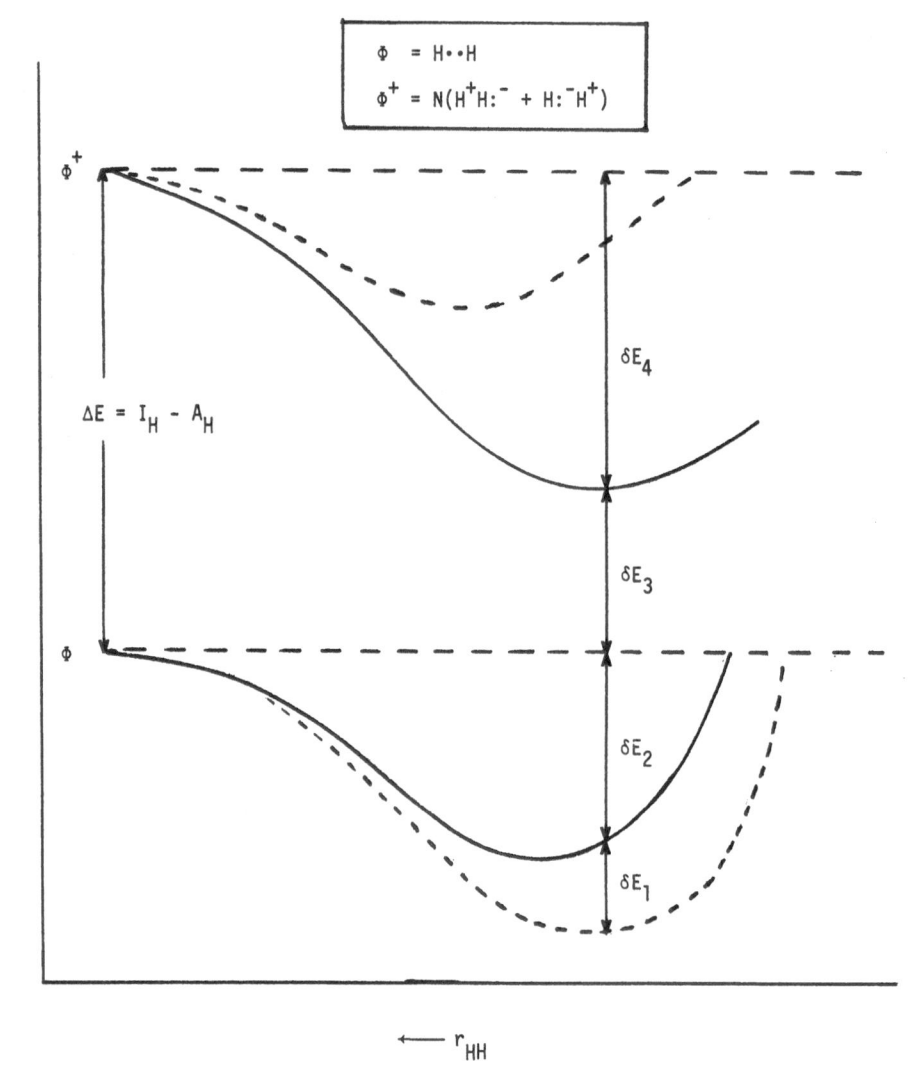

$$\Phi \;\; = H \cdot \cdot H$$
$$\Phi^+ = N(H^+H:^- + H:^-H^+)$$

Figure 3. The formation of the two-electron-two-orbital bond of H-H. The quantities ΔE and δE_n (n = 1-4) are defined in the text and in the Appendix. Solid lines indicate zero-order (diabatic) states and dashed lines the resulting final (adiabatic) states.

As a result, the bond dissociation energy of H_2 depends critically on two important quantities, namely, $I - A$ and β. These ideas can be generalized to any two electron-two orbital bond connecting identical or different atoms and they lead to the following statement: The bond dissociation energy of A-Z is a function of I_A-A_Z and β_{AZ}. The former is a measure of the tendency of the system to acquire low energy by electron transfer and the latter is an index of the tendency of the system to become stabilized through overlap interaction. In other words, we can say that there is a charge transfer and an overlap aspect which need to be considered when comparing the bonding abilities of a series of atoms.

In treating large polyatomic systems by MOVB theory, the role of the so-called Heitler-London (HL) CW's of VB theory (e.g., Φ in H_2) is most frequently assumed by the "perfect pairing", R, CW and the role of the "ionic" VB CW's by corresponding charge transfer CW's. The core-ligand bonds are now partly due to spin pairing in the R CW as well as the interaction of the R with higher lying CW's. Once again, there is a charge transfer and an overlap aspect to be considered, with I and A representing MO ionization potentials and electron affinities, respectively, for a given electronic configuration and with β_{tu} being replaced by the MO resonance integral, h_{mn}. The latter can be expanded to a <u>sum</u> of AO resonance integrals,[10] each one being differently weighted, so that, ultimately, the overlap interaction of core and ligands is dependent upon the <u>overlap binding ability</u> of the constituent atoms. Recognizing that charge transfer is, in a sense, a modulator of overlap interaction, we can now address the following question: How does the electronic nature of atoms determine the <u>magnitude</u> of core-ligand overlap interaction and, by extension, the stereochemistry of a molecule? In order to answer this question, we must first define a measure of the relative <u>overlap binding</u>, or, simply, the relative <u>overbinding</u> ability of atoms. As we shall see, the stereochemical changes which accompany the replacement of one atom with another one which is a stronger or weaker <u>overbinder</u> are fully predictable on the basis of MOVB theory and they define a target of investigation which is worth bringing to the attention of chemists and physicists.

Let us suppose that we want to compare the overbinding ability of two atoms X and Y with respect to A. That is to say, we are interested in the relative magnitudes of the components of the A-X and A-Y bond dissociation energies due to overlap interaction of the two atoms. A measure of the overbinding ability of X is β_{AX} and that of Y is β_{AY}. Hence, by computing these AO resonance integrals and comparing their magnitudes, we can determine which of X and Y is a stronger overbinder. In proposing such a computation, we recognize two problems:

a. The magnitudes of β_{AX} and β_{AY} are dependent on A. However, we can assume that, while <u>absolute</u> magnitudes will indeed change, the sign of $\beta_{AX}-\beta_{AY}$ will be independent of A, i.e., <u>the relative overbinding abilities of X and Y are reference invariant</u>. Test computations show that this is the case, at least in most problems of interest. Hence, we shall make the assumption that the relative overbinding abilities of X and Y can be inferred from the relative magnitudes of the β_{AX} and β_{AY}, where A is arbitrarily defined to be H.

b. The relative magnitudes of β_{HX} and β_{HY} are dependent upon the precise hybridization state of each of X and Y. In this case, we can assume that the hybridization of X and Y will be the one adopted by X and Y within the composite systems of interest and that it can be located within the bounds of the hybridization scale shown below:

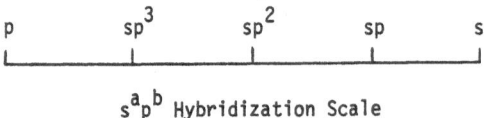

$s^a p^b$ Hybridization Scale

With the above in mind, we propose the following recipe:

a. Determine the hybridization of X and Y within the corresponding composite systems using the now familiar bond diagrammatic techniques of MOVB theory.

b. Calculate β_{HX} and β_{HY} using the appropriate hybrid AO's for X and Y. If we deal with "intermediate" hybridization, we can approximate the corresponding β by taking the average value of two β's computed for two "limiting" hybridizations. For example, if we determine that the bonding AO of X is intermediate between sp and s, we compute β_{HX} assuming sp hybridization, we repeat the calculation assuming s hybridization, and we take the average value:

$$\beta_{HX} = [\beta_{HX}(sp^3) + \beta_{HX}(s)]/2 \tag{5}$$

c. Compute each β by making use of the Wolfsberg-Helmholz approximation[11] which makes obvious the connection between the size of β and the electronegativities of the two atoms over which it is defined:

$$\beta_{tu} = K (I_{tt} + I_{uu}) s_{tu} \tag{6}$$

$$s_{tu} = <t|u> \tag{7}$$

$$I_{tt} = <t|H|t> \tag{8}$$

Since I_{tt} is directly related to the <u>atomic orbital electronegativity</u>, x_t, and the latter is related to the corresponding <u>atomic electronegativity</u>, x_A,[12] β_{tu} can be said to be a function of atomic electronegativities and AO overlap. I_{tt} is called the tth Valence Orbital Ionization Energy (VOIE) of atom X. Let us now carry out the procedure described above with the purpose of ranking the atoms of the first row of the Periodic Table with respect to their overbinding abilities.

The bond diagrams shown in Figure 4 reveal the hybridization of each of F, O, N, C, B, Be, and Li within the hydrides of most frequent occurrence. We note the following:

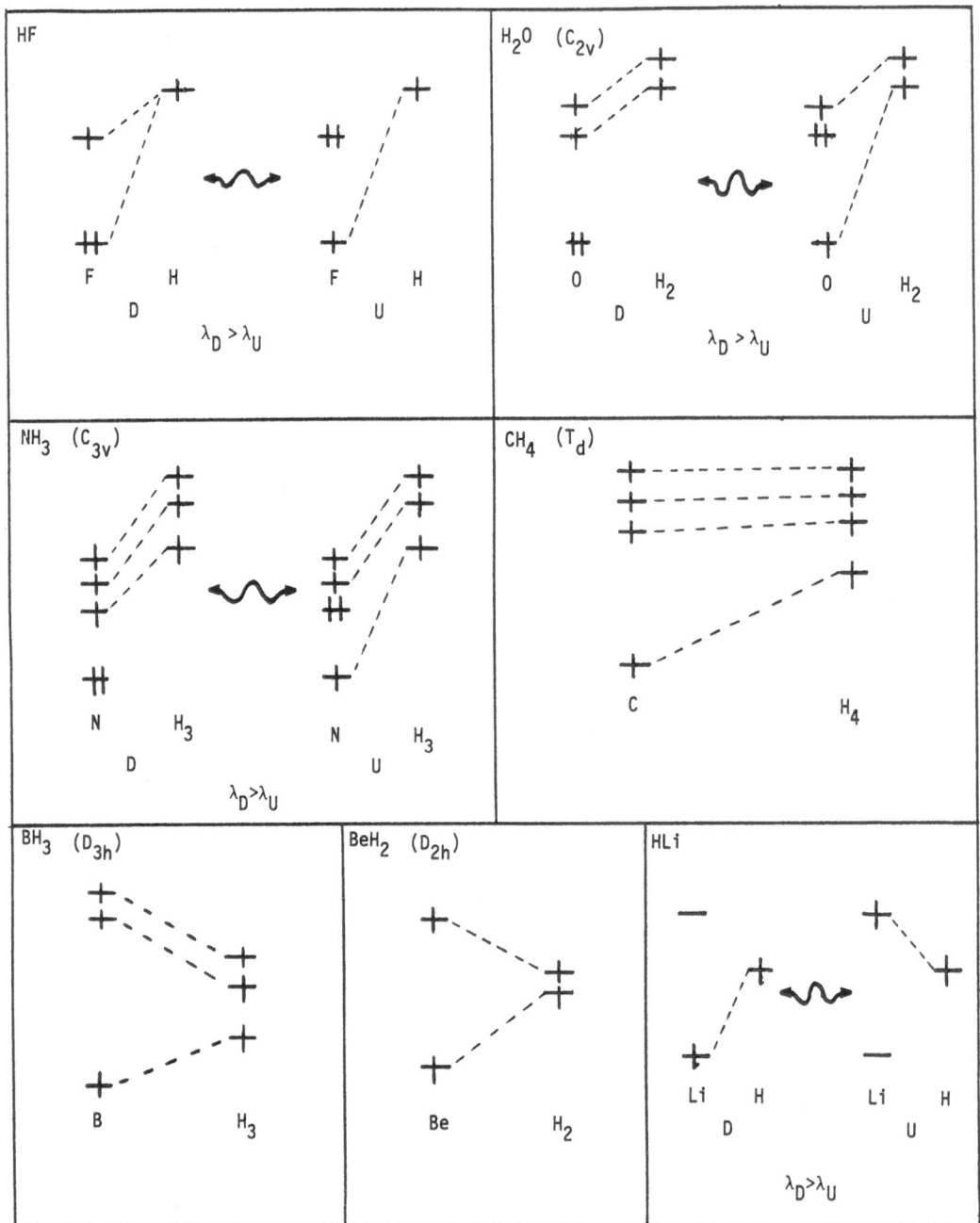

Figure 4. MOVB bond diagrams for the hydrides of most frequent occurrence of F, O, N, C, B, Be and Li showing only sigma orbitals, electrons, and bonds.

a. The single H-F bond in HF is made by major utilization of F2p and minor utilization of F2s because the large energy gap separating the F2s and F2p AO's dictates that the principal resonance bond diagram is D in Figure 4. Hence, the hybridization of F is towards the p end of the hybridization scale and inter-mediate between p and sp^3.

b. The two O-H bonds in H_2O are made by major utilization of the $O2p_x$ and $O2p_y$ and minor utilization of $O2p_x$ and O2s for the same reasons as in (a). Thus, the hybridization of O is towards the p end of the hybridization scale and intermediate between p and sp^3.

c. By using similar arguments as in (a) and (b), we can conclude that the hybridization of N in NH_3 is also between p and sp^3.

d. The hybridization of C in CH_4 is exactly sp^3.

e. The hybridization of B within BH_3 is exactly sp^2.

f. The hybridization of Be within BeH_2 is exactly sp.

g. The hybridization of Li within LiH is between s and sp.

The relative overbinding abilities of second row atoms can be determined using exactly the same procedure as above. The computed β_{HX}'s are given in Table 1 and the results can be very easily understood once one is aware of the following facts concerning the relative overbinding abilities of AO's of one and the same atom:

a. In <u>first row</u> atoms, the overbinding ability of 2s is much larger than that of 2p because of a more negative VOIE and, in most cases, a larger spatial overlap.

b. In <u>second row</u> atoms, the overbinding ability of 3s is no longer much different from that of 3p because the more negative VOIE of the former is counteracted by a greater spatial overlap of the latter.

Typical illustrations are given in Table 2. We hasten to add that, while absolute magnitudes are parametrization dependent, the trends are invariant, i.e., a 2s

Table 1. Overlap Bonding Ability of Atoms.[a,b,c]

	F	O	N	C	B	Be	Li
$-\beta_{X-H}/K$ (eV)	16.56 (21.61) [11.51]	16.98 (22.15) [11.82]	16.53 (20.84) [12.22]	19.45	18.20	15.81	8.07
	Cl	S	P	Si	Al	Mg	Na
$-\beta_{X-H}/K$ (eV)	16.62 (19.93) [13.31]	15.74 (18.67) [12.81]	15.20 (18.06) [12.33]	17.47	15.13	12.54	7.83

[a] For F, O, N, and Cl, S, P, β_{X-H} is the average of the values assuming sp^3 and p binding. For Li and Na, β_{X-H} is the average of the values assuming sp and s binding. The X-H bond lengths are those of the corresponding hydrides, i.e. r_{HF} is that of HF, r_{OH} that of H_2O, etc.

[b] Values in parenthesis are for sp^3 hybridization.

[c] Values in brackets are for p hybridization.

Table 2.   s versus p Overlap Bonding Abilities of First and Second Row Atoms.*

AO Resonance Integral	$-\beta/K$ (in eV)	$\Delta\beta/K$ (in eV)
$\langle 1s_H \vert \hat{H} \vert 2s_F \rangle$	23.29	
$\langle 1s_H \vert \hat{H} \vert 2p_F \rangle$	11.51	11.71
$\langle 1s_H \vert \hat{H} \vert 3s_{Cl} \rangle$	16.78	
$\langle 1s_H \vert \hat{H} \vert 3p_{Cl} \rangle$	13.31	3.47
$\langle 1s_H \vert \hat{H} \vert 2s_O \rangle$	23.82	
$\langle 1s_H \vert \hat{H} \vert 2p_O \rangle$	11.82	12.00
$\langle 1s_H \vert \hat{H} \vert 2s_S \rangle$	15.16	
$\langle 1s_H \vert \hat{H} \vert 2s_S \rangle$	12.81	2.35

* Calculated using H - X distances appropriate to H - F, H - Cl, H_2O, and H_2S.

is a stronger overbinder than a 2p AO and the differential binding ability of s and p AO's decreases as we go from a first to a second row atom within a column of the Periodic Table.

We will now "explain" the results in Table 1: Whether F, N, and O opt for maintaining their 2s AO's doubly occupied while making bonds using their weakly overbinding 2p AO's,or,whether they opt to hybridize in order to take advantage of the stronger overbinding ability of the 2s AO's at the expense of atom excitation,[13] the VOIE's and overlap integrals are such that F, O, and N end up having comparable overbinding strength. In making a transition from N to C, there is a dramatic change because tetravalent carbon <u>requires</u> hybridization at the expense of atom excitation. As a result, the fact that C is forced to be sp^3, while F, O, and N are allowed to be sp^b, with b > 3, becomes responsible for a significantly greater overbinding ability of C relative to any of F, O, or N. Once again, the situation changes dramatically as we shift from C to B. In this case, the reduction of the atomic VOIE's is such that it dominates the variation of the hybridization and the AO overlap integrals. As a result, B ends up being a weaker overbinder than C because of its smaller electronegativity. The same trend persists as we go from B to Be to Li. In summary, we can say that the overbinding ability of first row atoms remains initially constant (F, O, N), goes through a maximum (C), and then progressively declines as atomic electronegativity decreases until it reaches a minimum in the case of Li.

The overbinding abilities of second row atoms follow the same trend as those of first row atoms and need no separate discussion. However, the relative overbinding abilities of first and second row atoms deserve comment.

As we have already seen (Table 2) and discussed, the valence s has a much greater overbinding ability than the p AO of the first row atoms, while this trend tends to disappear in second row atoms. In fact, in the latter case the relative 3s and 3p overbinding abilities are parametrization dependent.[14]

An important consequence of this fact is that first row and second row atoms will be hybridized differently, whenever possible. For example, F, O, and N will tend to adopt sp^6 hybridization in order to exploit the very large overbinding ability of the 2s AO while Cl, S, and P will tend to use exclusively the 3p AO's since the 3s is no longer a far superior overbinder in comparison to the 2p AO. As a result, the relative overbinding abilities of F and Cl, O and S, and N and P can only be evaluated properly if due recognition of differential hybridization is made. No such problem exists in the case of C versus Si, B versus Al, and Be versus Mg, since hybridization is now fixed. Accordingly, the relative overbinding abilities of first and second row atoms are those shown in Table 3, in which β_{HX} for X = F, O, N has been computed by taking the average value of $\beta_{HX}(sp^3)$ and $\beta_{HX}(p)$ while β_{HX} for X = Cl, S, P has been computed by assuming p hybridization.

The principles discussed above are illustrated in Figure 5 which compares the bond diagrams for HF and HCl. Because of the much greater overbinding ability of the F2s relative to the F2p AO, F chooses to utilize the former AO at the expense of the required promotional energy and the bonding orbital of F becomes rich in s character. By contrast, the fact that the Cl3s and Cl3p AO's have comparable overbinding ability is responsible for the almost exclusive utilization of the latter AO for the purpose of bonding by Cl. Ultimately, the fact that the overbinding ability of F2s is much greater than that of F2p, Cl3s, or Cl3p becomes responsible for the fact that H-F is a stronger bond than H-Cl.

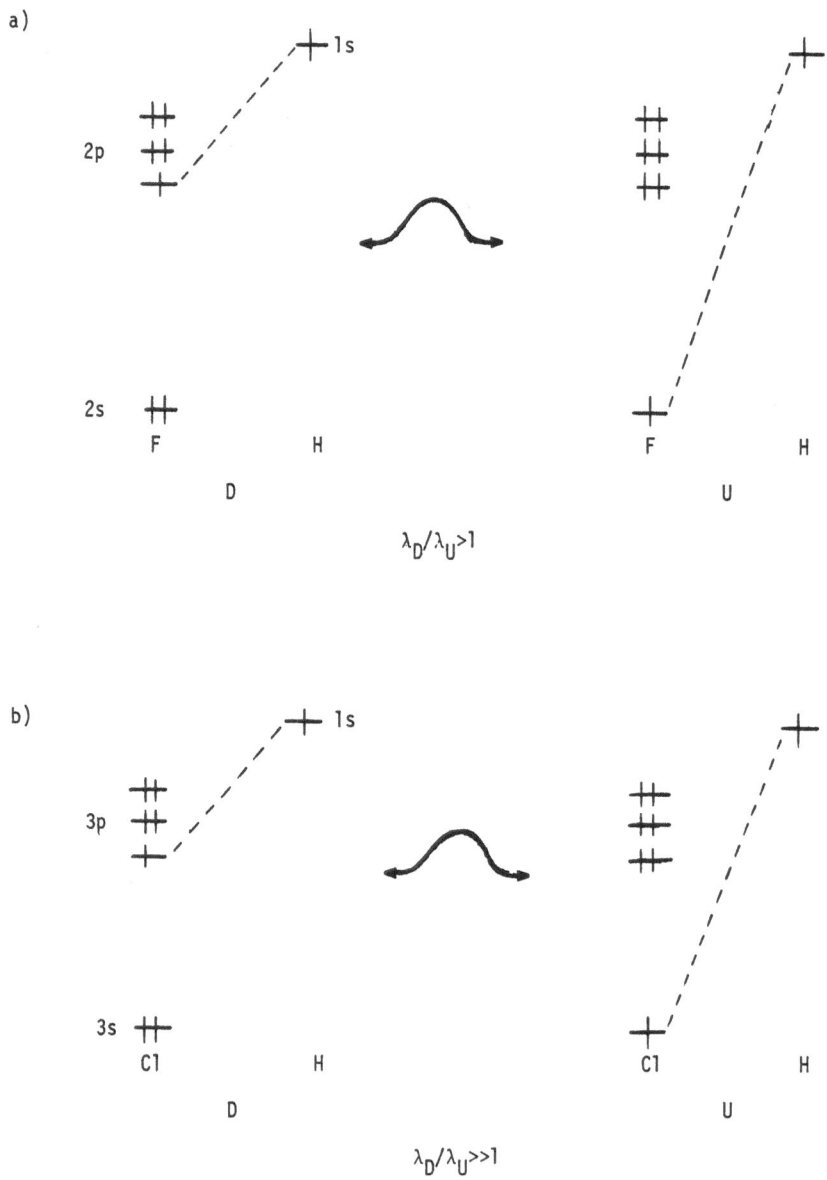

Figure 5. Bond diagrams for a) HF and b) HCl.

II. Orbital versus Atomic Electronegativities

With a clear understanding of the differences between atoms with regards to their overlap binding abilities, let us now turn our attention to another facet of atomic structure and seek to relate the VOIE's (I_{tt}'s) of the first and second row atoms <u>in the hybridization state defined above</u> with the corresponding atomic electronegativities. Table 4 shows the appropriate data with the atomic electronegativity numbers obtained from conventional sources and the VOIE data taken from an older paper of Hinze and Jaffe.[12] If we denote the atom electronegativity by x_A, it is evident that the following approximate relationship holds:

$$I_{tt,A} \ \alpha \ x_A$$

where t is the valence AO of atom A utilized for bonding with some other atom or group. The anomalous transition from N to C (and P to Si) is due to our arbitrary choice of p hybridization for N and P. Admixture of s character renders the I_{tt} of N equal or more negative than that of C.

Table 3. Relative Overlap Binding Abilities of First and Second Row Atoms $-(\beta_{X-H}/K).^a$

F	O	N	C	B	Be
16.56	16.98	16.53	19.45	18.20	15.81
Cl	S	P	Si	Al	Mg
13.31	12.81	12.33	17.47	15.13	12.54

[a] For F, O, and N β_{HX} is the average of $\beta_{HX}(sp^3)$ and $\beta_{HX}(p)$. For C and Si sp^3 hybridization is assumed, for B and Al sp^2 hybridization for Be and Mg sp hybridization and for Cl, S and P p hybridization.

Table 4. Pauling (x_p) and Mulliken (x_m) Electronegativities and Model
VOIE, I_{tt}, for 1st and 2nd Row Elements.[a]

	F	O	N	C	B	Be	Li
x_p	3.98	3.50	3.07	2.50	2.01	1.47	.97
x_m	20.90	15.08	14.585	12.376	8.626	9.13	6.21
$-I_{tt}$	20.86	17.28	13.94	14.61	11.29	8.58	5.39
	Cl	S	P	Si	Al	Mg	Na
x_p	3.16	2.58	2.19	1.90	1.61	1.31	.93
x_m	16.664	12.427	11.747	9.609	8.504	7.324	5.608
$-I_{tt}$	15.03	12.39	10.73	11.82	8.83	7.10	5.14

[a] x_m and x_p are atomic quantities, while I_{tt} is determined assuming the following hybridizations: F, O, N and Cl, S, P are p hybridized; C and Si are
sp^3 hybridized; B and Al are sp^2 hybridized. Be and Mg are sp hybridized;
and Li and Na are s hybridized.

III. The Effect of Atomic Replacement. Independent and Coupled Δx and $\Delta \beta$ Perturbations

Suppose that we were interested to find out how replacement of H by F would change the MOVB wavefunction of CH_2 if we were to neglect the fluorine lone pairs and consider only their sigma bonding orbitals. How can we tackle a fundamental problem of this type which time and again arises in the every-day practice of chemistry?

It is evident that we can view the replacement of H by F as a substitution modifying the β_{tu}'s and I_{tt}'s of the ligands. Does each of these changes affect all or some of the energy matrix elements in MOVB theory?

Let us recall the forms of the H_{ii} and H_{ij} matrix elements of MOVB theory:[6]

$$H_{ii} = F_i + G_i + X_i \tag{9}$$

$$H_{ij} = Q^0_{ij} \tag{10}$$

F_i is the energy of the isolated fragments, G_i is the "classical" interaction, and X_i the overlap interaction of the two fragments. The nature of Q^0_{ij} depends on the nature of the interacting CW's and, if attention is focussed exclusively on the monoelectronic charge transfer interaction of the CW's, then we obtain the equation shown below, where h_{mn} is the MO resonance integral over the MO's which constitute the origin and terminus of electron transfer and which belong to different fragments:

$$H_{ij} = d_{ij} \, \alpha \, h_{mn} \tag{11}$$

An example of a d_{ij} interaction matrix element is given schematically below:

$$y_2 \text{---} \text{---} y_3 \qquad y_2 \text{---} + y_3$$

$$\longleftrightarrow$$

$$y_1 \text{++} \text{++} y_4 \qquad y_1 + \text{++} y_4$$

$$\Phi_1 \qquad\qquad\qquad \Phi_2$$

$$H_{12} = d_{12} \, \alpha \, h_{13}$$

We now note two important facts:

a. F_i is independent of h_{mn} while both X_i and H_{ij} are dependent on h_{mn} and the latter is dependent upon interfragmental β_{tu}'s.

b. F_i is dependent upon h_{mn} which, in turn, depends on I_{tt}'s and, by extension, on atomic electronegativities, x_A's.

A plausible scenario for testing the effect of structural modification on molecular stereochemistry now unfolds as follows:

a. A molecule is viewed, in the usual way, as a composite of a core and ligands. In the previous example, C is the core and $X_2(X = H,F)$ the ligand fragment.

b. A replacement of core and/or ligands is tantamount to a perturbation of the original β_{tu}'s and/or I_{tt}'s. Accordingly, we can envision the following types of overlap and electronegativity perturbations:

1) An increase of the absolute magnitude of the β_{tu}'s. This will be called a $\Delta\beta^+$ perturbation. In a related sense, $\Delta\beta^-$ will symbolize a decrease of the same quantities.

2) An increase of ligand I_{tt}'s, or, in other words, a decrease of ligand electronegativity. This will be called a Δx^+ perturbation. The superscript projects the fact that the ligands become more electropositive. In a related sense Δx^- will symbolize a decrease of ligand I_{tt}'s, or, in other words, an increase of ligand electronegativity.

3) Simultaneous $\Delta\beta$ and Δx perturbations. We can define four types of coupled perturbations.

     i. $\Delta x^+ - \Delta\beta^+$

    ii. $\Delta x^+ - \Delta\beta^-$

   iii. $\Delta x^- - \Delta\beta^+$

    iv. $\Delta x^- - \Delta\beta^-$

A similar discussion and a similar set of definitions can be given in the case of perturbations introduced by core or core-and-ligand modifications.

How can one select the proper substituent so that it selectively brings into play one of the eight possible perturbations discussed above? By using Tables 1, 3, and 4 we can provide an answer in the format of Table 5 which shows the specific atomic replacements and the types of perturbations they represent. With a knowledge of what substitution corresponds to what type of perturbation, there is just one thing that remains to be done before we are able to predict the effect of a specific atomic substitution on the electronic structure of the parent substrate: We must understand the effect of each perturbation on the wavefunction of a composite C-L (C=core, L=ligands) system. In attempting to realize this goal, we now present a discussion of the effects of the various perturbations on the wavefunctions of prototypical systems in an effort to extract general concepts which can be ultimately applied to the problem of the stereochemistry of H_2O and its derivatives. For the purpose of space conservation, we restrict our attention to only four types of perturbations, namely, Δx^+, $\Delta\beta^-$, $\Delta x^+ - \Delta\beta^-$, and $\Delta x^- - \Delta\beta^-$ noting that Δx^-, $\Delta\beta^+$, and $\Delta x^+ - \Delta\beta^+$ have consequences opposite to those of Δx^+, $\Delta\beta^-$, and $\Delta x^- - \Delta\beta^-$, respectively.

Let us first consider the simple case of a two electron-two orbital C-L system described by the VB CW's shown in Figure 6 and let us inquire as to how the

Table 5. Examples of Perturbation Types.

		Perturbation Type	Heteroatom Substitution Effecting the Perturbation
Independent Perturbations		Δx^+ $(\Delta\beta\approx0)$	$F \rightarrow 0$, $S \rightarrow P$
		Δx^- $(\Delta\beta\approx0)$	$0 \rightarrow F$, $P \rightarrow S$
		$\Delta\beta^+$ $(\Delta x\approx0)$	$N \rightarrow C$
		$\Delta\beta^-$ $(\Delta x\approx0)$	$C \rightarrow N$
Coupled Perturbations		$\Delta x^+ - \Delta\beta^+$	$F \rightarrow C$, $Cl \rightarrow Si$
		$\Delta x^+ - \Delta\beta^-$	$F \rightarrow Cl$, $0 \rightarrow S$, $C \rightarrow Si$
		$\Delta x^- - \Delta\beta^+$	$Cl \rightarrow F$, $S \rightarrow 0$, $Si \rightarrow C$
		$\Delta x^- - \Delta\beta^-$	$C \rightarrow F$, $Si \rightarrow Cl$

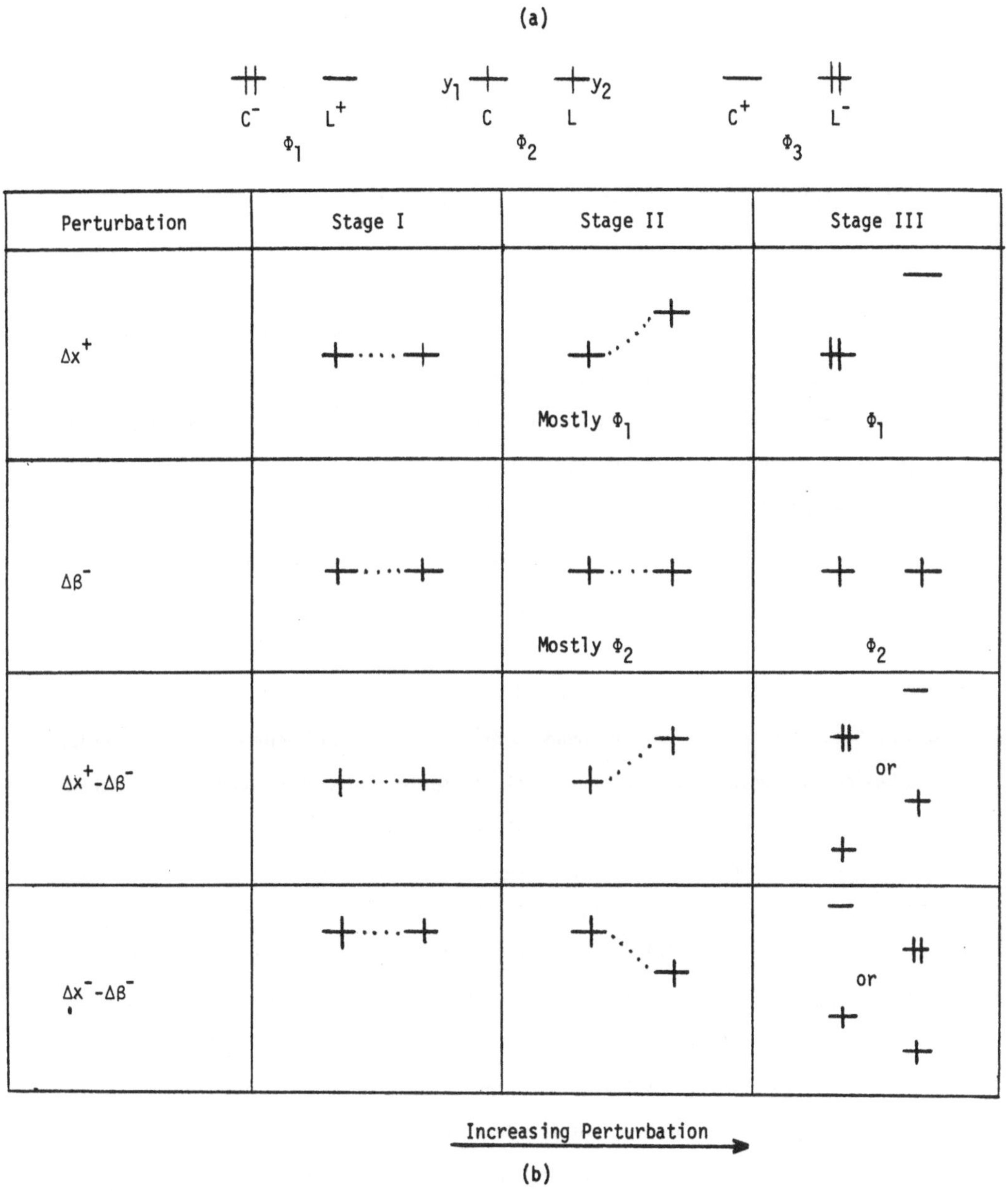

(a)

ϕ_1: C^- L^+ ϕ_2: y_1 C L y_2 ϕ_3: C^+ L^-

Perturbation	Stage I	Stage II	Stage III
Δx^+		Mostly ϕ_1	ϕ_1
$\Delta\beta^-$		Mostly ϕ_2	ϕ_2
$\Delta x^+ - \Delta\beta^-$			or
$\Delta x^- - \Delta\beta^-$			or

Increasing Perturbation

(b)

Figure 6. a. VB CW's for the two-electron-two-orbital C-L system. b. Bond diagrammatic representation of Δx^+, $\Delta\beta^-$, $\Delta x^+ - \Delta\beta^-$ and $\Delta x^- - \Delta\beta^-$ perturbations.

wavefunction responds to Δx^+ and $\Delta \beta^-$ perturbations. The first type of perturbation amounts to a progressive increase of the one-electron energy of orbital y_2, or, in other words, a progressive decrease of the electronegativity of L. The second type of perturbation amounts to a progressive diminution of the absolute magnitude of the interfragmental AO resonance integral, β_{12} ($\beta_{12} = \langle y_1 | \hat{H} | y_2 \rangle$). Clearly, Δx^+ renders the contribution of the Φ_1 CW increasingly important and preserves bonding due to spin pairing in Φ_2 and CW interaction until a limit (Stage III) is reached $[\varepsilon(y_2) = \infty]$ at which Φ_1 becomes the exclusive representative of the system. At this extreme, we say that C and L are ionically bound. On the other hand, $\Delta \beta^-$ progressively decreases the bonding due to spin pairing in Φ_2 as well as bonding due to CW interaction. As a result, the contribution of the "ionic" CW's Φ_1 and Φ_3 progressively decreases until, at the limit of $\beta_{12} = 0$, the wavefunction is represented exclusively by Φ_2.

A coupled Δx^+ - $\Delta \beta^-$ perturbation has the simultaneous effect of the independent Δx^+ and $\Delta \beta^-$ perturbations. That is to say, as the coupled perturbation increases, the wavefunction of the system becomes increasingly enriched in Φ_2 and Φ_1 until one of the two limits is reached first. Similarly, a coupled Δx^- - $\Delta \beta^-$ perturbation has the simultaneous effect of the independent Δx^- and $\Delta \beta^-$ perturbations. As the coupled perturbation increases, the wavefunction of the system becomes increasingly enriched in Φ_2 and Φ_3 until one of the two limits is reached first.

The effect of each of the four perturbations can be elegantly expressed in bond diagrammatic form as indicated in Figure 6. We say that VB theory with core-ligand dissection affords a pictorial way of representing heterolytic or homolytic bond cleavage, in general, and we note that Single Determinant (SD) MO theory can neither describe correctly homolytic bond cleavage nor can it provide a conceptually simple depiction of overbond destruction. What do we mean by the term "overbond"?

Let us first remind ourselves of the meaning of the word "bond" as used by most chemists. Specifically, the chemical equation shown below is taken to imply ("covalent") <u>bond</u> formation between the two hydrogen atoms.

$$H\cdot + H\cdot \longrightarrow H - H$$

Similarly, the chemical equation shown below is also taken to imply ("ionic") <u>bond</u> formation between Na^+ and Cl^-

$$Na\cdot + Cl\cdot \longrightarrow Na^+Cl^-$$

Thus, we say that both H_2 and NaCl are bound relative to their precursor atoms, <u>albeit by means of different bonding mechanisms</u>. This difference in bonding is obscured by mere usage of the term "bond". Recognizing that H_2 owes its ex-istence to the <u>overlap interaction</u> of the two hydrogen atoms while NaCl exists because of the "classical" interaction of Na^+ and Cl^- and it can be regarded as devoid of significant overlap induced bonding, we can say that the first of the two equations shown above denotes the formation of an <u>overbond</u>, the second one the formation of no new <u>overbonds</u> and we can argue that the imaginary conversion of H_2 to NaCl by means of atom replacement involves the destruction of an <u>over-bond</u>. The meaning of the term overbond is now self evident: It refers to the bonding component which is attributable to overlap interaction, i.e., spin pair-ing and CW interaction, and it is a strong function of the AO resonance integrals.

Using the data of Tables 3, 4, 5, we can offer the following examples of Δx^+, $\Delta\beta^-$, $\Delta x^+ - \Delta\beta^-$ and $\Delta x^- - \Delta\beta^-$ perturbations caused by atom replacement:

a. $H - F \longrightarrow H - NH_2$ (Δx^+)

b. $H - Li \longrightarrow H - Na$ $(\Delta\beta^-)$

c. $H - F \longrightarrow H - Cl$ $(\Delta x^+ - \Delta\beta^-)$

d. $H - CH_3 \longrightarrow H - OH$ $(\Delta x^- - \Delta\beta^-)$

An extremely important special case in which a $\Delta\beta^-$ perturbation is effected not by substitution but, rather, by conventional bond dissociation is illustrated below, where the double headed arrow indicates bonding due to spin pairing.

Chemical
Process: A — B →   A •••• B   →   A• + •B

Wavefunction
Change:

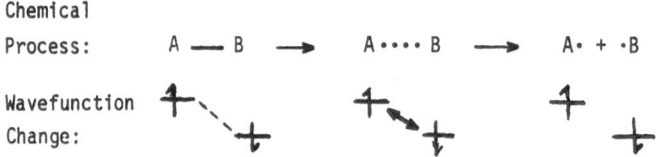

Let us now consider the effects of the four types of perturbations on
the wavefunctions of more complex systems. We first examine a four electron-
four orbital C = L system such as the one specified below:

$$y_2 \text{——} \cdots \text{——} y_3$$

$$y_1 \text{——} \cdots \text{——} y_4$$

$$\text{C} \qquad \text{L}$$

| $\beta_{13} = \beta_{24} = 0$ |
| $\beta_{14} = \beta_{23} < 0$ |

The VB or MOVB CW's necessary for the description of this system are shown in
Figure 7 and the bond diagrammatic representation of Δx^+, $\Delta \beta^-$, $\Delta x^+ - \Delta \beta^-$, and
$\Delta x^- - \Delta \beta^-$ are given in Figure 8 under the assumption that Δx^+ amounts to a progressive increase of the energies of y_3 <u>and</u> y_4, $\Delta \beta^-$ amounts to a progressive diminution of the absolute magnitude of the β_{14} <u>and</u> β_{23} resonance integrals, and $\Delta x^+ - \Delta \beta^-$ to a combination of these gradual changes ($\Delta x^- - \Delta \beta^-$
has analogous meaning). The drawings of Figure 8 can be easily understood by
thinking of C=L as a system of two independent two-electron bonds and noting
that an avoided surface crossing occurs before the limit of $\beta_{14} = \beta_{23} = 0$ is
reached. This example demonstrates that VB or MOVB theory with core-ligand
dissection affords a pictorial way of representing double bond cleavage and that
<u>this process can be simply viewed as electron deexcitation.</u>

Finally, we consider a four electron -three orbital C̈-L system such as the
one specified below.

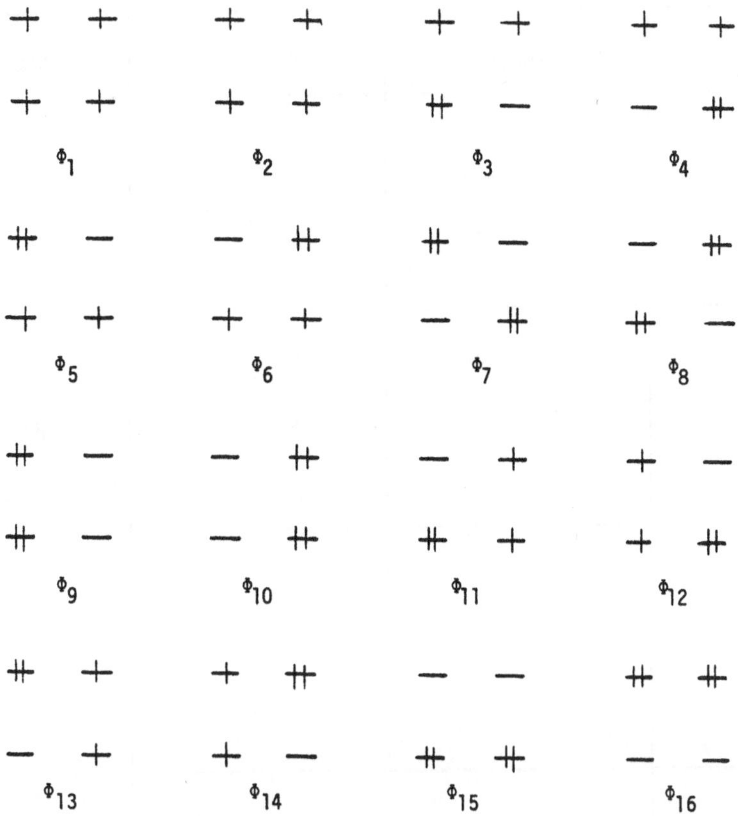

Figure 7. The 16 VB CW's for the four-electron-four-orbital C=L system.

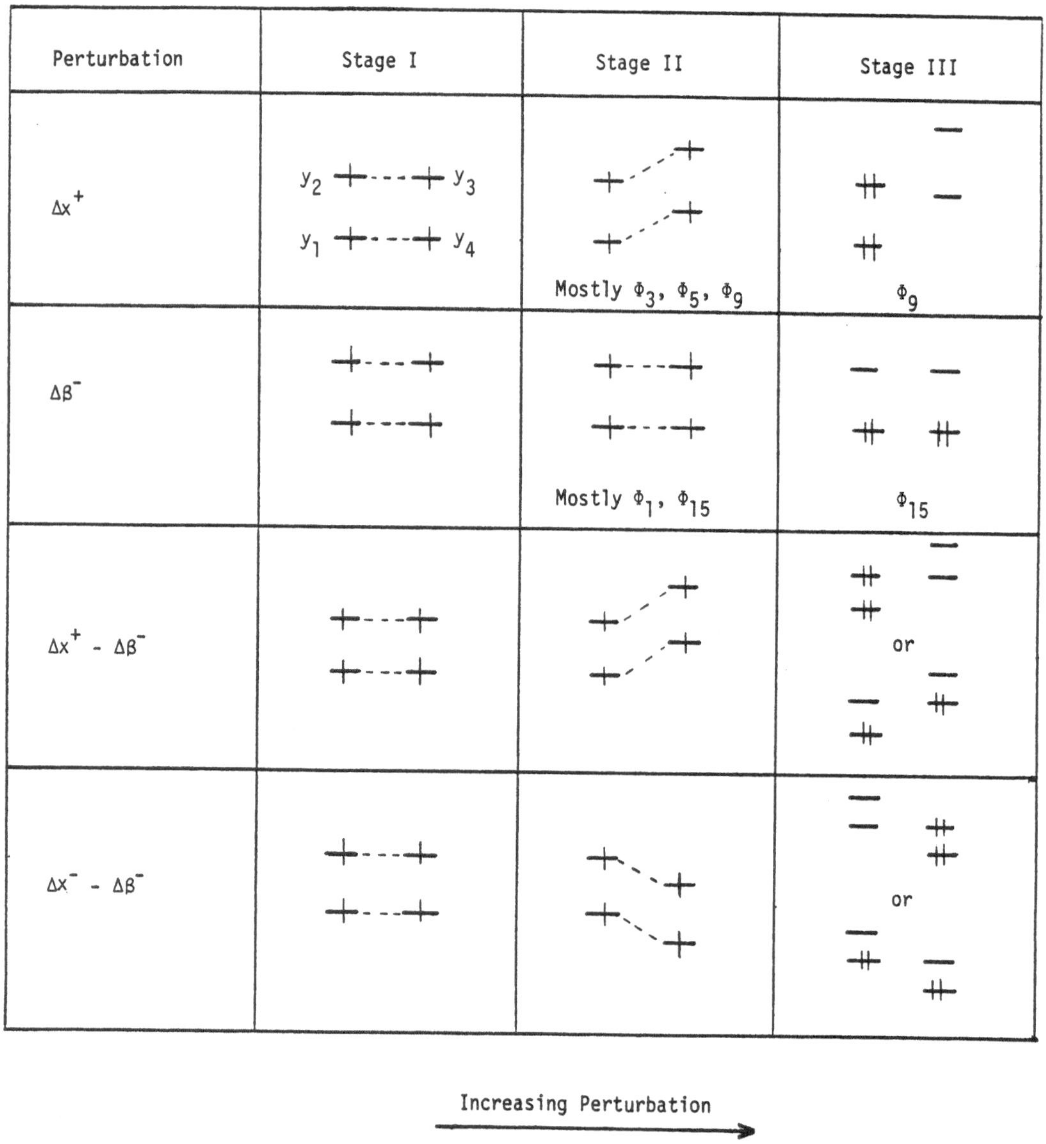

Figure 8. Bond diagrammatic representation of Δx^+, $\Delta \beta^-$, $\Delta x^+ - \Delta \beta^-$ and $\Delta x^- - \Delta \beta^-$ perturbations for the four-orbital-four-electron C=L system.

$$|\beta_{13}| \gg |\beta_{23}|$$

The VB or MOVB CW's required for the description of this system are shown in Figure 9 and the bond diagrammatic representations of Δx^{+}, $\Delta \beta^{-}$, Δx^{+}-$\Delta \beta^{-}$, and Δx^{-}-$\Delta \beta^{-}$ are given in Figure 10 under the assumption that Δx^{+} amounts to a progressive increase of the energy of y_3, $\Delta \beta^{-}$ to a progressive reduction of the absolute magnitudes of the β_{13} and β_{23} resonance integrals, and $\Delta x^{\pm}\Delta \beta^{-}$ to a combination of these gradual changes. The drawings of Figure 10 can be easily understood by recalling that the unperturbed Č-L system can be represented by a linear combination of resonance bond diagrams and noting that a perturbation acting on Č-L causes rehybridization, i.e., it causes a rebalancing of the weights of the resonance bond diagrams and ultimately projects a single CW as the sole descriptor of the system in question. This example demonstrates that VB or MOVB theory with core-ligand dissection affords a pictorial way of representing re-hybridization due to chemical modification, i.e., replacement of one atom with another having different orbital electronegativity.

There are two important aspects of Figure 10 which merit out attention. When both β_{13} and β_{23} are large in absolute magnitude, with the former greater than the latter, R* may lie below R due to stronger two-electron overlap attraction and weaker three-electron overlap repulsion despite the fact that the excitation energy of R* is larger than that of R. As both β_{13} and β_{23} decrease in absolute magnitude, a point is reached when R and R* become degenerate. Past this point, the lower promotional energy of R causes it to attain a lower energy than R*.

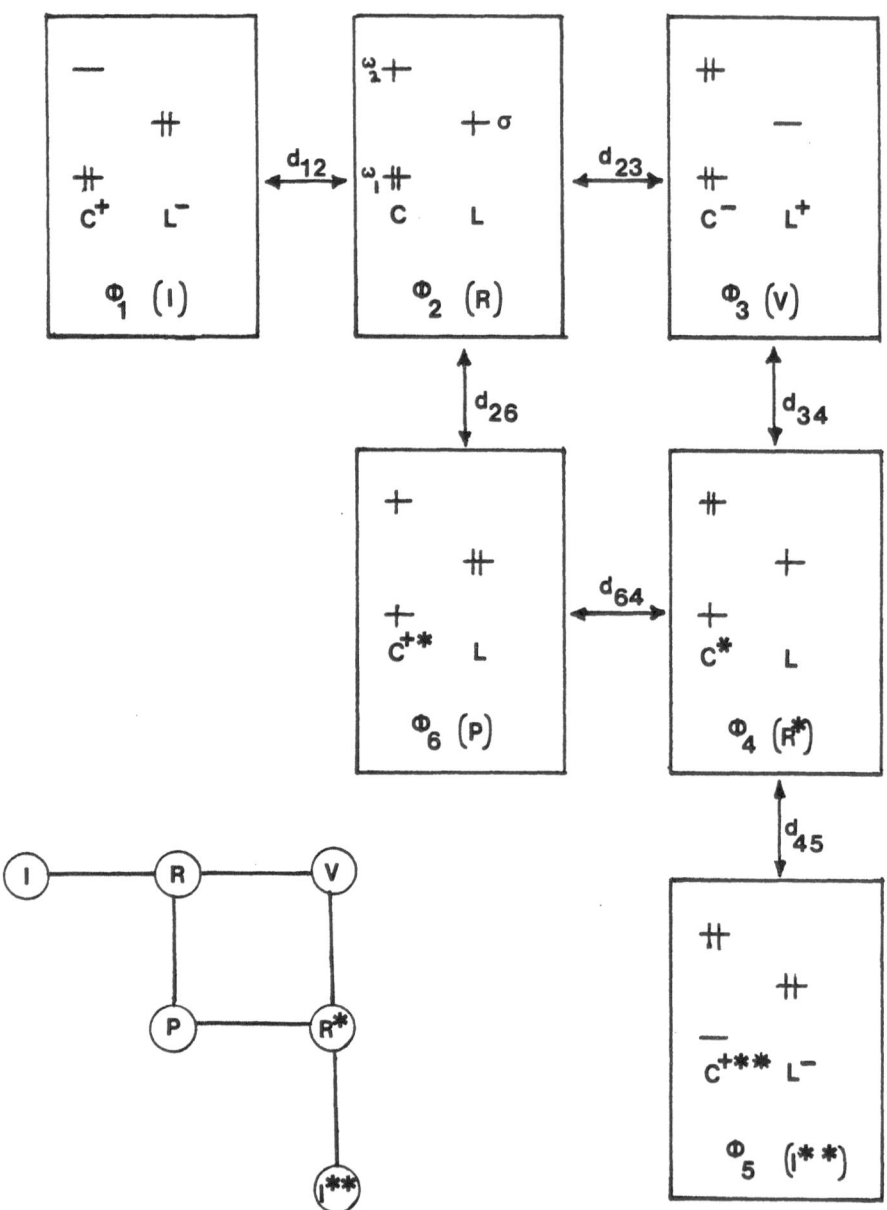

Figure 9. MOVB CW's Φ_i, for the treatment of a four-electron-three-orbital problem. C=Core fragment and L=Ligand fragment. Fragment charges and local excitation energies are indicated by the superscripts of C and L, with one asterisk implying single local excitation and a double asterisk double local excitation. CW nomenclarure is indicated in parenthesis. Double arrows indicate the monoelectronic CT interaction of the CW's through d_{ij} matrix elements. Orbital convention is spelled out in connection with the Φ_2 CW (ω_1 and ω_2 are orthogonal as they belong to the same fragment). At lower left, a schematic representation of the CT interaction of the six MOVB CW's is given with the lines connecting the circles implying d_{ij} matrix elements.

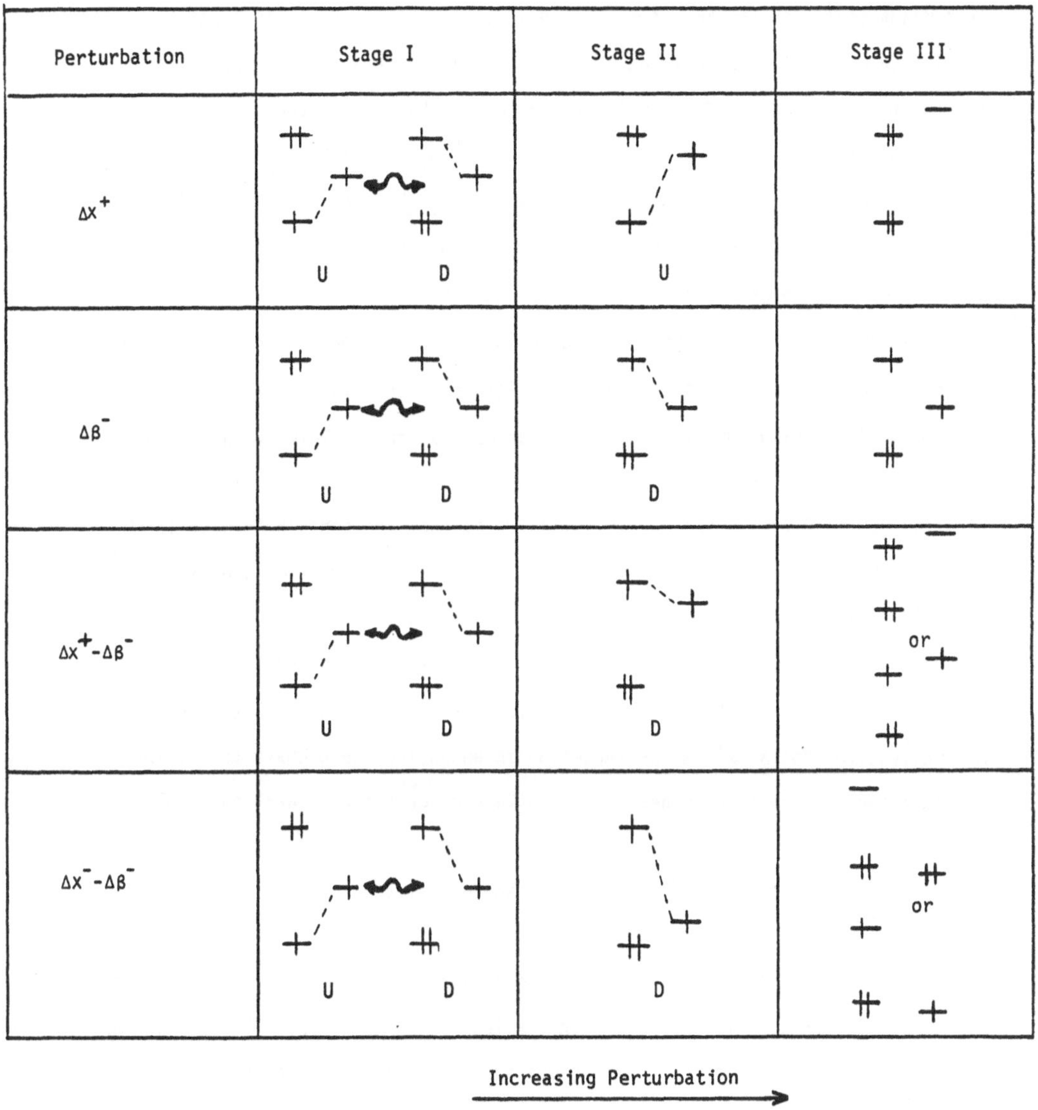

Figure 10. Bond diagrammatic representation of Δx^+, $\Delta \beta^-$, $\Delta x^+ - \Delta \beta^-$ and $\Delta x^- - \Delta \beta^-$ perturbations for the four-electron-three-orbital system.

This means that when the $\Delta\beta^-$ perturbation is turned "off", R* orchestrates bonding in combination with V and I** and the major resonance contributor is U. As $\Delta\beta^-$ increases, a point is reached at which the exact reverse occurs. namely, bonding is now due primarily to R in combination with V and I. In short, <u>$\Delta\beta^-$ causes a radical change of the way in which bonds are made in $\breve{C}$-L and gives advance warning of the potentially momentous stereochemical consequences of processes which amount to $\Delta\beta^-$ perturbations.</u>

A second aspect of Figure 10 which merits our attention concerns the way in which a coupled perturbation causes initial rehybridization consistent with an increased contribution of the D resonance bond diagram. This is due to the fact that the perturbation increasingly projects the R and V CW's which in combination define D type bonding, i.e., the two of them are contributors to the D resonance bond diagram. Similarly, a coupled $\Delta x^- - \Delta\beta^-$ perturbation causes initial rehybridization consistent with an increased contribution of the D resonance bond diagram because this perturbation increasingly projects the R and I CW's which in combination define D-type bonding. Obviously, a $\Delta x^- - \Delta\beta^-$ perturbation will be more effective than a $\Delta x^+ - \Delta\beta^-$ perturbation in enhancing the D character of the wavefunction because I describes only D while V describes D as well as U bonding.

The most important conclusions of the above analyses can now be stated as follows:

a. A Δx^+ perturbation causes rehybridization as described in previous works.

b. <u>A $\Delta\beta^-$ perturbation causes deexcitation of the core and/or ligands so that the bonds linking the two fragments change fundamentally in character in the sense that electron pairs are allowed to occupy the lowest energy orbitals of the core and/or ligands and bonds are made by utilization of the higher energy orbitals of the core and/or ligands.</u>

c. The effects of $\Delta x^+ - \Delta\beta^-$ and $\Delta x^- - \Delta\beta^-$ perturbations mimic the effect of a $\Delta\beta^-$ perturbation.

d. Since nature allows only very few substitutions which are tantamount to pure Δx perturbations, the most important type of perturbation, given (c), is the $\Delta\beta$ one. "New chemistry" can be discovered by judicious heteroatomic perturbations which constitute $\Delta\beta$ perturbations.

IV. A Reexamination of the Electronic Structure of H_2O Derivatives

Armed with the knowledge embodied in the above few paragraphs, we can now offer a diagrammatic representation of the effects of Δx^+, $\Delta\beta^-$, $\Delta x^+-\Delta\beta^-$ and $\Delta x^--\Delta\beta^-$ perturbations on the wavefunction of OX_2, where X is a univalent ligand. The orbital notation and the appropriate bond diagrams are shown in Figure 11. We can now proceed as follows:

a. Predict the effect of a given perturbation on the OX_2 angle by consulting Figure 11. It is assumed that only the ligands, X_2, are perturbed.

b. Associate the given perturbation with a specific atomic replacement using the information contained in Table 5.

c. On the basis of (a) and (b) predict the effect of substitution on the $X\hat{O}X$ angle.

Typical examples are briefly discussed below:

a. Δx^+ perturbation. According to Figure 11, Δx^+ will cause initial rehybridization consistent with increased U character of the wavefunction. As a result of the increased utilization of the O2s AO, the $X\hat{O}X$ angle with increase. We can find no example of such a type of atom replacement in the literature.

b. $\Delta\beta^-$ perturbation. According to Figure 11, $\Delta\beta^-$ will cause initial rehybridization consistent with increased D character of the wavefunction. As a result of the increased utilization of the O2p AO's, the $X\hat{O}X$ angle will decrease. Unfortunately, atomic replacements which constitute $\Delta\beta^-$ perturbations are difficult to find because most atoms differ significantly in electronegativity. Potential examples are OLi_2 and ONa_2 but both of these molecules represent the ionic limit of Figure 11 and both are linear in order to optimize the "classical" interaction brought into play by the $O^{-2}(X^{+1})_2$ type CW.

A bona fide example of $\Delta\beta^-$ perturbation is the actual dissociation process shown below:

$$H_2O \longrightarrow :\ddot{O} + H_2$$

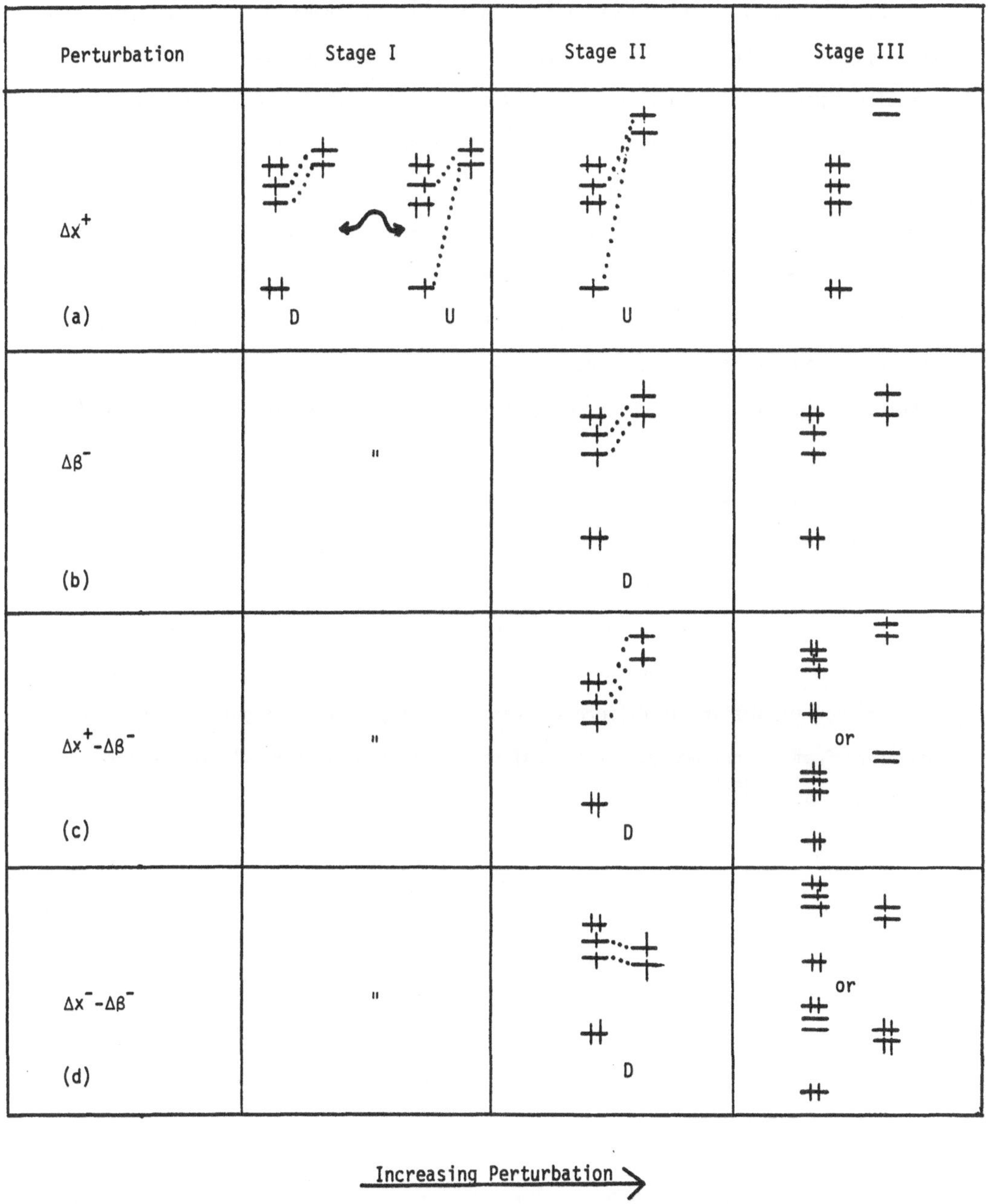

Figure 11. Bond diagrammatic representation of Δx^+, $\Delta \beta^-$, $\Delta x^+ - \Delta \beta^-$ and $\Delta x^- - \Delta \beta^-$ perturbations for H_2O.

c. Coupled Δx^- - $\Delta \beta^-$ perturbation. According to Figure 11, Δx^- - $\Delta \beta^-$ will cause initial rehybridization consistent with increased D character of the wavefunction. Hence, because of greater utilization of the O2p AO's, the $X\hat{O}X$ angle will decrease. The most illustrious example of such a perturbation is the comparison of the bond angles of $(CH_3)_2O$ and F_2O which shows that the angle of the former species shrinks as CH_3 is replaced by F approximately changing from 112° to 103°. The weaker overlap binding ability of F compared to CH_3 is, in part, due to the reasons discussed in this chapter with an additional indirect effect, that of core-ligand overlap repulsion due to the oxygen and fluorine electron pairs, also conspiring to the same end result. This latter aspect of the problem has been discussed in a previous work. Why F_2O, and related molecules which are made up of atoms having many lone pairs, are at all bound will be explained by usage of the concept of configuration aromaticity within the framework of VB theory (see Chapter 17). Indeed, the novel conclusion of this analysis is that OF_2 has a smaller angle than $O(CH_3)_2$ not only on account of the greater electronegativity of F, as commonly thought, but also because fluorine is a weaker overlap binder than CH_3 for two distinctly different reasons. This picture is significantly different from the one emerging from application of local (restrictive) models to the same problems.[15-18]

d. Coupled Δx^+-$\Delta\beta^-$ perturbation. According to Figure 11, Δx^+-$\Delta\beta^-$ will cause initial rehybridization consistent with increased D character of the wavefunction and, as a result of increased utilization of the O2p AO's, the $\widehat{XOX}$ angle will decrease. Replacement of CH_3 by SiH_3 is tantamount to such a perturbation. However, the experimental results shown below [3,19] make clear that the theoretical prediction is not borne out:

$$H_3C \overset{O}{\diagup\diagdown} CH_3 \qquad H_3Si \overset{O}{\diagup\diagdown} SiH_3$$

$$112° \qquad\qquad 144°$$

How can we rationalize this failure? Two different arguments can be made:

1. Replacement of CH_3 by SiH_3 brings into play a predominant Δx^+ perturbation and, as a result, the angle opens up.

2. Replacement of CH_3 by SiH_3 brings into play a Δx^+-$\Delta\beta^-$ perturbation but some other factor, e.g., nonbonded repulsion, plays a dominant role and the predicted angle shrinkage does not materialize despite the fact that the hybridization of O is as predicted by assuming that the substitution amounts to a Δx^+-$\Delta\beta^-$ perturbation. That the latter is a plausible interpretation is signaled by the experimental facts given below. [20-23] These imply that Walsh's rehybridization rule is not always obeyed and, since the predictions of this rule are the same as the ones arrived at on the basis of MOVB theory by assuming Δx perturbation dominance, a rationalization of its failure exists within the broader theoretical framework described in this work.

	$S(CH_3)_3$	$S(SiH_3)_3$
$\widehat{XSX}$ (X=C,Si)	99°	97°
	$P(CH_3)_3$	$P(SiH_3)_3$
$\widehat{XPX}$ (X=C,Si)	100°	100°

In the absence of well calibrated empirical data regarding the VOIE's and
AO overlap integrals of third, fourth, etc., row atoms, it is not possible to
determine whether the atomic overbinding ability continues to decrease as we move
down the column of the Periodic Table and past a second row atom or whether it
stays constant. However, we note that if a monotonic decrease of the overbinding
abilities does occur, the surprising results shown in Table 6 could be easily
rationalized in the same way as those given above. Specifically, one could argue
that the angle shrinkage predicted to accompany a Δx^+-$\Delta \beta^-$ perturbation begins to
manifest itself only when the ligands get far from each other, or, in other words,
as the bonds connecting them to the central atom become longer. In this way, nonbonded repulsion can interfere to a lesser extent with angle shrinkage. Of course,
other effects, such as d-orbital participation, may be responsible for the abrupt
change of geometry accompanying replacement of SiH_3 by GeH_3 in the molecules of
Table 6.

At this point, we digress in order to point out that the Frontier Orbital (FO)-
SOJT model affords a simple interpretation of the geometry of OH_2, NH_3 and their
derivatives in a way first pointed out by Levin [24] and Cherry et al.[25] For example, the bending tendency of H_2O is related to the strength of the HOMO-LUMO
interaction of the linear species upon bending. This in turn, is claimed to depend primarily on the energy separation of the two orbitals, the HOMO being a nonbonding O 2p AO and the LUMO a σ^* MO, as shown in the diagram below.

LUMO $\sigma^* =$ ◯ ⬤ ◯

$\qquad\qquad\quad s_1 \quad 2s \quad s_2$

Clearly, the HOMO-LUMO gap depends on the $2s$ - $(s_1 + s_2)$ interaction, or, in other
words, the HOMO-LUMO energy gap is a function of the overbinding ability of the
$2s$ valence AO of oxygen and the same is true for all the other systems. The fact

Table 6. Effect of Ligand Electropositivity on Bond Angles.

	X = CH_3	X = SiH_3	X = GeH_3
$X\widehat{O}X$	CH_3 O CH_3 $112°$[a]	SiH_3 O SiH_3 $144°$[b]	GeH_3 O GeH_3 $126°$[c]
$X\widehat{N}X$	CH_3 N CH_3 CH_3 $111°$[d]	SiH_3 N SiH_3 SiH_3 $120°$[e]	GeH_3 N GeH_3 GeH_3 $120°$[f]
$X\widehat{N}C$	$N = C = O$ | CH_3 $125°$[g]	$N = C = O$ | SiH_3 $180°$[h]	$N = C = O$ | GeH_3 $141°$[i]

a. Ref. 3.
b. Ref. 19.
c. Cradock, S.; Ebsworth, E. A. V.; Beagley, B. Inorg. Nucl. Chem. Lett. 1969, 5, 417.
d. Beagley, B.; Hewitt, T. G. Trans. Faraday Soc. 1968, 64, 2561.
e. Beagley, B.; Conrad, A. R. Trans. Faraday Soc. 1970, 66, 2740.
f. Glidewell, C.; Rankin, D. W. H.; Brockway, L. O. J. Am. Chem. Soc. 1940, 2935.
g. Eyster, E. H.; Gillette, R. H. Brockway, L. O. J. Am. Chem. Soc. 1940, 62, 3236.
h. Gerry, M. C. L.; Thompson, J. C.; Sugden, T. M. Nature 1966, 211, 846.
i. Murdock, J. D.; Rankin, D. W. H. Chem. Commun. 1972, 748.

that the HOMO-LUMO gap decreases as O is replaced by S reflects the fact that S can overbind through its 3s AO much more weakly than the oxygen through its 2s AO. Unfortunately, the SOJT model has restricted applicability, as pointed out by one of its originators,[17b] and it fails to give proper warning of the consequences of very weak overbinding of the type we will examine in the next section. These and other problems are traceable to the approximations which are involved, namely, the FO approximation, the utilization of low order perturbation theory in developing the argument, and the fact that the approach is essentially an SD-MO approach. MOVB theory is a rigorous qualitative theory of chemical bonding because none of these approximations has been made in developing the central concepts.

What is the most important message that the above discussions attempt to convey? Rehybridization due to a change of the electronegativity of core and/or ligands has been recognized many years ago and, in particular, since the publication of Walsh's paper[1] in 1947. On the other hand, rehybridization due to weakening of the overlap interaction of two fragments has never been considered explicitly before for two very simple reasons:

a. The initial apparent successes of the Walsh and related models created the impression that the only important thing which needs to be considered in attempts to predict the effect of substitution on stereochemisty is the change of core or ligand electronegativity while assuming that $\Delta\beta = 0$.

b. The effect of a Δx perturbation can be studied by SD MO theory. By contrast, the effect of a $\Delta\beta$ perturbation cannot be formally dealt with by SD MO theory since this brand of theory cannot properly describe bond weakening or breaking.[8] Even if it were capable of doing so, the conceptual intractability of MO theory is such that a clear and convincing dissection of the Δx and $\Delta\beta$ perturbations would have been impossible.

In summary, we can say that the effect of the Δx perturbation has been well understood for some time while the effect of the $\Delta\beta$ perturbation has eluded detection partly because we have been lulled to acceptance of the Walsh model and partly because SD MO theory, the major conceptual vehicle of the chemistry in the second half of this century, is incapable of handling this problem.

V. The Induced Deexcitation Model

Our knowledge of chemistry is mainly based on prototypical molecules containing strong overlap binders, e.g., hydrogen atoms bound to carbon cores in hydrocarbons. Very little is known about molecules which contain weak overlap binders, e.g., lithium or iodine atoms attached to carbon cores in perlithio and periodo alkanes. The shapes of molecules of the latter type have an excellent chance to be significantly different from those of molecules of the former type and, to a first approximation, they can be predicted by construction of bond diagrams in which electron pairs are allowed to occupy the lower energy orbitals of the core with bonding of core and ligands effected via utilization of the higher energy singly occupied orbitals of the core. A direct implication of this statement is that if a hydrocarbon exists in a given geometry which is consistent with U bonding (e.g., linear acetylene) replacement of the hydrogens by weak overlap binders, e.g., lithium atoms, will tend to favor a geometry consistent with $\overset{+}{H}$ bonding (e.g., trans acetylene), or, in more general terms, some geometry in which the core is allowed to be deexcited. In this light, we can propose a "recipe" for designing molecules which may adopt geometries in apparent "violation" of currently used stereochemical rules. The "recipe" is as follows:

a. Identify a molecule which exists preferentially in a geometry which is consistent with U bonding.

b. Replace the strong with weak binding core and/or ligands. The derivative molecule will non exist preferentially in a different geometry which is consistent with H bonding, or, in other words, some geometry which allows core and/or ligand deexcitation. This novel geometry can be predicted in a routine manner through construction of bond diagrams under the stipulation that nonbonding electron pairs occupy the lower energy orbitals of the core and/or ligands, where the term "nonbonding electrons" refers to electrons belonging to core or ligands which do not cause inter-fragmental bonding.

The network of ideas presented in this section defines what we will henceforth denote as the Induced Deexcitation (ID) model. Note that this is neither a model based on the collection of empirical facts nor an intuitive theoretical construct based on exclusive consideration of an allegedly dominant quantum chemical variable (e.g., overlap, coulomb repulsion, exchange repulsion, etc.). Rather it is a special application of qualitative MOVB theory as formulated in a previous work which seeks to project one simple yet fundamental principle: If a molecule is viewed as a combination of a core and a set of ligands, strong bonding justifies excitation of either or both components while weak bonding does not do so. This novel, yet self evident, in an a posteriori sense, viewpoint is one of the first rewards for our willingness to adopt a VB (as opposed to MO) way of thinking about chemical bonding.

The implications of the analysis of weak overbinding in molecules hardly need comment. The Periodic Table provides us with the strong overbinding atoms C, N, O, and F which in different combinations and numbers can define an infinite number of excited cores. The Periodic Table also provides us with weak and strong overbinding atoms which can serve as ligands of the aforementioned cores with a switch from the latter to the former nearly guaranteeing impressive stereochemical change, all predictable on the basis of MOVB theory, in a qualitative sense. Indeed, the research possibilities are endless and a systematic investigation is certain to be of great value.

References

1. Walsh, A.D. _Disc._ _Faraday_ _Soc._ _1947_, 2, 18.

2. Bent, H.A. _Chem._ _Revs._ _1961_, 61, 275.

3. $(CH_3)_2O$: Kimura, K.; Kubo, M. _J._ _Chem._ _Phys._ _1959_, 30, 151.

4. F_2O: Morino, Y.; Saito, S. _J._ _Mol._ _Spectrosc._ _1966_, 19, 435.

5. Cl_2O: Beagley, B.; Clark, A.H.; Hewitt, T.G. _J._ _Chem._ _Soc._ _1968_, 658.

6. Epiotis, N.D.; Larson, J.R.; Eaton, H. "Unified Valence Bond Theory of Electronic Structure" in Lecture Notes in Chemistry, Vol. 29, Springer-Verlag; New York and Berlin, 1982.

7. The expression of the total energy of a system as a sum of the energies of the individual subsystems is the direct result of the neglect of electron-electron interaction of these subsystems. This is done only for qualitative purposes. Indeed, the Independent Bond Model MOVB theory can be rigorously implemented in a quantitative _ab_ _initio_ sense without neglecting the subsystem bielectronic interaction. However, this is not one of the aims of this work.

8. Cotton, F.A. "Chemical Applications of Group Theory"; Wiley-Interscience: New York, 1971.

9. The meaning of the wiggly double arrow is discussed in reference 6.

10. This is so because β_{tu} and h_{mn} are interfragmental resonance integrals.

11. Wolfsberg, M.; Helmholz, L. _J._ _Chem._ _Phys._ _1952_, 20, 837.

12. Hinze, J.; Jaffé, H. _J._ _Am._ _Chem._ _Soc._ _1962_, 85, 540.

13. In a following paper, we shall see that F, O, and N actually have different hybridizations in the corresponding hydrides with N tending towards the sp^3 and F towards the p mark of the scale. For the time being, the assumption that all three have the same hybridization (Table 1) is safe for the purposes of this paper.

14. For example, according to the Wolfsberg-Helmholz approximation, 3s is a stronger overbinder, while, according to the Pople approximation, employed in CNDO calculations, 3p exceeds 3s in overbinding ability. See: Pople, J.A.; Beveridge, D.L. "Approximate Molecular Orbital Theory"; McGraw-Hill: New York, 1970.

15. Gillespie, R.J. "Molecular Geometry"; Van Nostrand-Rheinhold: London, 1972.

16. Bartell, L.S. J. Chem. Ed. 1968, 45, 754.

17. (a) Pearson, R.G. "Symmetry Rules for Chemical Reactions"; Wiley and Sons, Inc.: New York, 1971.

    (b) ibid., pp 211, 213.

18. DeFrees, D.J.; Levi, B.A.; Pollack, S.K.; Hehre, W.J.; Binkley, J.S.; Pople, J.A. J. Am. Chem. Soc. 1979, 101, 4085.

19. $(SiH_3)_2O$: Almennigen, A.; Bastiansen, O.; Ewing, V.; Hedberg, K.; Tratteberg, M. Acta Chem. Scand. 1963, 17, 2455.

20. $(CH_3)_2S$: Pierce, L.; Hayashi, M. J. Chem. Phys. 1961, 35, 479.

21. $(SiH_3)_2S$: Almenningen, A.; Hedberg, K.; Seip, R. Acta Chem. Scand. 1963, 17, 2264.

22. $(CH_3)_3P$: Lide, D.R.; Mann, D.E. J. Chem. Phys. 1958, 29, 914.

23. $(Si_3)_3P$: Beagley, B.; Robiette, A.G.; Sheldrick, G.M. J. Chem. Soc. 1968, 3002.

24. Levin, C.C. J. Am. Chem. Soc. 1975, 97, 5649.

25. Cherry, W.; Epiotis, N.D.; Borden, W.T. Acc. Chem. Res. 1977, 10, 167.

Chapter 2. Why do Organolithium Monomers have Strange Structures?

"Replacement of hydrogen atoms by lithium atoms may radically alter the stereo-chemistry of the parent hydrocarbon". Thus can be summarized some of the most important findings of the imaginative computational work of Schleyer and his coworkers.[1] What fundamental property of lithium is primarily responsible for the unexpected geometrical preferences of perlithio hydrocarbons? Is there some way to predict the geometry of these molecules? These and related questions can be dealt with within the framework of MOVB theory in a way which illustrates the basic utility of the Induced Deexcitation (ID) model presented in the previous chapter as well as the way in which MOVB theory[2,3] can be used in order to produce novel insights regarding the mechanism of vacant orbital participation in chemical bonding. The former illustrative application of the theory is made possible by the fact that Li is a weak overlap binding (overbinding) ligand which can readily induce core deexcitation while the latter is made possible by the fact that Li has low lying vacant 2p orbitals which can combine with doubly occupied orbitals to define new bonds or they can function as hybridization "holes" to promote more efficient "covalent" carbon-lithium bonding. At the out-set, we state that this work has been totally motivated by the calculational work of Schleyer and his collaborators[4] which was published at the time when we were in sore need of well established facts to test the central ideas of MOVB theory, such as the ones described in this and other chapters.

I. The Fundamental Bonding Differences Between Hydrogen and Lithium

Before we consider the various ways in which lithium differs from hydrogen and the problems which arise when we attempt to understand the stereochemical consequences of carbon-lithium bonding, let us briefly review some fundamental ground state bonding principles by reference to the hypothetical diatomic A_2, where A is an atom having one Atomic Orbital (AO), denoted by x, occupied by one electron.[5] In Valence Bond (VB) theoretical terms, the ground state wavefunction, Ψ, can be written as follows:

$$\Psi \ \alpha \ A\cdot \cdot A + \lambda(A^+ \ A\overline{\cdot} + A\overline{\cdot} \ A^+) \qquad\qquad \lambda \ll 1 \qquad\qquad (1)$$

The binding energy, E, of the system is defined as follows:

$$E = <\Psi|\hat{H}|\Psi>(at \ r_{AA} = r_{eq}) - <A\cdot\cdot A|\hat{H}|A\cdot\cdot A>(at \ r_{AA} = inf.)$$

or

$$E = <A\cdot\cdot A|\hat{H}|A\cdot\cdot A> (r_{eq}) + \Delta - <A\cdot\cdot A|\hat{H}|A\cdot\cdot A>(r_{inf}) \qquad (2)$$

where $\hat{H}$ is the nonrelativistic electrostatic Hamiltonian, r_{AA} is the interatomic distance, Δ represents the resonance interaction of $A\cdot\cdot A$ with $A^+ \ A\overline{\cdot}$ and $A\overline{\cdot} \ A^+$ at r_{eq}, and all wavefunctions are assumed to be normalised. The diabatic and adiabatic surfaces describing the combination of two A atoms are shown in Figure 1. By using the symbolism introducted in Figure 1, by employing perturbation theory for the evaluation of Δ, by setting $s^2 = 0$, and by neglecting differential overlap in computing the interaction term $<A\cdot\cdot A|\hat{H}|A^+A\overline{\cdot}>$, we obtain the following expression for the binding energy of A–A:

$$E = G + X - \underbrace{[2\beta - 2s <A\cdot\cdot A|\hat{H}|A\cdot\cdot A>]^2/J_{11} - J_{12}}_{\Delta} \qquad (3)$$

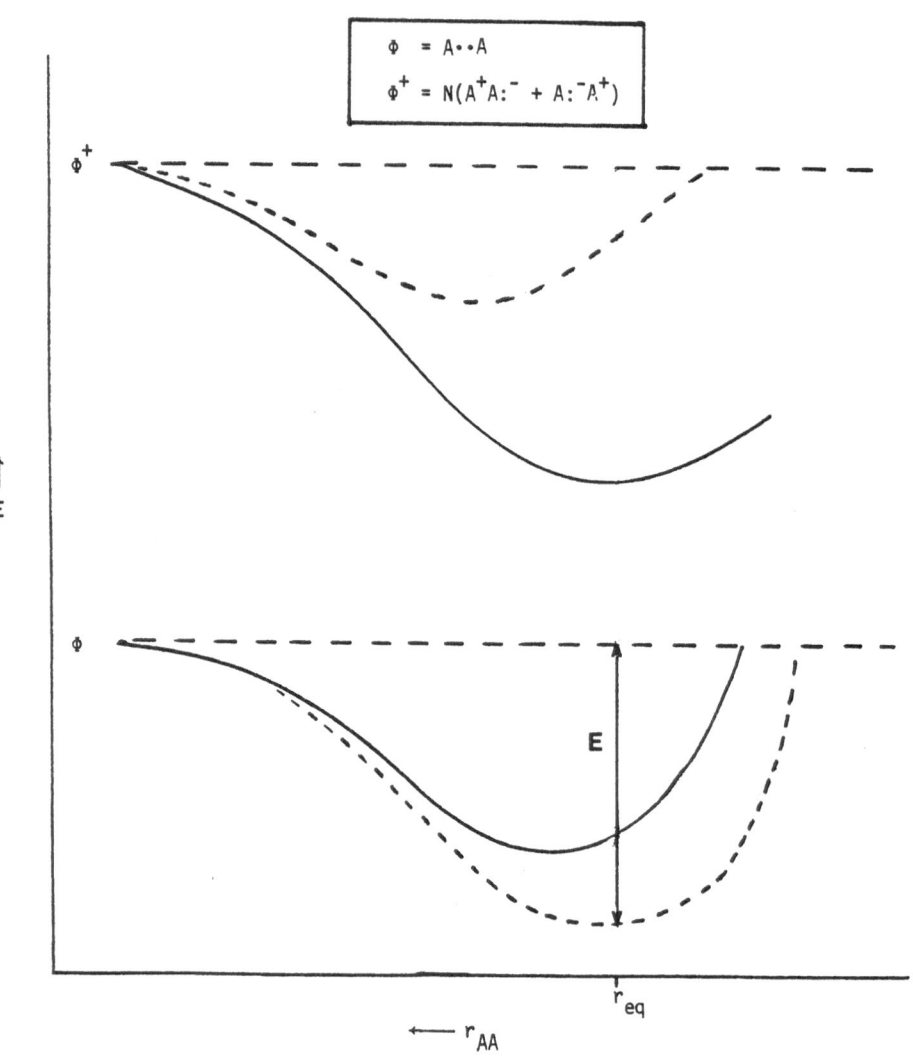

Figure 1. The formation of the two-electron-two-orbital bond of A - A. Diabatic surfaces are indicated by solid line and adiabatic surfaces by dashed line.

Appendix

NOTATION:
$$x_1 \underline{\qquad} \qquad \underline{\qquad} x_2$$
$$A_a \qquad\qquad A_b$$

x_1, x_2 : Orbitals

a, b : Nuclei

$s = \langle 1|2 \rangle$

$\epsilon_1 = \langle 1| -\frac{1}{2}\nabla^2 - \frac{Z_a}{r_a} |1\rangle$

$J_{12} = \langle 11| \frac{1}{r} |22\rangle$

$K_{12} = \langle 12| \frac{1}{r} |12\rangle$

$V_1 = \langle 1| -\frac{Z_b}{r_b} |1\rangle$

$V_N = \frac{Z_a Z_b}{r_{ab}}$

$\beta = \langle 1| -\frac{1}{2}\nabla^2 - \frac{Z_a}{r_a} - \frac{Z_b}{r_b} |2\rangle$

Table 1. Cohesive Energy (eV per atom) of H, Li and Na Clusters.

Species	H_6	Li_6	Na_6
D_{6h}	1.750	0.763	—
D_{3h}	1.267	0.870	0.590
O_h	1.025	0.820	0.573
D_{3d}	1.037	—	0.557
diatomic	2.374	0.525	0.375

where

$$G = V_1 + V_2 + J_{12} + V_N \tag{4}$$

and

$$X = 2\beta s + K_{12} \tag{5}$$

The definitions of the various integrals are given in the Appendix. We conclude that the binding energy of A-A is a function of G, X, and Δ, noting that both X and Δ are strong functions of the AO resonance integral β. Hence, we write:

$$E = G + X(\beta) + \Delta(\beta) \tag{6}$$

This equation, in conjunction with the discussion of the way in which β depends on the nature of the atoms involved in resonance interaction presented in a previous chapter, defines clearly how the nature of bonding in A-A changes as we replace H, a strong overlap binder, by Li, a very weak overlap binder. Specifically, $X(\beta)$ and $\Delta(\beta)$ decrease in absolute magnitude and the relative importance of overlap binding [$X(\beta)$ and $\Delta(\beta)$ terms] and "classical" coulomb binding (G term) changes. This has been recognized since the early days of the development of quantum chemistry[6] and it is responsible for some puzzling experimental and computational results which make abundantly clear that the rules of the binding game are not the same for H and Li:

a. In diatomics made up of strong overlap binding atoms, such as H, a two electron bond is stronger than a one electron bond. By contrast, in diatomics made up of weak overlap binding atoms, such as Li and Na, a one electron bond can be stronger than a two electron bond simply because a large part of the binding is due to "classical" (nonoverlap) coulomb attraction. Thus, H_2 has a larger bond Dissociation Energy (DE) than H_2^+.[7,8] By contrast, the bond DE of Na_2^+ is larger than that of Na_2 according to experiment.[8] Furthermore, computations indicate that

Li_2^+ is more strongly bound than Li_2[9,10,11] and $NaLi^+$ more strongly bound than $NaLi$.[11] These apparently peculiar trends have been noted in a monograph by Schaefer.[12]

DIATOMIC	DE(eV)
H_2	4.47
H_2^+	2.5
Li_2	0.62 (0.99)
Li_2^+	1.30 (>1.24)
Na_2	0.75
Na_2^+	1.02
$NaLi$	0.85
$NaLi^+$	0.92

b. Systems made up of strong overlap binders (e.g., H) tend to segregate because of net interbond overlap repulsion while systems made up of weak overlap binders (e.g., Li, Na) tend to aggregate. This is very nicely projected in Table 1 which contains data taken from recent important computation works.[13-15]

c. H_2 is very different from Li_2 and Na_2 insofar as the lowest triplet surface is concerned. This is purely repulsive in the case of H_2 while, by contrast, it displays a shallow minimum in the case of Na_2.[15]

What all these data are trying to tell us is that, in the case of strong overlap binders, trends are set by the overlap terms X and Δ while, in the case of weak overlap binders, the trends are set by the "classical" coulomb G term. Recognizing that the trends discussed above are valid not only for homonuclear diatomics of the type A_2 but also for heteronuclear diatomics of the type AX, where X is kept constant, we can now enumerate the key differences between H and Li insofar as their bonding to carbon atoms is concerned:

a. H is a strong and Li a weak overlap binder mainly because of the much greater electropositivity of Li. The Ligand Induced Deexcitation (LID) model is founded on the assumption that this is the principal difference between hydrogen and lithium. The stereochemical consequences of this particular difference are fully predictable within the confines of the LID model. Because of the fact that $|\beta_{HX}|$ is so much larger than $|\beta_{LiX}|$, the stereochemistry of H-containing molecules is much less dependent on "classical" coulomb interaction than the stereochemistry of Li-containing molecules. The latter is assumed to have no stereochemical implications within the confines of the LID model. However, we stress that this is only an assumption in need of computational testing since the stereochemical consequences of "classical" bonding cannot be predicted in any simple qualitative way. In this connection, the reader is alerted to the fact that point charge models, which implicitly aim at revealing the stereochemical consequences of "classical" bonding, are probably unrealistic since they do not reproduce the "classical" effect brought into play by the G term of equation (3).

b. Li has low lying vacant 2p AO's. On the other hand, there are no low lying vacant AO's of chemical significance in the case of H. How can the presence of the 2p AO's of Li affect the stereochemistry of a perlithio hydrocarbon? In order to answer this question, we must first digress and consider the MOVB theory of Vacant Orbital Participation (VOP) in chemical bonding.

II. The MOVB Theory of Vacant Orbital Participation

Let us consider the following fundamental and very common stereochemical problem: We are asked to consider two different geometries of one and the same molecular species such that in one geometry VOP is "on" while in the second it is "off". How are we to decide whether VOP is significant enough as to determine the stereochemistry of the molecule in question? MOVB theory gives a very direct answer to this question: If VOP enhances delocalization of existing core-ligand bond pairs, it is unimportant, while, if it introduces new core-ligand bond pairs, it is important and may be the underlying cause of a stereochemical preference. These concepts are illustrated in bond diagrammatic fashion in Figure 2. Let us see exactly what is involved by comparing the bond diagrams of Figure 2a and 2b and those of Figure 2c and 2d.

a. The bond diagram of Figure 2a shows a core and a ligand fragment joined by a single two-electron bond. Introduction of a vacant orbital in the ligand having symmetry compatible with that of the orbital of the core enhances the delocalization of the existing bond pair and lowers the energy of the system. The corresponding bond diagram is shown in Figure 2b.

b. The bond diagram of Figure 2c shows a core and a ligand fragment joined by a single two-electron bond, as in the previous case, only now the core posesses a lone pair. Introduction of a vacant orbital in the ligand having symmetry compatible with that of the orbital housing the core lone pair converts the latter into a bond pair and lowers significantly the energy of the system. The corresponding bond diagram is shown in Figure 2d.

Now, it can be easily demonstrated that the introduction of a new bond is energetically much more beneficial than enhancement of delocatization of an existing bond. Accordingly, we formulate the following VOP principle: VOP is operationally important only if it acts in a way which converts core or ligand lone pairs into core-ligand bond pairs.

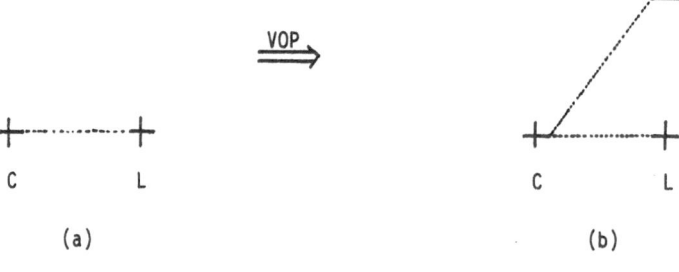

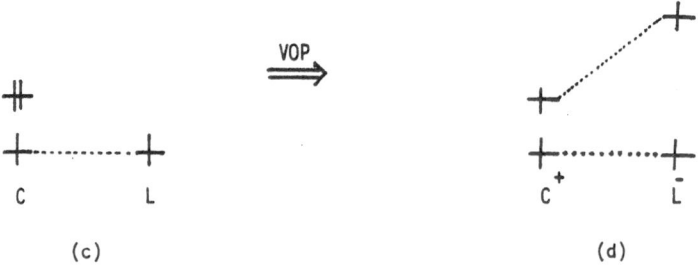

Figure 2. Illustration of Vacant Orbital Participation (VOP). Introduction of a vacant orbital into the system of (a) yields (b) which enjoys enhanced delocalization. Introduction of a vacant orbital in the system of (c) yields (d) which enjoys an extra bond.

Returning now to the problem at hand, we note that the Li 2s AO's can be thought of as the primary valence AO's of this atom and core-ligand bonds made by a hypothetical exclusive utilization of Li 2s AO's can be called primary bonds. Additional bonding due to the participation of the Li 2p AO's can then be referred to as VOP. As discussed above, we may distinguish two important cases of VOP:

1) The carbon core has no nonbonding electron pairs. In this case, the 2p AO's spanned by the ligand MO's will act as hybridization "holes".

2) The carbon core has nonbonding electron pairs. In this case, the ligand MO's spanning the 2p AO's will act in combination with the core MO's containing the nonbonding electron pairs to define new multicenter bonds connecting core and ligands.

According to the VOP principle, whenever there are core nonbonding electron pairs which can be converted to bonding electron pairs through Li 2p AO participation, the lithiated hydrocarbon will tend to adopt a geometry which makes an optimal conversion of this type possible. This will become clear as we work through the specific examples $XC \equiv CX$, $X_2C = CX_2$, and $X_3C - CX_3$ with X = H, Li.

III. Computational Tests of the MOVB Theoretical Treatment of Lithiated Hydrocarbons

We are now prepared to outline a rational method for unlocking the secrets of C-Li bonding and their implications for the stereochemistry of lithiated hydrocarbons. As a first step, the general principles of MOVB theory are applied to the problem at hand, starting with application of the ID model and considering additional electronic factors, such as "classical" interaction effects and low lying vacant orbital participation. As a second step, quantum chemical computations are carried out in order to test the validity of the MOVB analysis.

We will utilize the following four computational methods:

a. Extended HückelMO(EHMO) computations[16] with neglect of the Li 2p AO's, denoted by EH-p.

b. EHMO computations including the Li 2p AO's, denoted by EH+p.

c. Ab initio Self Consistent Field-MO (SCF-MO) computations[17] with neglect of the Li 2p AO's, denoted by SCF-p.

d. Ab initio SCF-MO computations including the Li 2p AO's, denoted by SCF+p. We "agree" that the EHMO computations be carried out using some arbitrary set of standard bond lengths and bond angles and the SCF-MO computations be carried out with geometry optimization.

We can define the following tabloid scheme:

EH-p	SCF-p
EH+p	SCF+p

Now, suppose that we examine two different geometries of a perlithio hydrocarbon, say, I and II. In a qualitative sense, there can be two possible computational results: Either I is found to have lower energy than II, or, vice versa. The first result can be denoted by an X mark and the second by the absence of an X mark. It is then obvious that in performing the four types of computations described above we may end up with any one of the following tabloid patterns, each repre-

senting schematically the results of four distinct computations:

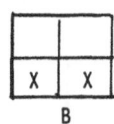 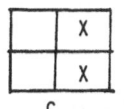 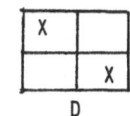

A   B   C   D

Needless to say, other possible tabloid patterns exist, but, for illustrative

purposes, we shall focus our attention on A, B, C, and D. A proper interpre-

tation of these tabloid patterns requires a clear recognition of the types of

bonding effects contained within each MO calculation. We note the following:

a. EH-p contains neither the "classical" nor the Li 2p AO participation

effect, but it does contain the "primary" valence AO overlap effect.

b. EH+p does not contain the "classical" effect, but it does contain the

Li 2p AO participation and the "primary" valence AO overlap effect.

c. SCF-p contains both the "classical" and "primary" valence AO overlap

effects but it leaves out the 2p AO participation effect.

d. SCF+p contains all effects.

We also require an understanding of the equivalence relationships between MO

and VB computational schemes. These have been discussed in a previous work[2] and

are displayed in Table 2. EHMO theory is equivalent to a brand of MOVB theory,

termed EHMOVB theory, which involves the same integral approximations as EHMO

theory. Because of these approximations, EHMOVB theory is an <u>overdelocalized</u>

version of MOVB theory. Similarly, EHMO is an overdelocalized version of SCF-MO-

Configuration Interaction(CI) theory. Also, SCF-MO theory is equivalent to a brand

of MOVB theory which makes use of the so-called constraint approximation

characteristic of monodeterminantal SCF-MO theory. Once again, this brand of MOVB

theory is an overdelocalized version of MOVB theory. The same relationship

Table 2. Approximate and Rigorous Types of Valence Bond and Molecular Orbital
Theories. Theories within a Row are Equivalent.

VB - Type Theory	MO - Type Theory	Approximation
VB or HL (delocalized AO's)	SCF - MO - CI	None
HL (localized AO's)	—	Truncation
—	3 x 3 SCF - MO - CI	Truncation
—	SCF - MO	Constraint
NDO - VB	—	Integral
—	NDO - SCF - MO	Integral and Constraint
EHVB (S ǂ 0)	EHMO (S ǂ 0)	Integral and Constraint
HVB (S = 0)	HMO (S = 0)	Integral and Constraint

exists between SCF-MO and SCF-MO-CI theory. What all this means is that, in performing EHMO and SCF-MO computations, we are not really testing MOVB theory, but, rather, we probe fundamental aspects of qualitative MOVB theory which can be reproduced approximately by EHMO and/or SCF-MO computations such as primary overlap bonding and low lying vacant orbital participation.

With these facts in mind, we can now interpret computational results, i.e., the obtained tabloid patterns, in the following way:

a. If tabloid pattern A is obtained, this can be taken as evidence of a controlling influence of primary AO overlap effects and a computational justification of the LID model. Note that this does not in any way mean that "classical" and Li 2p AO participation effects are unimportant. It merely implies that either these effects oppose but cannot overwhelm the valence AO overlap effects or that they operate in the same direction as the primary AO overlap effects.

b. If tabloid pattern B is obtained, this can be taken as evidence of a controlling influence of Li 2p AO participation, i.e., that "hole" hybridization and/or bond formation are the critical determinants of geometrical preference.

c. If tabloid pattern C is obtained, this can be taken as indicative evidence of a controlling influence of "classical" effects, the stereochemical consequences of which cannot be predicted in any simple qualitative manner.

d. If tabloid pattern D is obtained, this can be taken as evidence of a complex interplay of all effects.

We can now proceed with the examination of prototypical systems. Specifically, we shall examine whether and how replacement of H by Li alters the stereochemistry of $HC \equiv CH$, $H_2C = CH_2$, and $H_3C - CH_3$. The predictions will be generated by the LID model with additional consideration given to the role of **"nonclassical" Li** 2p AO participation effects. Computational results will be presented in connection with the MOVB analysis. The following preliminary comments, notes, and clarifications will render the subsequent discussions clear:

a. According to the LID model, replacement of strong by weak overlap-binding ligands will cause deexcitation of the core so that the stereochemistry of the derivative molecule will now be consistent with accommodation of nonbonding core electron pairs in the lower energy core MO's and core-ligand binding will be accomplished through utilization of the higher energy core MO's. If a molecule containing strong overlap binding ligands is U-bound, the corresponding derivative in which the strong have been replaced by weak overlap binding ligands will be H- or D-bound and it will exhibit a different geometrical preference. Bond diagrams constructed with the above stipulations can directly reveal the stereochemical change, if any, which a molecule will undergo due to replacement of H, a strong overlap-binder, by Li, a weak overlap-binder.

b. The stereochemical consequences of Li 2p AO participation can be predicted without constructing additional bond diagrams. All one needs to do is identify the core MO's which contain nonbonding electron pairs, if any, and match them with ligand MO's spanning the vacant ligand AO's according to the symmetry constraints imposed by the chosen geometry. The geometry which is consistent with the maximum number of strong core-ligand bonds generated in this fashion will tend to become the preferred geometry of the molecule as the vacant ligand orbitals attain increasingly lower energy.

c. In the EHMO computations of molecules containing H atoms, we have used the standard bond angles and C-H bond lengths recommended by Pople.[18] Finally, in the EHMO computations of molecules containing Li, we have used C-Li bond lengths and bond angles taken from the literature, whenever available.

d. The results of SCF-MO computations cited in this work have been taken from the literature. The vast majority of them have been obtained during the course of an extensive computational investigation of lithiated hydrocarbons by Schleyer and coworkers.[1]

IV. C_2H_2 versus C_2Li_2

We consider two different geometries of C_2X_2, a linear (L) and a rhomboidal (R) one.

$$X - C \equiv C - X$$

L   R

The bond diagrams for the case of X = H are shown in Figure 3. In the L geometry, the core (C_2) and ligands (X_2) are U-bound, that is to say the core is excited while, in return, there is maximum spatial overlap between the core and ligand MO's. By contrast, in the R geometry, the molecule is $D^{\ddagger}$-bound, where the dagger implies impaired spatial overlap between the core and ligand MO's. Here, the situation is the reverse of the one encountered in the L geometry. Specifically, the core is now deexcited at the expense of overlap. The loss of overlap is due to the fact that while the $\omega_2 - \sigma_L^*$ and $\omega_4 - \sigma_R^*$ overlap integrals are comparable in the two geometries, the $\omega_5 - \sigma_R$ overlap integral of the R geometry is much smaller than the $\omega_5 - \sigma_L$ overlap integral of the L geometry. This loss of overlap can be understood by closer inspection of the form of the core MO ω_5. Specifically, because of the way in which the 2s and 2p AO's contribute to this MO, the resulting lobes point outward in such a way so that the $\omega_5 - \sigma_R$ overlap integral becomes small in absolute magnitude and the corresponding multicenter bond very weak in the case of the R geometry. These considerations are illustrated in Figure 4. When X = H, i.e., when X is a strong overlap binder, the system opts for modest core excitation ($\omega_2 \rightarrow \omega_4$) in exchange for strong core-ligand overlap. A markedly different situation is encountered when X = Li, i.e., when X is a weak overlap binder. In this case, the interaction matrix elements become relatively small in absolute magnitude because of the highly electropositive nature of the ligands. Hence,

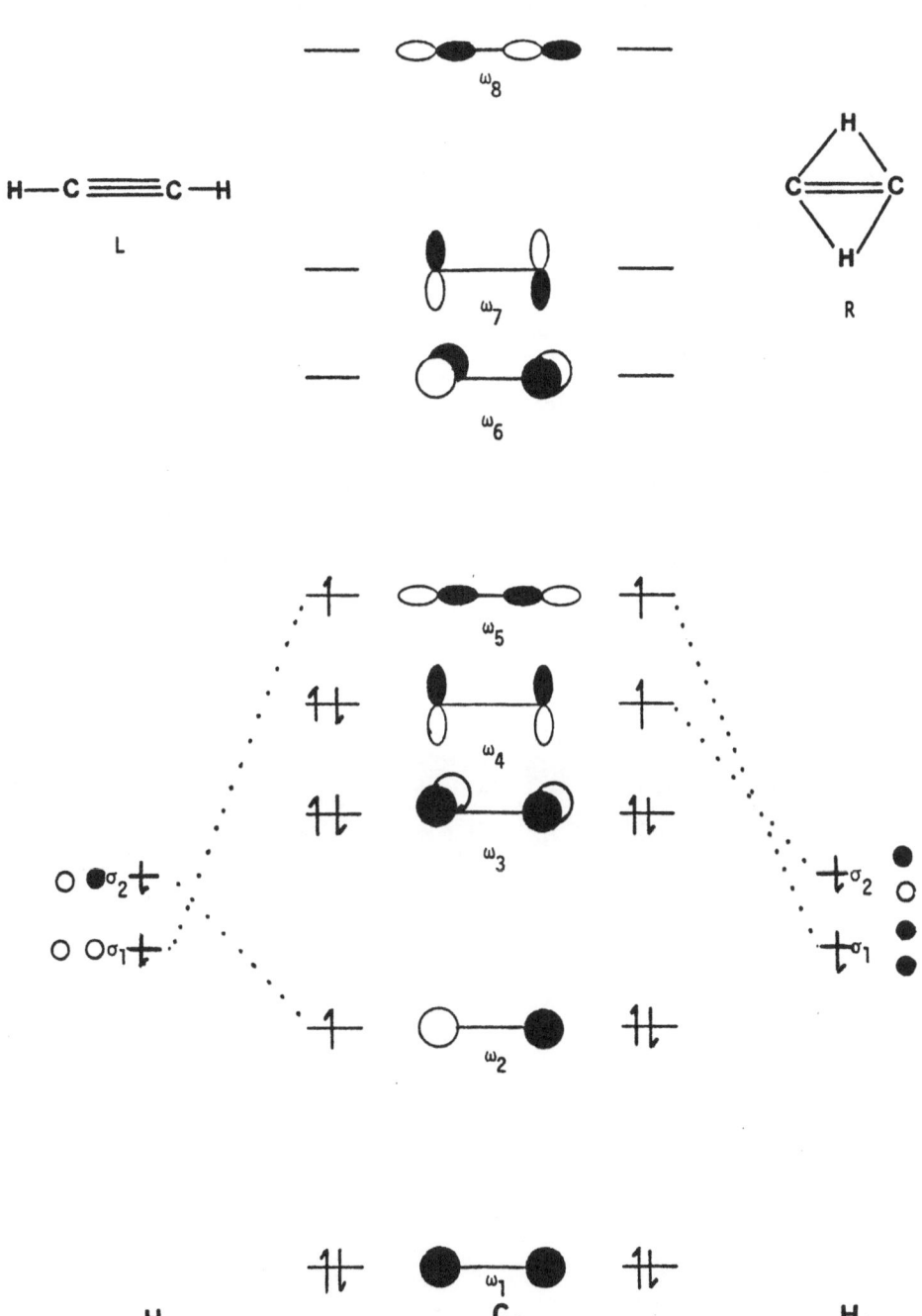

Figure 3. Compact bond diagrams for Linear (L) and Rhomboidal (R) C_2H_2.

a)

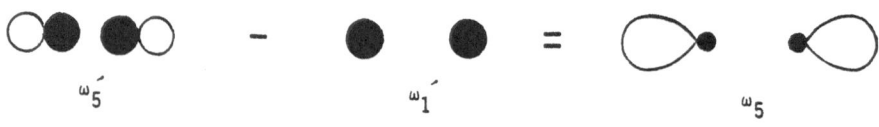

$$\omega_5' \qquad \omega_1' \qquad \omega_5$$

b)

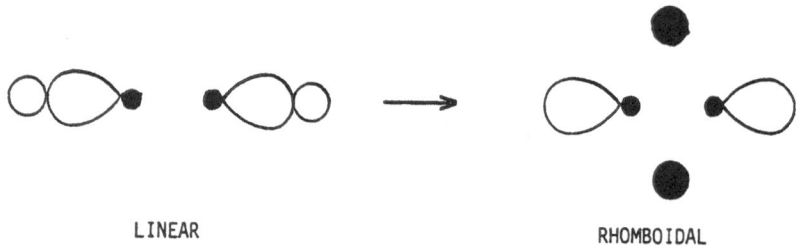

LINEAR   RHOMBOIDAL

Figure 4.   a.   Detailed form of the core orbital ω_5, formed by admixture of ω_1' and ω_5'. b.   Loss of $\omega_5 - \sigma_1$ overlap upon the L $\rightarrow$ R transformation of C_2H_2.

loss of overlap becomes less damaging than core excitation. Accordingly, C_2H_2 is predicted to be linear while C_2Li_2, in the hypothetical absence of any "classical" and 2p AO participation effects, is predicted to be rhomboidal. Actually, a more accurate statement is that the LID model forecasts a <u>tendency</u> of C_2H_2 for linearity and a <u>tendency</u> of C_2Li_2 to adopt a rhomboidal shape.

The above discussion is the first application of the LID model to a stereo-chemical problem. We arbitrarily considered two different geometries. In one of them, the core and ligands were U-bound (core-excited); in the the other, they were $D^{\ddagger}$-bound. The LID model predicts that replacement of strong by weak overlap-binding ligands will shift the preference from U to H or D core-ligand bonding, thus becoming responsible for a different geometrical preference of the derivative molecule. In the case of C_2X_2, it predicts that replacement of H by Li will generate a tendency for a rhomboidal geometry, which is consistent with $D^{\ddagger}$ bonding, rather than a linear geometry, which is consistent with U bonding.

The same problem could have been dealt with in an alternative and more direct way. Instead of arbitrarily choosing the L and R geometries for consideration, we could have begun with the recognition that C_2H_2 is linear, i.e., it exists in a geometry where core and ligands are U-bound. Then we would predict that replacement of H by Li will change the geometry in such a way so that the core becomes deexcited and the core-ligand multicenter bonds are formed through utilization of the higher energy MO's of the core. In other words, we could have attempted to identify a geometry which is consistent with a core configuration $\omega_1^2\omega_2^2\omega_3^2\omega_4^1\omega_5^1$. In turn, this means that we could have tried to find a geometry in which the symmetries of the ligand MO's σ and σ^* match correspondingly the symmetries of the core MO's ω_4 and ω_5. Two possible choices would have been a trans-bent geometry and the rhomboidal geometry we have already dealt with. The "match game" suggested by the LID model is exemplified in Figure 5.

The above discussion makes clear that we now have at our disposal a simple procedure, henceforth denoted as the LID Match procedure, for generating geometries which are likely to be the energy minima of perlithio compounds or any other compounds in which the ligands are weak overlap-binders. Before deciding, however, which of a number of candidate geometries is the global energy minimum in each case, we must consider the additional effects of "classical" bonding and VOP.

What is the effect of Li 2p AO participation on the stereochemistry of C_2Li_2? Consider the case of linear C_2Li_2 in the absence of Li 2p AO participation. The bond diagram will look like that for C_2H_2 (Figure 3) with the exception that the ligand MO's will lie higher in energy. Now, addition of the set of 2p AO's will generate additional ligand MO's, three of which can match the doubly occupied ω_1, ω_3 and ω_4 core MO's, thus generating three new multi-center core-ligand bonds. This matching procedure, henceforth denoted as the VOP Match Procedure, is illustrated in Figure 6.

The next question becomes: How do the new core-ligand bonds generated by converting core lone pairs into core-ligand bond pairs through VOP depend on geometry? As shown in Figure 6, all three bonds get stronger as a transition is made from the L to the R geometry because all corresponding overlap integrals become larger. Note that the ω_4 - X_4 bond in the L geometry has to be replaced by the ω_2 - X_4 bond in the R geometry due to the fact that one electron pair occupies ω_4 in the L and ω_2 in the R geometry. It is then clear that the rhomboidal geometry promotes three new core-ligand bonds which are stronger than the corresponding ones in the linear geometry. Accordingly, we conclude that Li 2p AO participation will favor a rhomboidal over a linear geometry, i.e., it will act in concert with the Li valence AO's in determining the stereochemical preference.

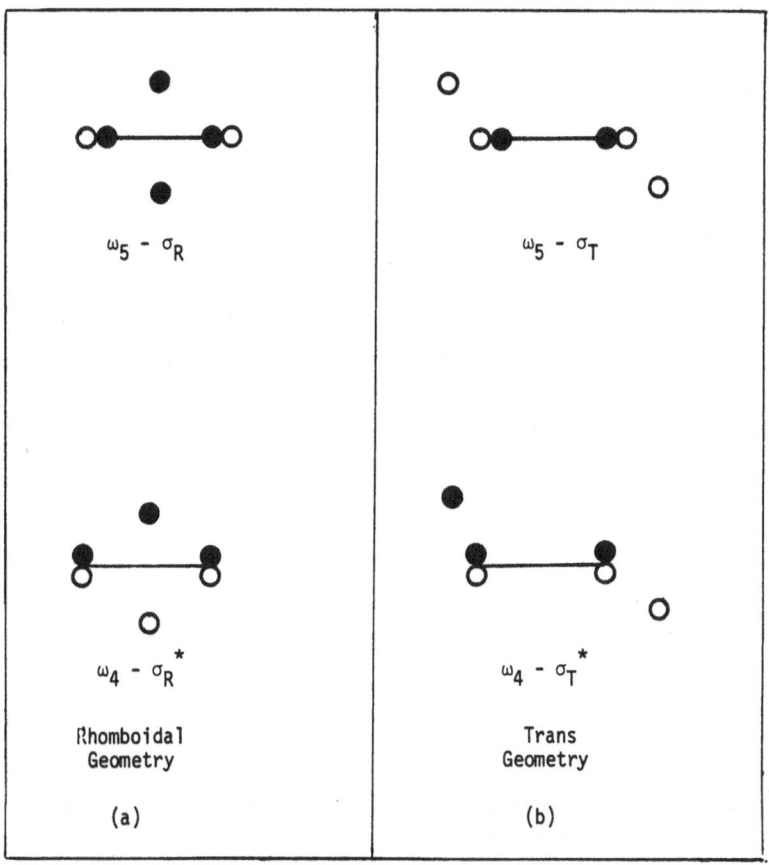

Figure 5. The LID Match Procedure for C_2Li_2. Both the trans bent and the rhomboidal geometries involve core-ligand bonding via the core orbitals ω_4 and ω_5.

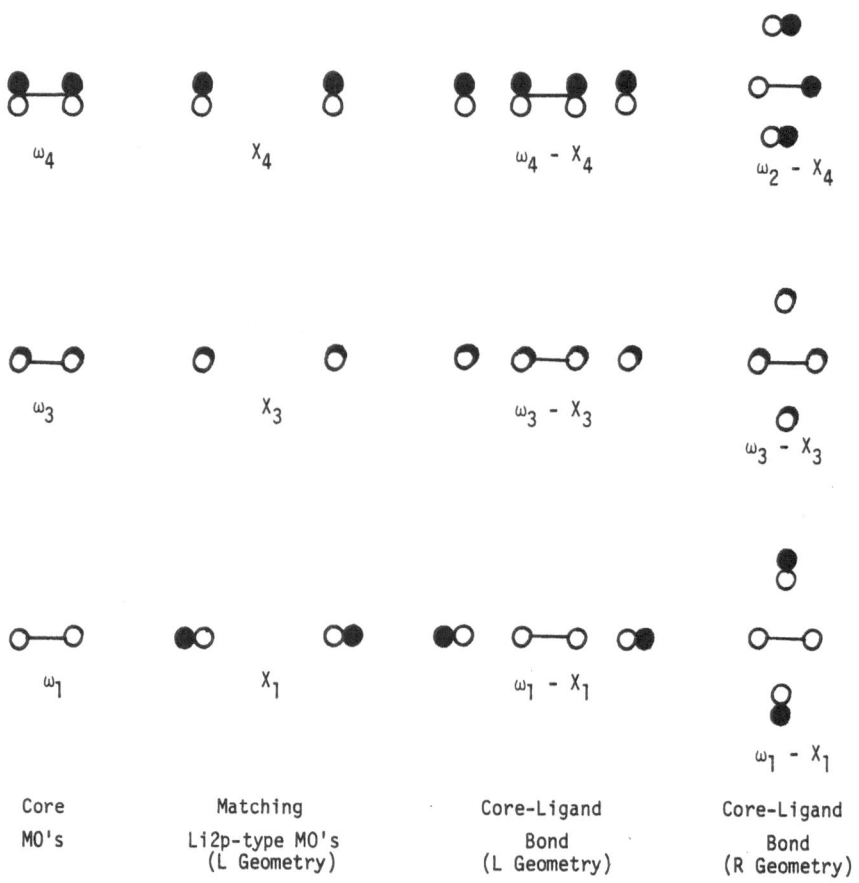

Figure 6. The VOP Match Procedure for C_2Li_2. Introduction of the ligand MO's spanning the Li 2p orbitals allows formation of 3 new bonds. Note that in the R geometry, X_4 matches ω_2 while in the L geometry X_4 matches ω_4.

In general, we will find that, due to the spatial extensions and phase properties of Li 2p AO's, their participation in bonding tends to favor bridged structures.

The final question is: How do "classical" effects operate insofar as dictating molecular stereochemistry? We are impotent to give a qualitative answer to this question. Thus, we must approach this problem by computation in a way we have already discussed.

The computational results shown in Table 3 make unambiguously clear the following trends:

a. The greater stability of L relative to R C_2H_2 is confirmed at both the EHMO and ab initio level, as it should be since the two geometries differ fundamentally in terms of overlap effects.

b. The greater stability of R relative to L C_2Li_2 is predicted at all levels of computation and regardless of whether lithium 2p AO's are included in the AO basis set. Thus, these computations define a tabloid pattern akin to A and constitute evidence in support of the LID model. Several results are worth emphasizing:

1) MOVB theory is a "perfect" theory of chemical bonding to the extent that all (nonrelativistic) electronic effects are explicitly included in the analysis though attention is paid exclusively to the electronic factors deemed most important in the case at hand. Because of this reason, i.e., due to the fact that MOVB theory does not make use of any of the standard approximations of Single Determinant (SD) MO theory, whether of the EHMO or SCF-MO type, the predictions of this brand of theory can formally be tested only by polydeterminantal computations. In this context, the computations cited in support of the LID model are valid indices of its success simply because the underlying principles of the LID model are "contained" within SD MO theory, these being overlap interaction principles. Furthermore, the prediction of

Table 3. Relative Energies (kcal/mole) of Linear and Rhomboidal C_2X_2 (X=H,Li).
at Different Levels of Approximation.

	Linear	Rhomboidal	Ref
X = H			
EHMO (K=1.75)	0	180	a
ab initio	0	197	b
X = Li (no 2p)			
EHMO K = 1.75	0	- 8	a
K = 1.50	0	-14	a
ab initio	0	- 3	c
X = Li (with 2p)			
EHMO K = 1.75	0	-32	a
ab initio	0	-21	c

[a] EHMO calculations performed using optimized geometries from **ab** initio
calculations and the indicated value of K in the Wolfsberg-Helmholz
approximation of β.

[b] Hehre, W. J.; Stewart, R. F.; Pople, J. A. J. Chem. Phys. 1969, 51, 2657.

[c] Ref. 1c.

Schleyer and his coworkers[1c] that Li adopts an R geometry was based on a thorough calculational study in which basis set and "correlation" effects were explicitly probed before the final conclusion was reached. Thus, their calculations constitute a true test of MOVB predictions.

2) The prediction of the LID model that replacing strong by weak overlap binders causes core deexcitation can also be probed by changing the constant K in equation (7) so that the interfragmental AO resonance integrals, β_{tu}'s, become smaller in absolute magnitude. Table 4 contains EHMO results which amply demonstrate the profound impact of β_{tu} reduction effected either by changing the VOIE of H or K in the equation:

$$\beta_{tu} = K (h_{tt} + h_{uu}) s_{tu}/2 \tag{7}$$

with

$$h_{tt} = <t|\hat{h}|t> \tag{8}$$

$$s_{tu} = <t|u> \tag{9}$$

3) The prediction that VOP favors the R over the L geometry is made evident by all computation types.

Table 4. Extended Hückel MO Relative Energies (kcal/mole) of Linear and Rhomboidal C_2H_2.

VOIE H (eV)	K [equation (7)]	Rel Energy [a] $E_{linear} - E_{rhomboidal}$
-13.6	1.75	-180
-13.6	1.50	-107
-3.6	1.75	+52

[a]Standard bond lengths (ref.18) were used throughout.

V. A Failed Experiment

Let us now use the MOVB concepts introduced in this work and try to identify potentially interesting molecules. We have seen that acetylene is linear because the molecule opts for spatial overlap maximization at the expense of modest core excitation. Replacement of the hydrogen ligands by weaker overlap-binding ligands, such as Li, "convinces" the molecule to opt for a different stereo-chemical arrangement which minimizes core excitation at the expense of spatial overlap which, in any event, is not so important because the ligands themselves are only weak overlap-binders. This immediately suggests that another interesting comparison would be that of C_2F_2 and C_2I_2 since the former contains strong and the latter weak overlap-binders. According to the LID model, the former molecule will be predisposed towards a linear and the latter towards a rhomboidal geometry. However, C_2I_2 may still prefer the linear geometry, whereas C_2Li_2 pre-fer the rhomboidal one, for one important reason: In the case of the iodo deriv-ative of acetylene, the ligands have doubly occupied AO's. This difference may become responsible for a different geometrical preference despite the fact that both Li and I are weak overlap binders. The way in which lone pair AO partici-pation in C_2I_2 differs from vacant AO participation in C_2Li_2 is illustrated in Figure 7 which shows the compact bond diagrams for L and R C_2I_2, including the iodine lone pairs. It can now be seen that the R geometry involves one less core-ligand bond than the L geometry. Hence, unlike the case of C_2Li_2, C_2I_2 will be dissuaded from adopting an R geometry despite the desire for core deexci-tation created by the presence of the weakly overlap-binding iodine ligands.

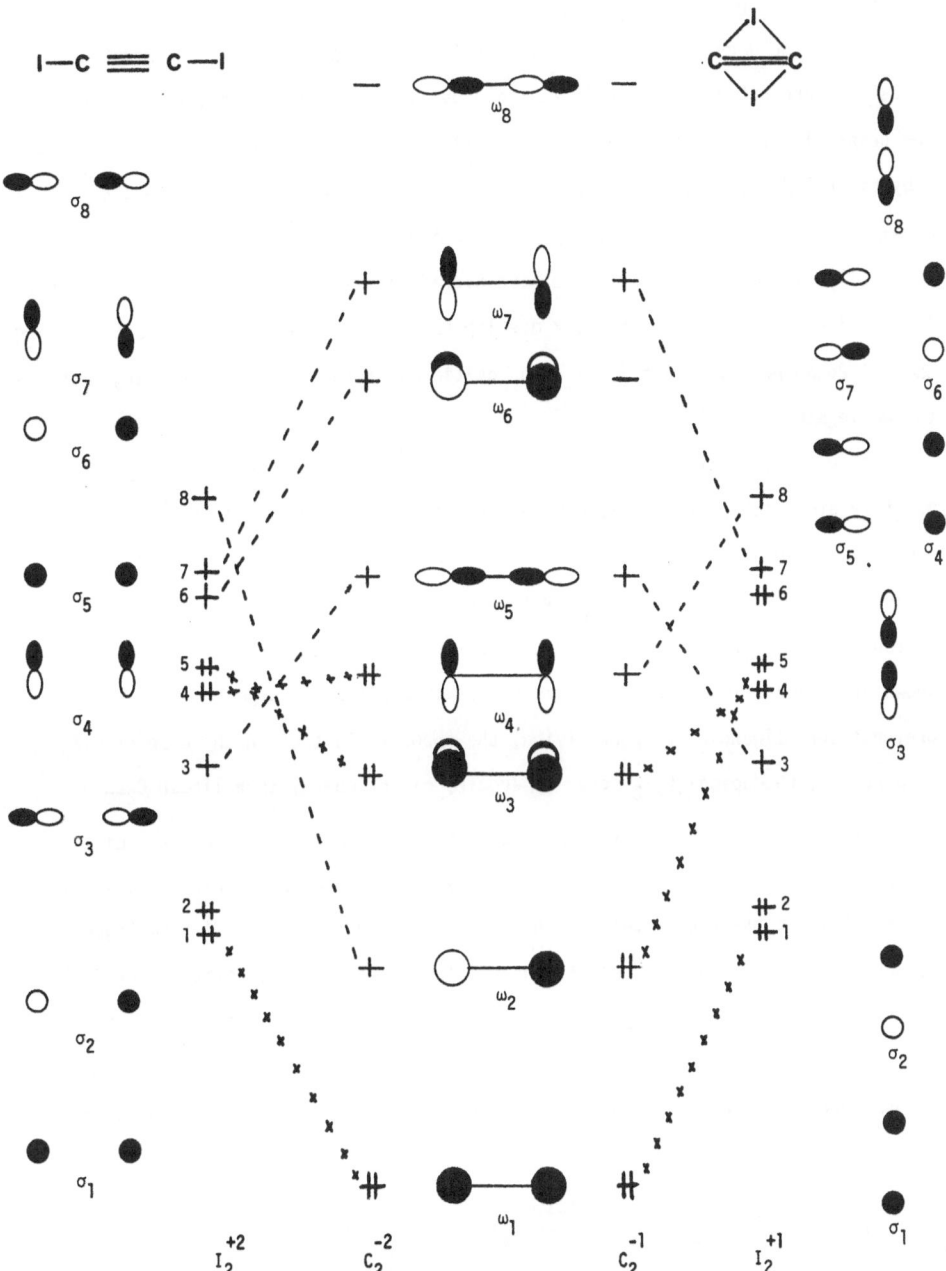

Figure 7. Compact bond diagrams for Rhomboidal (R) and Linear (L) C_2I_2.

VI. The MOVB Viewpoint of Bridging

One of the great advantages of theory is that comparisons of hypothetical situations are not only perfectly feasible but they can also be highly instructive. The transition from a linear to a rhomboidal structure which attends replacement of H by Li in C_2H_2 is an important phenomenon which is directly relevant to the mechanism of bridging itself, a topic of considerable interest among organic and, especially, inorganic chemists. We have already attributed the preference of C_2Li_2 for a bridged structure to the weak overlap bonding ability of Li, while recognizing that Li 2p AO participation is also an additional factor promoting bridging. Is this the whole story?

Let us consider the hypothetical molecule C_2Z_2 in which Z is a highly electronegative atom with weak overlap binding ability. Nature does not afford such an atom but we can use it for instructional purposes. The question now becomes: Would C_2Z_2 exhibit the same tendency for adopting a bridged structure as the C_2Li_2 molecule? The answer can be obtained directly by examination of the bond diagram shown in Figure 3 which makes evident that, while primary charge transfer in C_2Li_2 prevents core-ligand overlap repulsion, the opposite is true in the case of C_2Z_2. As a result, rhomboidal C_2Z_2 tends to acquire higher energy than linear C_2Z_2.

The didactic message of the above hypothetical comparison is clear: Li_2 bridges C_2 because Li is a weak overlap binder and also because delocalization is optimal because the electropositive nature of Li ensures primary CT from ligands to core without engendering core-ligand overlap repulsion. In addition, Li 2p AO participation favors bridging. Bridging is uncommon in organic chemistry simply because most organic compounds, in the narrow sense of the term, are made up of carbon cores and ligands which are either first row atoms or groups containing first row atoms, especially nitrogen, oxygen, and fluorine. Ligands like NR_2, OR, and F are strong binders; they direct interfragmental CT in a way which introduces core-ligand overlap repulsion; they do not have low lying vacant orbitals which can promote bridging; and they contain lone pairs which may interfere with bridging in the way illustrated for iodine.

VII. C_2H_4 versus C_2Li_4

In a previous paper, we developed and illustrated the general Induced Deexcitation model, a specific application of MOVB theory, hoping to draw attention to the way in which the strength of <u>overbonding</u> controls molecular geometry. In this work, we have gone further to formulate the MOVB theory of VOP, since low lying vacant orbitals are hallmarks of weakly overbinding electropositive ligands. Furthermore, we have developed useful procedures, such as the LID Match as well as the VOP Match procedures for routine applications of MOVB theory to molecules. Finally, we have agreed to infer the influence of "classical" effects from computations, in the absence of any simple way to predict them in a qualitative way. We shall now make use of these procedures in order to predict the stereochemical consequences of replacing H by Li in $H_2C = CH_2$.

Ethene ($H_2C = CH_2$) exists in a planar geometry in which an excited core (C_2) and ligands (H_4) are bound in the way indicated by the bond diagram in Figure 8. Replacement of H by Li will give rise to a tendency towards adoption of a new geometry in which a deexcited core binds the ligands. This deexcited core has now the reference electronic configuration $\omega_1^2\omega_2^2\omega_3^1\omega_4^1\omega_5^1\omega_6^1$. Accordingly, the qualifying geometries are the ones which match the ω_3, ω_4, ω_5, and ω_6 core MO's with four ligand MO's of appropriate symmetry. As shown in Figure 9, utilization of the LID Match procedure lead us to anticipate that the preferred geometry of C_2Li_4 will be the one shown below. No amount of intuition would suffice for predicting this extraordinary geometry as the potential ground state global minimum of C_2Li_4! The complete bond diagram for cis-X is given in Figure 8.

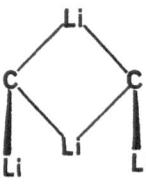

Cis-X

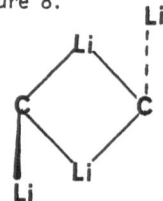

Trans-X

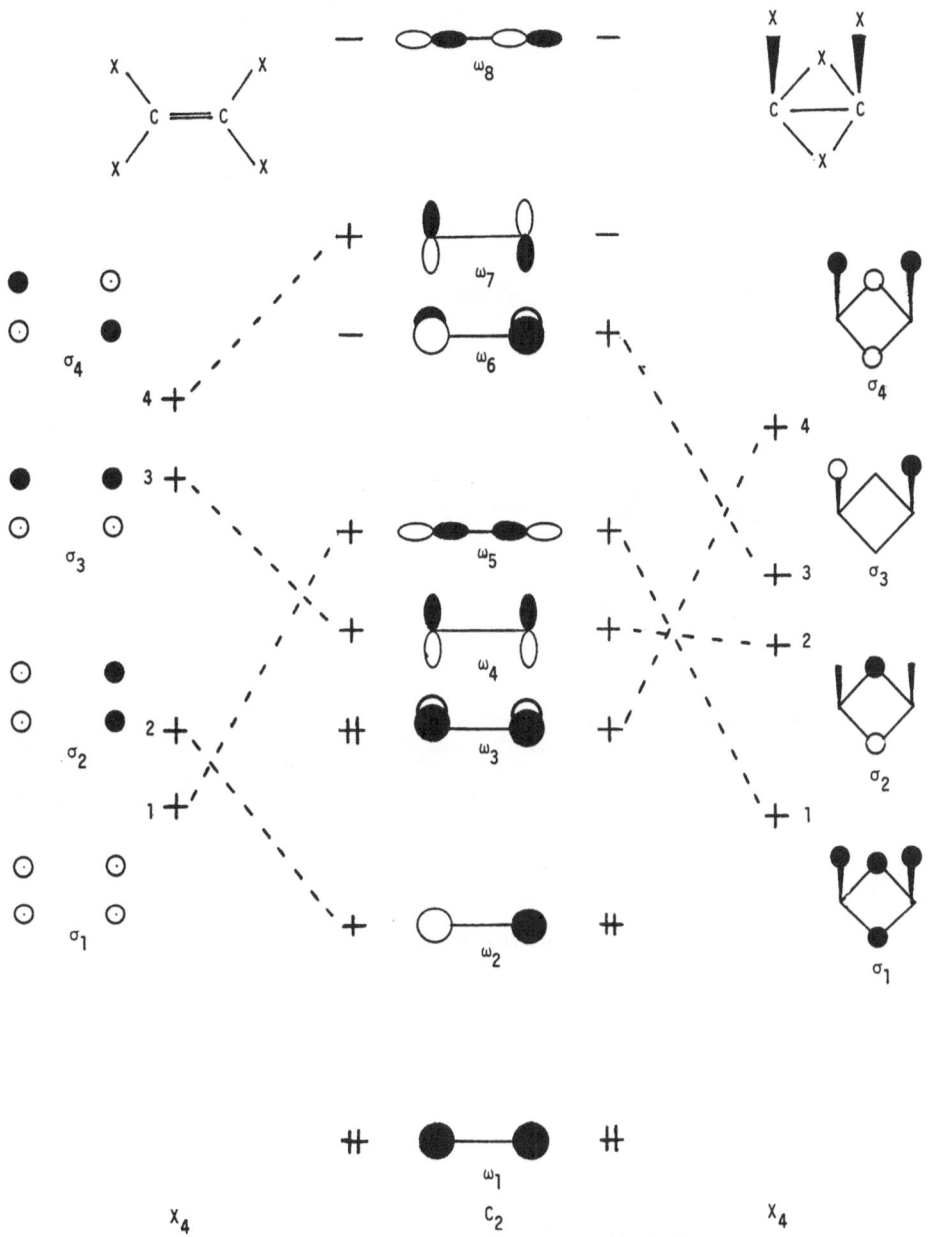

Figure 8. Bond diagrams for planar (D_{2h}) and cis-X (C_{2v}) C_2X_4.

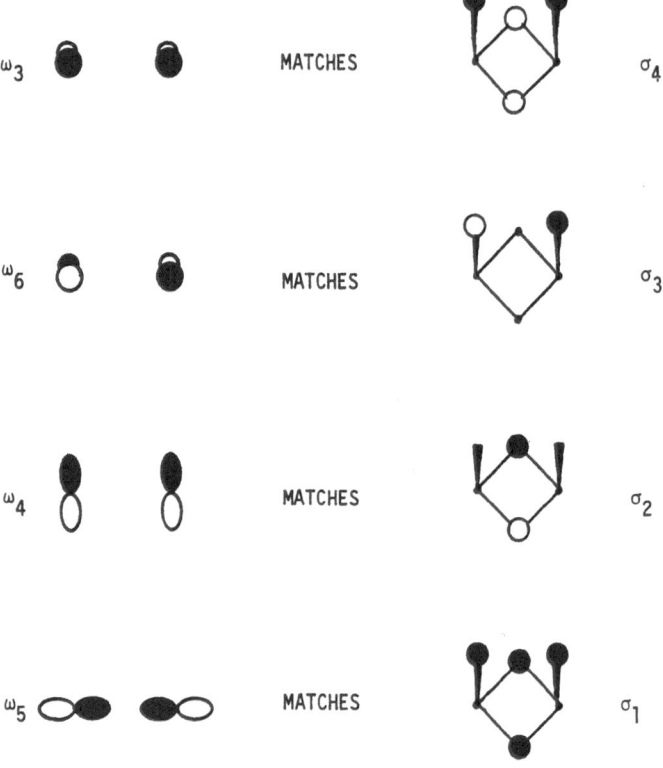

Figure 9. LID Match Procedure for C_2Li_4.

Utilization of the 2p AO Participation Match procedure reveals that an excellent match of the low energy core MO ω_2 (which houses the nonbonding electron in the
deexcited reference configuration of the core) and a ligand MO spanning the 2p AO's
is effected by the fully bridged geometry shown below:

Geometry B :

However, the bond diagram for this geometry shows that there is now <u>one less primary</u>
<u>bond</u> (Figure 10). Hence, the fully bridged geometry, B, is expected to be unfavorable
relative to cis-X predicted by the LID Match procedure. <u>Ab initio</u> calculations of
C_2Li_4 have not yet been reported.

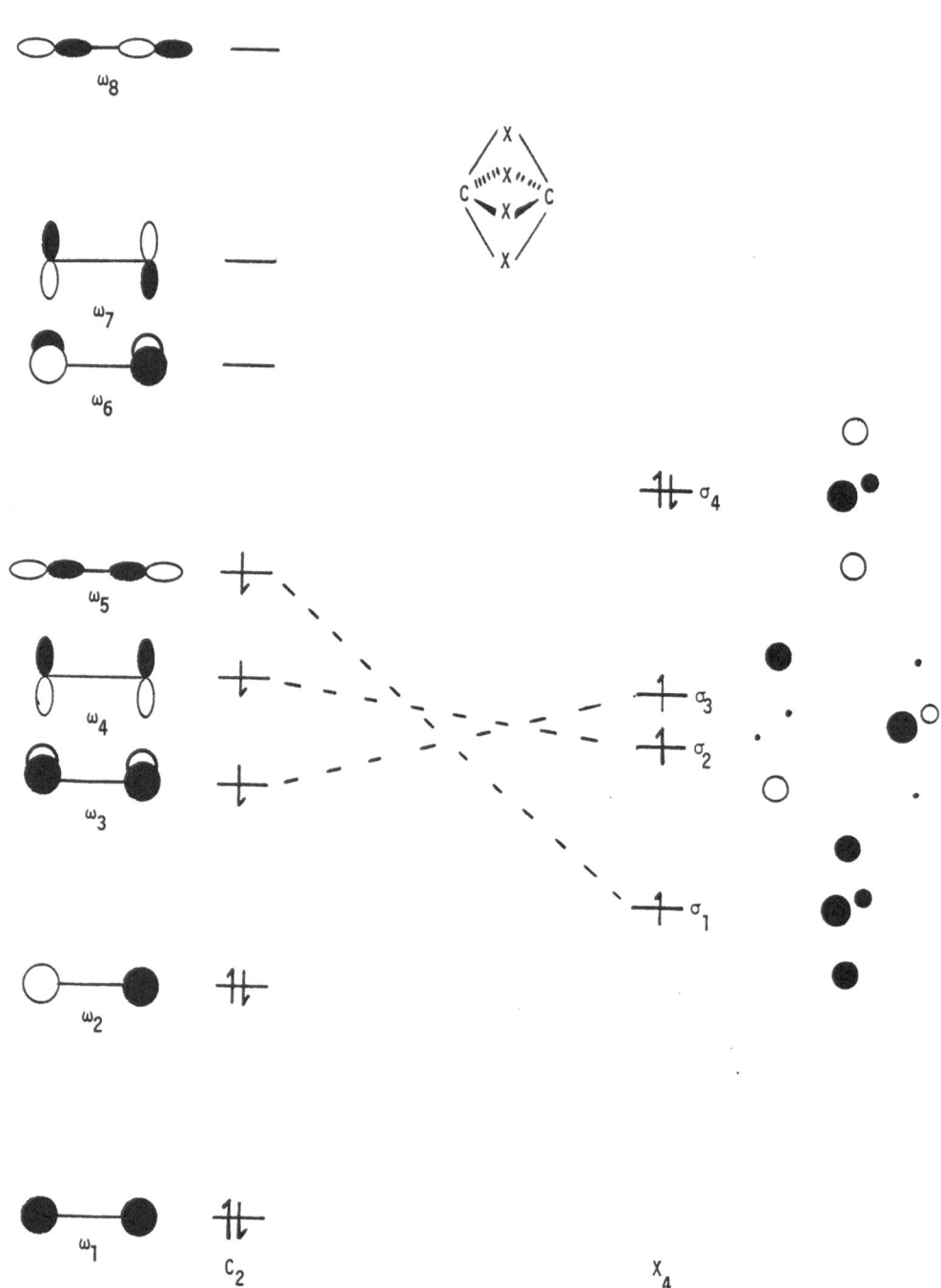

Figure 10. Compact bond diagram for the fully bridged geometry (B) of C_2X_4.

VIII. H_3C-CH_3 versus Li_3C-CLi_3

Ethane (CH_3CH_3) exists in a staggered geometry in which the core (C_2) and ligands (H_6) are bound in the way indicated by the bond diagram of Figure 11. It is immediately obvious that, neglecting the core electron pair which makes up most of the C-C bond, there are no core nonbonding electron pairs. Hence, there can be no core excitation or deexcitation and as a result the LID model predicts that replacement of H by Li will have absolutely no stereochemical consequences other than causing readjustment of the bond angles and C-C bond length. This readjustment of bond lengths is due to the fact that ω_1 and ω_5 are hybridized via σ_1 and the nature of ligands effects the energetic interrelationship of these three MO's and, hence, the hybridization of the corresponding subsystem in a way discussed before.[2] On the other hand, the VOP Match Procedure (see Figure 12) reveals that the optimum geometry for the formation of a core-ligand bond as a result of Li 2p AO participation is the fully bridged geometry shown below. However, much like in the case of C_2Li_4, there is now one less primary bond.

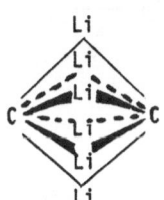

Accordingly, we predict that the preferred geometry of C_2Li_6 will be such as to maintain ethane-like bonding while simultaneously allowing for efficient VOP. The computational results of Schleyer and his coworkers[1k] bear out this expectation as they show that the most stable structure of C_2Li_6 is a distorted version of the staggered ethane structure which effectively allows good match of the doubly occupied core orbital with the ligand orbital spanning the 2p AO's of the lithium

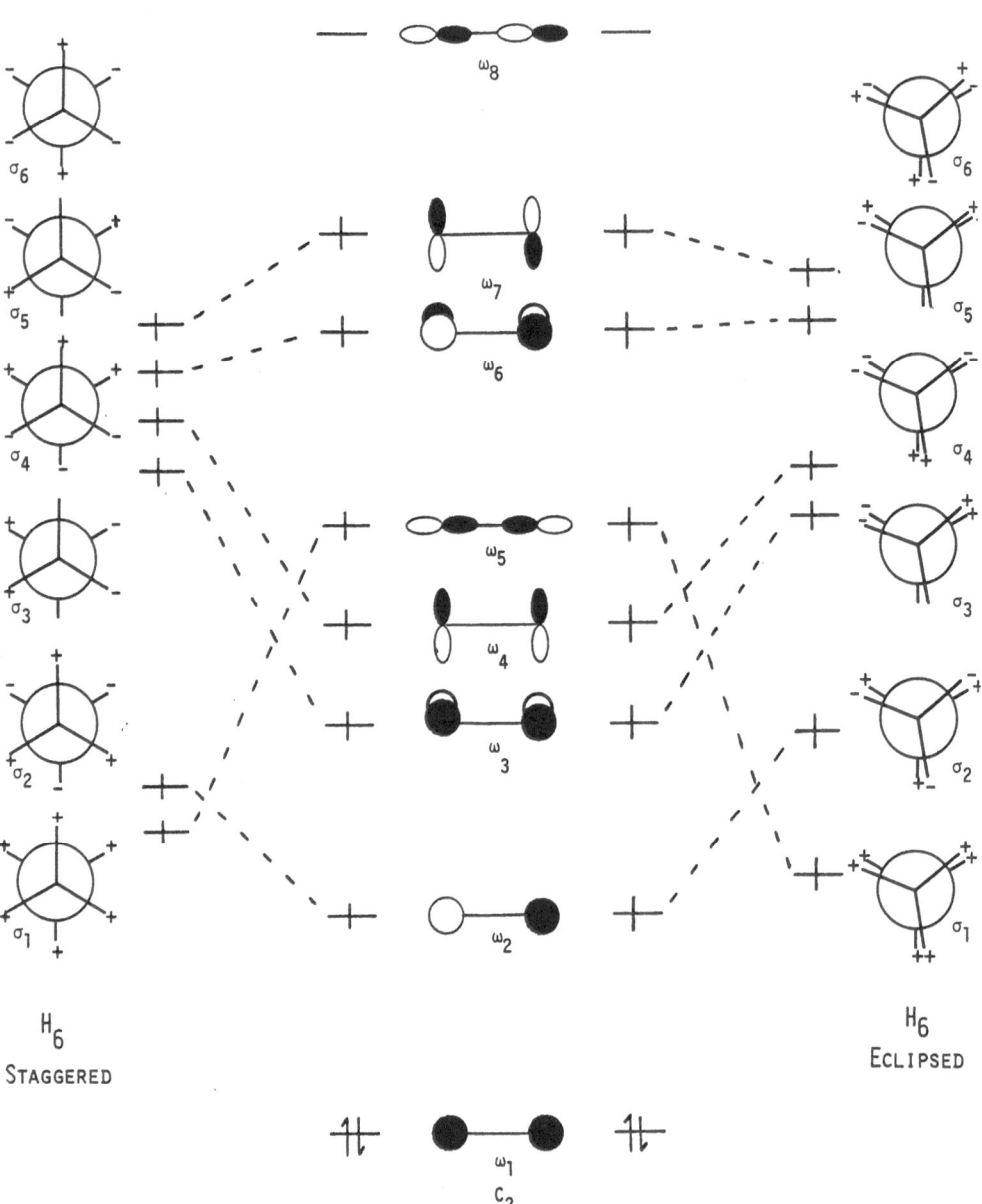

Figure 11. Compact bond diagram for staggered and eclipsed C_2H_6.

atoms (see Figure 11). In other words, the preferred geometry of C_2Li_6 can be regarded as a normal staggered ethane-like geometry which has undergone distortion in order to get the benefit of Li 2p AO participation.

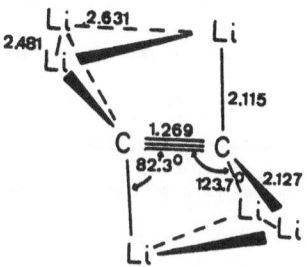

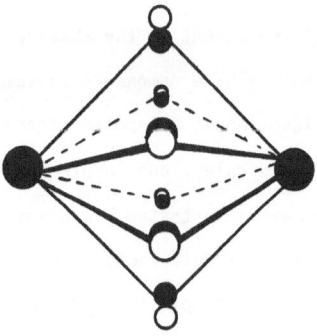

Figure 12. VOP Match Procedure for C_2Li_6.

Conclusion

A molecule can be viewed as a composite system of a core and a set of ligands. When the ligands are strong overlap binders, core excitation may enable the molecule to take advantage of the high overlap binding potential of the ligands. When the ligands are weak overlap binders, core excitation becomes energetically counter-productive. When the core is excited, the set of core orbitals allotted for core-ligand bond formation is different from the set of core orbitals allotted for the same purpose when the core is deexcited. As a result, strong and weak overlap bonding ligands create different geometrical preferences. Hence, we can say that the ovelap binding ability of ligands is a primary determinant of molecular stereochemistry. Replacement of strong by weak overlap binding ligands will always have profound stereochemical consequences unless the core connected to the strong overlap binding ligands is already deexcited, or, the stereochemical trans-formation of interest cannot produce core deexcitation. An example of the first type of exception is C_2H_6 in which the C_2 core makes bonds with the six H ligands by remaining deexcited. As a result, replacement of H by Li is not expected to change radically the geometry of the parent in the absence of Li 2p AO partici-pation. In short, we believe that we have demonstrated adequately the capability of MOVB theory to generate new ideas about molecular stereochemistry, define strategies for violating established rules, and inspire new experimental chemistry. In addition, we believe that we have illustrated how usage of MOVB theory may readily lead to the unraveling of electronic factors, e.g., overlap binding ability versus low lying vacant AO participation, and the roles they play in molecular sterochemistry. Finally, in the process of applying the LID model to hydrocarbons and lithiocarbon molecules we have defined conditions favorable for bridging.

None of the conclusions of this work could be arrived at on the basis of ordinary intuition, although, in an a posteriori sense, they seem understandable and reasonable. Thus, the fact that C_2H_2 is computed to have a rhomboidal geometry even when Li 2p AO's are omitted, the principal mode of operation of the low lying vacant 2p Li AO's, i.e., their action as converters of nonbonding to bonding electron pairs, the circumstances under which bridging can occur as opposed to those under which bridging is unfavorable, etc., can hardly be understood on the basis of present day qualitative MO theory in a simple and self-consistent way. Only space limitations prevent a more extensive application of MOVB theory and the LID model, in particular, to lithio derivatives of hydrocarbons. Thus, before ending this paper, we provide only one more instructive example of how MOVB theory can take us further along than mere intuition.

C_2Li_2 has been determined by computations to have a bridged structure. A chemist trained in MO theory might have intuitively guessed that this geometrical preference is due to the presence of vacant 2p AO's on lithium which become responsible for a double occupancy of an MO of the type shown below (something which is incorrect).

Reasoning by analogy, one might have also guessed that 1,2-dilithioethylene and 1,2 dilithioethane would have the bridged structures shown below:

Neither of these apparently resonable guesses turns out to be correct for reasons that are immediately understandable on the basis of MOVB theory. Specifically, a bridged dilithioethylene structure is unfavorable because it has one less core-ligand bond in comparison with the conventional planar structure. This conclusion is equally valid for any ethylene derivative. The argument can be readily understood by inspection of the bond diagrams of planar and bridged ethylene shown in Figure 13. Bridged 1,2 dilithioethane is also unfavorable relative to the conventional staggered structure for the same reason.

A final and important concluding remark: In all previous discussions, it has been implicitly assumed that lithium has sufficient overlap binding ability so that the global minimum of C_mLi_n is a species with the maximum number of core-ligand bonds, N. However, it must be pointed out that, as this ability tends to zero and depending on the relative electronegativities of core and ligand atoms, the global minimum of a ground state molecule may be a species containing any number from N to zero core-ligand bonds. Furthermore, the geometry of this molecule depends on how many core-ligand bonds are dictated by the overlap binding abilities of the core and ligand atoms. A detailed discussion of this principle is deferred to the next chapter. For the time being, we point out that different computational schemes effectively "give" different overlap binding abilities to an atom and may thus lead to different geometric predictions near the limit of weak core-ligand overlap binding. We have found this problem particularly acute with C_2Li_4 and C_2Li_6 and this is the reason why we did not rely on EHMO computations (which suffer from parametrization arbitrariness and integral and constraint approximations) in order to test the predictions of the LID model in these two cases.

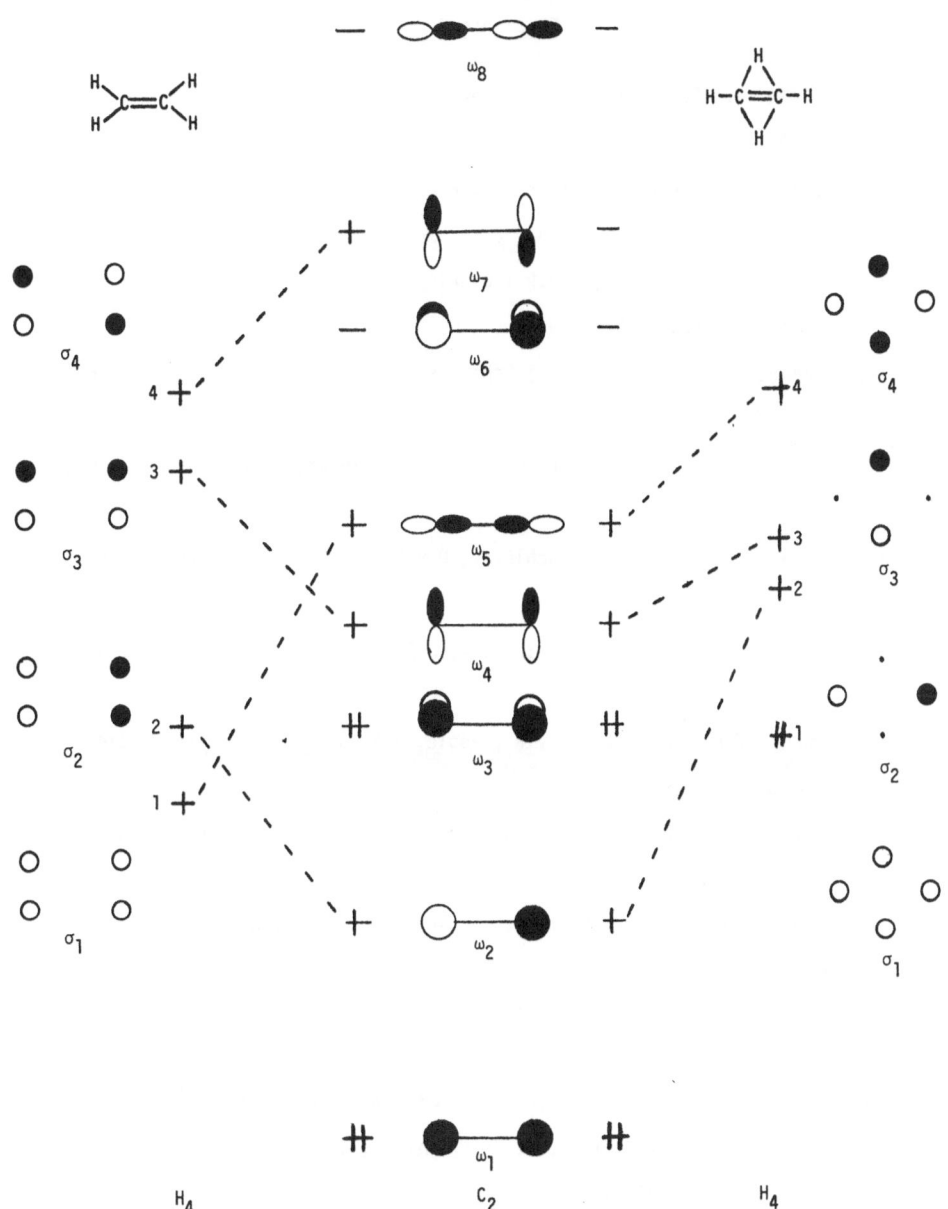

Figure 13. Compact bond diagrams for planar and bridged C_2H_4.

References

1. (a) Apeloig, Y.; Schleyer, P.v.R.; Binkley, J.S.; Pople, J.A. J. Am. Chem. Soc. 1976, 98 4332.

   (b) Collins, J.B.; Dill, J.D.; Jemmis, E.D.; Apeloig, Y.; Schleyer, P.v.R.; Seeger, R.; Pople, J.A. J. Am, Chem. Soc. 1976, 98. 5419.

   (c) Apeloig, Y.; Schleyer, P.v.R.; Binkley, J.S.; Pople, J.A.; Jorgensen, W.L. Tetrahedron Lett. 1976, 3923.

   (d) Jemmis, E.D.; Poppinger, D.; Schleyer, P.v.R.; Pople, J.A. J. Am. Chem. Soc. 1977, 99, 5796.

   (e) Rauscher, G.; Clark, T.; Poppinger, D.; Schleyer, P.v.R.; Angew. Chem. 1978, 90. 306.

   (f) Clark, T.; Jemmis, E.D.; Schleyer, P.v.R; Binkley, J.S.; Pople, J.A. J. Organomet. Chem. 1978, 150, 1.

   (g) Clark, T.; Schleyer, P.v.R.; Pople, J.A. J. Chem. Soc., Chem. Commun. 1978, 137.

   (h) Jemmis, E.D.; Schleyer, P.v.R.; Pople, J.A. J. Organomet. Chem. 1978, 154, 327.

   (i) Jemmis, E.D.; Chandrasekhar, J.; Schleyer, P.v.R. J. Am. Chem. Soc. 1979, 101, 537.

   (j) Jemmis,E.D.; Chandrasekhar, J.; Schleyer, P.v.R. J. Am. Chem. Soc. 1979, 101, 2848.

   (k) Kos, H.; Poppinger, D.; Schleyer, P.v.R.; Thiel, W. Tetrahedron Lett. 1980, 21, 2151.

2. Epiotis, N.D.; Larson, J.R.; Eaton, H. "Unified Valence Bond Theory of Electronic Structure" in Lecture Notes in Chemistry, Vol. 29; Springer-Verlag: New York and Berlin, 1982.

3. The complete MOVB theory of chemical bonding was first presented at the NATO Advanced Study Institute on "Topic in Theoretical Organic Chemistry" in Gargnano, Italy, in June, 1978.

4. Schleyer, P.v.R. Pure Appl. Chem., in press.

5. Slater, J.C. "Quantum Theory of Molecules and Solids", Vol. 1; McGraw-Hill: New York, 1963.

6. Glasstone, S.; Laidler, K.J.; Eyring, H. "The Theory of Rate Processes"; McGraw-Hill: New York, 1941.

7. Herzberg, G.; Monfils, A. J. Mol. Spectrosc. 1960. 5, 482.

8. Gaydon, A.G. "Dissociation Energies and Spectra of Diatomic Molecules"; Chapman and Hall: London, 1968.

9. James, H.M. J. Chem. Phys. 1935, 3, 9.

10. Zemke, W.T.; Lykos, P.G.; Wahl, A.C. J. Chem. Phys. 1969, 51, 5635.

11. Bertoncini, P.J.; Das, G.; Wahl, A.C. J. Chem. Phys. 1970, 52, 5112.

12. Schaefer, III, H.F. "The Electronic Structure of Atoms and Molecules"; Addison-Wesley: Reading, Massachusetts, 1972.

13. Dixon, D.A.; Stevens, R.M.; Herschbach, D.R. Faraday Discussion of Chem. Soc. 1977, 62, 110.

14. Pickup, B.T. Proc. Roy. Soc. A 1973, 333, 69.

15. Gelb, A.; Jordan, K.D.; Silbey, R. Chem. Phys. 1975, 9, 175.

16. (a) Hoffmann, R.; Lipscomb, W.N. J. Chem. Phys. 1962, 36, 2179, 3489; J. Chem. Phys. 1962, 37, 2873.

    (b) Hoffmann, R. J. Chem. Phys. 1963, 39, 1392.

17. Pople, J.A. Acc. Chem. Res. 1970, 3, 217.

18. Pople, J.A.; Beveridge, D.L. "Approximate Molecular Orbital Theory"; McGraw-Hill: New York, 1970.

Chapter 3. The Molecular Orbital-Valence Bond Theory of Excited States.

The original monograph introducing qualitative Valence Bond (VB) and Molecular Orbital Valence Bond (MOVB) theory to the chemical community was entitled "Unified Valence Bond Theory of Electronic Structure".[1,2] In this work, we begin to justify the use of the adjective "unified" by showing how the same MOVB concepts that are applicable to ground state chemistry can be applied to excited state chemistry. In particular, we shall use the MOVB theory and the accessory conceptual tools developed before in order to elucidate the energetic interrelationships of the low lying excited states of a given system and discover ways in which the energy ordering of these states can be altered. After reading the chapter describing the Induced Deexcitation model and this one, it is hoped that the reader will have no difficulty seeing that the energy ordering of <u>different molecular states at fixed geometry</u> and the energy ordering of <u>different ground state geometrical structures</u> are analogous problems which can be handled by the same MOVB concepts.

I. Theory

According to MO theory, the ground state of a closed shell n-electron molecule is constructed by utilization of the Aufbau Principle, i.e., by "feeding" electrons to the MO's starting from the lowest energy MO and proceeding up the ladder until the n/2 lowest energy MO's have been filled. Singly excited states are constructed by promoting one electron from an occupied to an unoccupied MO, doubly excited states by promoting two electrons, and so on. Of course, these are only approximate monodeterminantal descriptions which must be improved by inclusion of Configuration Interaction (CI). Nonetheless, this is the way in which a chemist who is trained in MO theory thinks, in a qualitative sense, about molecular states. According to MOVB theory with core-ligand dissection,[1] the ground state of a closed-shell n-electron system is constructed by generating the maximum number, P, of two-electron multicenter bonds linking core and ligands. MOVB excited states corresponding to the singly excited states of MO theory are of two types:

a. States having P-1 two-electron bonds plus one zwitterionic (singlet) or radicaloid (triplet) antibond.

b. States having P-2 two-electron bonds plus one one-electron and one three-electron bond consistent with singlet or triplet multiplicity.

MOVB excited states corresponding to the doubly excited states of MO theory belong to various types such as the following:

a. States having P-1 two-electron bonds plus a "delocalized antibond".

b. States having P-2 two-electron bonds plus a four-electron antibond.

c. States having P-4 two-electron bonds plus four bonds of the odd-electron variety.

The MOVB states defined in this way are only approximate molecular states.

The MOVB representation of molecular states is the main topic of this chapter and the description given above will be discussed, clarified, and illustrated in detail in a following section. For the time being the important thing to realize is that <u>in comparing approximate MOVB molecular states we shall be comparing numbers and types of core-ligand multicenter bonds</u>. Furthermore, one can now view each approximate molecular state as representing a particular valence state of the core and the ligand fragments. Hence, we shall refer to such molecular states as valence states for two reasons:

a. In order to project the fact that, according to MOVB theory, a study of ground and excited states is tantamount to the study of chemical valence.

b. In order to project the fact that valence states are approximate versions, and very good ones for that matter, of the true molecular states. Hence, we have chosen the letter V to represent a valence state, in general, with the number and type of multicenter bonds being specified in parenthesis next to the symbol V.

II. Ground C-L and Excited States

The simplest system which can be used to illustrate our approach is the two electron-two orbital CL system.[3] The three singlet and one triplet states of CL are shown in Figure 1. We note the following:

a. Figure 1a shows the ground valence state of CL represented by the appropriate bond diagram. Since only one two-electron bond connects the fragments C and L, this state is designated V(2), where the number 2 in parenthesis denotes one two-electron bond.

b. Figure 1b shows what in MO language is called the singly excited state of CL. The notation V(z) indicates that this is an antibonding zwitterionic state because the overlap term of the VB energy expression for V(z) is overall positive:

$$E[V(z)] = \frac{1}{1-s^2} \overbrace{[\epsilon_1 + \epsilon_2 + J_{11} + V_1 + V_2 - (2\beta s + K_{12})]}^{\text{overlap term}} + V_{nn}$$

$s = \langle 1|2 \rangle$ (AO overlap integral)

$\epsilon_1 = \langle 1|-1/2\nabla^2 -Z_c/r|1\rangle$

$\epsilon_2 = \langle 2|-1/2\nabla^2 -Z_\ell/r|2\rangle$ (one electron AO energy)

$V_1 = \langle 1|-Z_\ell/r|1\rangle$

$V_2 = \langle 2|-Z_c/r|2\rangle$ (nucleus-electron coulomb interaction)

$V_{nn} = Z_c Z_\ell/R$ (nucleus-nucleus coulomb interaction)

$J_{11} = \langle 11|11 \rangle$

$J_{12} = \langle 11|22 \rangle$ (electron-electron coulomb interaction)

$K_{12} = \langle 12|12 \rangle$ (electron-electron coulomb exchange interaction)

$\beta = \langle 1|-1/2\nabla^2 - Z_c/r - Z_\ell/r|2\rangle$ (AO "resonance integral")

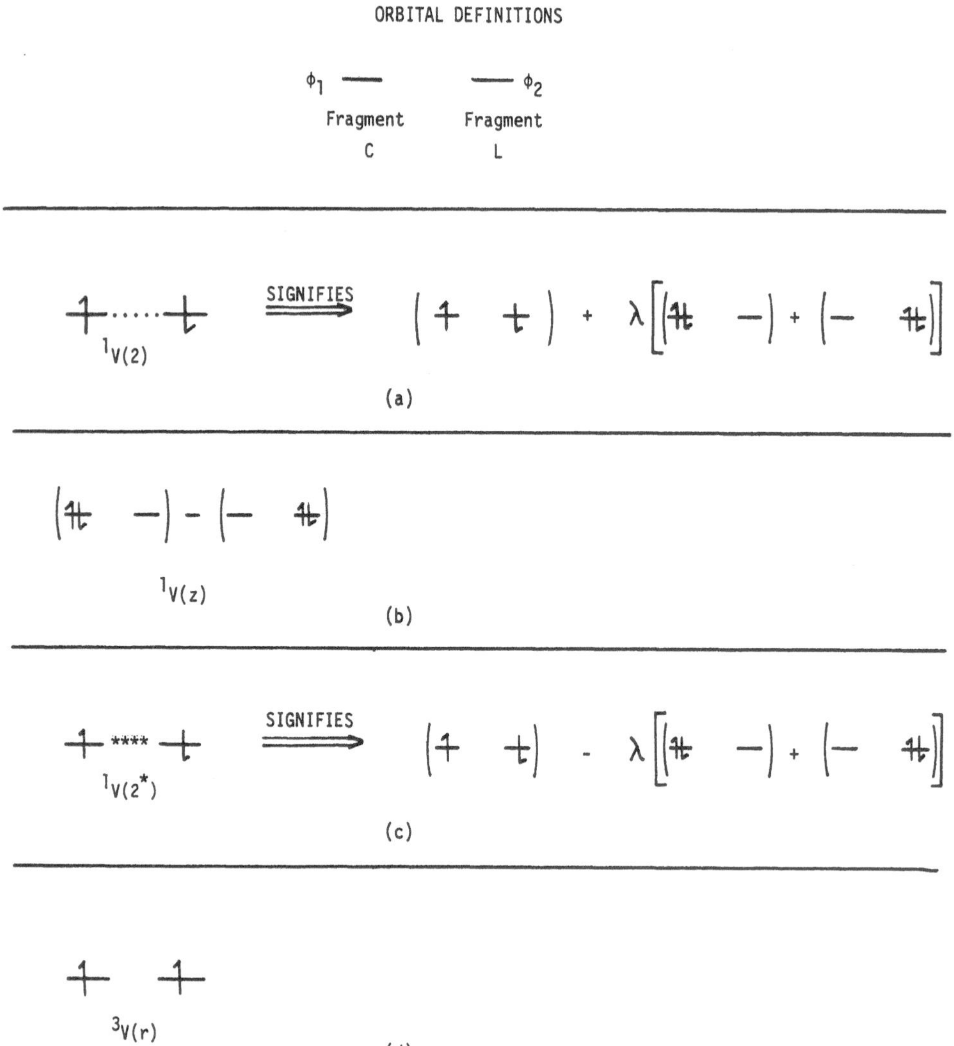

ORBITAL DEFINITIONS

ϕ_1 —   — ϕ_2

Fragment   Fragment
  C   L

SIGNIFIES

$\left(\begin{array}{cc} + & + \end{array}\right) + \lambda\left[\left(\begin{array}{cc} + & - \end{array}\right) + \left(\begin{array}{cc} - & + \end{array}\right)\right]$

$^1V(2)$

(a)

$\left(\begin{array}{cc} + & - \end{array}\right) - \left(\begin{array}{cc} - & + \end{array}\right)$

$^1V(z)$

(b)

SIGNIFIES

$\left(\begin{array}{cc} + & + \end{array}\right) - \lambda\left[\left(\begin{array}{cc} + & - \end{array}\right) + \left(\begin{array}{cc} - & + \end{array}\right)\right]$

$^1V(2^*)$

(c)

$^3V(r)$

(d)

Figure 1. Bond diagrams for valence states of the two-orbital-two-electron system. a. ^1V
(2) state enjoys a two-electron bond. b. ^1V (z) states suffers from an antibonding
zwitterionic interaction. c. ^1V (2*) state suffers from a two-electron antibond, here indicated by asterisks connecting the core and ligand orbitals. d. ^3V (r) state is an antibonding radicaloid state.

where C and L represent the two fragments, c and ℓ represent the corresponding effective nuclei, 1 and 2 stand for the two fragment orbitals ϕ_1 and ϕ_2, and r and R are the nucleus-electron and nucleus-nucleus distances, respectively.

c. Figure 1c shows what in MO language is called the doubly excited state of CL represented by what we may label an __antibond diagram__. Since now a two-electron antibond "connects" fragments C and L, this valence state is designated $V(2^*)$, where the asterisk implies an antibond.

d. Figure 1d shows the triplet state of CL. The notation $^3V(r)$ indicates that this is an antibonding radicaloid state because the overlap term of the VB energy expression for V(r) is overall positive:

$$E[V(r)] = \frac{1}{1-s^2} [\epsilon_1 + \epsilon_2 + J_{12} + V_1 + V_2 \overbrace{-(2\beta s + K_{12})}^{\text{overlap term}}] + V_{nn}$$

It is interesting to observe that the overlap terms of V(z) and $^3V(r)$ are equal.

With the above introduction to excited state electronic structure and having established the notation conventions to be used from now on, we can begin to deal with state manifolds of polyelectronic systems. Before we proceed to do so, we note parenthetically the lucid discussion and ingenius utilization of the "two electron-two orbital-four state" model of Salem and his collaborators.[4]

III. Ground C = L and Excited States

Consider two fragments, C and L, such that they both possess altogether four electrons, C has two nondegenerate orbitals ϕ_C and ϕ_C^*, and L has two nondegenerate orbitals ϕ_L and ϕ_L^*, where the subscript denotes the appropriate fragment and the asterisk indicates the orbital of higher energy within each fragment. Typical examples are shown in Figure 2. Furthermore, let us assume that the two fragments find themselves in such a geometry so that ϕ_C and ϕ_L are of one while ϕ_C^* and ϕ_L^* are of another symmetry species and that the overlap integral $<\phi_C|\phi_L>$ is larger than the overlap integral $<\phi_C^*|\phi_L^*>$. Indeed, this is the case in all examples of Figure 2. Now, the important thing to realize is that many reactive systems of interest can be adequately modeled by four electrons in four orbitals. Thus, the MOVB theory of the electronic states of the four-electron-four-orbital system has an extremely large range of applicibility.

The various valence states of the C=L system are shown schematically in Figure 3 along with the appropriate notation. The electronic nature of the valence states which involve zero, single, and double excitation is discussed below:

a. V (4,0) state. This is described by a single MOVB CW. In turn, this CW describes a four electron antibond due to the interaction of the doubly occupied ϕ_C and ϕ_L. Accordingly, the numbers in parenthesis indicate that this state has only one antibond.

b. ^{1}V (3,1) state. This is described by one bond diagram which, according to the Independent Bond Model, can be regarded as the sum of two uncoupled bond diagrams as illustrated below.

$$\phi_C^* \; \text{+} \cdots \text{-} \; \phi_L^*$$

and

$$\phi_C \; \text{++} \cdots \text{+} \; \phi_L$$

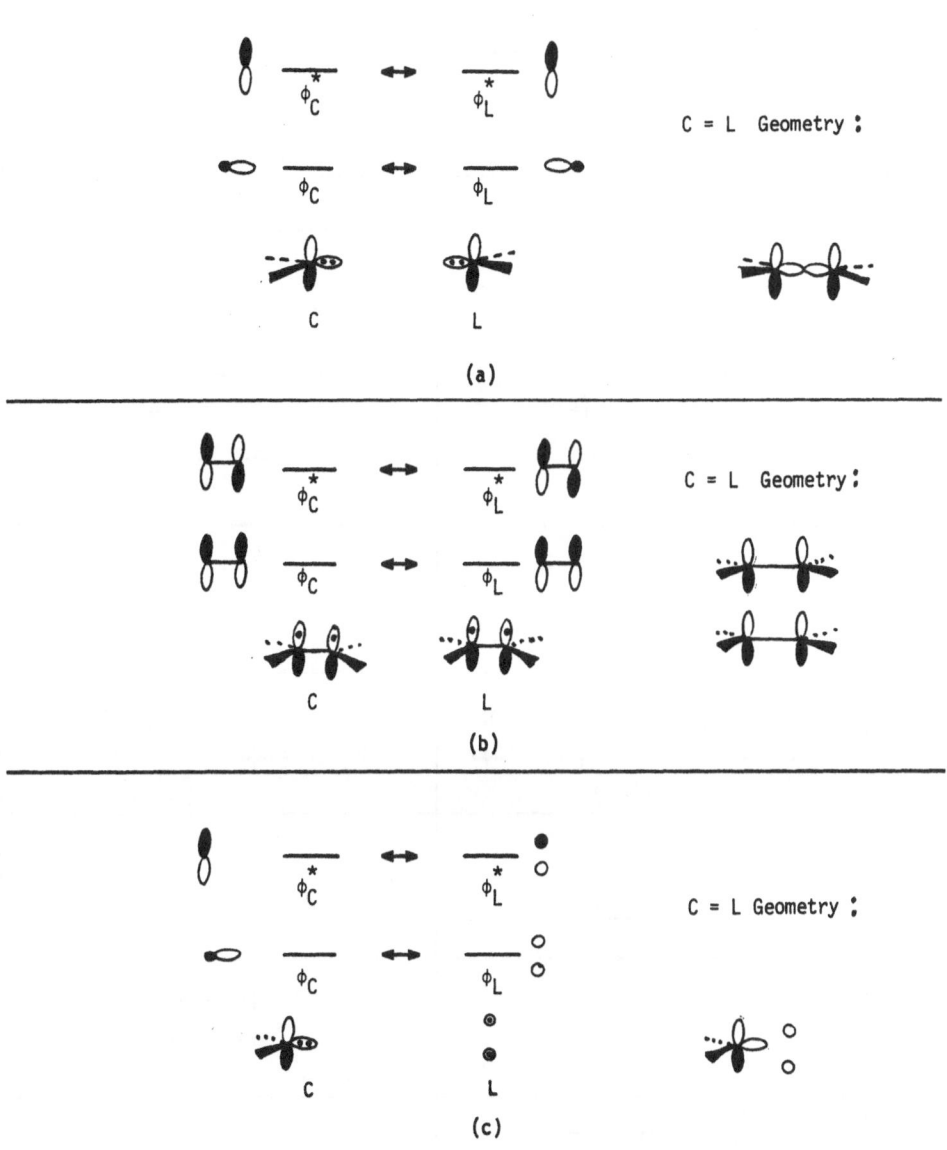

Figure 2. Examples of four-orbital-four-electron systems wherein the two lower energy orbitals, ϕ_C and ϕ_L, are of one symmetry type while the two higher energy orbitals, ϕ_C^* and ϕ_L^* are of another. In all cases $\langle\phi_C|\phi_L\rangle$ is greater than $\langle\phi_C^*|\phi_L^*\rangle$.

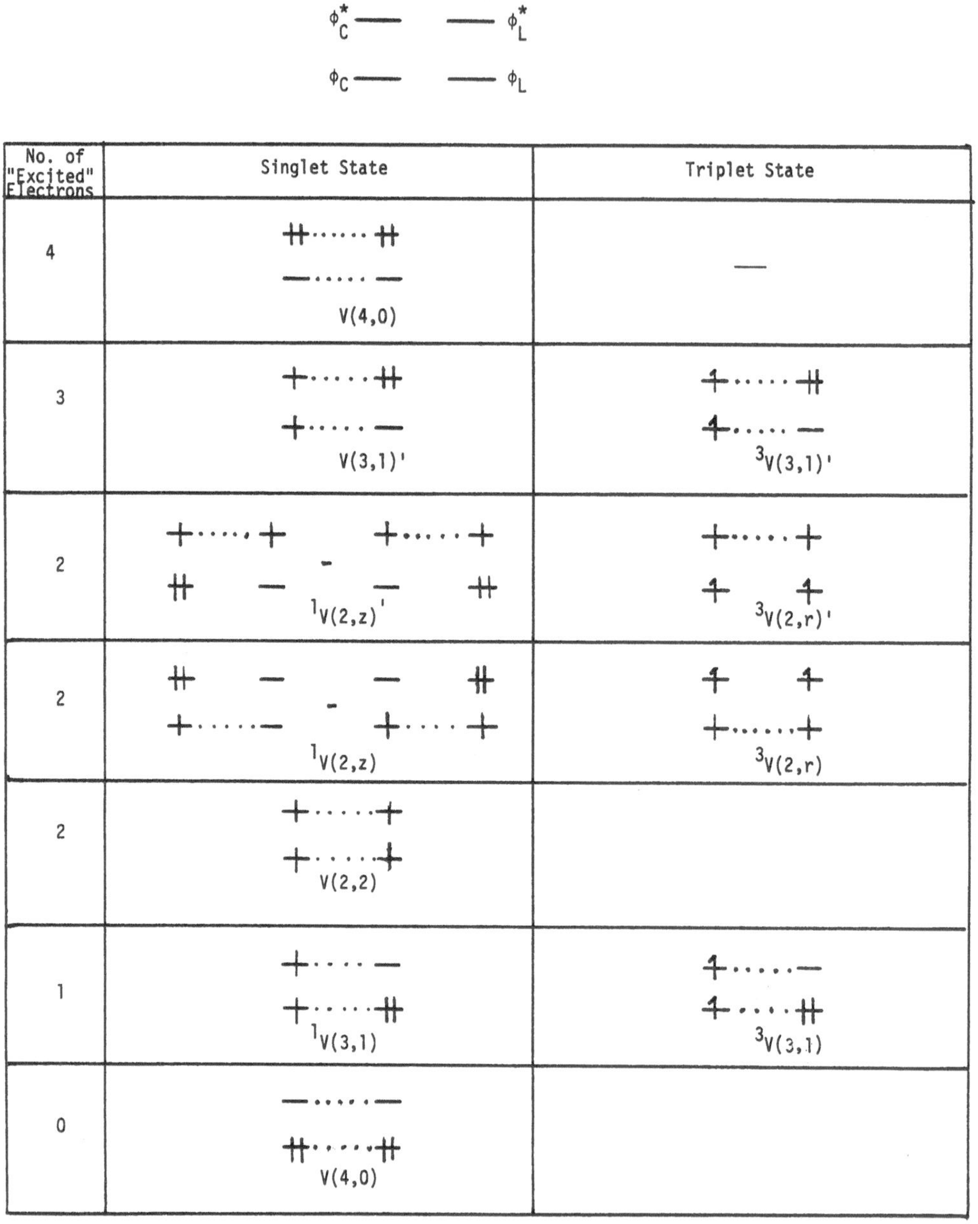

No. of "Excited" Electrons	Singlet State	Triplet State
4	╫ ⋯⋯ ╫ — ⋯⋯ — V(4,0)	—
3	+ ⋯⋯ ╫ + ⋯⋯ — V(3,1)'	+ ⋯⋯ ╫ ╄ ⋯⋯ — ^{3}V(3,1)'
2	+ ⋯⋯ + + ⋯⋯ + ╫ — — ╫ - ^{1}V(2,z)'	+ ⋯⋯ + ╄ ╄ ^{3}V(2,r)'
2	╫ — — ╫ + ⋯⋯ — + ⋯⋯ + - ^{1}V(2,z)	╄ ╄ + ⋯⋯ + ^{3}V(2,r)
2	+ ⋯⋯ + + ⋯⋯ ╄ V(2,2)	
1	+ ⋯⋯ — + ⋯⋯ ╫ ^{1}V(3,1)	╄ ⋯⋯ — ╄ ⋯⋯ ╫ ^{3}V(3,1)
0	— ⋯⋯ — ╫ ⋯⋯ ╫ V(4,0)	

Figure 3. Bond diagrams for the valence states of the four-orbital-four-electron C=L system.

Accordingly, this state can be viewed as containing one one-electron and one three-electron independent bonds and this fact is signified by the numbers in parenthesis following the symbol V.

c. V(2,2) state. This is described by one bond diagram which, according to the Independent Bond Model, can be decomposed into two uncoupled bond diagrams as illustrated below:

$$\phi_C^* + \cdots + \phi_L^*$$

and

$$\phi_C + \cdots + \phi_L$$

Accordingly, this state can be viewed as containing two two-electron independent bonds and this fact is signified by the numbers in parenthesis following the symbol V.

d. ^1V (2,z) state. According to the Independent Bond Model, the linear combination of the two bond diagrams can be decomposed as follows:

$$(\phi_C^* \uparrow\downarrow \cdots - \phi_L^*) - (- \cdots \uparrow\downarrow)$$

and

$$\phi_C + \cdots + \phi_E$$

Accordingly, we say that this state contains one two-electron bond and one zwitter-ionic antibond and this fact is signified by the number 2 and the letter z in parenthesis following the symbol V.

The electronic nature of the triplet valence states need not be discussed separately. Thus, $^3V(3,1)$ is entirely analogous to ^1V (3,1), the only difference being that the former has lower energy than the latter due to better coulomb correlation. Furthermore, ^3V (2,r) is entirely analogous to ^1V (2,z), the only difference being that the former has one two-electron bond and one radicaloid anti-bond, as compared to the latter having one two-electron bond and one zwitterionic

antibond, and that ^3V (2,r) profits from better coulomb correlation than ^1V (2,z)
does. This difference between singlet and triplet states will be discussed in a
following section.

In general, states which involve triple and quadruple electron excitation have
much higher energy than states of zero, single, or double electron excitation.
Furthermore, since ^1V (2,z) has lower energy than the related ^1V (2,z)' and ^3V (2,r)
lower energy than the related ^3V (2,r)', as a result of the assumption that $<\phi_C|\phi_L>$
is greater than $<\phi_C^*|\phi_L^*>$, we can focus our attention exclusively on the V (4,0),
^1V (2,z), V (2,2) singlets and the ^3V (3,1) and ^3V (2,r) triplets. Henceforth, we
shall refer to these as the low lying valence states of a four-electron-four-
orbital system.

The following clarifications are now in order:

a. Some of the V states shown in Figure 3 have the same and others have different
symmetry when compared in a pairwise sense. Thus V (4,0) and V (2,2) have the same
while V (4,0) and V (3,1) have different overall symmetry. Thus, in a formal sense,
V(4,0) and V (2,2) are zero order states since they mix in a two-electron sense.
The true electronic states are linear combinations of the type μ_1 V (4,0) + μ_2 V (2,2)
and ν_1 V (2,2) - ν_2 V (4,0). The symbol V serves the purpose of reminding us that
the conceptual building blocks of the MOVB theory of electronic states are valence
states which are approximate representations of the rigorous electronic states.
This approximation is entirely justified because the energy minima of valence states
come at different interfragmental distances, r_{CL} (see Figure 4). As a result, the
mixing of the valence states can be disregarded, in a first approximation. Typically,
the minimum energy of V (4,0) is achieved at large r_{CL} at which distance the energy
of V (2,2) is extremely high and, hence, the mixing of V (4,0) and V (2,2) small.
Of course, at distances other than equilibrium distances, one must consider explicitly

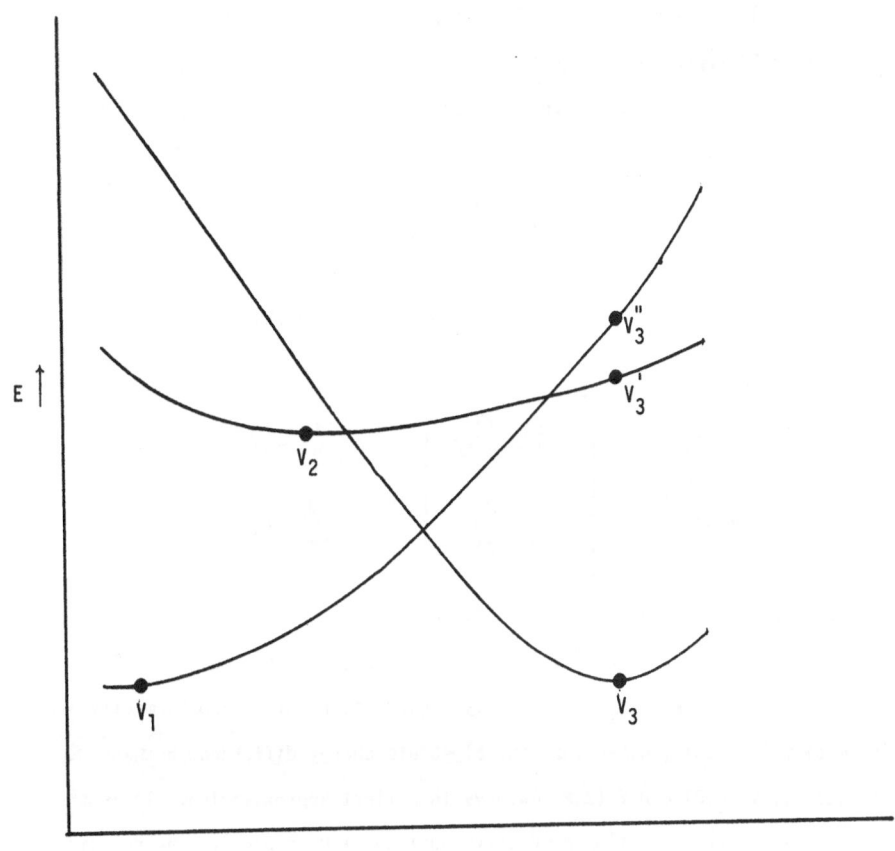

Figure 4. Energy of valence states as a function of interfragmental distance. V_1, V_2, and V_3 represent the energyminima of different valence states while V_3, V_3', and V_3'' represent one valence state and its related Franck-Condon states.

the mixing of valence states. Thus, for example, V (4,0) represents adequately the reactants, V (2,2) represents adequately the products, but only V (4,0) - V (2,2) is a good representation of the transition state of the so called "forbidden" reaction shown below:

C: + L: ⟶ [C: L: ↔ C = L] ⟶ C = L

V (4,0)   V (4,0) ↔ V (2,2)   V (2,2)

Reactants   Symmetrical   Products
                Transition State

A specific example is:

It is evident that what the chemist traditionally labels "reactants" and "products" are two different valence states, V (4,0) and V (2,2), of the same composite system. Furthermore, what the chemist traditionally regards as reaction exothermicity or endothermicity is nothing other than the algebraic energy difference between two valence states, V (4,0) and V (2,2), always in a first approximation. It is the purpose of this paper to develop predictive notions with respect to the relative energies of these and other low lying valence states.

   b. If the relative orientation of the two fragments A and B is fixed, the energy minima of different valence states occur at different interfragmental distances. A typical example is given in Figure 4. Now, frequently chemists are concerned with Franck-Condon excited states as illustrated in Figure 4. In this work, our concern is entirely different. Specifically, we are attempting to understand the factors which dictate the relative energies of V_1, V_2, and V_3, rather than the relative energies of one valence state and its related Franck-Condon states,

e.g., V_3, V_3', V_3''.

    c. The schematic MOVB and MO descriptions of low lying valence states of CL are shown in Figure 5. The generation of the MO from the MOVB description is particularly simple and it can be illustrated by reference to a specific example. Thus, the ground state of a two-electron-two-orbital system is represented by the MOVB bond diagram shown below. The corresponding MO representation is generated by summing the number of electrons occupying ϕ and x in the bond diagram and placing them in the MO resulting from the bonding interaction of ϕ and x.

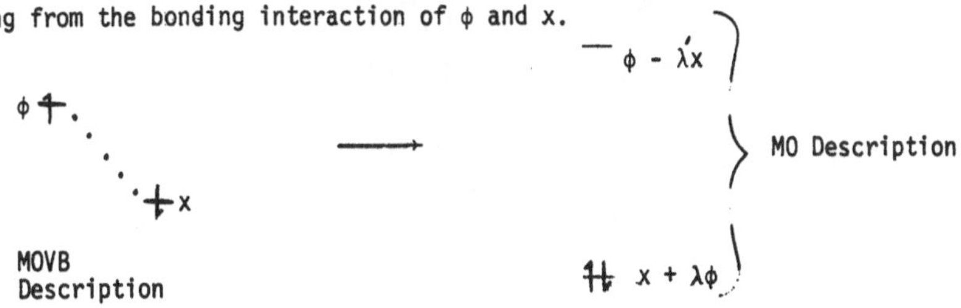

A second illustration can be given by reference to the specific case of V (4,0). The MOVB representation of V (4,0) is the single CW in which four electrons are associated with ϕ_C and ϕ_L. This means that the MO representation of V (4,0) will have four electrons associated with the MO's resulting from the interaction of ϕ_C and ϕ_L.

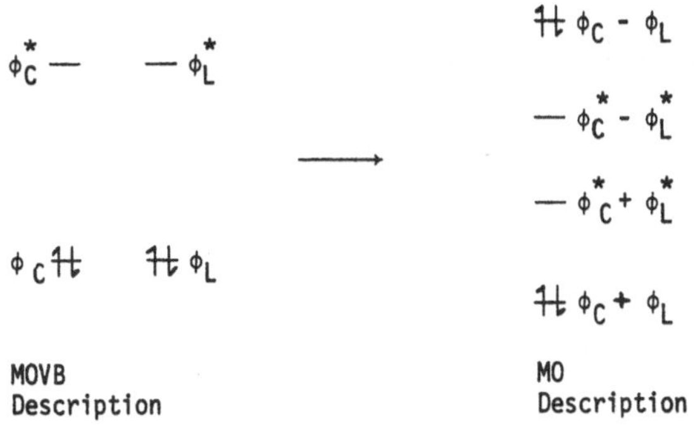

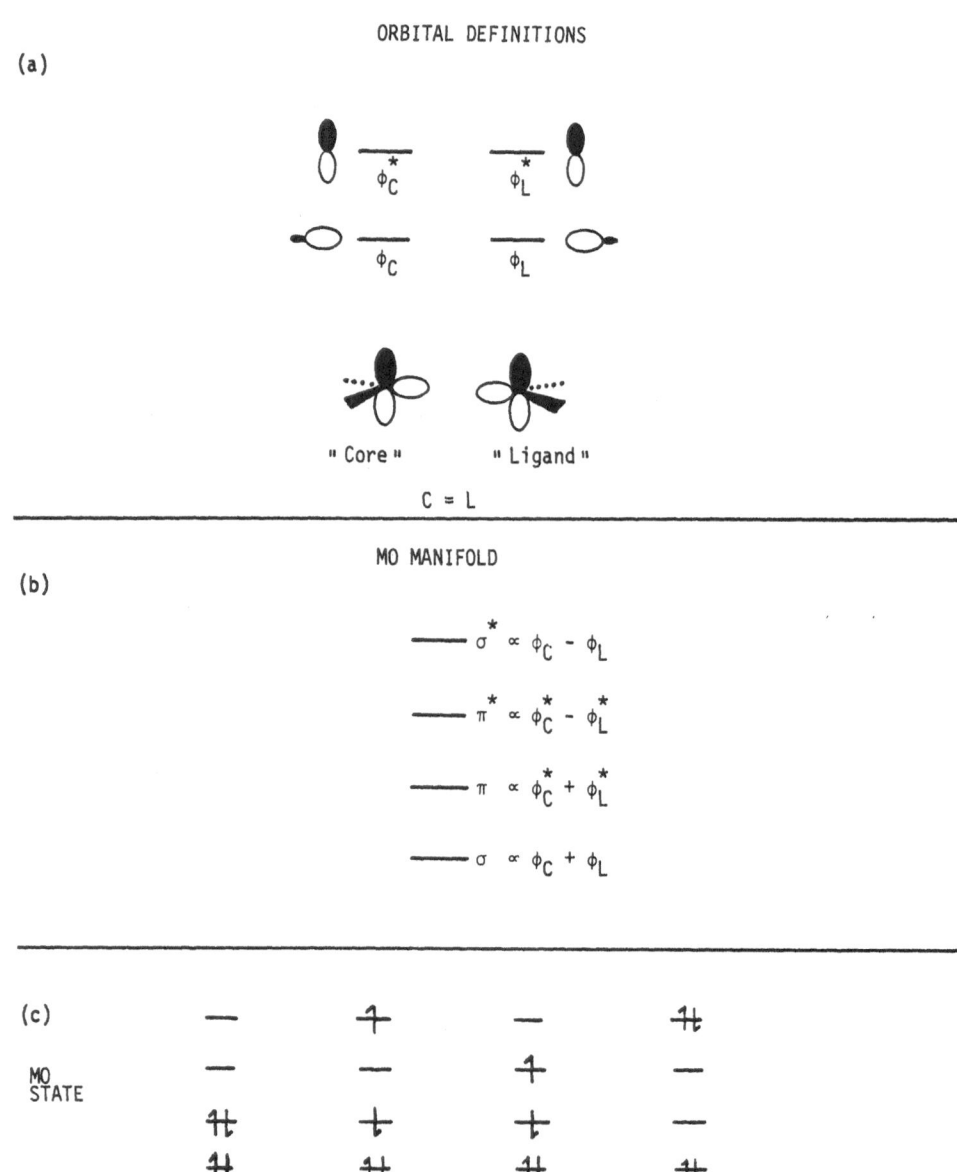

Figure 5. a. Orbital definitions for the four-electron-four-orbital system. b. MO manifold for the four-electron-four-orbital system. c. Schematic MO and MOVB representations of the low lying valence states of the four-electron-four-orbital system.

IV. Excitation and Overlap Binding: The Determinants of the Relative Energies of
     Molecular Electronic States

What are the factors that determine the relative energies of the various valence
states of CL shown in Figure 3? We recognize the following important variables:

a. Excitation energy. It is evident that different valence states involve
different degrees of electronic excitation as preparation for interfragmental bonding.
Thus, with respect to excitation energy, the stability order of the various low lying
valence states would be: ^1V (4,0) > ^1V (3,1) > ^1V (2,2), ^1V (2,z) and ^3V (3,1) >
^3V (2,r).

b. Bond strength. In addition to differing in terms of excitation energies,
the various states differ in terms of the number and types of core-ligand bonds due
to the overlap of core and ligand fragment orbitals. Thus, in the singlet manifold,
V (4,0) is characterized by a four-electron antibond, V (3,1) by a three-electron
bond (or antibond) and a one-electron bond, V (2,z) by a single two-electron bond
and a zwitterionic antibond and V (2,2) by two two-electron bonds. In the triplet
manifold, V (3,1) is characterized by one three-electron bond (or antibond) and one
one-electron bond and V (2,r) by one two-electron bond and a radicaloid antibond.

What are the relative energies of <u>sets</u> of different types of bonds? More
specifically, what are the relative energies of the following sets?

1) One set made up of two two-electron bonds, e.g., as in V (2,2), denoted as
the [2,2] set.

2) One set made up of one three-electron bond (or antibond) and one one-
electron bond, e.g., as in V (3,1), denoted as the [3,1] set.

3) One set made up of only one two-electron bond with an antibonding zwitterionic interaction neglected, e.g., as in V (2,z), denoted as the [2] set.
Now, the energy ranking of these sets depends on the nature of the interacting

fragment orbitals which, in turn, depends on the nature of the constituent atoms
of the two fragments. In a previous paper, we have seen that atoms or groups can
be classified according to their overlap binding (overbinding) ability. Specifically, if orbital x belonging to a variable atom or group A overlaps with orbital y
belonging to a fixed atom or group B and the corresponding resonance integral β
is large, in absolute magnitude, then we say that A is a strong overbinding group
or atom. If β is small, in absolute magnitude, we say that A is a weak overbinding
group or atom. As we have seen before, the relative overbinding abilities of first
and second row atoms are as shown in Table 1.

   With the above definitions in mind, we distinguish the following two extreme
situations:

   a. The constituent atoms are strong overbinders, i.e., they are located
towards the right of a period and the top of a column of the Periodic Table. In
this case, core-ligand bonding is due primarily to the overlap ("semiclassical"
and "nonclassical") interaction of the two fragments. Accordingly, the relative
energies of the [2,2], [3,1], and [2] sets will be determined by the relative
strengths of the one-, two-, and three-electron bonds. In order to determine
qualitatively the relative strengths of these bonds, it is useful to define the
following quantities:

   1) The Heitler-London (HL) CW's, Ψ°'s, for one-, two-, and three-electron
bonds shown below:

$$\text{+} \; \text{-} \qquad\qquad \text{+} \; \text{+} \qquad\qquad \text{+} \; \text{+}$$

   2) E° is the energy of the HL CW at infinite interfragmental distance (r=∞),
and E is the energy of the same CW at equilibrium interfragmental distance, r_{eq}.
On this basis, we define $\Delta E' = E - E^{\circ}$. If the "classical" coulomb interaction
terms are set equal to zero, $\Delta E'$ is positive in the case of a three electron bond,
zero in the case of one-electron bonds, and negative in the case of a two-electron
bond (see Table 2).

Table 1. Relative Overlap Binding Abilities of First and Second Row Atoms $-(\beta_{X-H}/K).$[a]

F	O	N	C	B	Be
16.56	16.98	16.53	19.45	18.20	15.81
Cl	S	P	Si	Al	Mg
13.31	12.81	12.33	17.47	15.13	12.54

[a]For F, O, and N β_{HX} is the average of $\beta_{HX}(sp^3)$ and β_{HX} (p). For C and Si sp^3 hybridization is assumed, for B and Al sp^2 hybridization for Be and Mg sp hybridization and for Cl, S and P p hybridization.

Table 2. Energy Expressions for One-, Two-, and Three-Electron Bonds.[a]

Interaction Term	Energy Expressions	
	$\Delta E'$	H_{ij}
One Electron Bond		
$< (\text{↑ —}) \|\hat{H}\| (\text{↑ —}) >$	0	-
$< (\text{↑ —}) \|\hat{H}\| (\text{— ↑}) >$	-	β
Two Electron Bond		
$< (\text{↑ ↑}) \|\hat{H}\| (\text{↑ ↑}) >$	$+2\beta s$	-
$< (\text{↑ ↑}) \|\hat{H}\| \phi^+ >$ $\phi^+ = (\text{↑↓ —}) + (\text{— ↑↓})$	-	$\frac{1}{\sqrt{2}}[2\beta+(\varepsilon_1+\varepsilon_2)s]$
Three Electron Bond		
$< (\text{↑↓ ↑}) \|\hat{H}\| (\text{↑↓ ↑}) >$	$-2\beta s$	-
$< (\text{↑↓ ↑}) \|\hat{H}\| (\text{↑ ↑↓}) >$	-	$-[\beta+(\varepsilon_1+\varepsilon_2)s+3\beta s^2]$

[a] Energy terms are defined in the text.

3) $\Delta E''$ is the energy depression resulting from the interaction of $\Psi°$ at r_{eq} with the rest of the CW's. The CW interaction in each of the three cases is shown schematically below. Each CW interaction matrix element, H_{ij}, is proportional to $k\beta$, where β is the AO resonance integral (see Table 2):

The energy gaps separating the interacting CW's favor a larger $\Delta E''$ in the cases of one- and three-electron bonds but the CW interaction matrix element favors a larger $\Delta E''$ in the case of a two-electron bond.

Piecing the above information together leads to the conclusion that a two-electron bond is very strong, a one-electron bond has intermediate strength, and a three-electron bond is either a weak bond or a weak antibond, depending upon the absolute magnitude of $\Delta E'$ and $\Delta E''$.

What is the dependence of one-, two-, and three-electron bond strength upon spatial overlap? Inspection of Table 2 reveals that, since β is directly proportional to s, an increase in spatial overlap will tend to leave $\Delta E'$ unaffected in the case of a one-electron bond while it will make $\Delta E'$ more negative in the case of a two-electron bond and more positive in the case of a three-electron bond. This will be referred to as the first order overlap effect. Next, we note that, in a perturbation theoretical sense, the interaction of two CW's is actually a function of $H_{ij} - E_i \cdot S_{ij}$, where H_{ij} and S_{ij} are the interaction matrix element and overlap integral of two CW's, Φ_i and Φ_j, respectively. As we have already seen, H_{ij} is proportional to $k\beta$. Similarly, S_{ij} is proportional to ks. Recalling that β increases in absolute magnitude as s increases, it is evident that an increase of spatial overlap will make H_{ij} more negative but it will also make $E_i S_{ij}$ more positive. The net result is generally a very "slow" increase of the term $H_{ij} - E_i \cdot S_{ij}$, in absolute magnitude, as s increases. <u>This will be referred to as the interaction overlap effect</u>.

The above discussion makes clear that the first order overlap effect is much more important than the interaction overlap effect. Thus, the former will dominate the latter even if the interacting CW's are degenerate. Hence, as s increases, two-electron bonds will tend to get stronger because the first order and interaction overlap effects operate in the same direction while three-electron bonds will tend to become three-electron antibonds because the two effects oppose each other with the former being dominant. Accordingly, we can make the following predictions:

1) The [2,2] set will have lower energy than either of the [3,1] and the [2] sets.

2) The relative energies of the [3,1] and [2] sets will depend on the case at hand.

Thus, the stability orders of the various low lying states in the case of strong core-ligand overlap binding will be:

$$^1V (2,2) > {}^1V (2,z) \gtrless {}^1V (3,1) > {}^1V (4,0)$$

and

$$^3V (2,r) \gtrless {}^3V (3,1)$$

The relative energies of $^1V (2,z)$ and $^1V (3,1)$, and well as those of $^3V (2,r)$ and $^3V (3,1)$, will depend on the ratio of the $<\phi_C|\phi_L>$ and $<\phi_C^*|\phi_L^*>$ overlap integrals.[5] As the latter remains constant and the former increases, the three-electron bonds

of ^1V (3,1) and ^3V (3,1) due to the overlap of ϕ_C and ϕ_L will become weaker and
eventually antibonds, while the two-electron bonds of ^1V (2,z) and ^3V (2,r), also
due to the overlap of ϕ_C and ϕ_L, will become stronger. Accordingly we can formulate
the following rule: As the $<\phi_C|\phi_L>/<\phi_C^*|\phi_L^*>$ ratio increases there will be a switch-
over of the relative stability of ^1V (2,z) and ^1V (3,1) as well as of ^3V (2,r) and
^3V (3,1). When the ratio equals one, the valence state with the [3,1] bond set will
tend to have lower energy while, when the ratio becomes large, the reverse will tend
to occur. We note that, if the antibonding interaction of the zwitterionic part of
^1V (2,z) and of the radicaloid part of ^3V (2,r) are taken into consideration, the
rule enunciated before is simply strengthened. This occurs because a decrease of
ϕ_C^* - ϕ_L^* overlap which causes the $<\phi_C|\phi_L>/<\phi_C^*|\phi_L^*>$ ratio to increase stabilizes
preferentially the ^1V (2,z) and ^3V (2,r) states relative to the ^1V (3,1) and
^3V (3,1) states, respectively, by weakening the zwitterionic or radicaloid antibonding interaction present in the former two and diminishing the strength of the
one-electron bond of the latter two.[6]

b. The constituent atoms are weak overlap binders, i.e., they are located
towards the left of a period and the bottom of a column of the Periodic Table. In
this case, core-ligand bonding is due equally to "classical" and overlap interaction
of the fragments and, because of the diminution of the importance of overlap binding,
one-, two-, and three-electron bonds tend to attain comparable energies. The
important thing to realize is that, as we proceed towards the limit of weak overlap
binding, the conventional ideas about bonding, inspired by studies of strongly
overlap-bound systems, are no longer valid and that overlap attraction and repulsion
cease to be dominant factors. Indeed, systems made up of weak overlap binding atoms
which demonstrate that a one-electron bond can be stronger than a two-electron bond,
for reasons which are easy to understand within the context of VB-type theory, are

known (see Table 3). In any event, at this limit, we expect one-, two-, and three-electron bonds to have comparable energies. Most importantly, we expect all bonds (or antibonds) regardless of type to be much weaker in systems made up of weak overlap binders (e.g., Li_2) than in systems made up of strong overlap binders (e.g., H_2). On the basis of the above considerations, we can now formulate the following guidelines:

Guideline 1. At the limit of strong overlap binding, the fragments will opt for making strong interfragmental bonds at the expense of excitation in which case the states which define interfragmental bonding will vary in energy in the order: ^1V (3,1) $\lessgtr$ ^1V (2,z) > ^1V (2,2) and ^3V (3,1) $\lessgtr$ ^3V (2,r). In other words, the relative energies will be dictated by the strength of interfragmental bonds. A nice example is provided by the union of two methylenes (CH_2) to form ethylene in the manner demonstrated in Figures 3a and 5. The energy order of the states of ethylene are as follows:

$$\sigma^2\pi\sigma^* \quad —— \quad [^1V\ (3,1)] \qquad\qquad \sigma^2\pi\sigma^* \quad —— \quad [^3V\ (3,1)]$$

$$\sigma^2\pi\pi^* \quad —— \quad [^1V\ (2,z)] \qquad\qquad \sigma^2\pi\pi^* \quad —— \quad [^3V\ (2,r)]$$

$$\sigma^2\pi^2 \quad —— \quad [^1V\ (2,2)]$$

Singlet   Triplet
States   States

Guideline 2. At the limit of weak overlap bonding, the fragments will opt for partial or no interfragmental bonding in favor of minimization of fragment excitation in which case the states which define interfragmental bonding will vary in energy in the order: ^1V (2,z) > ^1V (2,2) > ^1V (3,1) and ^3V (2,r) > ^3V (3,1). In other words, the relative energies will be dictated by the excitation energies.

Now it is not an exaggeration to say that the vast majority of chemists have been "reared" with molecules which are made up of strong overbinders, e.g., C_2H_4.

Table 3. Dissociation Energies of Some Alkali Metal Diatomics.

Diatomic	DE (eV)
Li_2	.62 (.99) [a,b,c]
Li_2^+	1.30 (>1.24)
Na_2	.75 [d]
Na_2^+	1.02
$NaLi$	.85 [c]
$NaLi^+$	.92

[a] James, H.M. J. Chem. Phys. 1935, 3, 9.

[b] Zemke, W.T.; Lykos, P.G.; Wahl, A.C. J. Chem. Phys. 1981, 51, 5635.

[c] Bertoncini, P.J.; Das, G.; Wahl, A.C. J. Chem. Phys. 1970, 52, 5112.

[d] Gaydon, A.G. "Dissociation Energies and Spectra of Diatomic Molecules";
Chapman and Hall: London, 1968.

Thus, the limit defined in Guideline 1 concerns state relationships which are familiar to most chemists. By contrast, the limit defined in Guideline 2 constitutes _terra_ _incognita_, at least for the most part. In this section, we have isolated the critical factors which are necessary for reversing conventional state orders and, as we shall see, there is ample evidence consistent with the expectations spelled out in Guideline 2.

V. Coulomb Correlation Differentiation of Valence States of Different Spin Multiplicity

Up until this point, we have restricted ourselves to making comparisons within a manifold of one type of spin multiplicity, i.e., we ranked singlet and triplet states separately because spin multiplicity effects cannot be discussed within the framework of the Independent Bond Model due to the fact that the electron-electron interaction of the bonds is neglected. By abandoning this model and reverting to complete MOVB theory, we can easily differentiate singlet and triplet states. For example, in considering the relative energies of singlet and triplet valence states of the same bonding type we recognize the following trends:

a. ^3V (3,1) has lower energy than ^1V (3,1) because of lower ionicity. This can be best understood in a qualitative way by inspection of the Hückel VB wavefunctions of the two states shown in tabular form in Table 4.[7] Clearly, ^3V (3,1) has greater radicaloid character and suffers less from electron repulsion.

b. ^3V (2,r) has lower energy than ^1V (2,z) because the former has greater radicaloid character than the latter, as the notation implies, while both suffer from comparable overlap repulsion, due to the interaction of the pair of electrons other than the one defining the two-electron bond.

Table 4 has great didactic significance for it shows that, although it appears that both ^1V (3,1) and ^3V (3,1) states have identical sets of three-electron and one-electron bonds, the two sets of bonds are created by the interaction of different sets of VB CW's. This becomes responsible for the fact that while the net overlap bonding due to the three- and one-electron bonds is the same in the two valence states, coulomb correlation is different. Thus, the true singlet and triplet states have entirely different wavefunctions and the triplet lies below the singlet state of corresponding bond-type.

Table 4. Hückel Valence Bond Wavefunctions for the ^{1}V(3,1) and ^{3}V(3,1) States.

^{1}V(3,1)

— + ╫ +	╫ — + +	+ ╫ + —	+ + — ╫	+ + ╫ —	╫ + — +	— ╫ + +	+ — + ╫	+ — ╫ +	╫ + + —
-.25	.25	-.25	.25	-.25	.25	-.25	.25	-.25	.25
+ ╫ — +	— + + ╫	╫ — ╫ —	╫ ╫ — —	— ╫ — ╫	— — ╫ ╫	— ╫ ╫ —	╫ — — ╫	+ + + +	+ + + +
-.25	.25	0	0	0	0	.35	.50	0	0

^{3}V(3,1)

— + ╫ +	╫ — + +	+ ╫ + —	+ + — ╫	+ + ╫ —	╫ + — +	— ╫ + +	+ — + ╫	+ — ╫ +	╫ + + —
-.25	.25	.25	-.25	-.25	-.25	.25	.25	-.25	-.25

+ ╫ — +	— + + ╫	+ + + +	+ + + +	+ + + +
.25	.25	.35	.20	-.29

VI. Unification of Ground and Excited State Stereochemistry

With the background of the previous sections and the previous chapters, let us now consider how a conceptual unification of ground and excited state chemistry can be effected. First, we recognize that the conceptual poles of MOVB theory are two:

a. Fragment excitation.

b. Interfragmental overlap bonding.

When the fragments contain strong overlap binders, the system opts for strong bonds at the expense of excitation energy. Conversely, when the fragments contain weak overlap binders, the reverse choice is made.

Next, we recall the situation with CH_2, C_2H_2, and C_2Li_2 insofar as the ground state geometries of these molecules are concerned.[8] In the case of CH_2 [viewed as C(core) + H_2(ligands)], the system opts for no core excitation and weak core-ligand bonds and CH_2 adopts a bent, rather than a linear geometry which would require a very large core excitation. In the case of C_2H_2 [viewed as C_2(core)+ H_2(ligands)], the system opts for strong core-ligand bonds and modest core excitation and C_2H_2 adopts a linear, rather than a bent, geometry. Finally, replacement of strong overlap binders (H atoms) by weak overlap binders (Li atoms) in C_2H_2 tips the balance in favor of no core excitation and weak core-ligand bonding and a rhomboidal geometry is now adopted. The latter is nothing other than an extreme trans-bent geometry, so to speak. In summary, these three examples illustrate that bending goes hand in hand with one type of system "decision" (option for core excitation minimization) and linearity with another type of system "decision" (option for core-ligand bond strength maximization through maximization of core-ligand overlap). In this light, let us now consider the relative energies of two different valence states, say ^1V (2,2) and ^1V (4,0). Clearly, when the fragments contain strong overlap binders, the system will opt for bond formation at the expense of fragment excitation. Accordingly, ^1V (2,2) will have lower energy than ^1V (4,0) for exactly the same reason that C_2H_2 ended up linear and not bent. By contrast, when the fragments contain weak binders, the system will opt for exactly the opposite: ^1V (2,2) will now have higher energy than ^1V (4,0) for exactly the same reason that CH_2 and C_2Li_2 are bent and not linear.

The above discussion makes clear that whenever we compare a set of ground states, e.g., two (or more) different geometries of a molecule in one and the same valence state (linear versus bent ground state CH_2, C_2H_2, C_2Li_2), or, a set of valence states, e.g., two (or more) different valence states of a molecule, we are using the same conceptual devices. It is exactly the different choices of different systems with regards to excitation energy and bond strength optimization which are the basis of "chemical diversity" in nature.

We can now apply the ideas developed on the basis of MOVB theory to pairwise comparisons of valence states of diverse systems. Three types of comparisons are of great importance for chemistry:

a. The ^1V (4,0) versus ^1V (2,2) comparison. This is tantamount to a thermo-dynamic analysis of any chemical reaction which leads to a generation of two new bonds. Organic chemists are familiar with insertion reactions, inorganic chemists deal routinely with oxidative additions, and so on. Typical examples are given below.

$$\ddot{C}H_2 + H \!-\! H \longrightarrow H_2C \begin{smallmatrix} \nearrow H \\ \searrow H \end{smallmatrix} \qquad \text{"Insertion"}$$

$$M\colon + X \!-\! X \longrightarrow M \begin{smallmatrix} \nearrow X \\ \searrow X \end{smallmatrix} \qquad \text{"Oxidative Addition"}$$

Thus, the comparison of ^1V (4,0) and ^1V (2,2) is extremely interesting since it is relevant to a vast number of synthetically important bond-making reactions.

b. The ^1V (2,2) versus ^3V (3,1) comparison. Chemists, through the influence of the Pauling "qualitative" VB theory and monodeterminantal MO theory of whatever brand(Hückel MO, Extended Hückel MO, Self Consistent Field MO), are extremely fond of even-electron bonds. On the other hand they have been very slow to appreciate the reality and importance of odd electron bonds. Thus, while V (even) ground states are very familiar to chemists, V (odd) excited states are novelties. The latter constitute spectroscopic Franck-Condon states or high energy intermediates in chemical reactions. The analysis presented in this work raises the distinct possibility that <u>V (odd) states can actually become ground states when the constituent fragments contain weak overlap binders</u>. This raises the possibility that a large part of our current understanding of molecular structure may have been based on incorrect assumptions. For example, some inorganic systems MX_2 which are thought to be bound as in Figure 6a [a picture corresponding to a V (2,2) state] may actually be bound as in Figure 6b [a picture corresponding to a ^3V (3,1) state].

c. The ^1V (3,1) versus ^1V (2,z) comparison. This comparison illustrates the overlap ratio rule stated before.

We now consider specific examples.

A. V (4,0) versus V (2,2)

Examples of V (2,2) preference over V (4,0), and vice versa, are given in Table 5 along with the relevant fragment excitation energies. **Importantly, note the** dramatic **change of the relative** energies of the two valence states as one goes from 1 and 2 to 3 is the result of the fact that the CH_2 and C_2H_4 fragments are strong while the Be fragment is a weak overbinder.

B. ^1V (2,2) versus ^3V (3,1)

Examples of systems with ^1V (2,2) having lower energy than ^3V (3,1) and vice versa, are given in Table 6. It is easily seen that replacement of a strong by a weak overlap binding group causes ^3V (3,1) to come under ^1V (2,2).

Table 5. Systems with ^1V (4,0) and ^1V (2,2) Ground States.

V (4,0) $\longrightarrow$ V (2,2)		ΔE (kcal/mole)	Singlet-Triplet Energy Gap (eV)
1. $:CH_2$ + $:CH_2$ $\longrightarrow$	$H_2C = CH_2$	-188.9 [a]	.75 [b]
2. $CH_2 = CH_2$ + $CH_2 = CH_2$ $\longrightarrow$	□	-19.0 [a]	8.6 [c]
3. Be + Be $\longrightarrow$	Be_2	~ 0 [e]	5.4 [d]

[a] Stull, D. R.; Westrum, Jr., E. F.; Sinke, G. C. "Chemical Thermodynamics of Organic Compounds"; John Wiley and Sons: New York, 1969.

[b] Frey, H.M. J. Chem. Soc. Chem. Commun. 1972, 1024.

[c] Mulliken, R.S. J. Chem. Phys. 1977, 66, 2449. (Triplet has D_{2h} symmetry)

[d] Moore, C.E. "Atomic Energy Levels"; National Bureau of Standards: Washington, D.C., 1949.

[e] Be_2 is slightly bound because of coulomb polarization, discussed in Chapter 20. See: Whiteside, R.A.; Krishnan, R.; Pople, J.A.; Krogh-Jespersen, M.- B.; Schleyer, P.von R.; Wenke, G. J. Comput. Chem. 1980, 1, 307. In the absence of this bonding mechanism, the sign of ΔE is positive.

Table 6. Systems with ^1V (2,2) and ^3V (3,1) Ground States.

Core-Ligand System	Lower Energy Valence State
$:CH_2$ + ⟩⟨	^1V (2,2)
Ni + ⟩⟨	^1V (2,2) [a]
Be + ⟩⟨	^3V (3,1) [b]
$:CH_2$ + Li_2	^3V (3,1) [c]

[a] Ref. 10.

[b] Ref. 9.

[c] Ref. 11 and 12.

c. ^1V (3,1) versus ^1V (2,z)

Ethylene (C_2H_4) can be conceived as the product of the union of two triplet methylenes (CH_2) as indicated in Figure 5. Since $<\phi_C^*|\phi_L^*>$ is much smaller than $<\phi_C|\phi_L>$, ^1V (2,z) must have lower energy than ^1V (3,1). Indeed, the former state ($\pi\pi^*$ in MO parlance) is known to have lower energy than the latter state ($\pi\sigma^*$ in MO parlance). By contrast, cyclobutane (C_4H_8) can be conceived of as the product of the union of two triplet ethylenes (C_2H_4) as indicated in Figure 2b. If non-nearest neighbor overlap is neglected, $<\pi|\pi> = <\pi^*|\pi^*>$. Hence, in this case, ^1V (3,1) must have lower energy than ^1V (2,z).

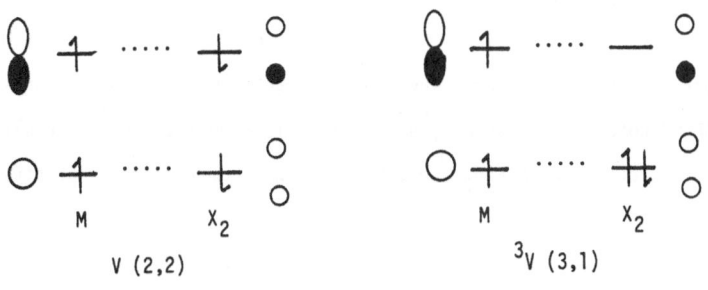

$$V\ (2,2) \qquad\qquad {}^3V\ (3,1)$$

Figure 6. Valence states of MX_2:

    a. V (2,2)   b. ^3V (3,1)

VII. A Closer Look at V (3,1)-type States

In our work, we have sought to develop a conceptual framework suitable for the following purposes:

a. The prediction of chemical trends in the absence of known facts.

b. The rationalization of known facts which cannot be explained on the basis of normal intuition.

c. The rational analysis of computational results and the evaluation of computational theory itself.

In MOVB theory we have found a tool which appears capable of performing admirably in these three domains. In this section we show how this theory can be used as an analytical tool for decoding the information contained within good quality computations, thus making them relevant to a broad spectrum of chemists. A fascinating paper by Swope and Schaefer[9] provides the data "to be explained" and the opportunity to further illustrate the application of MOVB theory to chemistry.

How does ground singlet and triplet Be interact with acetylene and ethylene? In order to answer this question, Swope and Schaefer performed a series of mono-determinantal SCF-MO computations which yielded the following interesting results:

a. Ground Be interacts with C_2H_2 and C_2H_4 in a C_{2v} geometry in a repulsive manner.

b. Triplet Be interacts with C_2H_2 and C_2H_4 in a C_{2v} geometry in a very different manner depending upon the electronic configurations of the atom. Using the C_{2v} point group classification the three possible electronic configurations of triplet Be, depending on whether a $2p_x$, a $2p_y$, or a $2p_z$ AO is occupied, are:

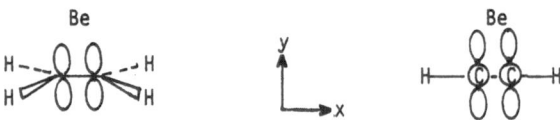

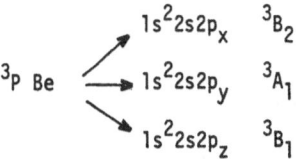

The results for the $^3Be-C_2H_4$ interaction are shown in Figure 7. The same trends are found for the $^3Be-C_2H_2$ system. Thus, we can ask the question: Why is the interaction of ^3Be with a pi bond so dependent on the configuration of ^3Be? The answer is obtained immediately by construction of the appropriate bond diagrams shown in Figure 7:

a. $^3B_2$ Be interacting with a pi bond generates a one-electron plus a three-electron bond, and this leads to bonding.

b. $^3B_1$ Be interacting with a pi bond generates only one three-electron bond, which lies in the gray region of a bond or antibond (<u>vide supra</u>). As a result, this leads to a nonbonding interaction of the two fragments.

c. $^3A_1$ Be interacting with a pi bond generates two coupled three-electron bonds. However, each of the three defining CW's suffers from twice (roughly speaking) the overlap repulsion of each of the CW's which define the single three-electron bond in the $^3B_1$ species. As a result, the shapes of the Be-C_2H_4 ($^3A_1$) and Be-C_2H_2 ($^3A_1$) curves are similar to those which characterize inert gas - inert gas inter-action. Charge redistribution and other interesting structural details can also be anticipated on the basis of the bond diagrams shown in Figure 7. However, space

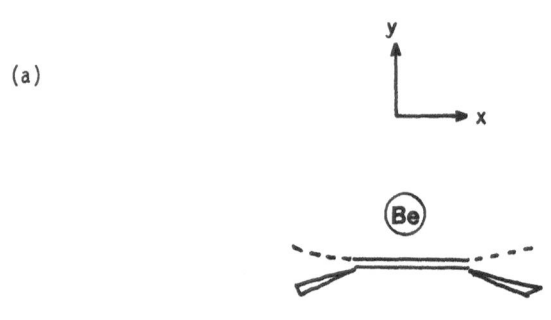

(a)

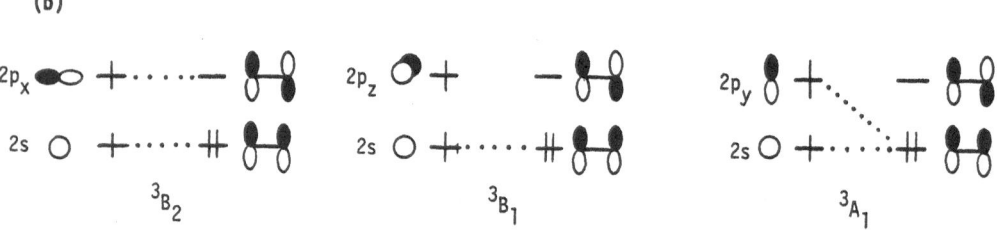

(b)

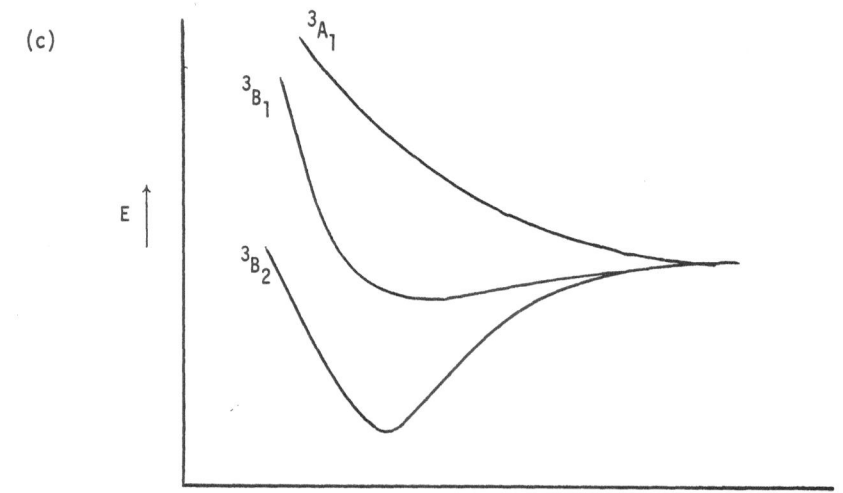

(c)

Figure 7.   a.   Geometry of the Be - C_2H_4 complex.   b.   Bond diagrammatic representation of Be - C_2H_4 interaction for $^3B_2$, $^3B_1$, and $^3A_1$ configurations of Be.   c.   Energies of $(^3B_2)$ - C_2H_4, Be $(^3B_1)$ - C_2H_4, and Be $(^3A_1)$ - C_2H_4 complexes as a function of interfragmental distance.

limitations do not permit a fuller discussion, especially of the importance of polarization holes, e.g., the role of the 2p Be AO's in the bonding of ^{3}Be-C_2H_4 and ^{3}Be-C_2H_2.

The Swope-Schaefer study is of critical importance not only for the information it provides regarding the mode of binding of ^{3}Be with a closed-shell molecule but also because it illustrates the consequences of weak overlap binding predicted by MOVB theory. Because Be is a weak overlap binder, the ^{1}V (2,2) state lies much above the ^{3}V (3,1) state, which,in turn, lies well above the ^{1}V (4,0) state. The Dewar-Chatt-Duncanson model is undefined and it cannot allow a distinction between molecular states. Once made, this distinction opens new vistas in our appreciation of molecular structure and reactivity.

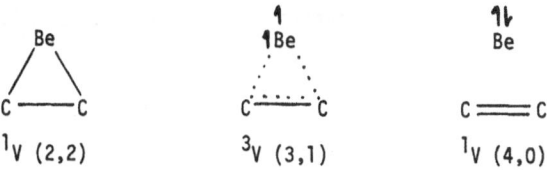

Recent calculations of the Ni-C_2H_2 complex by Goddard, et al.[10] and good-quality calculations of CH_2Li_2 by Schleyer, et al.[11] and Schaefer, et al.[12] can be analyzed similarly,the low lying triplet complexes predicted by those studies being ^{3}V (3,1) species. The low energy of the ^{3}V (3,1) relative to the other valence states is again due to the weak overbinding ability of Ni and Li.

Having understood exactly how valence state reversal can be effected in four electron-four orbital systems, we can use the same concepts in analyzing the valence state manifolds of polyelectronic systems. This is illustrated by a comparative study of the electronic states of C_2 and Si_2. The choice is deliberate since C_2 is inherently interesting in that it is one of the simplest organic molecules and a component fragment of a great many organic molecules. Also, the bonding of ground state C_2 illustrates MOVB principles which lead to conclusions that may surprise many of the readers. Finally, the C_2-Si_2 comparison is important in view of the recent great interest in organosilicon chemistry.

C_2 and Si_2 can be viewed as composite systems made up of C plus C and Si plus Si atomic fragments, respectively. Because the AO's of atoms are orthogonal, the union of the fragments to make up the total system can be described theoretically by using MOVB theoretical concepts. If we symbolize C and Si by A and consider only valence electrons, the valence state of A having the greatest number of two electron bonds is represented by the detailed bond diagram of Figure 8a and the compact bond diagram of Figure 8b. These imply a quadruple bond connecting the two atoms, i.e., A_2 can be symbolized by A≡A. However, because of the direction of primary charge transfer, interatomic overlap repulsion generated by the ω_1-σ_2 and ω_2-σ_1 bonds keep the interatomic distance longer than expected for a quadruple bond. This is made evident by the approximate representation of A_2 given in Figure 8c. This implies a double bond connecting the two atoms, i.e., A_2 can now be symbolized by ꟾA=Aꟾ. This is the way in which most chemists think about an eight valence electron diatomic A_2. Which of the two representations is closer to the truth? If we adopt the representation of Figure 8b we imply significant electron delocalization from an ns AO of one atom to a np AO of the other. If we adopt the representation of Figure 8c, we imply zero electron delocalization of this type. We now show that the operationally correct representation of A_2 is A≡A and not ꟾA=Aꟾ.

Let us compare the valence state of A_2 with essentially four two-electron bonds (Figure 8b) with a valence state of A_2 in which the four original two-electron bonds have been essentially replaced by two three-electron bonds and two one-electron bonds (Figure 8e) by demoting two electrons from the np to the ns AO's, i.e., effecting fragment deexcitation. Of course, another more accurate

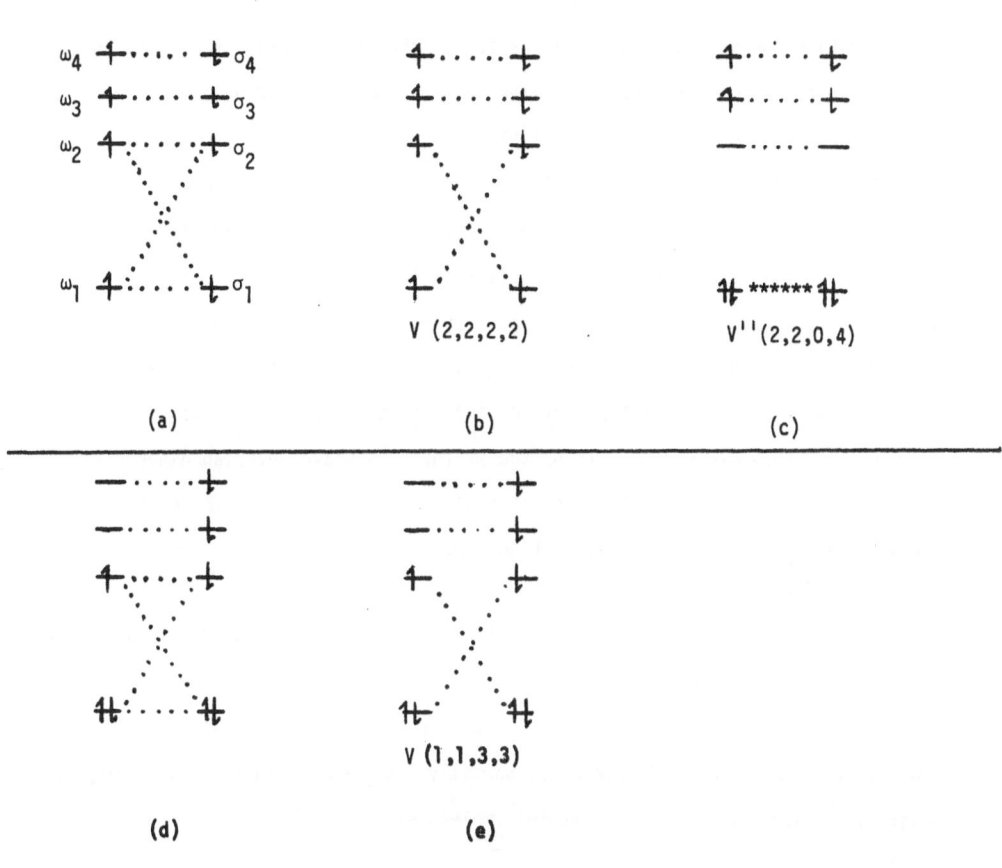

Figure 8. Detailed (a) and compact (b) bond diagrams of ground A_2 (e.g., A=C) when A is a strong overlap binder. When A is a weak overlap binder (e.g., A=Si), the ground state is that indicated by the detailed and compact bond diagrams (d) and (e) respectively. V (2,2,2,2) has larger (average) CW excitation but stronger bonds compared to V(1,1,3,3).

way of describing the same process would be to say that two original two-electron-two-orbital bonds and one original four-electron-four-orbital bond have been replaced by two new one-electron-two-orbital bonds and a new six-electron-four-orbital bond with the profit of fragment deexcitation. In short, we reduced core-ligand binding in exchange for deexcitation. The approximate MOVB representations are given in Figures 8d and 8e. On the basis of this description, we predict the following result: As the overlap binding ability of A decreases, there will be an energy switchover of the V (2,2,2,2) and V (1,1,3,3) states, with the former being the ground state when A is a strong overlap binder (e.g., A=C) and the latter becoming the ground state when A attains the status of a weaker overlap binder (e.g., A=Si). Indeed this is what experiment demonstrates![13] The MOVB and MO descriptions of C_2 and Si_2 are shown in Figure 9.

We summarize our findings:

a. The interplay between bond strengths and excitation energy becomes responsible for valence state switchovers as we go from C_2 to Si_2.

b. The ground state of C is properly represented by the Lewis formula C≣C, and not by IC≡CI. Thus, the quadruple bond, normally the preoccupation of the inorganic chemist, is a staple of primordial organic molecules!

MO

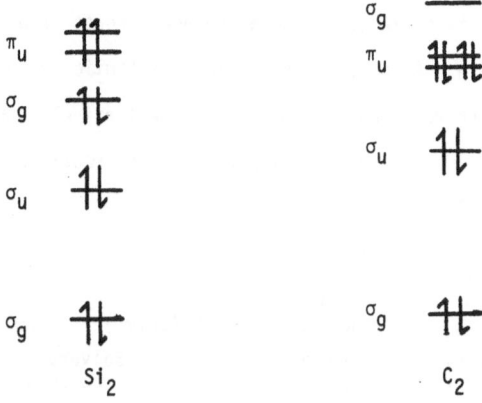

MOVB

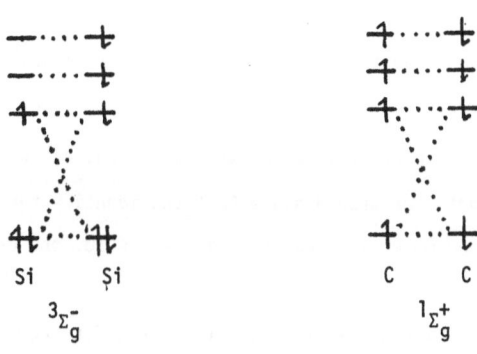

Figure 9. The MO and MOVB descriptions of bonding in C_2 and Si_2.

VIII. Excited States and Chemical Intermediates

Let us now take advantage of the clear and chemically meaningful MOVB description of excited states in order to define new targets of experimental and theoretical investigation. As an example, we focus attention on unimolecular nucleophilic substitution reactions which have been extensively studied during the last half of the century. The general mechanism of reactions of this type is the one shown below.[14]

$$
R - X \longrightarrow \underset{\substack{\text{Tight} \\ \text{Ion Pair}}}{R^+ X^-} \longrightarrow \underset{\substack{\text{Solvent} \\ \text{Separated} \\ \text{Ion Pair}}}{R^+||X^-} \longrightarrow \underset{\substack{\text{Solvated} \\ \text{Free Ions}}}{R^+(\text{solv.}) + X^-(\text{solv.})} \longrightarrow \text{Products}
$$

with Products arising from both the Tight Ion Pair and Solvent Separated Ion Pair stages.

The dipolar species can be viewed as excited states of the precursor alkyl halide (R-X) and the question becomes: Are the tight ion pair (TIP) and the solvent separated ion pair (SIP) one and the same electronic species involving two different solvation shells or are they two different electronic species which are also differently solvated? In order to anwer this question, all we have to do is construct the bond diagram for ground state CH_3F and identify the low lying dipolar excited species of the molecule. As shown in Figure 10, one can identify three types of dipolar excited states:

a. The Z state which arises from electron transfer from the sp hybrid AO of C to the valence AO of F. This excited state dissociates probably as shown below.

$$
Z \longrightarrow [\overset{+}{C}H_3||F:^-] \longrightarrow \overset{+}{C}H_3(\text{solv.}) + F:^-(\text{solv.})
$$

b. The singlet and triplet Y states which arise from electron transfer from the H_3 ligand orbitals to the valence AO of F. These excited states dissociate probably as follows:

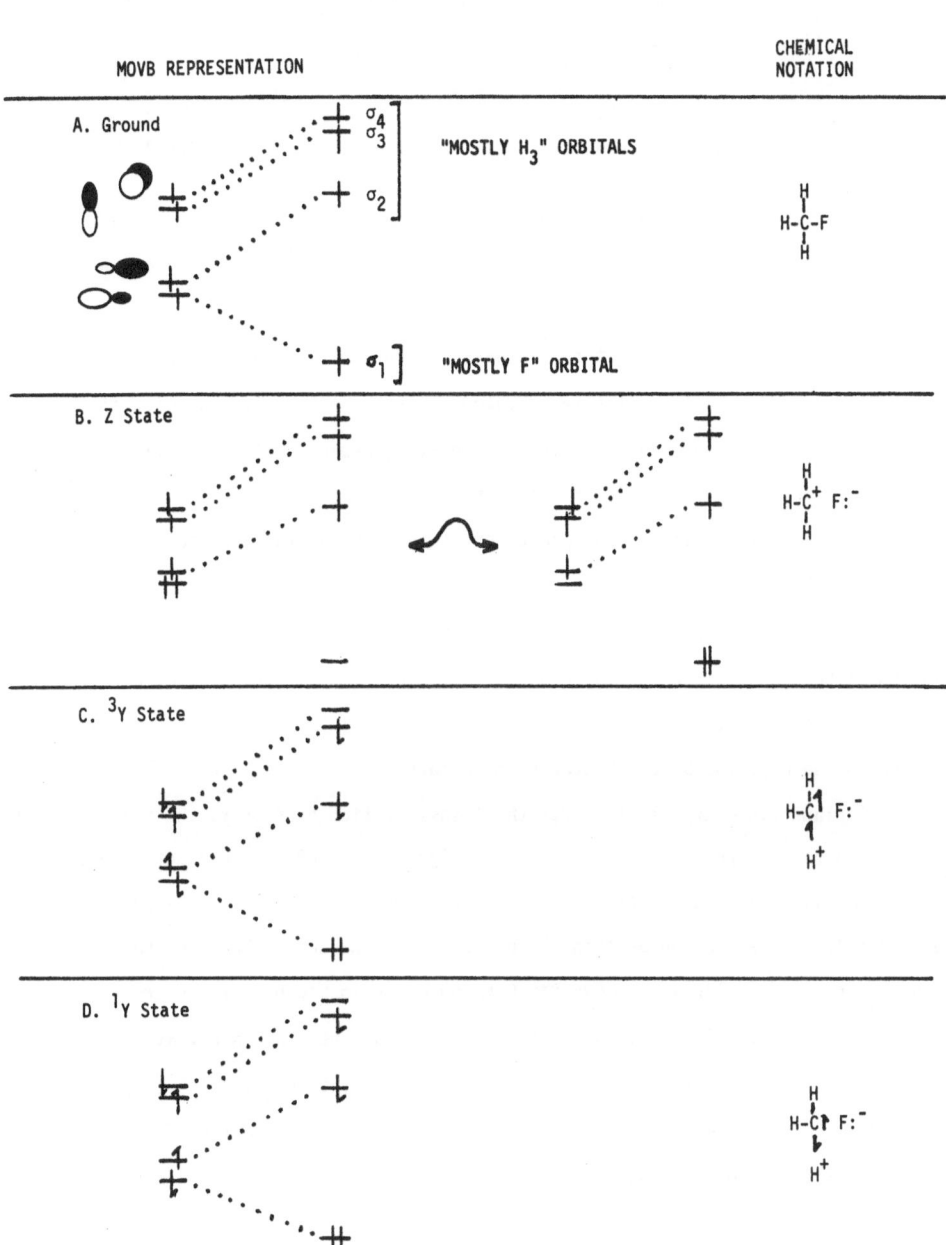

Figure 10. MOVB bond diagrams for the ground and Z,^{3}Y and ^{1}Y excited states of CH_3F.

$$^1Y \longrightarrow [^1CH_2H^+F:^-] \longrightarrow [^1CH_2H^+||F:^-] \longrightarrow {}^1CH_2 + H^+(solv.) + F:^-(solv.)$$

$$^3Y \longrightarrow [^3CH_2H^+F:^-] \longrightarrow [^3CH_2H^+||F:^-] \longrightarrow {}^3CH_2 + H^+(solv.) + F:^-(solv.)$$

Note that $[\overset{+}{CH_3}||F:^-]$ is a <u>closed shell</u> whereas $[^*CH_2H^+F:^-]$ and $[^*CH_2H^+||F:^-]$ are <u>open shell</u> molecules. The brackets signify that the enclosed species may or may not be an energy minimum.

Because of the different localization of the positive hole in the $\overset{+}{CH_3}$ fragment, the Z and Y states must be differentially solvated. Also, the relative energies of the Z and Y states are dependent on the electron donating capacity of the alkyl groups in R_3CX. We can now raise the following possibilities:

   a. The singlet Y state is the precursor of the TIP and the Z state the precursor of the SIP.

   b. A low lying triplet Y state may be chemically accessible if a spin orbit coupling mechanism exists which could permit the conversion of the singlet precursor R_3CX to triplet Y. The likelihood of "seeing" such a triplet dipolar species will increase as the R groups become better sigma donors.

   c. A plausible energy surface for the transformation of R_3CX to the solvent separated in pair $R_3\overset{+}{C}||X:^-$ is the one shown in Figure 11a. At the reactant stage, the Y states lie below the Z state because charge transfer from the R_3 ligands is energetically more advantageous than charge transfer from the carbon because the ligand σ_3 and σ_4 MO's lie well above the C sp AO and also because of the greater amount of interelectronic repulsion in the Z state. On the other hand, at the product stage, the greater stability of $\overset{+}{C}R_3$ relative to the lowest lying triplet and singlet CR_2R^+ species goes hand in hand with another very important factor which furthers the energetic advantage of $\overset{+}{C}R_3$. Specifically, the positive charge

a)

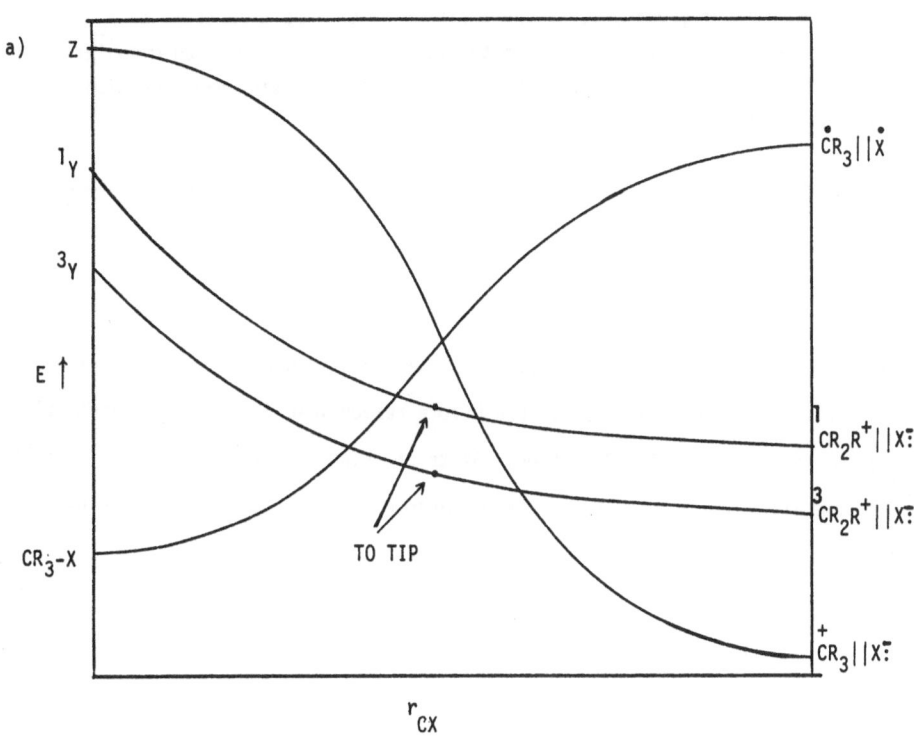

b)

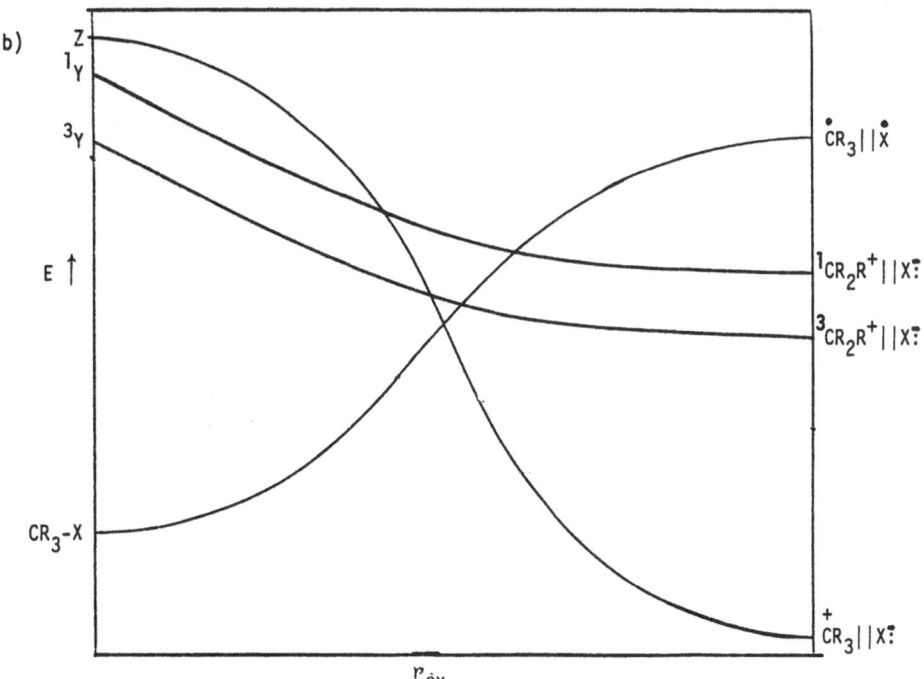

Figure 11. Energies of ground,^{3}Y, ^{1}Y and Z states of CR_3F as a function of r_{CX} for the unimolecular solvolysis reaction. a. State crossing near the reaction midpoint can lead to intermediate (TIP) formation. b. Lack of state crossing near reaction midpoint does not allow intermediate formation.

is localized on C in $\overset{+}{C}R_3$ but delocalized in the ligands in the case of the higher lying CR_2R^+ species and this factor can become responsible for more efficient solvation of the closed shell $\overset{+}{C}R_3$ cation. State crossing near the "reaction midpoint" can become responsible for a third energy minimum, the tight ion pair, singlet or triplet.

References

1. Epiotis, N.D.; Larson, J.R.; Eaton, H. "Unified Valence Bond Theory of Electronic Structure" in Lecture Notes in Chemistry, Vol. 29; Springer-Verlag: New York and Berlin, 1982.

2. Epiotis, N.D. Pure Appl. Chem., in press.

3. For lucid discussion of the MO and VB theory of such a primordial system, the reader is referred to the work of J. C. Slater.

4. Dauben, W.G.; Salem, L.; Turro, N.J. Acc. Chem. Res. 1975, 8, 41.

5. The following clarifying comments with regards to the overlap rule are in order. Since there are two bonds in ^1V (3,1) and ^3V (3,1), a one and a three electron bond due to $\phi_C - \phi_L$ and $\phi_C^* - \phi_L^*$ overlap respectively, and since there is only one bond, a two electron bond, in ^1V (2,z) and ^3V (2,r) due to $\phi_C - \phi_L$ overlap, neither of the two overlap integrals by itself is an appropriate index of the relative energies of ^1V (3,1) and ^1V (2,z) and of ^3V (3,1) and ^3V (2,r). Only the ratio of the two overlap integrals constitutes a valid index.

6. Hückel MO (HMO) theory tends to emphasize two-electron bonding over one- or three-electron bonding, thus, unreasonably favoring ^1V (2,z) over ^1V (3,1) and ^3V (2,r) over ^3V (3,1). The reason for this is that the HL CW's are correctly perceived as degenerate but the ionic CW's are incorrectly perceived as degenerate due to the integral approximation of HMO theory. As a result, the contributions of the interaction of the HL and the ionic CW's to three-electron and one-electron bond formation are relatively well reproduced while the contribution of the interaction to two electron bond formation is exaggerated.

7. For details on the calculation of HVB wavefunctions, see Reference 1.

8. See Chapter 1 and ref. 1.

9. Swope, W.C.; Schaefer, III, H.F. J. Am. Chem. Soc. 1976, 98, 7962.

10. Upton, T.H.; Goddard, III, W.A. $\underline{J}$. $\underline{Am}$. $\underline{Chem}$. $\underline{Soc}$. $\underline{1978}$, 100, 321.

11. Chandrasekhar, J.; Pople, J.A.; Seeger, R.; Seeger, U.; Schleyer, P.V.R. $\underline{J}$. $\underline{Am}$. $\underline{Chem}$. $\underline{Soc}$., in press.

12. Laidig, W.D.; Schaefer, III, H.F. $\underline{J}$. $\underline{Am}$. $\underline{Chem}$. $\underline{Soc}$. $\underline{1978}$, 100, 5972.

13. The ground state of gaseous C_2 is the singlet $^1\Sigma_g^+$ state with the triplet $^3\Pi_u$ state lying 1.74 kcal higher in energy:

    (a) Ballik, E.A.; Ramsay, D.A. $\underline{J}$. $\underline{Chem}$. $\underline{Phys}$. $\underline{1958}$, 29, 1418.

    (b) Ballik, E.A.; Ramsey, D.A. $\underline{Astrophys}$. $\underline{J}$. $\underline{1963}$, 137, 61, 84.

    The ground state of gaseous Si_2 is a $^3\Sigma_g^-$ triplet with all indications being consistent that the lowest energy singlet lies <u>significantly</u> above the ground triplet state. For review, see: Bürger, H.; Enjen, R. $\underline{Top}$. $\underline{Curr}$. $\underline{Chem}$. $\underline{1974}$, 50, 1.

14. Winstein, S.; Robinson, G.C. $\underline{J}$. $\underline{Am}$. $\underline{Chem}$. $\underline{Soc}$. $\underline{1958}$, 80, 175.

Chapter 4. The "Forbidden" World of Chemistry.

Hückel's rule[1], the Woodward-Hoffmann rules[2], and, in a broader sense, Hückel Molecular Orbital (HMO) theory[3] have had profound impact on chemistry in a way which is now well recognized and admired. Perhaps, the most important and time-lasting contribution of HMO theory has been the revelation of what in Valence Bond (VB) terms we call parity control of stereoselection[4], or, in more familiar language, the revelation of the fact that a ground state molecule, a transition state, or, any molecular system, in general, can be thought of as the product of a "forbidden" or "allowed" union of two component fragments with the latter mode of union being energetically more favorable than the former one. We can summarize what most chemists know about "forbidden" and "allowed" unions in Frontier Orbital (FO)[5] Hückel Perturbation MO (HPMO)[6] language as follows[7]:

a. A "forbidden" union occurs when the Highest Occupied MO's (HOMO's) of the two fragments are of one symmetry species and the Lowest Unoccupied MO's (LUMO's) of the same two fragments are of a second symmetry species so that there is HOMO-HOMO and LUMO-LUMO overlap while HOMO-LUMO and LUMO-HOMO overlap is zero. This pattern guarantees net antibonding when interfragmental overlap is weak and double bonding when interfragmental overlap is strong, the switch from the former to the latter situation being effected by virtue of composite system MO crossing. The spatial overlap properties of the FO's in a "forbidden" union are shown schematically below, with the double arrows indicating orbitals with congruent symmetry.

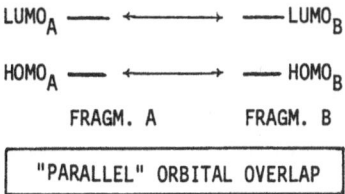

b. An "allowed" union occurs when the HOMO of fragment A and the LUMO of fragment B are of one symmetry species and the LUMO of fragment A and the HOMO of fragment B of a second symmetry species so that there is HOMO-LUMO and LUMO-HOMO overlap while HOMO-HOMO and LUMO-LUMO overlap is zero. This pattern guarantees net double bonding regardless of the magnitude of interfragmental overlap, i.e., there is no longer composite system MO crossing. The spatial overlap properties of the FO's in an "allowed" union are shown schematically below.

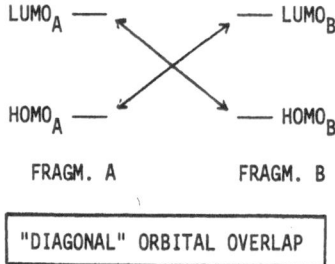

FRAGM. A   FRAGM. B

"DIAGONAL" ORBITAL OVERLAP

c. A hybrid "allowed"-"forbidden" union is called a "nonaromatic" union and it occurs when all FO's are of the same symmetry species but orbital overlap is impaired relative to that in the corresponding "forbidden" or "allowed" union. A schematic illustration is given below. Each orbital overlap integral is roughly one-half of each overlap integral of (a) and (b).

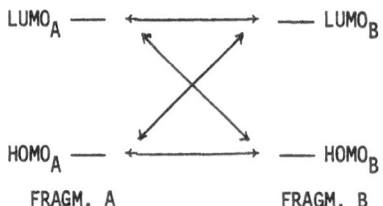

FRAGM. A   FRAGM. B

In short, we can say that a four-electron-four-orbital system can be bound in three different ways symbolized by the letters F, A, and N denoting "forbidden", "allowed", and "nonaromatic" fragment union, respectively. Furthermore, we note that the hallmark of a "forbidden" union is "parallel" fragment orbital overlap while the earmark of an "allowed" union is "diagonal" fragment orbital overlap. We fully expect that larger systems also can be thought of as products of "forbidden" or "allowed" fragment unions having analogous characteristic properties as far as fragment orbital overlap is concerned. For example, we can conceive of two different eight-electron-eight-orbital isomeric systems wherein symmetry dictates the fragment overlap pattern shown below.

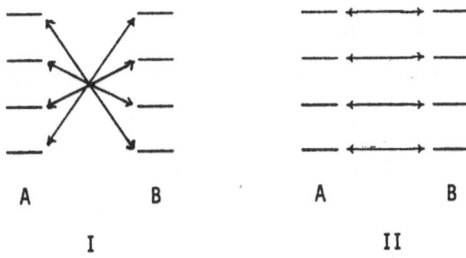

A   B   A   B

I   II

In addition, we can conceive of intermediate possibilities such as the ones shown below.

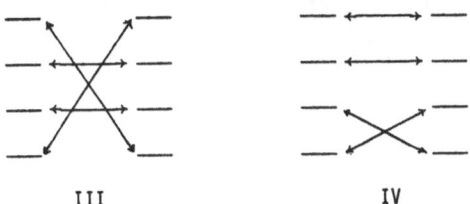

III   IV

We can say that isomer I is fully "allowed", isomers III and IV partly "forbidden", and isomer II fully "forbidden".

How is the problem of chemical stereoselection reformulated when we shift from approximate HPMO to rigorous MOVB theory? First, we recall that MOVB theory has the great advantage of combining rigor with simplicity. The first attribute is due

to the fact that the theory itself has been developed starting from the basic
equations of VB theory[8,9] and that the approximations involved are not only less
than drastic but they can be identified and corrected, if necessary, by retracing
the path defined by a series of equations which have taken us from a two-electron-
two-orbital model system to a real polyatomic. The second attribute is due to the
fact that MOVB theory makes use of the construct of the chemical bond and the
resulting "thinking philosophy" is not foreign to the vast majority of chemists who
have been previously exposed to "resonance theory",[10] which is actually an approximate form of VB theory and which also makes use of the construct of the chemical
bond. The formal basis of MOVB theory and the way in which it can be implemented
in a routine way in order to "solve" qualitatively any problem involving organic
or inorganic, ground or excited, small or large molecules have been discussed in a
monograph.[4] Restricting our attention to the problem at hand and focusing on systems
which are comprised of two strongly overlapping fragments, such as ground state
molecules, we can say that a four-electron-four-orbital system can be bound in
three different ways as illustrated below by means of MOVB bond diagrams.[4]

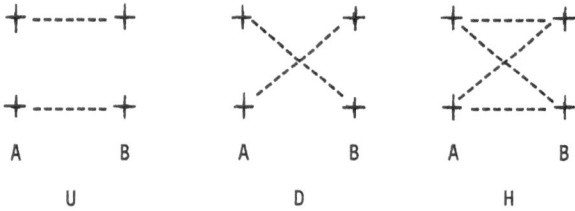

In a sense, U bonding in MOVB theory is analogous to F bonding in MO theory, D
analogous to A, and H analogous to N.

The characteristic property of a U-bound system is the fact that the
Perfect Pairing (R) Configuration Wavefunction (CW) is the leading contributor
to the total MOVB wavefunction. This is a direct consequence of "parallel overlap" and it means that many stereochemical features of a U-bound system can be
understood by reference to the corresponding R CW and many useful indices can be
extracted from it. In our model four-electron - four-orbital system, the R CW
is Φ_1 shown below.

By contrast, the characteristic property of a D-bound system is that there now
exists a CW other than the R CW which places electrons in the lowest energy
orbitals of the two fragments thus attaining equal, if not higher, status. This
is a direct consequence of "diagonal overlap" and it means that the stereochemistry
of D-bound systems can only be understood by recognizing the important role of
this low energy CW (Φ_2 in our model four-electron - four-orbital system). The
MOVB description of the eight-electron - eight-orbital isomers is given below.

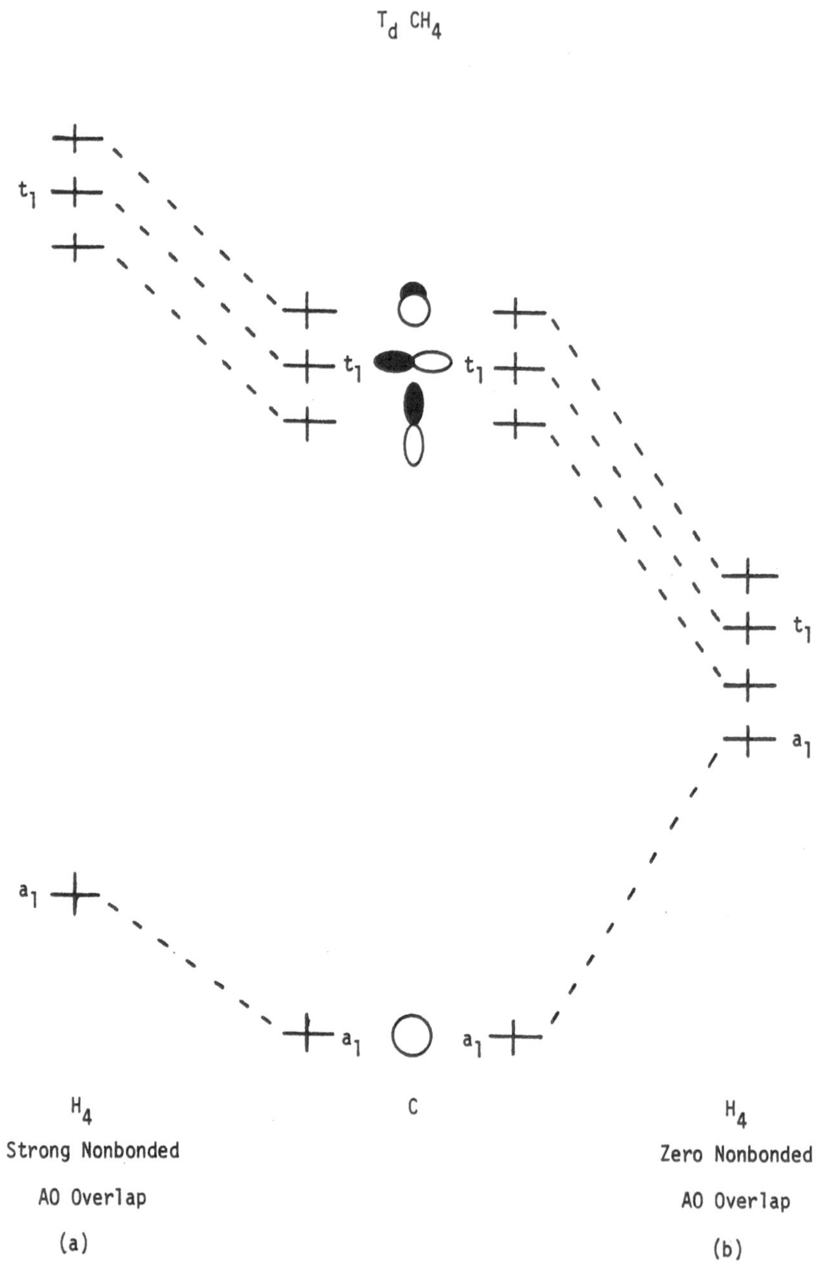

T_d CH_4

t₁

t₁ t₁

t₁

a₁

a₁

a₁

a₁ a₁

H₄

C

H₄

Strong Nonbonded

Zero Nonbonded

AO Overlap

AO Overlap

(a)

(b)

Figure 1. Bond diagrams for tetrahedral methane under the assumptions of strong and zero nonbonded AO overlap.

Anticipating following developments, we can say that if we were to discover that the vast majority of ground state _organic_ molecules can be viewed as composite systems each of which is made up of a core and a ligand fragment that are U-bound in a _reference geometry_, we would be entitled to call these molecules "forbidden" and to claim that chemistry itself is the science partly devoted to the study of "forbidden" relationships, by keeping in mind that the term "forbidden", though a misnomer, is being used for maintainance of conceptual continuity and unambiguous specification of interfragmental overlap characteristics. That this improbable scenario is actually what nature has concocted can be illustrated by a simple example: The ground state molecule methane (CH_4), one of the simplest molecules of what people call "organic chemistry". The bond diagram of CH_4, viewed as C plus H_4, is shown in Figure Ia and the "parallel" overlap of the fragment orbitals is immediately noticeable. Clearly, CH_4 can be regarded as a "forbidden" molecule, i.e., a molecule in which a core (C) and a ligand (H_4) fragment are U-bound. To the extent that CH_4 is a prototype of a vast number of molecules (alkanes) and to the extent that _similar conclusions are reached regardless of_ _the nature of core and ligands and subject only to the condition that ligand_ _orbital overlap is nonzero_, we can claim that almost every organic molecule can be viewed as the product of a "forbidden" union of core and ligands. This implication of MOVB theory is hidden in the elementary texts of undergraduate chemistry in a way explained below.

Let us consider two polyelectronic fragments A and B and inquire as to whether their overlap characteristics are consistent with an "allowed", or, a partly or totally "forbidden" overlap interaction. We know that an increase of the number of nodes parallels an increase of the energy of an MO. This means that the lowest energy MO of fragment A will have zero nodes and the same thing will be true of the lowest energy MO of fragment B. As a result, the two MO's will have "matching symmetry". The same trend will persist, either totally or partly, going up the MO ladders of fragments A and B with the net result being that the MO's of the two fragments will overlap, either totally or partly, in a "parallel" sense thus defining a "forbidden" composite system. The bond diagram of CH_4 is the simplest illustrator of this simple idea which is based on the elementary fact that the nth energy level of a fragment corresponds to an orbital with n-1 nodes and that nodes are the determinants of the overall symmetry of the orbital itself.

The reader may now ask: How is one to reconcile "forbidden" molecules with the fact that preferred reaction paths are "allowed" unless some special constraint is operative? The answer is that FO-MO theory permits only a partial visualization of a multidimensional problem. The success of the Woodward-Hoffmann rules, founded on such a brand of theory, is due to the fact that they correctly forcast the relative "forbiddenness" of two distinct stereochemical paths. Thus, there is nothing in the MOVB viewpoint of molecules which negates the validity of these and related rules despite the appearance that, by painting a picture of chemistry as the science of "forbidden" relationships, one fails to give recognition to "allowed" relationships. In summary, there are fundamental reasons which dictate that organic molecules are actually totally or partly "forbidden" composites and that the superiority of one isomer over another can be due to the fact that one is simply less "forbidden" than the other, e.g., isomer III (or IV) is energetically more favorable than II above though both are "forbidden" to some extent according to the overlap criterion to which the terms "forbidden" and "allowed" are connected. The momentous implication of this analysis is that organic stereo-chemistry, in the broadest sense of the term, can be studied by making use of concepts pertinent to "forbidden" reactions and strategies designed to remove reaction "forbiddenness", where by "reaction" we mean any hypothetical or actual union of elementary chemical fragments. That these salient points have never been made before is the natural consequence of two things:

a. The crude nature of qualitative MO theoretical models and, in particular, the FO approximation which indirectly provides the basis for a rigid dichotomy between "allowed" and "forbidden" unions. For example, note that in making a transition from the four-electron-four-orbital system, which can be adequately treated by FO-MO theory since the four orbitals are actually the four FO's, to an eight-electron-eight-orbital system, which can no longer be properly treated by FO-MO theory, we discovered that there are no "allowed" and "forbidden" unions but, rather, a spectrum of unions of varying degrees of "forbiddenness" and that molecules, in general, almost never qualify as totally "allowed" entities.

b. The incomplete understanding of the role of nonbonded interaction and especially that between atoms or groups attached to one and the same central atom.

The first obstacle does not exist in MOVB theory. It has also been recognized and bypassed by other workers who use MO theory for qualitative analyses of molecular stereochemistry.[11] The second obstacle, though it has been effectively removed in section II.I of ref. 4, needs renewed exposure because failure to recognize the central role of nonbonded interaction precludes the proper under-standing of many interesting stereochemical trends at any level of theory. Accordingly, a rather detailed discussion of this subject is presented in the following section.

I. Why Nonbonded Interaction is Anything But Qualitatively Unimportant

Consider the union of C and H_4 to produce CH_4 as depicted by the bond diagram of Figure 1a. We note that the a_1 and t_1 ligand orbitals are separated by a substantial energy gap, a consequence of the fact that, in constructing the bond diagram, we have assumed that the $1s_H - 1s_H$ nonbonded overlap is nonzero. It is exactly this assumption of nonbonded interaction which entitles us to view methane as a "forbidden", or, in MOVB terminology, as a U-bound system. This point is further clarified by examining a second bond diagram, shown in Figure 1b, which now depicts the formation of CH_4 from C and H_4 under the assumption of zero $1s_H - 1s_H$ nonbonded overlap. In this case, all ligand orbitals are degenerate, no distinction between U and D binding can be made, and CH_4 can no longer be called a "forbidden" molecule. It is then evident that, depending on which of the two bond diagrams comes closer to the "truth", two different sets of bonding concepts can be developed. This dilemma is easily resolved by inspection of the computational results given below which demonstrate that geminal nonbonded overlap can be of the same order of magnitude as bonded overlap. Hence, we conclude that Figure 1a contains the correct representation of CH_4 and thus that this molecule can be justifiably called a "forbidden" or U-bound molecule.

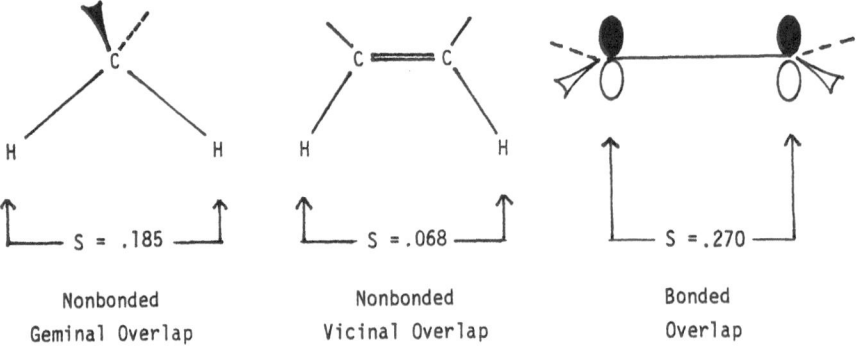

| Nonbonded | Nonbonded | Bonded |
| Geminal Overlap | Vicinal Overlap | Overlap |

S = .185   S = .068   S = .270

The reader who is familiar with qualitative MO theory will have no difficulty recognizing the ominous forebodings of the analysis presented above for we have hardly ever incorporated nonbonded interaction effects in qualitative MO models[7a] except when discussing problems such as the rotational barrier of ethane, where consideration of nonbonded interaction is actually inevitable since the staggered and eclipsed conformers are differentiable only if vicinal nonbonded interaction is assumed to be nonzero.

We can summarize our major conclusions as follows:

a. Nonbonded overlap is very large when the nonbonded atoms are geminally disposed, appreciable when they are vicinally related, and negligible when they are attached to atoms separated by intervening fragments.

b. As a result of nonbonded overlap, most organic molecules can be formulated as systems in which the core and the ligands are U-bound in a reference geometry. This realization then leads to the following predictions:

1) A molecule which is U-bound in a reference geometry will tend to adopt a different geometry so that U is replaced by H or D bonding. In other words, a molecule will "search" for a geometry which best relieves its "forbiddenness" without reducing the number or significantly affecting the strength of already existing core-ligand bonds.[12]

2) A structural modification which changes the bonding from U to H or D will have stereochemical consequences which can be easily anticipated from mere inspection of the pertinent diagrams.

These statements are strictly valid when the atoms of the core belong to the first row of the Periodic Table. A replacement of first by second row component atoms causes a decrease of nonbonded overlap in a way that has well defined chemical consequences which will be discussed in a subsequent section.

The way in which stereochemical transformations effect chemical rebonding (e.g., a change from U-type to H- or D-type bonding) has been discussed elsewhere.[4] In this paper, we focus exclusive attention on the stereochemical consequences of structural modifications which effectively destroy a symmetrical electron distribution in the ligand fragment and thus cause U $\rightarrow$ H rebonding of core and ligands. In order to conserve space and maximize the simplicity of the MOVB analysis, we shall illustrate our approach by reference to the prototypical systems AH_2, AHX, AX_2, AHF, and AF_2, where A is a divalent atom containing one sp and one p AO and X is a hypothetical monovalent ligand devoid of any electron pairs with significantly larger electronegativity than H. For example, methane can be viewed as the product of the "forbidden" union of CH_2 and H_2, and CH_2 is a divalent fragment which can be represented by A. Accordingly, instead of demonstrating concepts by drawing bond diagrams for methane and its derivatives, we can accomplish exactly the same thing by constructing bond diagrams for AH_2 and its derivatives. Furthermore, we can isolate sigma bonding effects in the absence of lone pair participation by examining the hypothetical systems AHX and AX_2. Subsequently, the role of lone pairs can be understood by examining the real AHF and AF_2 and comparing their electronic structure with that of the model AHX and AX_2 (with X having been assigned electronegativity equal to that of F) molecules.

II. Conditions for Orbital Symmetry Annihilation

Consider the two-electron-two-orbital diatomic A-A and the associated orbital manifold made up of the σ and σ^* MO's which have the form shown below.

$$\sigma \;=\; c_1 \, x_1 + c_2 \, x_2 \tag{1}$$

$$\sigma^* \;=\; c_1^* \, x_1 - c_2^* \, x_2 \tag{2}$$

Since $c_1 = c_2$ and $c_1^* = c_2^*$, we say that electron density is symmetrically distributed in each of the two MO's, or, that each MO is symmetrical with regards to the electron density distribution. We now seek the conditions for maximum MO electron density asymmetry. These are suggested by simple perturbation theory[6] which tells us that the mixing of an orbital x_2 into an orbital x_1 is proportional to their interaction matrix element and inversely proportional to the energy gap separating the two orbitals. Hence, in order to localize electron density on one center in σ and on the other center in σ^*, one must diminish the absolute magnitude of $\langle x_1 | \hat{H} | x_2 \rangle$, henceforth denoted by β_{12}, and increase the energy gap $\varepsilon_2 - \varepsilon_1$, henceforth denoted by $\Delta\varepsilon$. This simple analysis suggests that there are two coupled conditions for obtaining substantial electron density asymmetrization of σ and σ^* in A-A. Specifically, we must replace A by a more electronegative or electropositive atom, Y, so that $|\Delta\varepsilon|$ will change from zero to some arbitrarily large value, while increasing the A-Y interactomic (or interfragmental) distance, r_{AY}, so that $|\beta_{12}|$ will decrease since β_{12} is proportional to s_{12}, the overlap integral of x_1 and x_2, which is a function of r_{AY}.

At this point, we open a parenthesis in order to emphasize that maximum electron density asymmetrization is obtained by replacing A by a more electropositive Y while simultaneously increasing the r_{AY} simply because β_{12} depends on the orbital energies of x_1 and x_2, commonly taken to be the Valence Orbital

Ionization Energies (VOIE's), as well as the corresponding overlap integral according to the Wolfsberg-Helmholz approximation:[13]

$$\beta_{12} = K (\varepsilon_1 + \varepsilon_2) s_{12}/2 \tag{3}$$

As a result, maximization of $|\Delta\varepsilon|$ by raising, rather than lowering, the energy of, say, x_2 is much more effective in simultaneously reducing β_{12}. In short, equation (3) tells us that maximum electron density asymmetrization is obtained by replacing A-A by "stretched" A-Y where Y is much more electropositive than A. The reader is referred to a previous chapter dealing with the stereochemical consequences of weak overlap binding,[14] i.e., interfragmental interaction due to small, in absolute magnitude, β_{mn}'s.

With the above in mind, we formulate the following rule: A substantial asymmetrization of the MO electron densities of a homonuclear core or ligand fragment can be achieved by replacement of one or more atoms or groups by more electronegative or electropositive ones while simultaneously increasing their distance. As an example, consider the hypothetical molecule AY_2 viewed as a composite of core (A) and ligand (Y_2) fragments. Different types of Y substitution as well as different geometrical arrangments of the three atoms cause differing asymmetrizations of the σ and σ^* ligand MO's. Specifically, the electron density of these two MO's becomes asymmetrical as Y is replaced by a more electronegative Z and as bent AYZ is transformed to linear AYZ so that the distance between Y and Z increases. A more drastic asymmetrization is obtained by replacement of Y by a more electropositive Z and linearization of AYZ.

III. Stereochemical Consequences of "Forbiddenness" Reduction in Ground State

     Molecules

Consider the hypothetical molecule AH_2 described by the bond diagram of
Figure 2. It is evident that AH_2 is a U-bound molecule. Replacement of one of
the two hydrogen atoms by X (as defined before) will break the symmetry of the
ligand MO's and will reduce interfragmental spatial overlap. As a result, the
derivative molecule AHX will be $H^\ddagger$-bound. Again, this is illustrated in Figure 2.
Furthermore, we observe that the $U \rightarrow H^\ddagger$ rebonding as a result of substitution of
H by X allows electrons formerly forced to occupy the high lying orbitals p and
σ^* to "drop" to the low lying orbitals sp and σ as a result of symmetry breaking.
In a formal sense, we say that annihilation of orbital symmetry in the ligand
fragment causes the Configuration Wavefunction (CW) shown below to become a very
important contributor to the total MOVB wavefuction and this has two very important
stereochemical consequences:

    a. The ligand interaction becomes attractive because of the double occupation
of the bonding σ ligand MO.

    b. The core-ligand interaction becomes repulsive because of four electron
overlap destabilization.

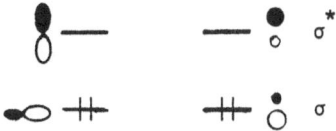

We conclude then that replacement of H by X in AH_2 will cause angle shrinkage as
well as elongation of the A-H bond in AHX relative to the same bond in the U-bound
AH_2 molecule. By using the same line of reasoning, we can also conclude that AHX
will have a smaller angle and a longer A-X bond than AX_2.

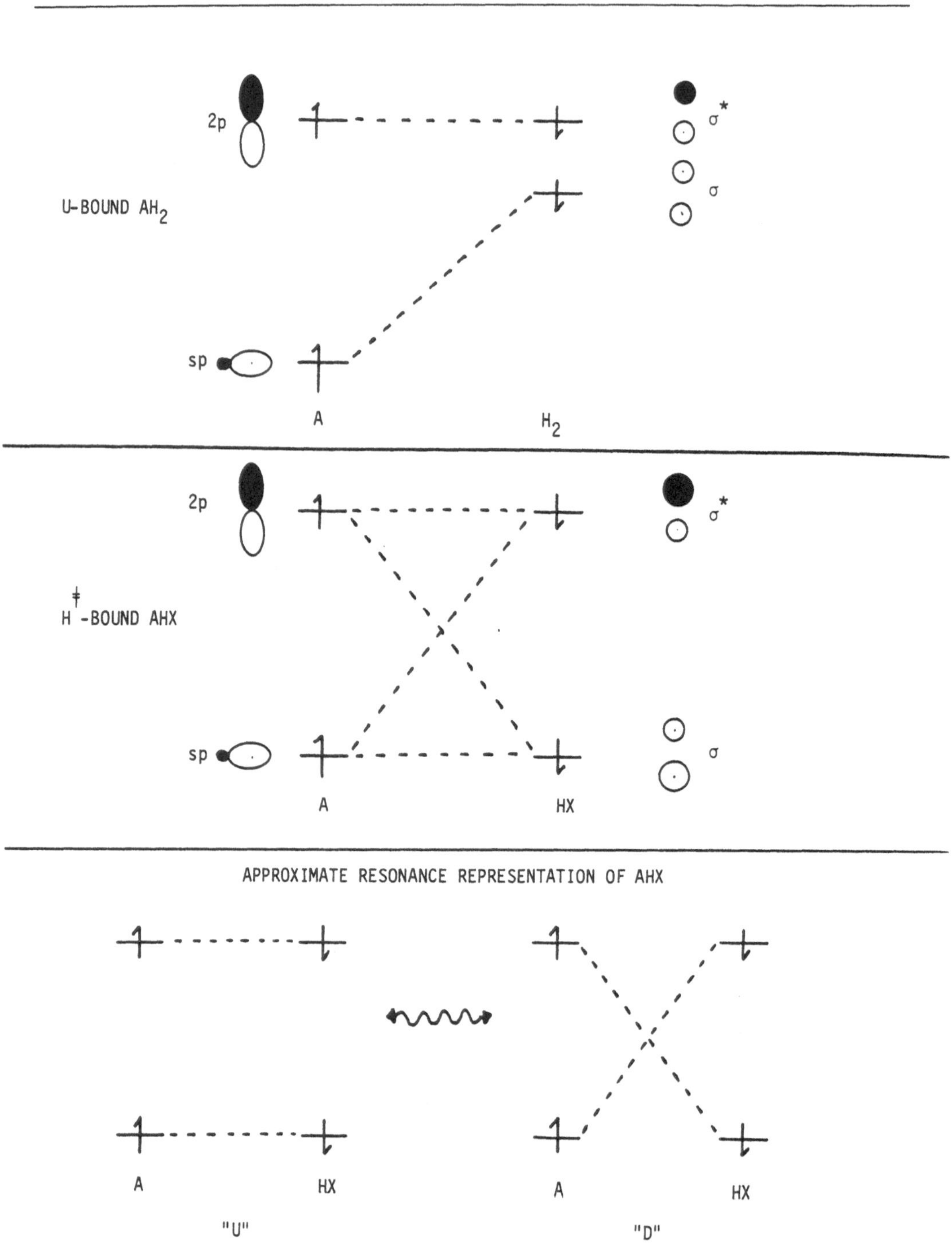

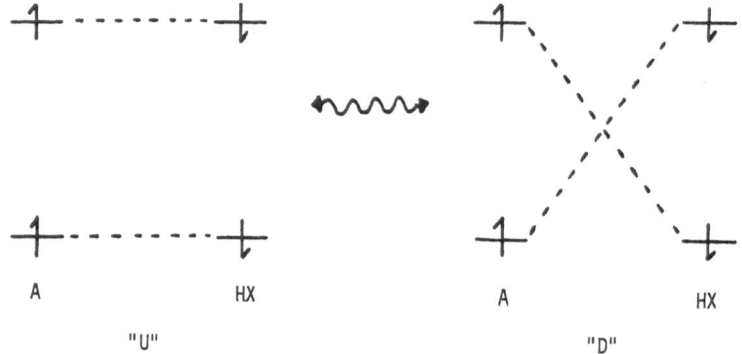

Figure 2. Bond diagrammatic representations of AH_2 and AHX. The relative contribution of "U" and "D" depends on the degree of σ and σ^* asymmetrization in AHX.

Next, let us consider the molecule H_2O in its equilibrium bent geometry. Figure 3 shows the MOVB representation of this molecule as well as the MOVB representation of the HOX derivative where X is again more electronegative than H, e.g., F treated as a univalent ligand with neglect of its lone pairs. Recognizing that the major contributor bond diagram of H_2O is $\equiv_2$ and those of HOX $\equiv_2 - \equiv_4$, realizing that the CW shown below is a major contributor to the total MOVB wavefunction of HOX, and using the same arguments as before, we would predict that HOF must have a smaller angle than either H_2O or F_2O and that the O-H and O-F bonds in HOF must be longer than the O-H bond in H_2O and the O-F bond in F_2O, respectively. This is indeed what experimental investigations have revealed[14] and the pertinent data are shown below.

$$
2p \left\{ \begin{array}{l} \text{\LARGE ⧺} \\[6pt] \text{\LARGE —} \\[6pt] \text{\LARGE ⧺} \end{array} \right. \qquad \qquad \text{—} \quad \sigma^* \text{ (H-like)}
$$

$$
\text{⧺} \quad \sigma \text{ (F-like)}
$$

$$
2s \quad \text{⧺}
$$

Dominant CW of HOX

	Bond Angle	r(OH)	r(OF)
H-O-H	105.5°	0.959	-
H-O-F	97.2°	0.964	1.442
F-O-F	103.1°	-	1.405

We have now completed a sketch of the scenario to unfold and we are prepared to go ahead with the major task: The demonstration of the potential of MOVB theory to explain naturally many puzzling trends of chemistry for which apparently

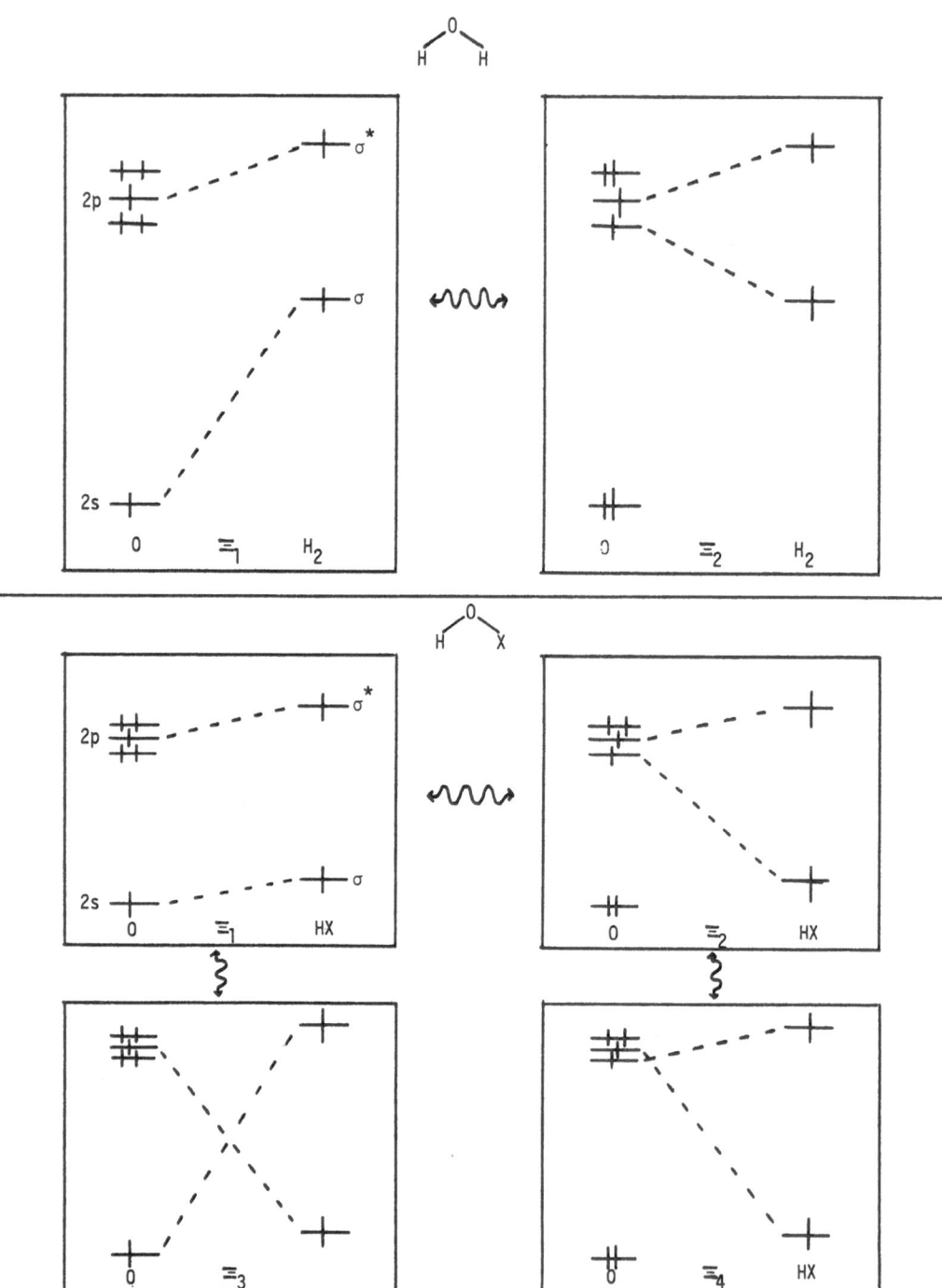

Figure 3. Bond diagrammatic representations of HOH and HOX. Ξ_i are resonance bond diagrams.

successful, intuitive rationalizations exist, but, which, once scrutinized by MOVB
theory, turn out not to contain the essential factor which is the basis, in part
or in toto, of the chemical phenomena itself! In this work, we confine our
attention to the problem of symbiosis in organic molecules.

IV. The MOVB Theory of Symbiosis on Carbon Cores

Let us consider the isodemic reaction[15] shown below and inquire as to the sign

of the energy difference between reactants and products, $E_P - E_R$, henceforth denoted

by ΔE, under two different sets of conditions:

$$AH_2 + AX_2 \longrightarrow AHX + AHX \qquad\qquad (I)$$

a. The nonbonded overlap of the ligand AO's is zero.

b. The nonbonded overlap of the ligand AO's is nonzero and the entire molecule

can be viewed as the product of the "forbidden" union of core and ligands.

A direct answer to the question posed indirectly above can be given by constructing

the bond diagrams for reactants and products and evaluating their relative energies.

The isodesmic reaction is recast in bond diagrammatic form as shown in Figure 4

with the assumption that the nonbonded overlap of the ligand AO's is zero. A mere

inspection of the one-electron energies of the major contributor CW's is sufficient

to convince one that 2AHX will have lower energy than $AH_2 + AX_2$. In other words,

we can say that in AHX electrons are permitted to occupy the lowest lying orbital

of the core and the low lying orbital of the electronegative ligand X while this

is no longer possible in AH_2.

The conclusion reached above on the basis of the zero nonbonded AO overlap

assumption is somewhat altered if we now assume that indeed the ligand AO's overlap.

In this case, we can write the same isodesmic equation in bond diagrammatic form as

shown in Figure 5. We now say that the fact that AH_2 and AX_2 are "forbidden" while

AHX partly "forbidden" is responsible for products being more stable than reactants.

At the same time, we recognize that the magnitude of ΔE depends on the λ_D/λ_U ratio

because occupation of the low lying core and ligand orbitals is possible only in

the "D" but not the "U" resonance bond diagram of AHX.

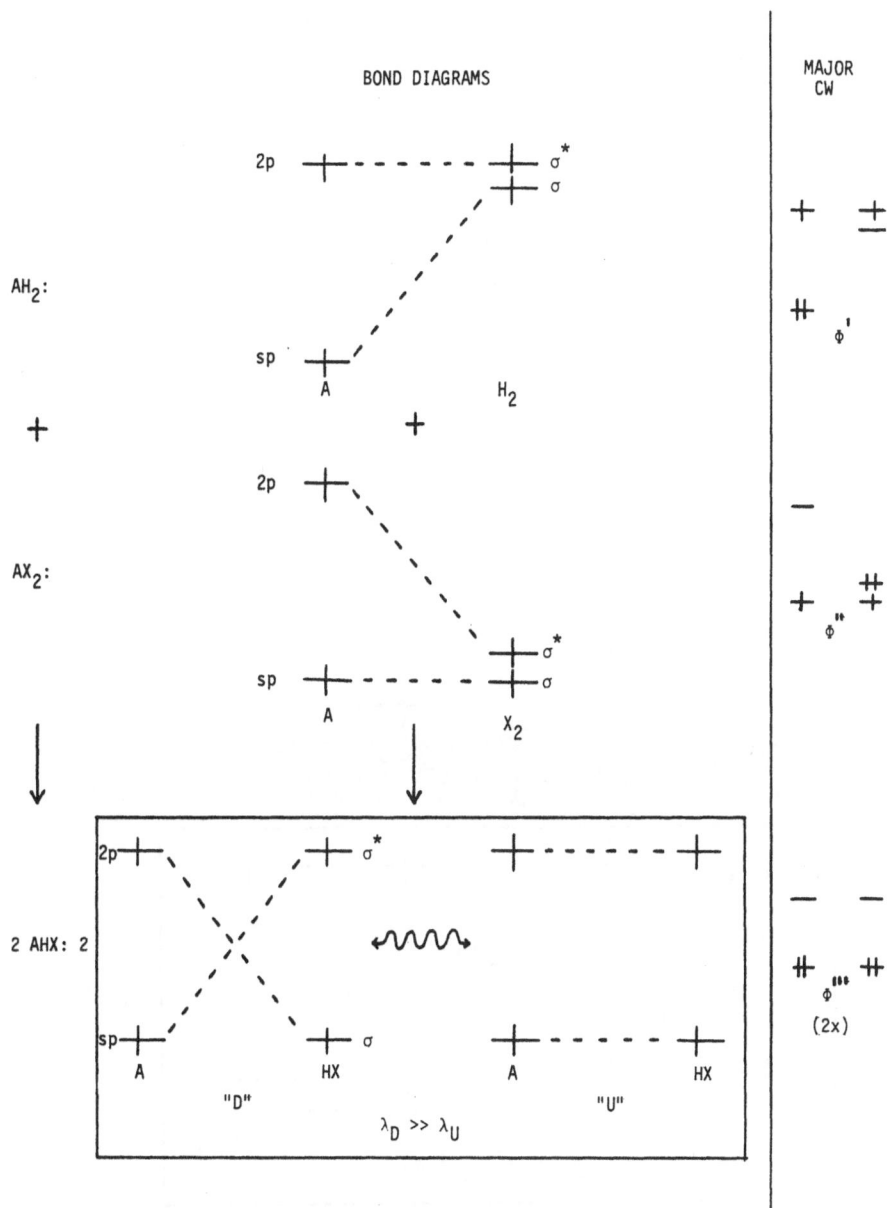

Figure 4. Bond diagrammatic representation of the reaction $AH_2 + AX_2 \rightarrow 2AHX$ based on the assumption that nonbonded AO overlap is zero.

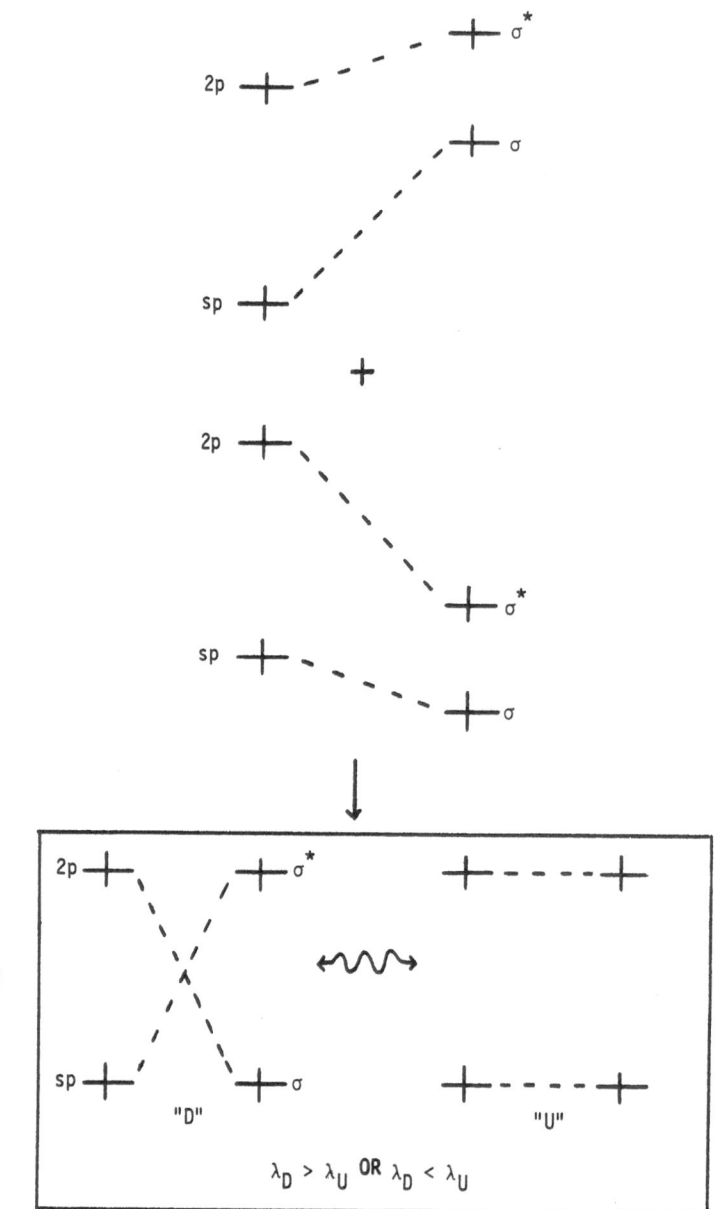

Figure 5. Bond diagrammatic representation of the reaction $AH_2 + AX_2 \rightarrow 2AHX$ based on the assumption of appreciable nonbonded AO overlap.

The conclusions reached above are fully supported by quantum chemical computations of different degrees of sophistication. Thus, for example, we have computed the energy of the reaction shown below where D is an electropositive and A an electronegative monovalent ligand in a variety of ways and have always obtained the same qualitative answer, namely, that the less symmetrical products are more stable than the more symmetrical reactants. The results shown below[16] are convincing evidence that the effect we have uncovered is a fundamental one and, consequently, reproducible at all levels of MO theory.

$$CD_4 + CA_4 \longrightarrow 2\ CD_2A_2 \qquad\qquad (II)$$

Calculation Type	ΔE (kcal/mol)
EHMO	- 8
Ab initio SCF-MO	- 49
Ab initio SCF-MO-CI	- 48

The conclusion is inescapable: If only sigma bonding effects are considered, there is ligand segregation, or, ligand antisymbiosis, i.e., ΔE in equation (I) as well as (II) is negative.

Next, let us consider the isodesmic equation shown below, which differs from the previous one only to the extent that the monovalent electronegative ligand X has been replaced by F, a real ligand having the same electronegativity as X but possessing, in addition, lone pairs. The question now becomes: Will the presence

$$AH_2 + AF_2 \longrightarrow AHF + AHF \qquad\qquad (III)$$

of the lone pairs on F change the sign of ΔE predicted in their absence? Before we can answer this question, let us refresh our memory with regards to the conditions for optimal lone pair delocalization in molecules by consulting Figure 6 which shows that lone pair delocalization and, hence, overall stabilization, depends on the direction of primary Charge Transfer (CT) which, in turn, depends upon the energetic interrelationship of core and ligand MO's.[4] Figure 6 shows three cases in which the mode of primary bond formation determines whether lone pair delocalization will be turned off or on. Since extensive discussions of these fundamental principles can be found in the original work, we merely point out that the lone pair delocalization becomes increasingly favorable as primary CT occurs towards the fragment carrying the lone pair, which in our example is the ligand fragment. With this information at hand, we can now rephrase the question posed above as follows: Is lone pair delocalization more effective in two AHF molecules than in one AF_2 molecule?

Fluorine has seven valence electrons and, in bonding to other atoms, it makes primary use of a singly occupied 2p AO with the other electron pairs occupying the other 2p and the 2s AO's. In the case of the bent AHF and AF_2 species, the pi type lone pair occupying, say, the $2p_z$ AO can be neglected.[17] The 2s lone pair approaches being truly a lone pair to the extent that the large 2s-2p energy gap hardly encourages its delocalization and so it can also be neglected. Thus, for our purposes, we can think of F as being sigma-bound to A as a three electron atom. With this convention, we can now draw the bond diagrams for AHF and AF_2 and decide whether 2p lone pair delocalization is more favorable in the former than in the latter, or vice versa, under two different assumptions:

    a. Zero nonbonded ligand AO overlap.

    b. Nonzero nonbonded ligand AO overlap.

The bond diagrams of Figure 7 can be interpreted in a straightforward manner without recourse to lengthy discussions. In particular they reveal the following:

    a. Regardless of whether one assumes that nonbonded ligand AO overlap is zero or nonzero, one lone pair is well delocalized and the other is modestly delocalized in AF_2.

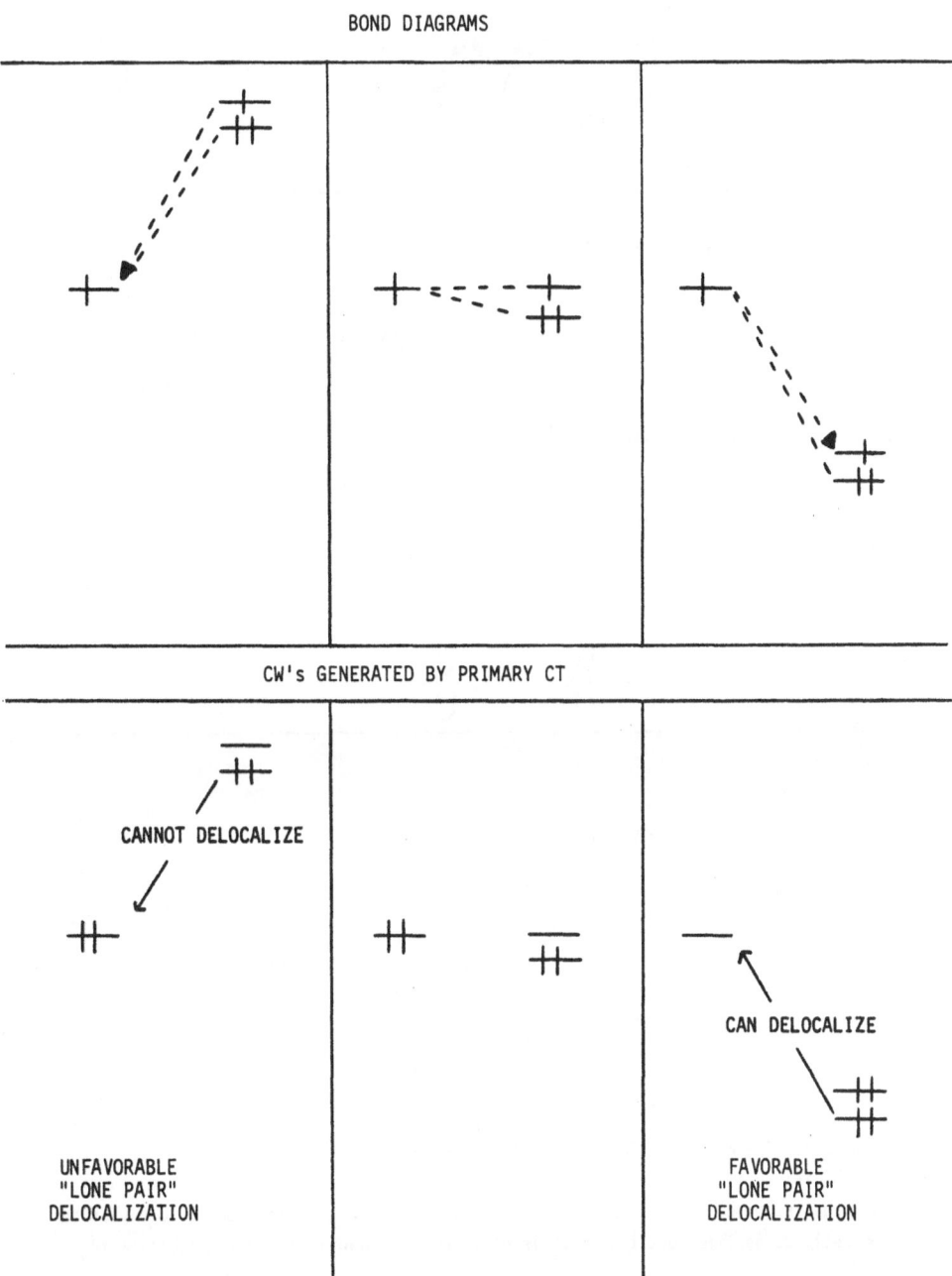

Figure 6.   The way in which the direction of primary charge transfer (indicated by arrow) triggers or prevents "lone pair" delocalization.

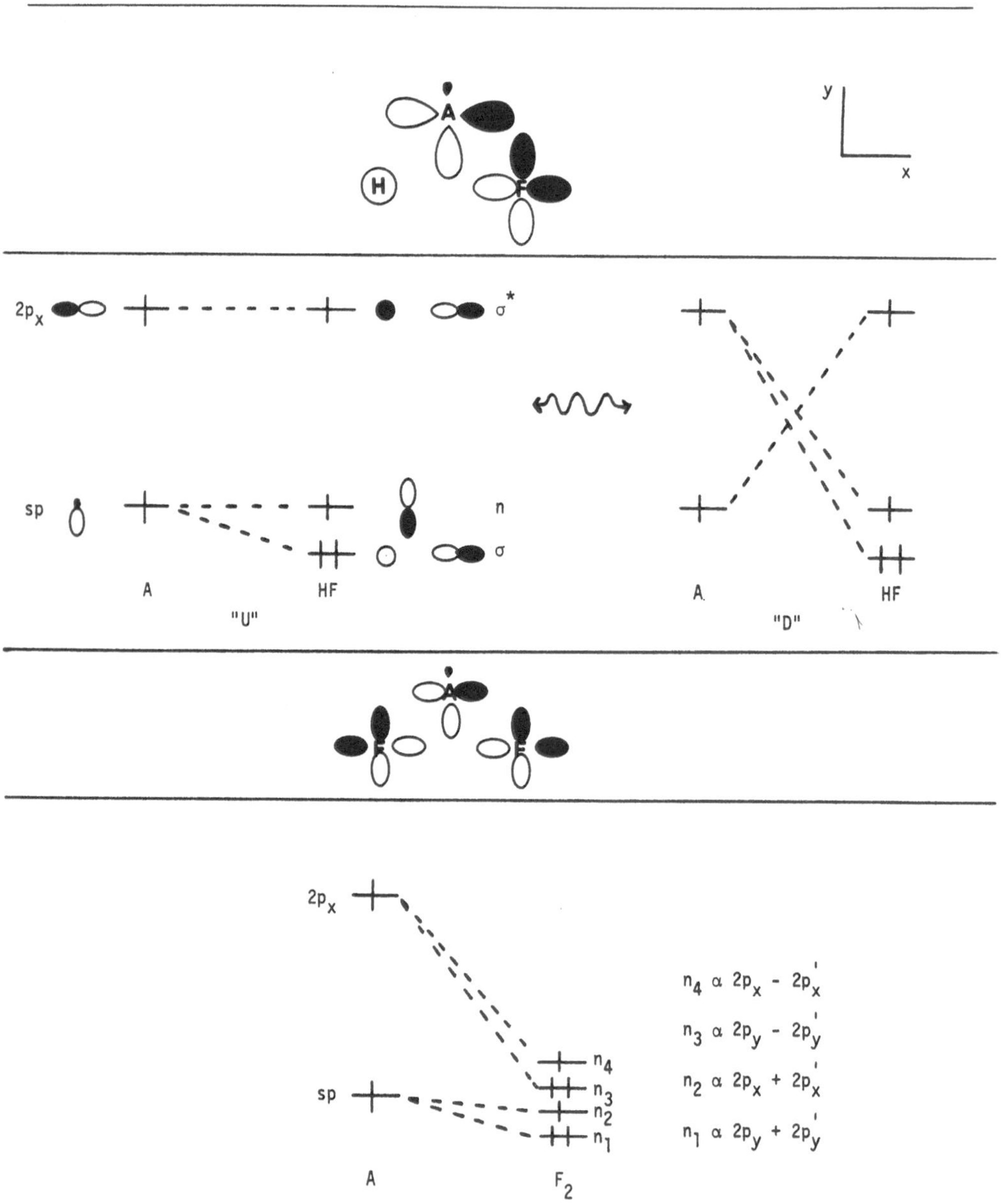

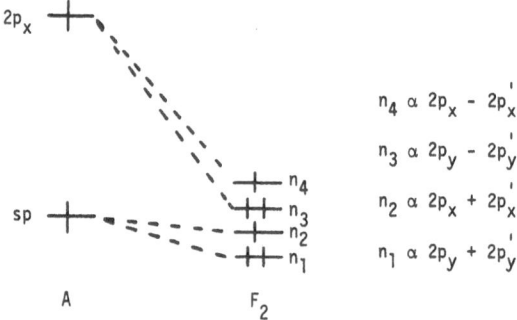

y
x
2px
σ*
sp
n
σ
A
HF
"U"
A
HF
"D"
2px
sp
n4
n3
n2
n1
A
F2
n4 α 2px - 2px'
n3 α 2py - 2py'
n2 α 2px + 2px'
n1 α 2py + 2py'

Figure 7. The mode of F lone pair delocalization in AHF and AF_2. In the first case, the dominance of "U" (due to strong H---F AO overlap) becomes responsible for modest delocalization of the single F lone pair (see Figure 5). In AF_2, one F lone pair is modestly delocalized but the other is given an excellent chance for delocalization.

b. If one assumes that nonbonded ligand AO overlap is zero, then $\lambda_D \gg \lambda_U$ and, as a result, there is good lone pair delocalization in AHF. By contrast, if now one assumes that there is substantial nonbonded ligand AO overlap, then $\lambda_U > \lambda_D$ and, as a result, there is modest lone pair delocalization in AHF.

This analysis leads to the following important conclusion: If molecules are viewed as "forbidden" entities, then lone pair delocalization is more favorable in AF_2 than in two AHF's and this will tend to reverse the sign of ΔE predicted by neglecting the lone pairs. If, by contrast, nonbonded ligand AO overlap is negligible, i.e., if molecules do not "deserve" the description "forbidden", then lone pair delocalization is more favorable in two AHF's rather than AF_2, a fact which will reinforce the conclusion arrived at by neglecting the lone pairs. Now, all types of MO computations[18] as well as experimental studies[19] demonstrate that the reaction shown below is endothermic.

$$\underset{\underset{\text{F F}}{/ \backslash}}{\text{CH}_2} \quad + \quad \underset{\underset{\text{H H}}{/ \backslash}}{\text{CH}_2} \quad \longrightarrow \quad \underset{\underset{\text{H F}}{/ \backslash}}{\text{CH}_2} \quad + \quad \underset{\underset{\text{H F}}{/ \backslash}}{\text{CH}_2} \qquad\qquad \text{(III)}$$

	ΔE (kcal/mol)
EHMO	26.8
<u>Ab initio</u> SCF-MO	11.5
Experiment:	24.9

Two things now become obvious:

a. Lone pair delocalization favors fluorine symbiosis and its effect overrides the competing sigma bonding effect discussed before. This implies that methane-like molecules deserve the title "forbidden".

b. All alleged explanations of symbiosis in methane, i.e., all explanations of the fact that 2 CH_3Y are less stable than CH_2Y_2 plus CH_4 and related observations which are not explicitly based on the notion that ground state molecules are "forbidden" entities are necessarily heuristic interpretations because, <u>in the absence of nonbonded ligand AO overlap, ΔE would be negative rather than positive</u>. A correct interpretation demands that the role of nonbonded ligand AO overlap is properly taken into account. So, here we are after half a century of intense preoccupation with a fundamental problem now discovering that previous explanations have only been illusions, with the correct interpretation becoming only apparent once we properly understand what "allowed" and "forbidden" really means in many electron-many-orbital systems.

Next, let us consider the following question: What happens to the endothermicity of equation (III) as F is sequentially replaced by OH, NH_2, and CH_3? The answer to this question is straightfoward once it is recognized that each of these groups is effectively a monovalent heteroatom Y with one singly occupied AO and three doubly occupied orbitals which can be thought of as the containers of three "lone pairs". Thus, in sweeping from F to CH_3 we are effectively changing the ligand Y electronegativity by making it more electropositive relative to a fixed core. The consequences of this are straightforward:

a. Interfragmental CT is redirected to the extent that CT from core to ligands becomes less and less favorable.

b. As a result of (a) "lone pair" delocalization becomes less and less <u>intrinsically</u> favorable (see Figure 6).

c. As a result of (b), the bias in favor of the left hand side of equation (III) due to lone pair delocalization tends to disappear and so does the reaction endothermicity.

$$CH_4 \quad + \quad CH_2Y_2 \longrightarrow 2 \ CH_3Y \tag{IV}$$

Schleyer and coworkers have calculated the ΔE of reaction (IV) for a variety of Y's and their computational results are shown in Table 1. It is clear that, with the exception of OH, there is a trend of decreasing ΔE with decreasing Y electro-negativity. The "anomaly" of OH is probably due to the "hydrogen bonding" - type interaction of two OH's in $CH_2(OH)_2$. Like everything else, this effect can be demonstrated by constructing more detailed bond diagrams but this exercise is left to the reader since it is not directly relevant to our pursuits.

We can now summarize the most important conclusions of our analysis as follows:

a. A ground state composite system made up of a core (comprised of first row atoms) and ligand fragments which contain nondegenerate orbitals is a partly or totally "forbidden" entity with the nondegeneracy of the ligand orbitals being due to appreciable overlap of the orbitals of the nonbonded ligands.

b. In molecules of the type specified above and in the hypothetical absence of ligand and core lone pairs, there is chemical segregation, i.e., the sign of the ΔE of equation (I) is negative. The presence of ligand or core electron pairs can reverse this trend and promote chemical symbiosis, i.e., it can become responsible for changing the sign of the ΔE of equation (I) from negative to positive.

c. If nonbonded ligand orbital overlap is assumed to be zero, then a correct theoretical analysis of the electronic structure of the reactants and products of equation (I) leads to the conclusion that chemical segregation must exist uniformly and without exception. Hence, all analyses based on this very assumption but leading to an opposite conclusion are incorrect.

d. As the electronegativity of a ligand containing an effective lone pair decreases, while the core fragment is kept constant, lone pair delocalization becomes less important and ligand symbiosis is progressively abolished.

As we have discussed before, all common heteroatomic groups which adorn "organic" compounds, i.e., F, OH, NH, and CH_3 itself, have real or "effective" lone

pairs. Furthermore, when the central atom is carbon and the core-ligand bonds
are relatively short, nonbonded ligand orbital overlap is appreciable so that the
composite system can be regarded as being primarily U-bound in which case lone
pair delocalization will promote symbiosis. Accordingly, we can state the
following hypothesis: <u>Whenever there is appreciable nonbonded ligand overlap and
the ligands carry lone pairs, there will be cooperative ligand interaction promoting aggregation or symbiosis as well as stronger core-ligand binding in the
symbiotic state with the effect decreasing as the electronegativity of the ligand
changes in a way which discourages lone pair delocalization, i.e., as the ligands
are made increasingly electropositive relative to the core.</u> That this hypothesis
has been arrived at by a sequence of "intricate" arguments which, nonetheless, can
be cast in pictorial form, is made evident in Scheme 1 which, in effect, explains
why the statement just made is a hypothesis and not a definite prediction.

SCHEME 1.

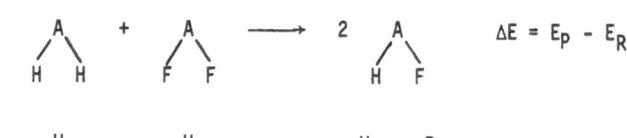

$$\Delta E = E_P - E_R$$

Bonding
Type: U   U   U $\longleftrightarrow$ D

Condition	Trend
Lone Pairs "Off"	$\Delta E < 0$ if $\lambda_U > \lambda_D$
	$\Delta E \ll 0$ if $\lambda_U < \lambda_D$
Lone Pairs "On"	$\Delta E > 0$ if $\lambda_U > \lambda_D$
	$\Delta E < 0$ if $\lambda_U < \lambda_D$

$\lambda_D > \lambda_U$ if nonbonded ligand AO overlap is zero.

This work has been inspired by known and well publicized experimental facts.[20]
Thus, to feign amazement at the fact that the stated hypothesis is consistent with
experimental results would be inappropriate. Rather, we can say that MOVB theory
provides a plausible interpretation of some remarkable trends which have long
fascinated chemists and which constitute, in part, the basis of the concept of
chemical symbiosis. In this connection, we mention that many explanations of
fluorine symbiosis can be found in the chemical literature of the past one half
century. None of them amounts to a proper interpretation of these trends spoken
of above and the reason is very simple: Previous investigators had not been
alerted as to the consequences of "orbital symmetry", so to speak, and, thus,
they did not search for the right answer in the right place. The fact that no
clear and convincing interpretation of fluorine symbiosis exists in the literature was fully recognized by Chambers[20] who commented: "...the very range of
explanations that have been offered is some indication in itself of the uncertainty which exists in explaining this very fundamental and interesting aspect
of fluorine chemistry and casts a pointer at the level to which we can hope to
discuss the chemistry of fluorocarbon systems at this time."

Table 1. The computed $CH_2X_2 + CH_4 \longrightarrow 2\ CH_3X$ Transformation
Energy (ΔE).[a]

X :	Li	BeH	BH$_3$	CH$_3$	NH$_2$	OH	F
ΔE : (Kcal/mol)	-3.1	+7.8	+4.1	+1.2	+8.3	+15.2	+11.5

[a] Data taken from Dill, J.D.; Schleyer, P.v.R.; Pople, J.A. J. Am.
Chem. Soc. 1973, 98, 1663.

References

1. Hückel, E. Z. Physik. 1931, 70, 204; ibid. 1932. 76, 628. Hückel, E. Z. Electrochem. 1937, 43, 752.

2. (a) Woodward, R.B.; Hoffmann, R. "The Conservation of Orbital Symmetry"; Verlag Chemie: Weinheim, 1970.

    (b) The independent recognition of orbital symmetry control of reaction stereochemistry by a number of brilliant investigators is documented in a recent article: Epiotis, N.D.; Shaik, S.; Zander, W "Rearrangements in Ground and Excited States", De Mayo, P., Ed., Vol. 2; Academic Press, Inc.: New York, 1980.

3. (a) Streitwieser, Jr., A. "Molecular Orbital Theory for Organic Chemists"; John Wiley and Sons, Inc.: New York, 1961.

    (b) Heilbronner, E.; Bock, H. "Das HMO-Modell und Seine Anwendung"; Verlag Chemie, Gmbh: Weinheim, 1968.

4. Epiotis, N.D.; Larson, J.R.; Eaton, H. "Unified Valence Bond Theory of Electronic Structure" in Lecture Notes in Chemistry, Vol. 29; Springer-Verlag: Heidelberg, 1982.

5. The "father" of the FO approximation in "qualitative" MO theory is K. Fukui. Fukui, K.; Yonezawa, T.; Shingu, H. J. Chem. Phys. 1952, 20, 722, Fukui, K.; Yonezawa, T.; Nagata, C.; Shingu, H. J. Chem. Phys. 1954, 22, 1433.

6. Dewar, M.J.S. "The Molecular Orbital Theory of Organic Chemistry"; McGraw-Hill: New York, 1969.

7. For applications of the FO-PMO model to problems of structural chemistry and chemical reactivity, see:

    (a) Epiotis, N.D.; Cherry, W.R.; Shaik, S.; Yates, R.L.; Bernardi, F. Top. Curr. Chem. 1977, 70, 1.

7. (b) Ref. 6.

    (c) Hudson, R.F. Angew Chemie, Int. Ed. Engl. 1973, 12, 36.

    (d) Klopman, G. in "Chemical Reactivity and Reaction Paths", Klopman, G.,
         Ed.; Wiley-Interscience: New York, 1974.

    (e) Fukui, K. "Theory of Orientation and Stereoselection"; Springer-Verlag:
         New York and Berlin, 1975.

    (f) Fleming, I. "Frontier Orbitals and Organic Chemical Reactions"; John
         Wiley: New York, 1976.

8. (a) Heitler, W.; London, F. Z. Physik. 1927, 44, 455.

    (b) Slater, J.C. Phys. Rev. 1931, 38, 1109.

    (c) Eyring, H.: Kimball, G.E. J. Chem. Phys. 1933, 1, 239, 626.

    (d) Eyring, H.: Frost, A.A.; Turkevich, J. J. Chem. Phys. 1933, 1, 777.

    (e) Pauling, L. J. Chem. Phys. 1933, 1, 280.

    (f) Van Vleck, J.; Sherman, A. Rev. Mod. Phys. 1935, 7, 168, 200.

    (g) McWeeny, R. Proc. Roy. Soc. A. 1953, 233, 306.

    (h) Bobrowicz, F.W.; Goddard, III, W.A. in "Modern Theoretical Chemistry",
         Schaefer, III, H.F., Ed.; Plenum Press: New York, 1976.

9. Some important texts which contain discussions of fundamental VB theory are
    the following:

    (a) Pauling, L.; Wilson, E.B. "Introduction to Quantum Mechanics"; McGraw-Hill
         Book Co., Inc.: New York, 1935.

    (b) Glasstone, S.; Laidler, K.J.; Eyring, H. "The Theory of Rate Processes";
         McGraw-Hill: New York, 1941.

    (c) Eyring, H., Walter, J., Kimball, G.E. "Quantum Chemistry"; Wiley and
         Sons: New York, 1954.

9. (d) Slater, J.C. "Quantum Theory of Molecules and Solids", Vol. 1; McGraw-Hill: New York, 1963.

    (e) Sandorfy, C. "Electronic Spectra and Quantum Chemistry"; Prentice-Hall: Englewood Cliffs, NJ, 1964.

10. Wheland, G.W. "Resonance in Organic Chemistry"; John Wiley and Sons: New York, 1955.

11. For example, the work of R. Hoffmann and his collaborators in the area of inorganic chemistry, illustrates the importance of all orbitals in ground state molecular stereochemistry and the necessity for utilization of high order perturbation theory in dealing with problems of this type.

12. Molecules which shun deexcitation, i.e., "forbiddenness" relief, and exist as U-bound species are known and examples have been discussed in Ref. 4. The reason for U over H or D preference is maintenance of strong interfragmental overlap at the expense of excitation. A typical illustrator is acetylene which is U-bound in the linear $D_{\alpha h}$ geometry. This geometry is preferred over the trans-bent one although the latter involves H-bonding.

13. (a) Wolfsberg, M.; Helmholz, L. J. Chem. Phys. 1952, 20, 837.

    (b) Hoffmann, R.; Lipscomb, W.N. J. Chem. Phys. 1962, 36, 2189.

    (c) Hoffmann, R. J. Chem. Phys. 1963, 39, 1397.

14. Callomon, J.H.; Hirota, E.; Kuchitsu, K.; Lafferty, W.J.; Maki, A.G.; Pote, C.S. in Landolt-Bornstein, "Numerical Data and Function Relationships in Science and Technology" Vol. 7, New Series, "Structure Data of Free Polyatomic Molecules", Hellwege, K.H., Ed.; Springer-Verlag: Berlin, 1976.

15. Hehre, W.J.; Ditchfield, R.; Radom, L. ; Pople, J.A. J. Am. Chem. Soc. 1970, 92, 4796.

16. The C-D and C-A bond lengths were taken to be equal to the C-H bond length and tetrahedral geometries were used throughout. In the EHMO computation, the VOIE of D was set at -11.6 eV and that of A at -15.6 eV. In the ab initio computations, we used an STO-3G AO basis and set the nuclear charge of D at 0.9 and that of A at 1.1. The CI computations were carried out using a program kindly provided by Professor E. R. Davidson of this department.

17. The vicinal pi nonbonded overlap of two F's is small and, hence, any pi conjugation in AF_2 will produce an effect which is roughly the same as any pi conjugation in two AHF's, i.e., the pi conjugation effect of the two F's in AF_2 is additive.

18. The EHMO calculations were carried out using standard bond angles and bond lengths. The ab initio value is taken from the work of Schleyer et al. cited in Table 1.

19. The experimental ΔE has been computed using the thermochemical data contained in: Cox, J.D.; Pilcher, G. "Thermochemistry of Organic and Organometallic Compounds"; Academic Press: New York, 1970.

20. Chambers, R.D. "Fluorine in Organic Chemistry"; John Wiley and Sons: New York, 1973. A compilation of data of bond dissociation energies and bond lengths of haloalkanes can be found in Table 7.1, p. 138.

Nonbonded repulsion, commonly referred to as "steric effect", is one of the priceless weapons of the conceptual arsenal of the chemist, especially the organic chemist. It is repeatedly invoked to rationalize why the most "crowded" arrangement of atomic nuclei is often the least stable one.[1] The simplicity of the concept, an intuitively obvious one, has much to do with the popularity it enjoys. While excellent for a posteriori rationalization, the concept of nonbonded repulsion has no predictive value. For example, ethane is staggered but water is non-linear. If we did not know the actual physical facts, we would predict ethane to be staggered and water linear on the basis of "steric effects". Furthermore, the concept of nonbonded repulsion is really an empirical one for it was originally developed by experimentalists in order to codify a subset (not a complete set) of experimental observations which indicated that "crowding" of nonbonded atoms is energetically unfavorable. Thus, the concept of nonbonded repulsion is both nonpredictive and nontheoretical (heuristic).

At this stage the reader may object: Is it not true that nonbonded repulsion has been placed on a theoretical basis and is it not true that it is well understood as a result of many quantum chemical studies? The answer to this question depends on the level of comprehension of chemical bonding one desires. For example, it can be said that much is known about nonbonded repulsion through local theoretical models, i.e., models which are applicable to one type of problem but not another. Thus, we can rationalize why water is bent and not linear with the Second Order Jahn Teller (SOJT) model.[2a,b] Similarly, we can provide an a posteriori explanation of why ethane is staggered and not eclipsed through the hyperconjugation model.[2c-g] However, the hyperconjugation model is not applicable to angular distortion problems and the SOJT model does not "work" for most conformational problems! Thus, we can hardly claim to understand nonbonded repulsion (ethane problem), or, its antipode, nonbonded attraction (water problem). Indeed, we realize that nonbonded repulsion and attraction are not concepts but descriptions of the optimal bonding

<u>resolutions made by the system in question</u>. We can claim an understanding of why some molecules opt for an "unconjested" and others for a "crowded" geometry only within the context of a general theory of bonding which can foretell which choice the molecule will likely make.

The above serves as an introduction necessitated by the traditional education which prepares one to become a researcher in chemistry and which shapes attitudes and credos which are not founded on solid ground. For example, the undergraduate student of chemistry is introduced to nonbonded repulsion, most often via the examples of rotational isomerism of ethane and geometric isomerism of 2-butene[3], while no mention is made of nonbonded attraction, e.g., in 1,2 difluroethane and 1,2 difluoroethylene. Later on, during his graduate career, he learns that 1,2 difluoroethane is gauche [4] and 1,2 difluoroethene cis[5]. Finally, he may encounter literature articles on "steric attraction" or "nonbonded attraction"[6] which rationalize preferences of diverse systems for "crowded" geometries on the basis of <u>local</u> theoretical models. However, by that time, the following unjustified opinions have been formed:

    a. Nonbonded repulsion is the rule and nonbonded attraction the exception.

    b. Nonbonded repulsion is "intuitive" while nonbonded attraction is "counterintuitive".

    c. Systems dominated by nonbonded repulsion are not interesting while those dominated by nonbonded attraction are novel.

Thinking along these lines demonstrates that one has lost track of what is a true concept and what is an empirical descriptor. That nonbonded repulsion and attraction are descriptions of two possible modes of operation of nonbonded atoms, with neither being the rule or the exception, is vividly demonstrated by compilations of data pertaining to the relative stability of 1,2 disubstituted ethenes.[7] Depending on whim, one can easily pronounce either nonbonded repulsion or attraction the rule or the exception! For example, alkyl substituents generate a preference for the trans while halogen substituents create a preference for the cis geometry in 1,2 substituted ethylenes.

With the above in mind, we can state that the purpose of this paper is to show that there exists an electronic mechanism which is present in the vast majority of organic polyatomic molecules and which, if not overridden by other factors, it causes them to adopt "uncrowded" geometries. This mechanism is natural nonbonded repulsion and it is the direct consequence of the fact that ground state organic molecules are (partly or totally) "forbidden" entities, i.e., they are the result of (partly or totally) "forbidden" unions of cores and ligands.

I. Theory

Consider the hypothetical four-electron-four-orbital CL_2 molecule which can be viewed as the product of the union of a core(C) and a ligand (L_2) fragment so that there are two core-ligand multicenter bonds, the molecule is bent, and the ligand orbitals do not overlap with each other. There exist two different bond diagrams which adequately describe such a molecule and these are shown on the left of Figure 1. Note that because of the degeneracy of $\sigma(\lceil_a)$ and $\sigma^*(\lceil_b)$ it is irrelevant whether ω_1 is of $\lceil_a$ and ω_2 of $\lceil_b$ symmetry, or _vice versa_.

Next, let us consider what happens if we "push" the ligands towards each other so that the $\widehat{LCL}$ angles shrinks without any concommitant change of the core-ligand AO resonance integrals. In this case, the degeneracy of σ and σ^* will be lifted and, depending upon the symmetry labels of ω_1 and ω_2, the system will be converted either to a U- or a D-bound species. This now projects the fact that the symmetry labels of ω_1 and ω_2 are critical because they will determine whether angle shrinkage is an energy raising or energy reducing process. We distinguish two possibilites:

a. Symmetry dictates that ω_1 overlaps with σ and ω_2 with σ^* ("_parallel overlap_"). In this case, angle shrinkage will cause U-bonding which is unfavorable.

b. Symmetry dictates that ω_1 overlaps with σ^* and ω_2 with σ ("_diagonal overlap_"). In this case, angle shrinkage will cause D-bonding which is favorable.

In other words, we can say that (prior to angle shrinkage) we have a nascent U [in case (a)] and a nascent D [in case(b)] four-electron-four-orbital system. By using brackets to symbolize this nascent state, we can say that ligands will tend to repel each other in [U] and attract each other in [D] systems. Now, for reasons which we have explained before, the vast majority of organic molecules can be viewed as the result of a "forbidden" union of core and ligands, or, in

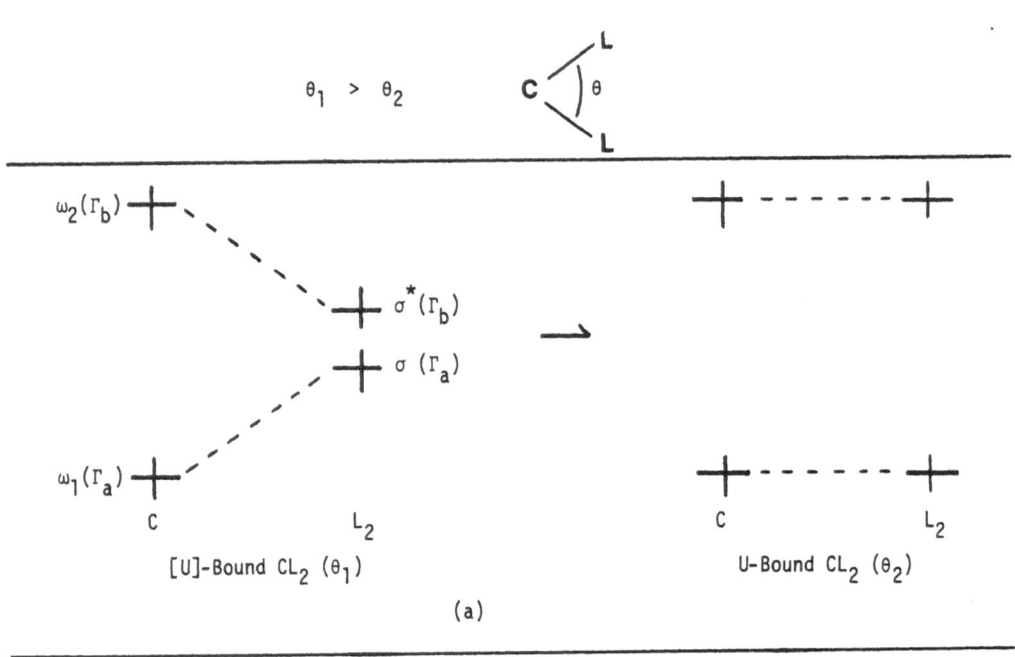

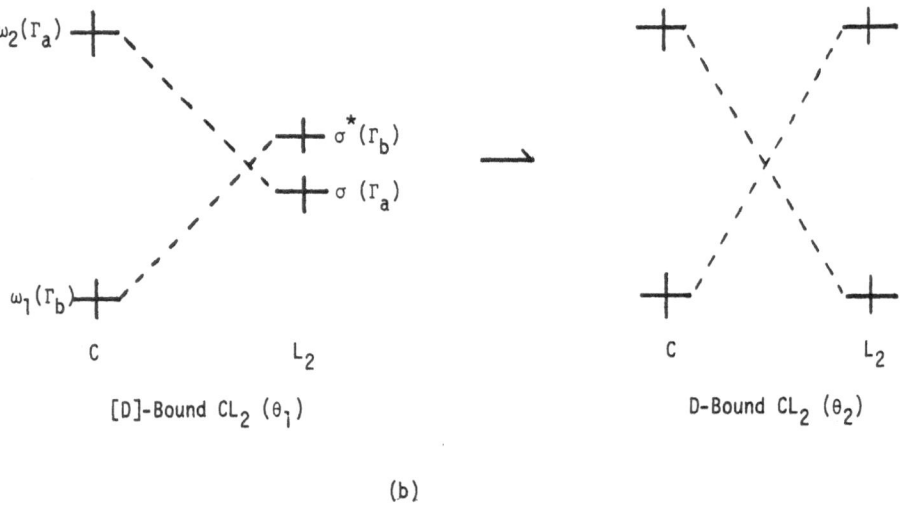

Figure 1. Definitions of [U] and [D] binding.

other words, organic molecules involve "parallel overlap" and they constitute
U-bound species. As a result, it follows that, in all such systems, ligands will
tend to avoid each other in an effort to render U-bonding less unfavorable. Thus,
natural nonbonded ligand repulsion is the direct result of the nodal structure of
orbitals which dictates the "parallel overlap" relationships spoken of above. For
example, when the core is carbon and the ligands hydrogens, we have the prototypical
CH_2 unit of organic chemistry in which the two ligands tend to stay away from each
other so that they avoid as much as possible U-bonding while still maintaining
strong core-ligand overlap. The bond diagram is shown below:

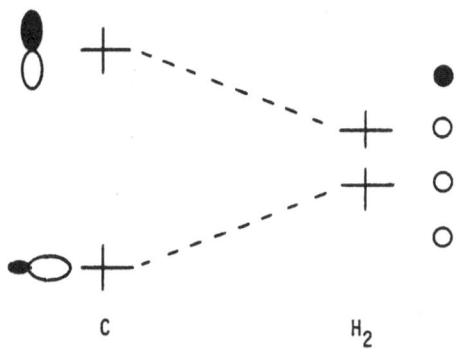

C H_2

Up until this point, all arguments were developed without ever requiring that
overlap is included in the computation of the symmetry adapted core and ligand
orbitals. As a result, we have identified only one component of natural Nonbonded
Ligand Repulsion (NLR). Let us now see what would happen had we included overlap
in the analysis by reference to the prototypical U-bound four-electron-four-orbital
systems of Figure 1a. We recognize two facts:

     a. The leading term of the MOVB wavefunction of a U- or [U]-bound system
is the perfect pairing CW.

     b. Inclusion of overlap, becomes responsible for the unsymmetrical splitting
of σ and σ^*.

It follows that the perfect pairing CW and, hence, the total wavefunction, will have higher energy in the case of a U- as compared to the case of a [U]-bound species exactly because inclusion of overlap causes σ^* to rise more than σ is lowered in energy and both orbitals are singly occupied in the perfect pairing CW.

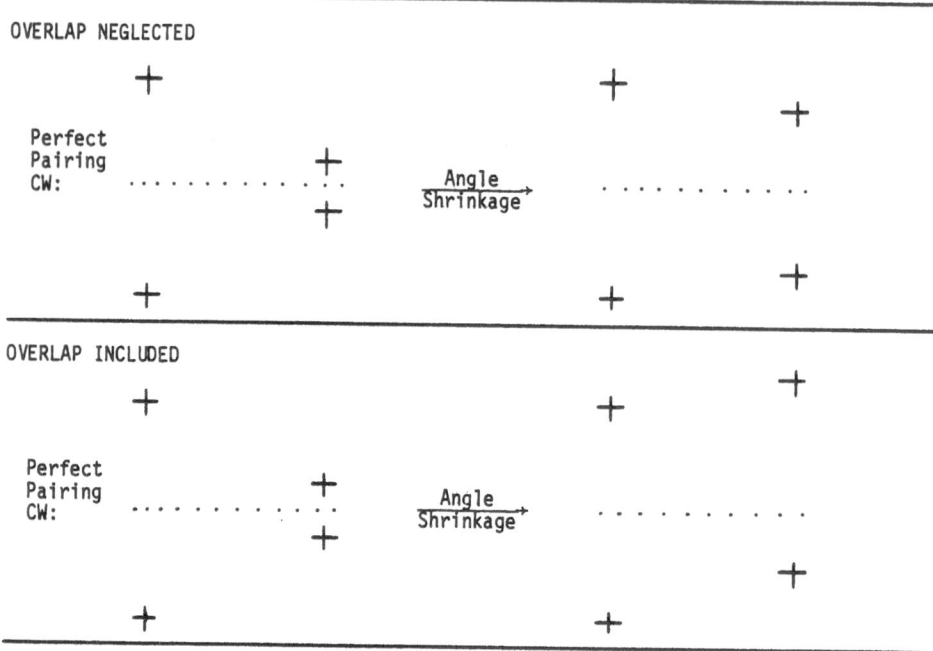

Thus, even a one-CW theory of ethane conformational isomerism would predict that staggered is more stable than eclipsed ethane, if overlap were included in the analysis (vide infra).

Let us now recall the VB description of bound systems and draw analogies with the above analysis. At the level of VB theory, a two-electron-two -orbital bond is described by the wavefunction shown below.

We say that bond formation is due to two effects:

    a. Spin pairing within the Heitler-London (HL) CW, Φ_1, which is the main contributor to the total VB wavefunction.

    b. Electron delocalization due to the interaction of Φ_1 with the higher lying charge transfer CW's, Φ_2 and Φ_3.

Thus, a specific phenomenon, in this case bond formation, is attributed to the cooperative action of _two_ electronic mechanisms: spin pairing and electron delocalication.

    A four-electron-four-orbital system having two bonds, a-d and b-c, which are in the vicinity of each other is described by the wavefunction shown below.

$$
\Psi \;=\; \begin{matrix} x_a \uparrow \;\; \downarrow x_d \\[4pt] x_b \uparrow \;\; \downarrow x_c \end{matrix} \;+\; \lambda \left(\begin{matrix} \uparrow\downarrow & - & -\uparrow\downarrow \\[4pt] \uparrow\;\downarrow & + & \uparrow\;\downarrow \end{matrix} \right) \;+\; \text{etc.}
$$

$$
\Phi_1 \qquad\qquad\qquad \Phi_2 \quad\; \Phi_3
$$

The HL CW, Φ_1, now describes a-d and b-c bonds due to spin pairing as well as a-b, c-d, a-c, and d-b antibonds due to overlap repulsion with the bonds outweighing the antibonds. We can say that the origin of both bonds and antibonds is spin correlation. Finally, the interaction of Φ_1 with higher lying charge transfer CW's further enhances a-d and b-c bonding. We can now generalize by saying that the energy of a system of aggregated bonds is due to two effects:

    a. Spin correlation.

    b. Electron delocalization.

    In MOVB theory, the role of the lowest energy HL CW in U- and [U]-bound systems is played by the perfect pairing CW. As a result, the stereochemistry of such systems is determined by the joint operation of two effects, in exact analogy with the situation encountered in VB theory:

a. Spin correlation in the perfect pairing CW.

b. Electron delocalization due to the interaction of the perfect pairing CW with higher lying MOVB charge transfer CW's.

Accordingly, <u>natural NLR is the consequence of the operation of both of these mechanisms</u>. In current qualitative theory, <u>natural NLR due to spin correlation is called bond-bond overlap repulsion and natural NLR due to electron delocalization is termed hyperconjugation.</u>

CH_2 is a model organic fragment. What happens as we make a transition to large molecules comprised of polyatomic cores and many ligands? The answer is that the arguments enunciated on the basis of consideration of the model CL_2 system remain essentially unaltered. Thus, a large molecule is made up of a number of subsystems which can be classified as [U] or [D] and with the nature of core-ligand overlap being such that there are more [U] than [D] subsystems, at least in most cases. A beautiful example is ethane which can be dissected into two [U] and one [D] sub-systems as shown in Figure 2 with the complete bond diagrams shown in Figure 3. Because staggered ethane is made up of two [U] and one [D] subsystem, rotation to the eclipsed conformation is energetically counterproductive because it brings the vicinal hydrogens closer to each other and this has the consequence depicted in Figure 2, i.e., the two [U] are converted to two U sybsystems and the one [D] to a D subsystem, with the net result being natural NLR. Note that both staggered and eclipsed ethane have the same number of core-ligand multicenter bonds, core-ligand spatial overlap is identical in the two conformations if ligand AO overlap is neglected, no promotion or demotion of electrons or holes in the core or ligands occurs to first approximation, and the only difference is the spatial relationships of the hydrogens attached on different carbons.

How do bond lengths change as rotation occurs and the eclipsed form is trans-formed to the staggered form? Mere inspection of the diagrams of Figure 2 tells us that the consequences are the following:

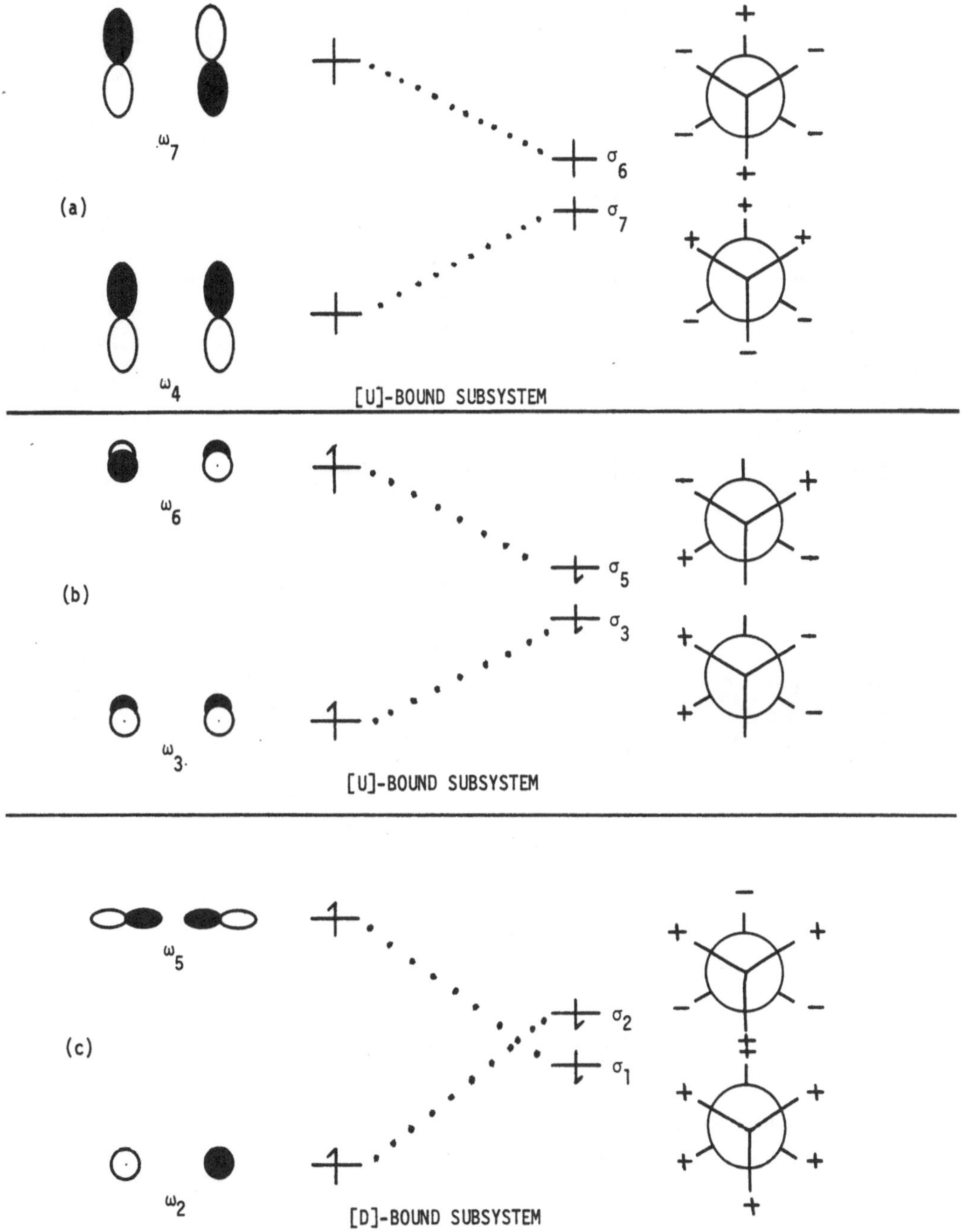

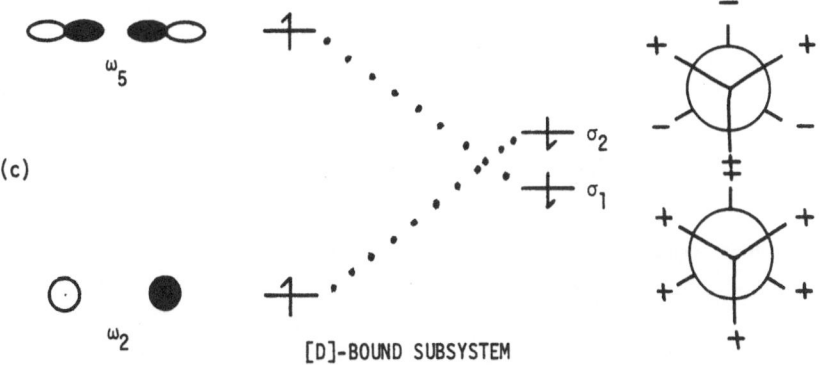

Figure 2. The three subsystems of CH_3 - CH_3. The low lying ω_1 (responsible for C - C sigma bonding) and the high lying ω_8 are assumed to play no key role.

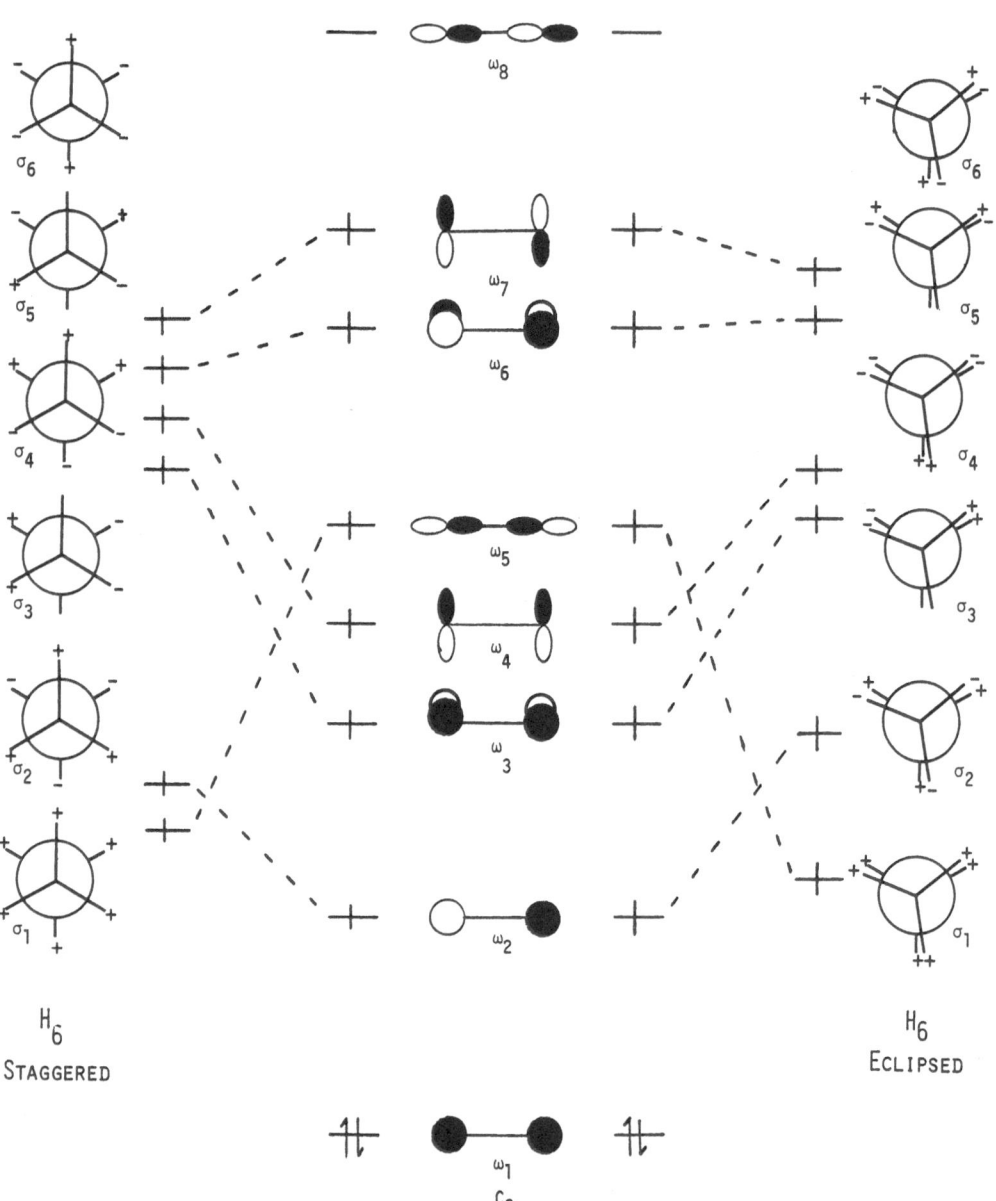

Figure 3. Bond diagrams of staggered and eclipsed ethane.

Subsystem 1	:	Depopulate ω_1 and populate ω_4
Subsystem 2	:	Depopulate ω_6 and populate ω_3
Subsystem 3	:	Depopulate ω_2 and populate ω_5

We predict that there will be net shortening of the C-C bond, something which calculations have already shown to be the case.[8]

Ethane has been the target of many theoretical investigations. The results of them can be summarized as follows: Regardless of the level of theory, staggered is always predicted to be more stable than eclipsed and the magnitude of the rotational barrier is correctly reproduced at the monodeterminantal MO level. To put it crudely, it is nearly impossible to get ethane "wrong" no matter what computational MO method one employs, something which cannot be said for many "simpler" molecules, e.g., H_2O and NH_3.[9] Why is this so? Recalling that the lower energy of staggered as compared to eclipsed ethane is due to the joint operation of two mechanisms, spin correlation and electron delocalization, and, further, recognizing that neglect of overlap will annihilate only the first but not the second mechanism, we predict that any MO computation will reproduce qualitatively the ethane internal barrier regardless of whether it is based on an ab initio or a Zero Differential Overlap (ZDO) method. Indeed, CNDO[10] computations get ethane "right", although overlap is neglected in the computations, because they "see" the electron delocalization aspect of natural NLR. EHMO[11] computations are also successful because they reproduce both spin correlation as well as electron delocalization. Ab initio SCF-MO[12] computations do the same, in a more realistic sense.

In summary, ethane is staggered because the nodal properties of MO's dictate that most organic molecules are partly or totally U-bound. An equivalent statement is that most organic molecules contain a greater number of U or [U] than D or [D] subsystems. If we implement rigorous MOVB theory with inclusion of overlap and we restrict the CW basis to a single perfect pairing CW, the barrier is due to spin correlation: This is bond-bond overlap repulsion in the chemist's language. If we augment the basis, the barrier is due additionally to delocalization differences reflected in the MOVB labels, U, H, and D and this is now called hyperconjugation in the chemist's language. The net effect is properly called natural LNR. How much is due to bond-bond overlap repulsion and how much to hyperconjugation is a moot point because the ratio of the effects depends on whether one implements orthogonal, approximate-orthogonal (Neglect of Overlap or Neglect of Differential Overlap), or, nonorthogonal MOVB theory. Ourselves, we prefer to think of partitioning of effect at the level of nonorthogonal MOVB theory because it is chemically more convenient. However, this decision is purely arbitrary. Thus, we conclude that the explanations of rotational barriers based on either bond-bond overlap repulsion or hyperconjugation are defensible and that the analyses of the problem by Lowe,[13] Hoffmann,[14] Weinhold,[15] and others are entirely compatible though seemingly different. The novel contribution of this work is to argue that the most general concept is natural LNR, that this is due to the nodal properties of core and ligand orbitals, and that it operates in most organic molecules which can be thought of as products of the "forbidden" union of core and ligands, sometimes dominated by other factors and sometimes exerting the sole major influence on molecular stereochemistry.

The natural tendency of ligands to avoid each other can be overridden by other rebonding processes which also "aspire" to reduce the "forbiddenness" of the union of core and ligands, or, it can simply become energetically counterproductive if it entails loss of core-ligand spatial overlap. In general, we say that natural LNR can be overridden because of two reasons:

    a. Minimization of natural LNR runs counter core-ligand overlap maximization.

    b. Minimization of natural LNR runs counter core or ligand deexcitation.

An example of the first type is the structure of ethylene, planar rather than staggered, and an example of the second type is water, bent rather than linear.

II. Indices of Natural Nonbonded Ligand Repulsion

A simple-to-compute and very useful index of natural NLR in every U-bound-like system is the one-electron energy of the perfect pairing CW at infinite inter-fragmental distance. This is a measure of natural NLR due to spin correlation and it is denoted by R. If in a series of comparisons the core remains constant, then the value of R is given by Constant + R^0, where R^0 stands for the one-electron energy of the ligand fragment within the perfect pairing CW. R and R^0 can be easily computed at the level of Extended Hückel MO (EHMO) theory[11] through calcu-lation of the core and ligand symmetry adapted MO's and construction of the appropriate bond diagram which directly identifies the perfect pairing CW of lowest energy. As an example, the R and R^0 indices of the CL_2 systems of Figure 1a are given below. The less negative R_2 and R_2^0 quantities tell us that the second geometry is sterically unfavorable relative to the first because of natural NLR.

$$\left.\begin{array}{l} R_1 = \varepsilon(\omega_1) + \varepsilon(\omega_2) + \varepsilon_1(\sigma) + \varepsilon_1(\sigma^*) \\ R_1^0 = \varepsilon_1(\sigma) + \varepsilon_1(\sigma^*) \end{array}\right\} \theta_1$$

$$\left.\begin{array}{l} R_2 = \varepsilon(\omega_1) + \varepsilon(\omega_2) + \varepsilon_2(\sigma) + \varepsilon_2(\sigma^*) \\ R_2^0 = \varepsilon_2(\sigma) + \varepsilon_2(\sigma^*) \end{array}\right\} \theta_2$$

and

$$R_2 > R_1$$
$$R_2^0 > R_1^0$$

if

$$\theta_1 > \theta_2$$

Once again, we emphasize that the R and R^0 indices ought to be used only when a molecule can be legitimately viewed as the product of the "forbidden" union of core and ligand fragments, i.e., when there are no high lying orbitals of one fragment coupled with low lying orbitals of the other in defining a bond, and the Perfect Pairing CW is the major contributor to the total MOVB wavefunction. Henceforth, we shall implicity assume that this condition is met whenever we make usage of the R and R^0 indices.

Let us now illustrate these ideas by an appropriate example: Conformational isomerism in ethane. In this case, we can use EHMO theory in order to compute the ligand (H_6) symmetry orbitals for the staggered and eclipsed conformations, assuming standard bond angles and bond lengths. The sum of the orbital energies is then the R^0 index. These calculations can be routinely carried out even in the most primitive theoretical laboratory. We find that the EHMO R^0 indices of the two conformers are as shown below. We conclude that eclipsed ethane suffers from greater natural NLR and that this is the cause (in part) for its higher energy. Using the index R^0 (eclipsed) - R^0 (staggered) we predict a rotational barrier of 3.458 kcal/mol to be compared with the experimental value of 2.928 kcal/mol.[16]

	Eclipsed Ethane	Staggered Ethane
R^0(eV):	- 77.43	- 77.58

III. A Broader Look at Natural Nonbonded Ligand Repulsion

The way in which the perfect pairing CW determines natural NLR can be excellently illustrated by comparing four different four-electron-four-orbital systems such as the ones shown below by employing the fragment dissections also indicated below.

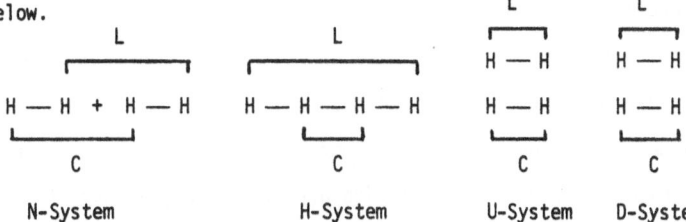

The bond diagrams shown in Figure 4 make evident the following trends:

a. The perfect pairing CW makes the dominant contribution to N, H, and U and only in D can an alternative CW, which has much smaller excitation energy (but forms no core-ligand bonds by spin pairing) and generates comparable coulomb repulsion, become as important as the perfect pairing CW. This is the CW shown below.

b. Assuming that (a) is valid, it follows that the total energy should increase in going from N to H to U because the energies of the corresponding perfect pairing CW's increase in this direction. Indeed, this is what computations aptly demonstrate for the case of N, H, and U H_4 complex.

c. Hückel MO (HMO) theory is totally inadequate for handling aggregation stereoselection and it predicts that H which will be more stable than N which will be as stable as U (if overlap is neglected). The reason for this is that "ionic" CW's are as important as the perfect pairing CW at this level of theory because of neglect of interelectronic repulsion. Since deexcitation by one electron transfer starting from the perfect pairing CW is more effective in H than N or U, pi butadiene, for example, is predicted to be more stable than two pi ethylenes or pi cyclobutadiene. This single electron transfer generates a CT CW (see below) which involves high interelectronic repulsion not "seen" at the level of HMO theory. The reader will now begin to appreciate how VB and MOVB theory makes us appreciate coulomb repulsion through pictures and without the necessity of explicitly evaluating electron repulsion integrals.

Single
Electron
Transfer :
CW's

$\underline{N}$ H$_4$
(H$_2$ + H$_2$)

$\underline{H}$ H$_4$
(H-H-H-H)

$\underline{U}$ H$_4$
$\begin{pmatrix} H - H \\ | \quad | \\ H - H \end{pmatrix}$

All decompose to "ionic"
VB CW's: H H$^-$ H$^+$ H, etc.

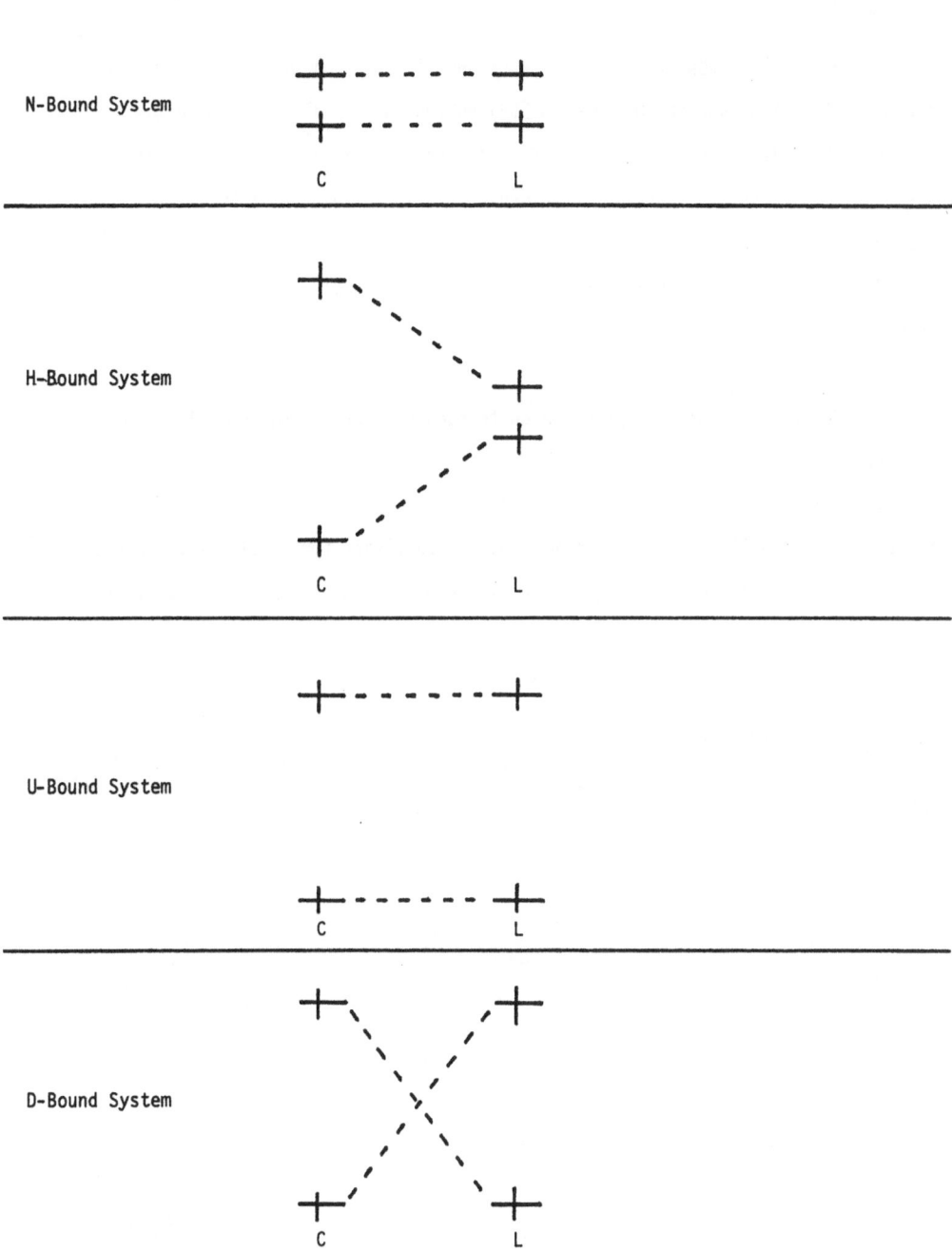

Figure 4.   Four different types of core-ligand binding in a four-electon-four-orbital system.

d. Since the staggered ⟶ eclipsed ethane transformation actually involves two H ⟶ U (H=[U]) and one H ⟶ D (H=[D]) rebonding and since HMO correctly predicts the relative energies of H-, U- and D-bound four-electron-four-orbital systems, we say that the barrier of internal rotation of ethane is "within" HMO theory.

In summary, we can say that the nonbonded repulsion of ligands can be due to three factors:

a. "Classical" coulomb repulsion.

b. The presence of lone pairs which through overlap define four-electron antibonds.

c. Natural NLR.

Henceforth, we shall reserve the term nonbonded repulsion for a collective description of all thee factors, while referring to them individually when in need to do so.

References

1. Newman, M.S. "Steric Effects in Organic Chemistry"; John Wiley: New York, 1956.

2. (a) Bartell, L.S. J. Chem. Ed. 1968, 45, 754.

   (b) Pearson, R.G. "Symmetry Rules for Chemical Reactions"; Wiley and Sons, Inc.: New York, 1976.

   (c) Wheland, G.W. J. Chem. Phys. 1934, 2, 474.

   (d) Pauling L.; Springall, H.E.; Palmer, K.J. J. Am. Chem. Soc. 1939, 61, 927.

   (e) Mulliken, R.S. J. Chem. Phys. 1939, 7, 339.

   (f) Mulliken, R.S.; Rieke, C.A.; Brown, W.G. J. Am. Chem. Soc. 1941, 63, 41.

   (g) Dewar, M.J.S. "Hyperconjugation"; Ronald Press Co.: New York, 1962.

3. Morrison, R.T.; Boyd, R.N. "Organic Chemistry", 3rd Ed.; Allyn and Bacon, Inc.; Boston, 1973.

4. Van Schaik, E.J.M.; Geise, H.J.; Mijlhoff, F.C.; Renes, G. J. Mol. Struct. 1973 16, 23.

5. Whipple, E.B.; Stewart, W.E.; Reddy, G.S.; Goldstein, J.H. J. Chem. Phys. 1961, 34, 2136.

6. (a) Hoffman, R.; Olofson, R.A. J. Am. Chem. Soc. 1966, 88, 943.

   (b) Epiotis, N.D. J. Am. Chem. Soc. 1973, 95, 3087.

   (c) Epiotis, N.D.; Yates, R.L. J. Am. Chem. Soc. 1976, 98, 461.

7. (a) Fieser, F.L.; Fieser, M. "Reagents for Organic Synthesis"; John Wiley: New York, 1967.

   (b) Waldron, J.T.; Snyder, W.H. J. Am. Chem. Soc. 1973, 95, 5491

   (c) Epiotis, N.D.; Cherry, W.R.; Shaik, S.; Yates, R.L.; Bernardi, F. Top. Curr. Chem. 1977, 70, 1.

8. Veillard, A. Theoret. Chim. Acta 1970, 18, 21.

9. H_2O is predicted to be linear by EHMO theory for a "reasonable" choice of empirical parameters and NH_3 is predicted to be planar by SCF-MO theory if "poor" AO basis sets are used in the calculations.

10. Pople, J.A.; Beveridge, D.L. "Approximate Molecular Orbital Theory"; McGraw-Hill: New York, 1971.

11. (a) Wolfsberg, M.; Helmholz, L. J. Chem. Phys. 1952, 20, 837.

    (b) Hoffmann, R.; Lipscomb, W.N. J. Chem. Phys. 1962, 36, 2189.

    (c) Hoffmann, R. J. Chem. Phys. 1963, 39, 1397.

    (d) An excellent introduction to applied quantum chemistry and, in particular, to the "how to do it" aspects of HMO and EHMO theories can be found in McGlynn, S.P.; Vanquickenborne, L.G.; Kinoshita, M.; Carroll, D.G. "Introduction to Applied Quantum Chemistry"; Holt, Rinehart, and Winston, Inc.: New York, 1972.

12. (a) Pitzer, R.M.; Lipscomb, W.N. J. Chem. Phys. 1963, 39, 1995.

    (b) Veillard, A. Chem. Phys. Letters 1969, 3, 128.

    (c) Stevens, R.M. J. Chem. Phys. 1970, 52, 1397.

    A large body of computations suggests that the magnitudes of rotation and inversion barriers are satisfactorily predicted by single determinant Hartree-Fock MO theory: Allen, L.C. Ann.Rev. Phys. Chem. 1969, 20, 315.

13. Hoffmann, R. Pure Appl. Chem. 1970, 24, 567.

14. Lowe, J.P. Science 1973, 179, 527.

15. Brunck, T.K.; Weinhold, F. J. Am. Chem. Soc. 1979, 101, 1700.

16. Weiss, S.; Leroi, G.E. J. Chem. Phys. 1968, 48, 962.

Chapter 6. Conformational Isomerism of N_2H_4 and Derivatives. The Stereochemical
Consequences of "Forbiddenness" Removal.

According to Molecular Orbital-Valence Bond (MOVB) theory,[1] the vast majority
of organic molecules in a reference geometry can be viewed as the result of a
"forbidden" union of a core(C) and a ligand(L) fragment.[1-3] The transition from
the reference geometry to the lowest energy geometry of the molecule is accompanied
by an energy reduction which is a reflection of "forbiddeness" removal.[4] The purpose
of this paper is to focus on one and only one type of system, namely, A_2X_4 with
fourteen valence electrons, in order to demonstrate how radically MOVB theory has
changed our view of molecular stereochemistry. In pursuing this goal we shall ask
the following two key questions:

a. Why is N_2H_4 gauche?

b. Why does fluorination of N_2H_4 have entirely different stereochemical consequences than fluorination of H_2O_2?

Since current qualitative MO theoretical models seemingly provide answers to these
questions, we shall be able to compare the conclusions of MOVB theory with the
conclusions of these models and, thus, define the present limits of our understanding of chemical bonding and the new horizons opened by MOVB theory. Our
first task is then to describe exactly what the currently popular models of chemical
bonding predict with regards to the stereochemistry of N_2H_4, H_2O_2 and their derivatives. In more specific terms, we want to first examine how the hyperconjugation
model[5] can be applied to the problem at hand and how the answers it provides match
known experimental facts.

Like aromaticity,[6] hyperconjugation is nearly a half-century old idea. It
refers to the "mechanism" by which the interaction of an electron pair bond with a
vicinal doubly occupied, singly occupied, or vacant AO, or, with a vicinal electron
pair bond confers stabilization to the total system in question. The original concept of hyperconjugation was founded on intuitive VB theory and it was presented

in the pictorial way illustrated in Figures 1a and 1b. Mulliken placed the idea of hyperconjugation on an MO theoretical basis in 1941. In recent times, the hyper-conjugation argument has been enunciated in the laguage of Frontier Orbital (FO)[7] Perturbation MO (PMO) theory[8] as illustrated in Figures 1c and 1d. A number of investigators have made imaginative use of the concept of hyperconjugation in rationalizing and predicting chemical phenomena.[9] The gauche preference of H_2O_2 as well as that of N_2H_4 can be rationalized by hyperconjugation FO-PMO-style as illustrated in Figure 2. After some time of intense fascination with these ideas and extensive exploration of their applicability, we began to realize that the successes of hyperconjugation were rivaled by failures for which no easy rational-ization could be offered. A typical example is provided by the comparative study of conformational isomerism of H_2O_2 and N_2H_4. As first pointed out by Bauer and Yokozeki,[10] the gauche preference of H_2O_2 seems to be augmented while the same preference of N_2H_4 is reduced upon replacement of the hydrogens by fluorines despite the fact that hyperconjugation ascribes the gauche preference of both H_2O_2 and N_2H_4 to the same stabilization "mechanism" and, thus, forecasts an identical response to substitution of the ligands or the core. The hyperconjugation argument, simply put, is that H_2O_2 and N_2H_4 are gauche because the stabilizing interaction of a doubly occupied AO of O or N with a vicinal O-H or N-H vacant antibonding orbital is maximized in the gauche geometry as illustrated in Figure 2. Replacement of H by F is expected to lower the energy of the antibonding orbital and to change the mag-nitudes of the AO coefficients so as to enhance the hyperconjugative stabilization of the gauche form in both systems. This very straightforward prediction is not borne out by the experiment. On one hand, the dihedral angle of H_2O_2 indeed shrinks upon replacement of H by F, a fact consistent with the idea of enhanced hyperconju-gation. On the other hand, the gauche form is found to be <u>destabilized</u> relative to the trans form upon substitution of H by F in N_2H_4. The key data are presented below.[11-14]

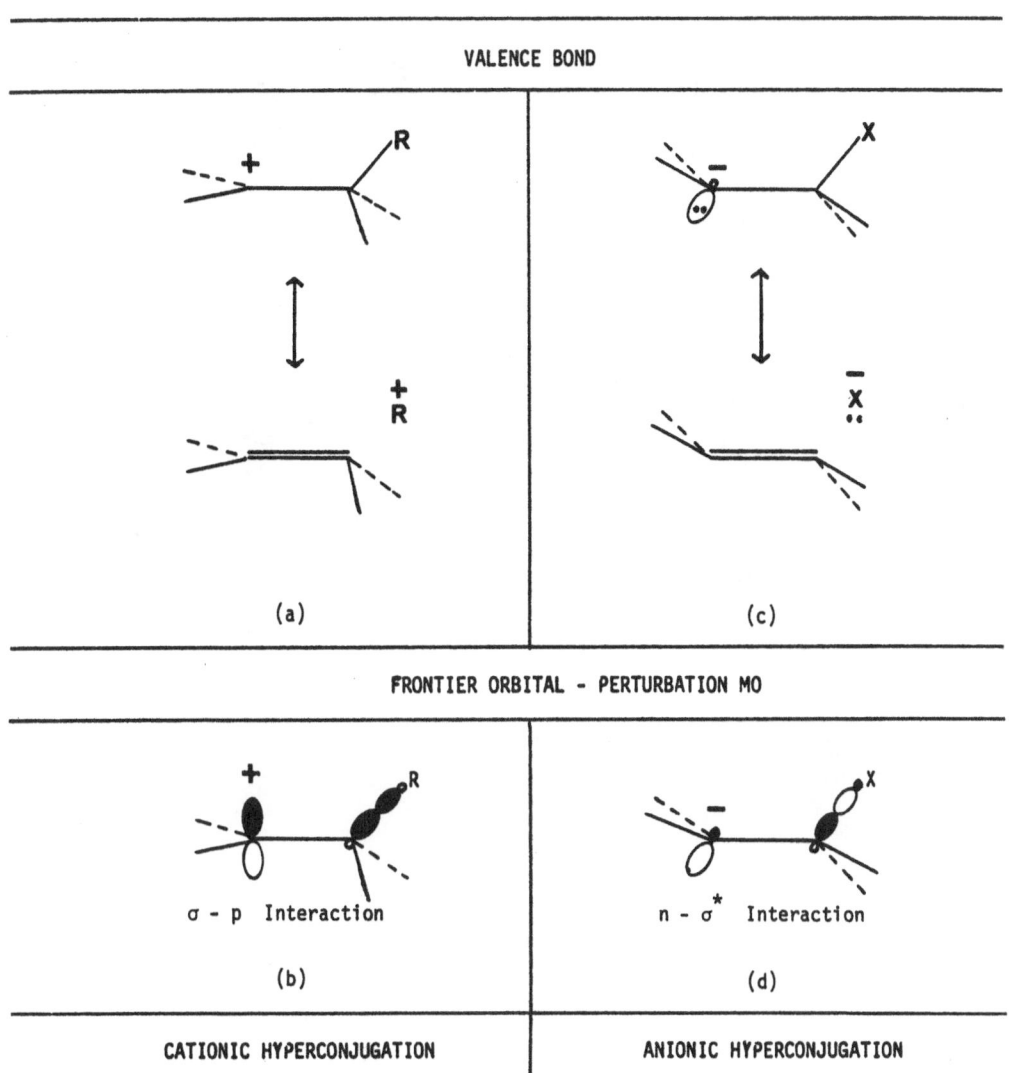

Figure 1. Valence Bond and Frontier Orbital-Perturbation MO formulations of hyperconjugation.

H₂O₂ Geometry

Stabilizing Interaction:
$$P_O - \sigma^*_{OH}$$

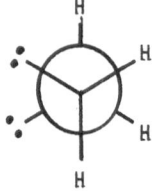

N₂H₄ Geometry

Stabilizing Interaction:
$$sp^3_N - \sigma^*_{NH} \ (anti)$$

Figure 2. Geometry control of lone pair delocalization into and adjacent σ^* antibond in H_2O_2 and N_2H_4.

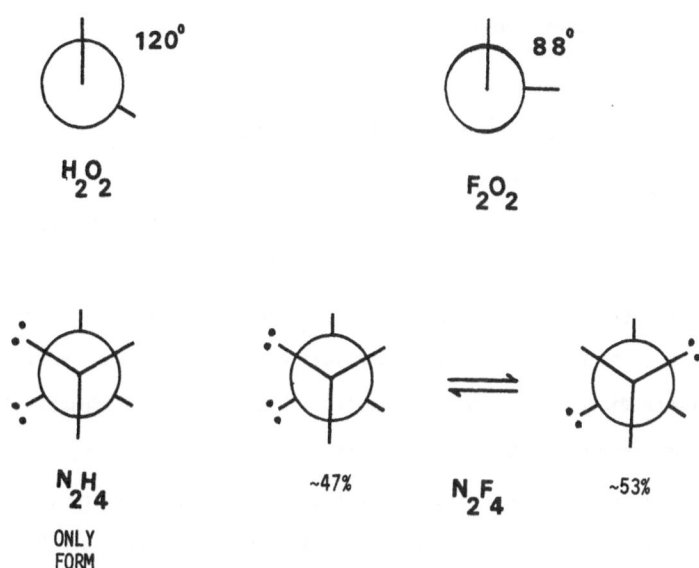

These observations raise the question: Why does hyperconjugation fail in rationalizing qualitatively substituent effects? In this work we will demonstrate that
H_2O_2 and N_2H_4 are qualitatively unrelated systems, the gauche preferences of which
are due to entirely different factors. As a result, the effect of substituents on
conformational isomerism is entirely different in the two systems. In other words,
hyperconjugation works by accident in rationalizing the preferred conformation of H_2O_2
and N_2H_4 and it fails naturally in rationalizing the effects of substituents upon conformational preference.

I. Conformational Isomerism of N_2H_4

The preferred geometry of N_2H_4 can be predicted (or, better, anticipated) by MOVB theory in a very simple and straightforward manner. Thus, one may consider three likely geometries, planar (P), trans (T), and gauche (G) and compare their electronic structure by reference to the corresponding bond diagrams. These can be constructed by assuming a core-ligand dissection in which N_2 plays the role of the core and H_4 the role of the ligand fragment, developing the core and fragment symmetry orbitals according to well known perturbation theoretical procedures,[15] and identifying the core-ligand bonds dictated by orbital symmetry. Now, as we have emphasized before, one cannot understand molecular stereochemistry unless the role of nonbonded interaction is properly appreciated. Thus, though all operations involved in the construction of a bond diagram are trivial, it is instructive to consider how the symmetry orbitals of the fragments, and, in particular, the symmetry orbitals of the ligand fragment, are constructed.

The symmetry orbitals of H_4 for the <u>planar</u> geometry of N_2H_4 can be constructed by usage of FMO theory and recognition of the following two general trends:

a. Geminal nonbonded overlap of the hydrogen 1s AO's is very large. For example, using $r_{NN} = 1.47$ Å, it is computed to be 0.1542 in P N_2H_4.

b. Vicinal nonbonded overlap of the hydrogen 1s AO's is much smaller but far from negligible. For example, it is computed to be 0.0338 in P N_2H_4. Statements (a) and (b) are strictly valid whenever the core is made up of first row atoms. When the core is made up of second, third, etc., row atoms, the increasing core atom-ligand atom bond lengths cause the ligands to move further away from each other. The resulting reduction in nonbonded ligand AO overlap has profound chemical consequences which will be identified later on. Using the AO numbering system shown in Figure 3,

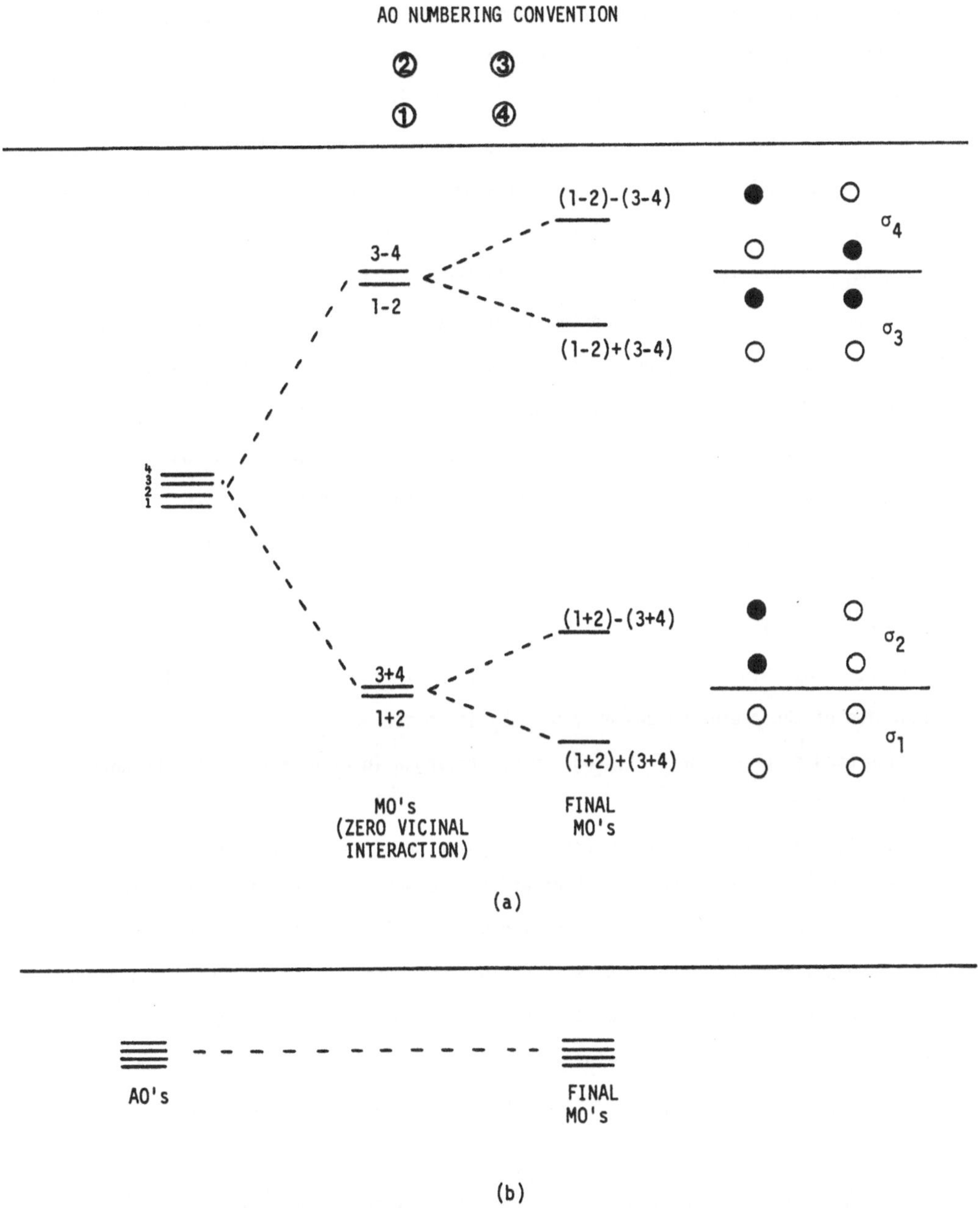

AO NUMBERING CONVENTION

② ③

① ④

(1-2)-(3-4)

3-4

1-2

(1-2)+(3-4)

σ_4

σ_3

4
3
2
1

(1+2)-(3+4)

3+4

1+2

(1+2)+(3+4)

σ_2

σ_1

MO's
(ZERO VICINAL
INTERACTION)

FINAL
MO's

(a)

AO's

FINAL
MO's

(b)

Figure 3. a. The perturbation theoretical derivation of the H_4 symmetry adapted MO's for planar N_2H_4 assuming strong geminal and small vicinal H---H AO overlap. b. The same construction assuming zero nonbonded H---H AO overlap.

we can the predict that the bonding and antibonding MO's spanning hydrogens 1 and 2 and those spanning 3 and 4 will be separated by a large energy gap due to the strong nonbonded overlap of 1 with 2 and 3 with 4. This is illustrated in Figure 3. Now, the subsequent interaction of the resulting four MO's will be weaker due to smaller nonbonded overlap of 1 with 4 and 2 with 3. As a result, the final symmetry MO's of H_4 will separate into two bands as illustrated in Figure 3a. This pattern will persist in all three geometries of N_2H_4 (P, T, G) due to the fact that geminal nonbonded overlap remains always large. As a result, electron excitation from σ_1 or σ_2 to σ_3 or σ_4 will always be deemed costly and the corresponding deexcitation beneficial regardless of geometry.

The qualitative energy levels of H_4 appropriate to P, T and G N_2H_4 would be radically different had we assumed that nonbonded overlap of the hydrogen 1s AO's is zero. In such an event, we would have obtained the symmetry orbitals shown in Figure 3b. Note that now excitation or deexcitation can no longer be defined simply because all four MO's are degenerate. We will see that assumption of one or the other H_4 MO patterns makes a "night-day" difference insofar as the final prediction of the preferred geometry of N_2H_4 is concerned.

The bond diagrams for P, T, and G N_2H_4 are shown in Figures 4 and 5. We note the following:

a. In comparing P and T, we first recognize that the former is the maximum overlap form.[2] Hence, a dagger must be affixed to whatever label is appropriate for T. The transition P → T is accompanied by U → $H^{\dagger}$ rebonding. In more descriptive language, we can say that it is accompanied by core deexcitation of the $\omega_6 \rightarrow \omega_2$ type. Since the energy separation of the ω_2 and ω_6 orbitals is very large, we expect that deexcitation will dominate loss of spatial overlap and T will attain lower energy than P. This problem is entirely analogous to the problem of ammonia pyramidalization.[16]

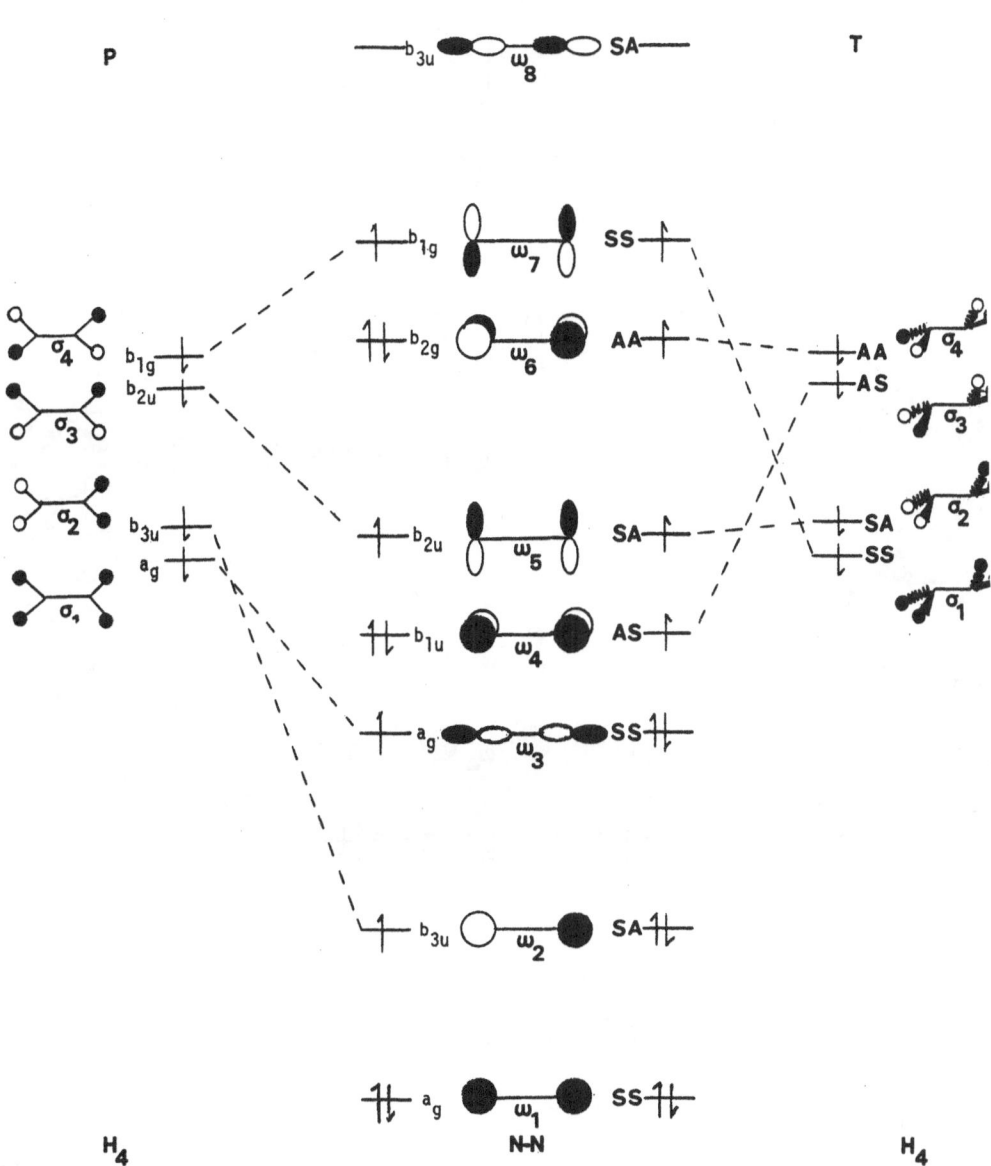

Figure 4. Compact bond diagrams of planar (P) and trans (T) N_2H_4. The sigma ω_3 and the pi ω_4 and ω_5 MO's of N_2 are approximately degenerate.

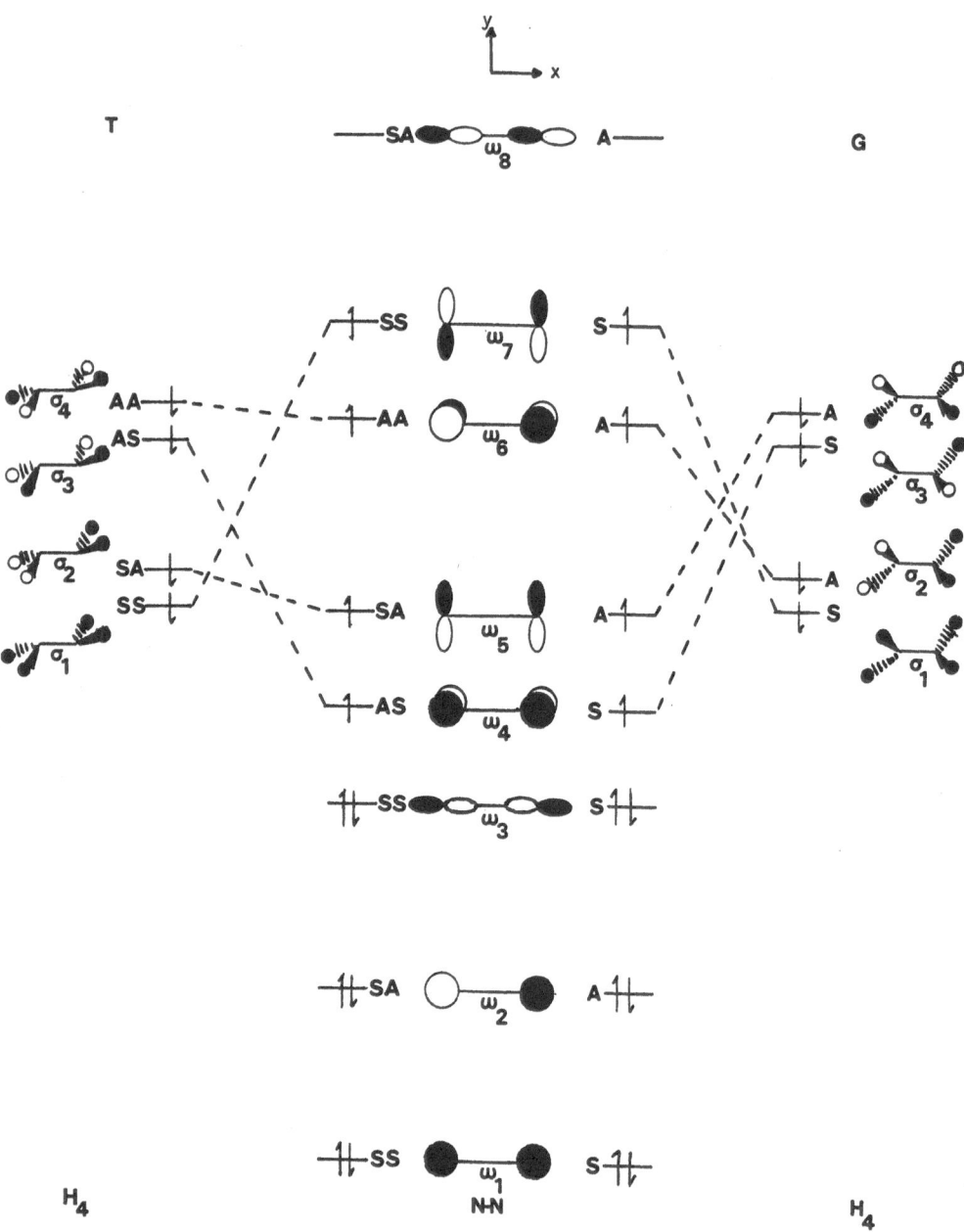

Figure 5. Compact bond diagrams of trans (T) and gauche (G) N_2H_4.

b. In comparing T and G, we recognize that, now, the former is the maximum overlap form. The transition $T \rightarrow G$ is again accompanied by $U \rightarrow H^{\ddagger}$ rebonding. In more descriptive terms, we can say that it is accompanied by core plus ligand deexcitation of the $\sigma_4 \rightarrow \sigma_2$ and $\omega_6 \rightarrow \omega_5$ type. Since the energy separation of σ_4 and σ_2 as well as the energy separation of ω_6 and ω_5 are large, we expect again that deexcitation will dominate the loss of spatial overlap and G will attain lower energy than T. Note that, if σ_1 to σ_4 were degenerate, the rebonding would be simply of the $N \rightarrow N'$ type, i.e., rebonding accompanied by bond weakening due to loss of spatial overlap without any deexcitation benefit. In this case, T would end up being energetically more favorable than G. Since whether σ_2 and σ_4 are degenerate or not depends on the strength of geminal nonbonded interaction, we can see that two different predictions can be arrived at depending on whether geminal nonbonded overlap is deemed important or not:

1) If geminal nonbonded interaction is appreciable, σ_2 and σ_4 will be substantially split in energy, the $T \rightarrow G$ transformation will be accompanied by $U \rightarrow H^{\ddagger}$ rebonding, and G will be the lowest energy form.

2) If geminal nonbonded interaction is negligible, σ_1 to σ_4 will be degenerate, the $T \rightarrow G$ transformation will be accompanied by $N \rightarrow N'$ rebonding and T will be the lowest energy form.

Since it is alway possible to replace N and/or H by some other atom or group so that geminal nonbonded overlap tends to zero and determine how the conformational equilibrium is affected by such a change, it is evident that MOVB theory lead to hard testable conclusions.

What happens to the N-N and N-H bond lengths as we go from P to T to G hydrazine? Consider the $U \rightarrow H^{\ddagger}$ rebonding, which accompanies the $P \rightarrow T$ transformation and which is shown in Figure 4. The rebonding consequences are clear cut: An antibonding core MO is depopulated and a nonbonding one[17] is populated while overlap repulsion is generated between core and ligand mainly through the CW shown in Figure 6a. Next, consider the $U \rightarrow H^{\ddagger}$ rebonding which accompanies the $T \rightarrow G$ transformation and which is shown in Figure 5. Again, rebonding consequences are straightforward: Antibonding MO's are depopulated and bonding ones are populated while overlap repulsion is generated between core and ligands through the CW shown in Figure 6b. It is then self-evident that in going from P to T to G geometry, the N-N bond must be shortened while the N-H bonds must become elongated.

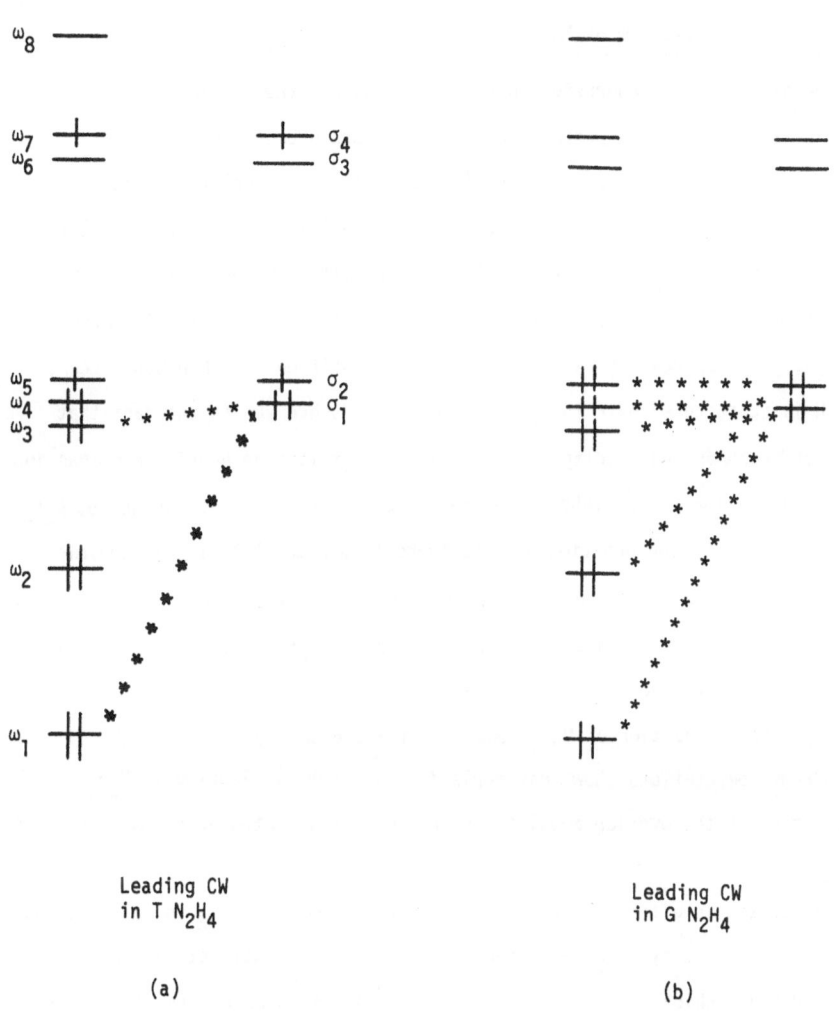

Leading CW
in T N_2H_4

(a)

Leading CW
in G N_2H_4

(b)

Figure 6. Leading CW's of trans and gauche N_2H_4 showing that the latter form accommodates four electron in the π_y and π_z core MO's while the former keeps three electrons in π_y and π_z and one in π_y^{*z}. Asterisked lines indicate four-electron antibonds.

II. Conformational Isomerism of N_2X_4

Once we have properly understood why N_2H_4 is gauche, the reason for the fact that N_2F_4 shows no comparable preference for the gauche geometry becomes self-evident. Specifically, because of the smaller spatial extension of the fluorine valence AO's, geminal nonbonded overlap between F and F in N_2F_4 is much smaller than between H and H in N_2H_4 (see Table 1). As a result, the motivation for the T → G transformation largely ceases to exist and N_2F_4 tends to adopt a T, rather than a G, geometry for reasons explained above. In addition, the N-N bond becomes longer as H is substituted by F in N_2H_4.[18a] Both trends are due to the fact that diminution of nonbonded AO overlap has reduced the deexcitation benefit accompanying a T → G transformation. One might be tempted to argue that N_2H_4 is gauche but N_2F_4 tends to adopt a trans conformation because there is greater "classical" coulomb ligand repulsion in the case of N_2F_4. However, this is unlikely because the electro-negativity difference between N and F is comparable to that between N and H. In addition, various facts suggest that such an electrostatic effect is not the main reason behind the disposition of N_2F_4 towards a trans geometry.

a. SCF-MO computations show that replacing hydrogen by fluorine in N_2H_4 changes drastically the overlap populations in a way anticipated by the MOVB analysis (vide infra).

b. Electrostatic effect cannot explain the preference of N_2H_2 for a trans and of N_2F_2 for a cis geometry.[18b] In fact, these results suggest that net ligand repulsion is greater in N_2H_2, which is "forced" to be trans while N_2F_2 adopts the "naturally" preferred cis geometry.

c. Electrostatic effects cannot account for the stereochemical trends which are encountered as N is replaced by atoms of the same column of the Periodic Table.

Table 1. Geminal Nonbonded Ligand AO Overlap Integrals.

(a) H $\cdots$ H Nonbonded AO Overlap in P N_2H_4.

	1s'
1s	.1943

(b) H $\cdots$ H Nonbonded AO Overlap in P P_2H_4.

	1s'
1s	.0641

(c) F $\cdots$ F Nonbonded AO Overlap in P N_2F_4.

	2s'	$2p_x'$	$2p_y'$	$2p_z'$
2s	.0060	.0044	.0077	0
$2p_x$	.0044	.0020	.0062	0
$2p_y$	.0077	.0062	.0092	0
$2p_z$	0	0	0	.0016

(d) F $\cdots$ F Nonbonded AO Overlap in P P_2F_4.

	2s'	$2p_x'$	$2p_x'$	$2p_z'$
2s	.0021	.0016	.0027	0
$2p_x$	.0016	.0008	.0022	0
$2p_y$	.0027	.0022	.0034	0
$2p_z$	0	0	0	.0005

III. Conformational Isomerism of P_2X_4 and As_2X_4

Consider the molecule AH_n in which A is a first row atom. Replacement of A by a second row atom of the same columnn of the Periodic Table will cause an increase of the H---H distance, r_{HH}, and a very significant reduction of nonbonded AO overlap (see Table 1). This suggests that upon replacing N by P in N_2X_4, we must observe a dramatic reduction of the energy difference between the G and the T form and, in fact, a reversal of its sign. That is to say, we expect that the diminution of the geminal nonbonded interaction will tend to convert the rebonding accompanying the T → G transformation from $U \rightarrow H^{\ddagger}$ to N → N' and, thus, reverse the order of relative stability of T and G. The trend will become even more pronounced as P is replaced by As. The experimental results collected in Table 2 leave no doubt that as the covalent radius of A increases the trans becomes increasingly more stable relative to the gauche conformer. This trend is exactly opposite to the one encountered in ethane-like A_2X_6 molecules and it has been noted by several authors.[19,20]

Table 2. The Conformations of Gaseous N_2X_4, P_2X_4, and As_2X_4 Molecules.

Molecule	% Gauche	% Trans	Reference
N_2H_4	100	0	a, b, c
$N_2(CH_3)_4$	100	0	d
$N_2(CF_3)_4$	100	0	e
N_2F_4	47	53	f
P_2H_4	90	10	g
$P_2(CH_3)_4$	~ 60	~ 40	h
	16	84	i
$P_2(CF_3)_4$	10	90	i
P_2F_4	0	100	j, k
As_2H_4	-	-	
$As_2(CH_3)_4$	12	88	i
$As_2(CF_3)_4$	0	100	i
As_2F_4	-	-	

References

a. Morino, Y.; Iijima, R.; Murata, Y. Bull. Chem. Soc. Japan 1960, 33, 46.

b. Yamaguchi, A.; Ichishima, I.; Shimanouchi, T.; Mizushima, S.I. Spectro. Chim. Acta. 1960, 16, 1471.

c. Kasuya, T.; Kojima, T. J. Phys. Soc. Japan 1963, 18, 364.

d. Durig, J. R.; MacNamee, R. W.; Knight, L. B.; Harris, W. C. Inorg. Chem. 1973, 12, 804.

e. Durig, J. R.; Thompson, J. W.; Witt, J. D. Inorg. Chem. 1972, 11, 2477.

f. Cardillo, M. J.; Bauer, S. H. Inorg. Chem. 1969, 8, 2086.

g. Elbel, S.; Dieck, H.; Becker, G.; Ensslin, W. Inorg. Chem. 1976, 15, 1235.

h. Durig, J. R.; MacNamee, R. W. J. Mol. Struct. 1973, 17, 426.

i. Cowley, A. H.; Dewar, M. J. S.; Goodman, D. W.; Padolina, M. C. J. Am. Chem. Soc. 1974, 96, 2648.

j. Rudolph, R. W.; Taylor, R. C.; Parry, R. W. J. Am. Chem. Soc., 1966, 88, 3729.

k. Hodges, L.; Bartell, L. S. Fourth Austin Meeting on Molecular Structure, Austin, Texas, 1972.

IV. H_2O_2 versus N_2H_4

In a previous work, we discussed the structure of H_2O_2 and its derivatives from the standpoint of MOVB theory. The bond diagrams of linear (L), trans (T), and gauche (G) H_2O_2 are shown in Figures 10 and 11. We conclude that H_2O_2 is gauche because the L → T → G transformation is accompanied by sequential $U \rightarrow H^{\ddagger}$ rebondings. In more descriptive language, the L → T → G transformation is accompanied by sequential core deexcitations. Replacement of H by F was shown to lead to rehybridization which increases the gauche preference. Finally, the problem of the stereochemistry of H_2O and its derivatives was shown to be entirely analogous to the problem of H_2O_2 and its derivatives. It is now evident that H_2O_2 and N_2H_4 are both gauche for the same fundamental reason, i.e., maximal "removal of forbiddenness" of the L H_2O_2 and P N_2H_4 structures in the G H_2O_2 and G N_2H_4 geometries, respectively. However, this reduction of "forbiddenness" is accomplished by two different mechanisms in the two molecules: In H_2O_2 by core deexcitation and in N_2H_4 by core and ligand deexcitation. Now, replacement of H by F affects indirectly core deexcitation and directly ligand excitation. In the former case, we say that replacement of H by F causes rehybridization of the core. In the latter case, we say that substitution of H by F eliminates largely the necessity for "forbiddenness removal" by driving the ligand MO's towards degeneracy and, thus, rendering T superior to G N_2F_4 by changing the rebonding accompanying the T → G transformation from $U \rightarrow H^{\ddagger}$ to $N \rightarrow N'$.

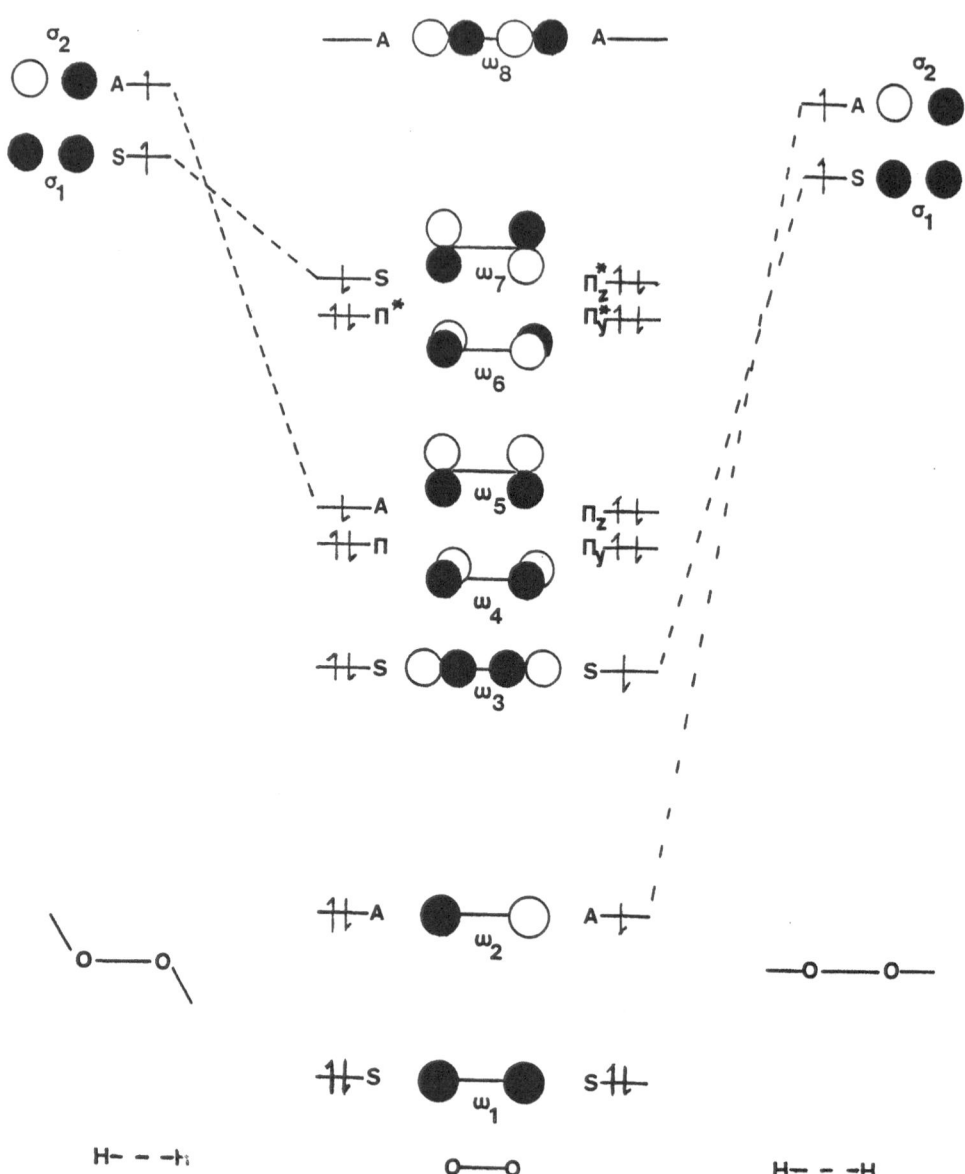

Figure 10. Compact MOVB bond diagrams of linear and trans H_2O_2.

227

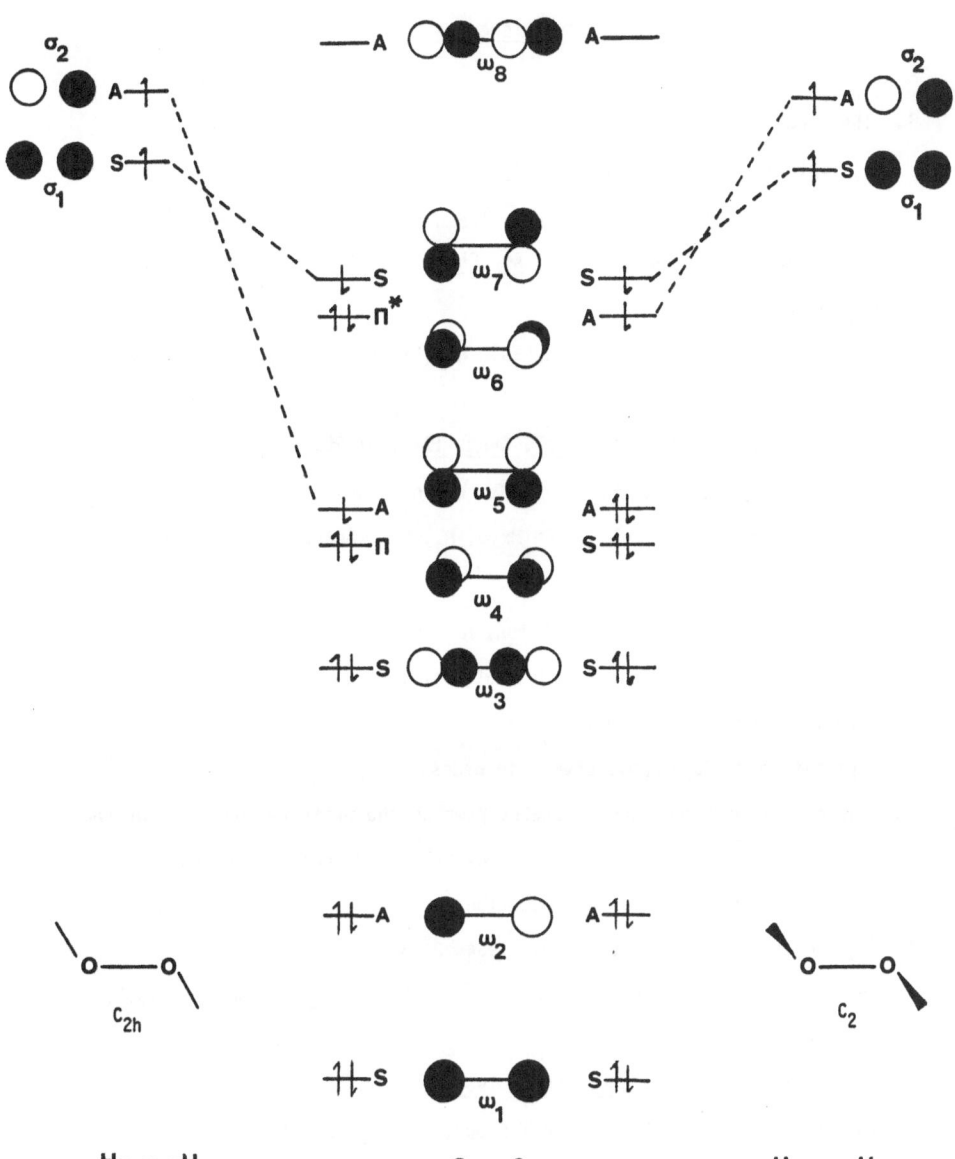

Figure 11. Compact bond diagrams of trans and gauche H_2O_2.

References

1. The "seeds" of the formal MOVB theory can be found in:

    (a) Epiotis, N.D. Angew Chemie, Int. Ed. Engl. 1974, 13, 751.

    (b) Epiotis, N.D.; Shaik, S. Prog. Theor. Org. Chem. 1977, 2, 348.

    (c) Epiotis, N.D.; Shaik, S. J. Am. Chem. Soc. 1978, 100 1, and subsequent
        papers.

    (d) Epiotis, N.D. "Theory of Organic Reactions"; Springer-Verlag: Berlin
        and New York, 1978.

    (e) Epiotis, N.D. Pure and Appl. Chem. 1979, 51 203

    (f) Epiotis, N.D.; Shaik, S.; Zander W. in "Rearrangements in Ground and
        Excited States", Vol. 2, De Mayo, P., Ed.; Academic Press: New York,
        1980.

2. Epiotis, N.D.; Larson, J.R.; Eaton, H. "Unified Valence Bond Theory of
    Electronic Structure" in Lecture Notes in Chemisty, Vol. 29; Springer-Verlag:
    New York and Berlin, 1982.

3. Epiotis, N.D. Pure Appl. Chem., in press.

4. The reference "forbidden" geometry remains the preferred geometry of the
    system if "forbiddenness" removal may take place only at the expense of
    substantial spatial overlap reduction.

5. (a) Wheland, G.W. J. Chem. Phys. 1934, 2, 474.

    (b) Pauling, L.; Springall, H.S.; Palmer, K.J. J. Am. Chem. Soc. 1939, 61,
        927.

    (c) Mulliken, R.S. J. Chem. Phys. 1939, 7, 339.

    (d) Mulliken, R.S.; Rieke, C.A.; Brown, W.G. J. Am. Chem. Soc. 1941, 63, 41.

    (e) Dewar, M.J.S. "Hyperconjugation", Ronald Press Co.: New York, 1962.

6. The "father" of the concept of aromaticity is E. Hückel. Hückel, E.
    Z. Physik. 1931, 70, 204; ibid. 1932, 76, 628. Hückel, E. Z. Electrochem.
    1937, 43, 752.
    Its applicability to problems of chemical reactivity and, in particular,
    pericyclic reactions was recognized independently, under different theoretical
    disguises, by M. G. Evans, M. J. S. Dewar, and H. E. Zimmerman:
    (a) Evans, M.G. Trans. Faraday Soc. 1939, 35, 824.
    (b) Dewar, M.J.S. Angew. Chem., Int. Ed. Engl. 1971, 10, 761.
    (c) Zimmerman, H.E. Acc. Chem. Res. 1971, 4, 272.

7. The father of the FO approximation in "qualitative" MO theory is K. Fukui.
    (a) Fukui, K.; Yonezawa, T.; Shingu, H. J. Chem. Phys. 1952, 20, 722.
    (b) Fukui, K.; Yonezawa, T.; Nagata, C.; Shingu, H. J. Chem. Phys. 1954, 22. 1433.

8. In most practical applications, PMO theory is implemented with the integral
    approximations of Hückel MO theory. For early formulation and application,
    see: Dewar, M.J.S. "The Molecular Orbital Theory of Organic Chemistry";
    McGraw-Hill: New York, 1969.

9. Epiotis, N.D.; Cherry, W.R.; Shaik, S.; Yates, R.L.; Bernardi, F. Top. Curr.
    Chem. 1977, 70, 1, and references therein.

10. Yokozeki, A.; Bauer, S.H. Top Curr. Chem. 1975, 93, 289.

11. H_2O_2: Redington, R.L.; Olson, W.B.; Cross, P.C. J. Chem. Phys. 1962, 36, 1311.
    Oelfke, W.C.; Gordy, W. J. Chem. Phys. 1969, 51, 5336.

12. F_2O_2: Jackson, R.H. J. Chem. Soc. 1962, 4585.

13. N_2H_4: Kasuya, T. Sci. Papers Inst. Phys. Chem. Res. Tokyo 1962, 56, 1.

14. N_2F_4: Cardillo, M.J.; Bauer, S.H. Inorg. Chem. 1969, 8, 2086.

15. An excellent text: Jorgensen, W.L.; Salem, L. "The Organic Chemist's Book of
    Orbitals"; Academic Press: New York, 1973.

16. Epiotis, N.D.; Larson, J.R.; Eaton, H., submitted for publication.

17. The detailed nature of ω_2 is such that it makes it deserving of the epithet nonbonding. For more details, see ref. 2.

18. (a) The N-N bond lengths of N_2H_4 and N_2F_4 are 1.453 $\overset{\circ}{A}$ and 1.489 $\overset{\circ}{A}$, respectively.

    (b) Colburn, C.B.; Johnson, F.A.; Kennedy, A.; McCallum, K.; Metzger, L.C.; Parker, C.O. J. Am. Chem. Soc. 1959, 81, 6397. Armsgrong, G.T.; Marantz, S. J. Chem. Phys. 1963, 38, 169.

19. Durig, J.R.; Gimarc, B.M.; Odom, J.D. in "Vibrational Spectra and Structure", Vol. 2, Durig, J.R., Ed.; Marcel Dekker: New York, 1973.

20. Cowley, A.H.; Dewar, M.J.S.; Goodman, D.W.; Padolina, M.C. J. Am. Chem. Soc. 1974, 96, 2648.

Chapter 7. Geometric Isomerism: The Simplest Illustrator of Orbital Symmetry Control of Molecular Stereochemistry.

The thermal conversion of 1,3-butadiene to cyclobutene may occur in a conrotatory or a disrotatory fashion. In the former case, an axis of symmetry is maintained along the reaction coordinate while in the latter case a plane of symmetry is preserved during the conversion or reactants to products. This difference with respect to the existing symmetry elements becomes responsible for a difference in the symmetry labels of reactant and product orbitals. In turn, this becomes responsible for the existence of a barrier in the case of disrotation and the absence of a barrier in the case of conrotation at the level of Hückel MO theory. This is clearly revealed by the Longuet—Higgins-Abrahamson-Woodward-Hoffmann MO correlation diagrams[1] for con- and dis-rotatory ring closure of 1,3-butadiene. The conrotatory ring closure of 1,3-butadiene is termed a symmetry "allowed" and the disrotatory ring closure of the same molecule is termed a "forbidden" reaction.

The approach to electrocyclizations highlighted above has been more or less accepted by chemists for the following reasons:

a. A large body of experimental evidence is consistent with the predictions of the theory.[2]

b. Exceptions can be rationalized by making a reasonable invokation of "steric effects".

c. The theoretical argument is unambiguous and clear cut to the extent that two different stereochemical pathways are associated with two different symmetry elements which, in turn, become responsible for two different sets of reactant and product orbitals, insofar as the symmetry labels are concerned.

The purpose of this paper is to point out that the problem of cis-trans isomerism in 1,2-disubstituted ethylenes, $CHX = CHX$, is very analogous to the problem of conrotation-disrotation in 1,3-butadiene and that geometric isomerism, examined from the vantage point of MOVB theory,[3] is one of the simplest stereochemical problems one is likely to encounter, a rather surprising statement in view of the controversy with which it has been surrounded.[4]

I. Sigma Rebonding in Cis and Trans CHX = CHX

In order to determine the relative stability of cis and trans CHX = CHX, we need to examine the electronic structure of the uncoupled sigma and pi systems as revealed by the appropriate bond diagrams. Assuming that H and X differ substantially in electronegativity the sigma bond diagrams are those in Figure 1. We note the following:

a. The ligand MO's separate into two sets, one set being composed of two MO's localized predominantly on the H's and a second set being composed of two MO's localized mainly on the X's.

b. There exists a plane of symmetry in the cis and an axis of symmetry in the trans isomer. Accordingly, the core and ligand MO's of the cis form can be classified as either symmetric (S) or antisymmetric (A) with respect to the σ_v plane and those of the trans form can be classified as either S or A with respect to the C_2 axis. It is evident that because of the localization of the ligand MO's, these are geometry invariant with respect to the symmetry labels. By contrast, the symmetry classification of the C_2 core MO's is geometry dependent in the way emphasized below.

$\boxed{A}$ ✛ ω_5	$\boxed{S}$ ✛ ω_5
S ✛ ω_4	S ✛ ω_4
$\boxed{S}$ ✛ ω_3	$\boxed{A}$ ✛ ω_3
A ✛ ω_2	A ✛ ω_2
S ✚ ω_1	S ✚ ω_1
CIS	TRANS

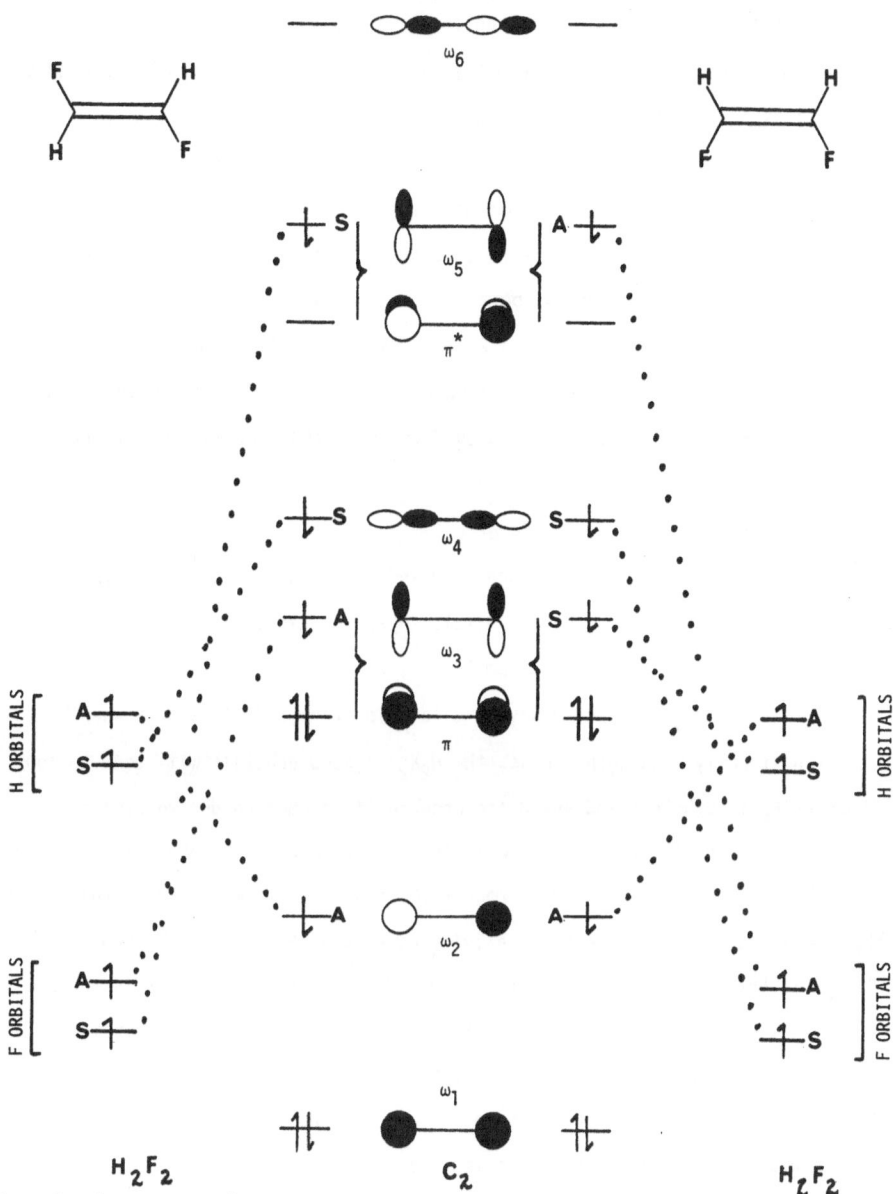

Figure 1. Compact sigma bond diagrams of cis and trans 1,2-difluoroethylene assuming that the F_2 and H_2 ligand MO's are segregated. Core pi MO's are shown for calibration.

Note that the presence of a symmetry plane in cis and a symmetry axis in trans is responsbile for a switchover of the symmetry labels of the two "pi-type" sigma MO's, ω_3 and ω_5. This is the critical difference between the cis and trans isomers insofar as orbital symmetry is concerned.

The interpretation of the sigma bond diagrams of Figure 1 is straightforward. Since core-ligand spatial overlap is kept constant, the cis and tans isomers differ in terms of bond type with the former being D- and the latter U-bound. This difference is better projected in Figure 2, with the dotted lines indicating differing bonds and the faded dashed lines marking similar bonds. Accordingly, the trans → cis transformation is accompanied by favorable rebonding and the cis isomer will tend to be more stable than the trans one.

Some clarifying remarks are now in order:

a. The electronegativity difference of H and X and their spatial overlap are assumed to be such so that the ligand orbitals separate into two sets, one set localized mostly on the H's and the other localized mostly on the X's. This assumption breaks down when X and H have similar electronegativity and/or when H and X overlap strongly. In such a case, the H_2X_2 ligand orbitals will tend to look like those of H_4 in ethylene and any difference in sigma bonding due to orbital symmetry will tend to disappear. This will leave unencumbered the operation of nonbonded ligand repulsion which will render the trans isomer more stable than the cis isomer. Since group electronegativity decreases as we change X from F to OR, to NR_2, to CR_3 and since the geminal nonbonded H-X overlap increases in the same direction, we expect that the cis preference in CHX = CHX will progressively diminish as X is varied in this way, if only sigma effects are considered.

b. The assignment of the labels U and D to the trans and cis isomers, respectively, depends on the energy gap separating the core MO's, ω_3 and ω_4. If the gap is appreciable, then a distinction between U and D bonding can be made. If the

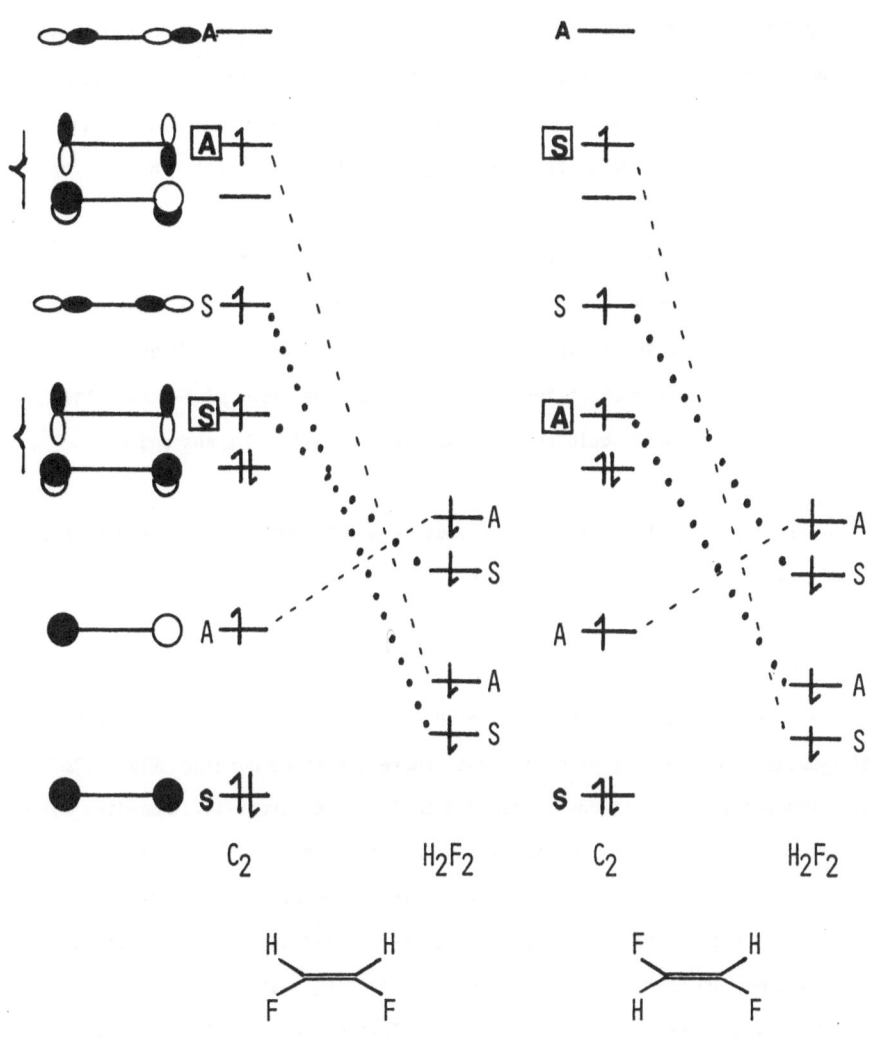

Figure 2. The key sigma bonding difference between cis and trans 1,2 difluoroethylene.

gap is very small or zero, then the two isomers belong to the same bond type. Now, it turns out that the $\omega_3-\omega_4$ energy gaps for a variety of diatomic cores calculated by the Extended Hückel MO (EHMO) method[5], are as shown in Table 1. The results indicate (not prove) that the differential sigma bonding indeed acts in favor of cis CHX = CHX.

At this point the reader may express doubts: While the energy gap separating the 2s and 2p AO's of oxygen may undoubtedly be termed "large", being of the order of 17 eV, the energy gap separating ω_3 and ω_4 of the C_2 diatomic core of XHC = CHX seems rather "small" in a comparative sense, being of the order of 1-2 eV. The question then arises: What would be the relative stability of trans and cis XHC = CHX isomers if ω_3 and ω_4 were degenerate.

In order to answer this question, we must recall the most general recipe for the qualitative application of MOVB theory. This has been spelled out before[3f] and it is the following:

a. In comparing two isomers, one constructs the compact MOVB diagram for each one of them and determines the type of bonding imposed by core and ligand orbital symmetry in each case recalling that there are three bonding "flavors": D bonding which permits electrons to descend to low lying orbitals, U bonding which confines some of the electrons to high lying orbitals, and H bonding which represents a hybrid of D and U bonding accompanied by impairment of core-ligand spatial overlap, at least in most cases of interest. Loss of spatial overlap is indicated by affixing a dagger superscript to the appropriate latter, most often H. The letter (H, $H^{\ddagger}$, and D) assignment is always made in a relative sense and the selection rules are: D is always betterthan $H^{\ddagger}$ and U bonding but $H^{\ddagger}$ may be superior or inferior to U bonding depending on whether deexcitation is more important than loss of spatial overlap or _vice versa_. This latter principle constitutes the quantum mechanical rationalization of stereochemical diversity in nature.

Table 1. EHMO $\omega_3(\pi_u)$ - $\omega_4(\sigma_g)$ Energy Gap of Diatomic Cores.

X - X	r_{XX}	$E(\omega_4) - E(\omega_3)$ (eV)
C = C	1.34	1.53
C - C	1.54	0.86
N - N	1.45	0.61
O - O	1.48	-0.65
F - F	1.42	-1.01

Table 2. Enthalpies of Cis ⟶ Trans Gas Phase Isomerization of
CHX = CHX.[a]

X	ΔH^0 (Kcal/mol)
CH_3O	1.445
F	0.928
Cl	0.650
Br	0
I	0

[a] Data taken from the compilation of Waldron, J.T.; Snyder, W.H.
J. Am. Chem. Soc. 1973, 95, 5491.

Table 3. Enthalpies of Cis ⟶ Trans Gas Phase Isomerization of
CHR = CHR.[a]

R	ΔH^0 (Kcal/mol)
CH_3	-1.00
CH_3CH_2	-1.63

[a] Computed by using standard heat of formation data for the two
isomers.

b. If the compact bond diagrams of two different isomers belong to the same bond type (e.g., both involve D-bonding, to a first approximation), one proceeds to examine the consequences of the principal direction of interfragmental Charge Transfer (CT) in each of the core-ligand bonds of the two isomers. The way in which primary and secondary CT differentially affect two isomers has been extensively discussed in the original work[15]. This last procedure effectively amounts to a one-to-one energetic comparison of the compact bond diagrams (Ξ_i's) the linear combinations of which constitute the approximate hybrid diagrammatic representations of the two species.

Let us now interpret the compact bond diagrams of Figure 1 using this procedure and by assuming that ω_3 and ω_4 are degenerate. In such a case, both cis and trans isomers can be said to involve the same type of core-ligand binding, e.g., both can be regarded as D-bound. However, if we focus on the direction of primary CT in the core-ligand bond which will undoubtedly be more "ionic", namely, the $\omega_5^1 - \sigma_a^1$ bond (a=1 in trans and a=2 in cis), then we immediately discover that CT from core to ligand will render σ_a doubly occupied thus engendering $\omega_1^2 - \sigma_a^2$ core-ligand overlap repulsion in trans but not in cis due to symmetry constraints (superscript denotes number of electrons). Hence, we again predict that there still exists an orbital-symmetry factor which will tend to render cis more stable than trans XHC = CHX even if the two isomers are , in a first approximation, similarly bound. If indeed ω_3 and ω_4 are nearly degenerate, the reason why cis is more stable than trans FHC = CHF becomes the same as the reason why cis is more stable than trans FN = NF.

The analysis presented above is in excellent agreement with various well known trends:

a. A large body of experimental evidence, some of which is summarized in Table 2, suggest that the cis is more stable than the trans isomer. The data of Table 2 can be said to define a cis preference rule in CHX = CHX molecules.

b. Exceptions to the a cis preference rule arise when the substituents are alkyl groups (Table 3). The trans preference may be ascribed to "steric effects" in the absence of significant sigma bond type differentiation due to the fact that H and C have comparable electronegativity.

c. In cis CHF = CHF (see the diagram of Figure 1), ω_4 and ω_5 are primarily responsible for core-fluorine bonding and ω_2 and ω_3 for core-hydrogen bonding. accordingly, each fluorine is bound to the core, C_2, via carbon orbitals which are intermediate between p and sp because ω_5 is an out-of-phase combination of two carbon p AO's and ω_4 is an in-phase combination of two sp-like carbon AO's.

Detailed ω_4:

Also, each H is bound to the core in a similar fashion because ω_3 is an in-phase combination of two carbon p AO's and ω_2 is an out-of-phase combination of two sp-like carbon AO's.

Detailed ω_2:

Thus, we expect cis CHF = CHF to have the standard structure shown below.

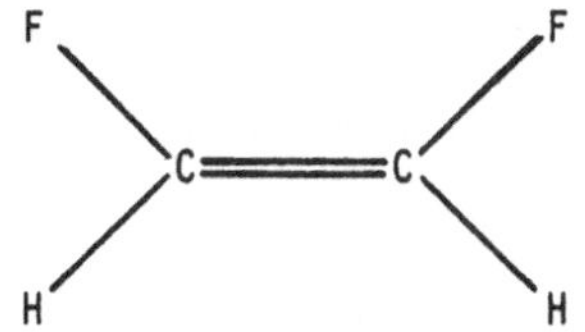

By contrast, in trans CHF = CHF (see Figure 1), it is now ω_5 and ω_3 which are primarily responsible for core-fluorine binding and ω_2 and ω_4 for core-hydrogen binding. This means that the fluorines tend to be bound via carbon p AO's and the hydrogens by carbon sp-like AO's. The shape of trans CHF = CHF would thus tend to be as shown below.

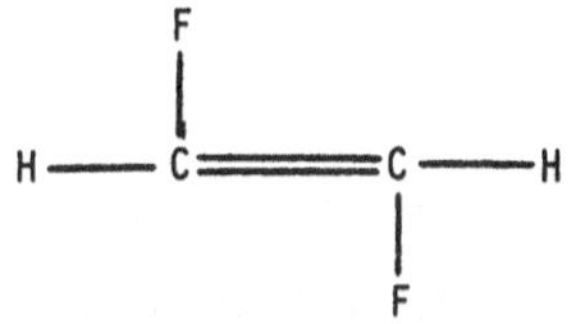

This analysis suggests that while cis CHF = CHF will tend to have standard bond angles, trans CHF = CHF will adopt a geometry which can be visualized as the result of conrotation of the CHF groups of the hypothetical standard trans CHF = CHF structure. The experimental results given below clearly show that these expectations are met.[6]

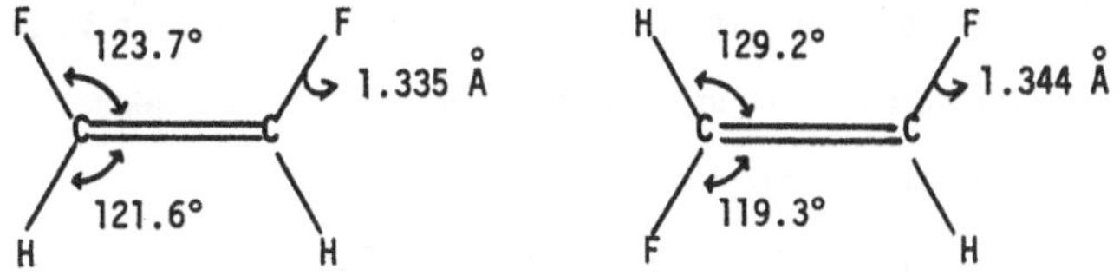

An important difference between cis and trans 1,2-difluoroethylene, insofar as geometry is concerned, is the fact that the C-F bonds are <u>shorter</u> in the cis geometry. This is so because the carbon hybrid AO's directed towards the two fluorines have greater s character in the cis isomer because the C_2 core binds the fluorines through primary utilization of ω_4 and ω_5 in the cis isomer and mainly via ω_5 and ω_3 in the trans isomer and only ω_4 is a linear combination of two sp^q carbon AO's, the other two core MO's being pure pi type MO's. As we have already seen,[7] an s AO is a stronger overlap binder than a p AO in first row atoms. Finally, the dependence of hybridization and, by extension, the relative stability of the geometric isomers on ligand electronegativity is further illustrated by the fact that the $=C-CH_3$ bond lengths and $\widehat{CCC}$ bond angles of cis and trans 2-butene define trends which are entirely different from those encountered in cis and trans 1,2 difluoroethylene. Thus, the $=C-CH_3$ bond lengths are nearly the same and the $\widehat{CCC}$ angles are larger than 120° in <u>both</u> isomers.[8]

The reader may now ask: How are the conclusions regarding the relative stabilities of the cis and trans isomers of 1,2 difluoroethylene modified if we take into consideration the presence of sigma fluorine sigma lone pairs? We can rephrase the question as follows: Are the sigma fluorine lone pairs better delocalized in the cis or in the trans geometry? Clearly, the way in which the core and ligands are connected by the sigma bonds will determine the extent of the lone pair delocalization. Since D-bonding directs primary CT from core to ligands more strongly than U-bonding, it follows that secondary CT from ligand to core, i.e., lone pair delocalization, will be more favorable in the cis geometry. This is a general phenomenon: The sigma lone pairs of electronegative ligands which are D-bound to a core will be more delocalized than if they belonged to electronegative

ligands which are U-bound to the core, thus enhancing the energetic advantage of
the D-bound system predicted by neglecting the effect of the lone pairs and by
considering only the primary sigma bonds. This general principle is illustrated
pictorially below.

U-Bound

  System:

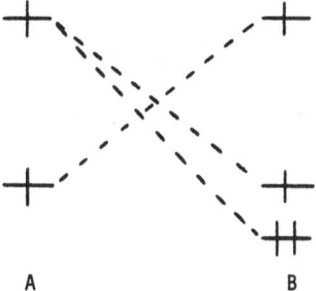

D-Bound

  System:

This analysis makes evident the fact that as the electronegativity of X increases relative to that of C, the sigma electron pairs of X become increasingly delocalized in a way which progressively enhances the stability of cis relative to trans CHX = CHX. Conversely, as the difference in electronegativity between C and X decreases, the delocalization of the sigma electron pairs of X is turned off and these same electron pairs begin to destabilize the cis isomer by nonbonded overlap repulsion. Thus, the sigma lone pairs of F stabilize the cis isomer of CHF = CHF but the doubly occupied bonding orbitals of $-CH_3$ of sigma symmetry stabilize the trans isomer of $CH(CH_3) = CH(CH_3)$. We conclude that primary sigma bonding and sigma X electron pair delocalization both dictate that the cis isomer of CHX = CHX will be increasingly favored as X becomes more electronegative relative to C.

II. Pi Bonding Control of the Relative Stability of Geometric Isomers

The effect of pi-type electron pairs of X on the relative stability of cis and trans CHX = CHX molecules can be illustrated by reference to the specific case of cis and trans CHF = CHF. The pi bond diagrams shown in Figure 3 require little comment. There is no pi rebonding accompanying the cis → trans conversion. The only one-electron difference is to be found in the normalization factors of the ligand MO's, a difference which renders the $n^*-\pi^*$ multicenter bond stronger and the n-π four-electron antibond weaker in the cis isomer. Since this effect has been analyzed and tested before in an MO theoretical frame,[9] no further discussion is needed. The important point is that the cis → trans conversion of CHF = CHF is accompanied by sigma but not pi rebonding. Hence, the more profound sigma effects are most likely the cause for the greater stability of the cis isomer of CHF = CHF and related molecules.

At this point, it is instructive to consider the type of bonding in the sigma and pi framework of CHF = CHF. Since, cis is D- and trans U-bound, we say that the former isomer has sigma bonds of higher "ionicity" than the latter one. By contrast, because of the stronger F---F interaction and the larger (due to the normalization factor) $<n^*|\hat{H}|\pi^*>$ matrix element (in absolute magnitude) in the cis isomer, the single pi core-ligand bond is more "ionic" in the trans isomer. We shall use this information when we attempt to predict qualitatively the effect of "electron correlation" on the relative stability of geometric isomers.

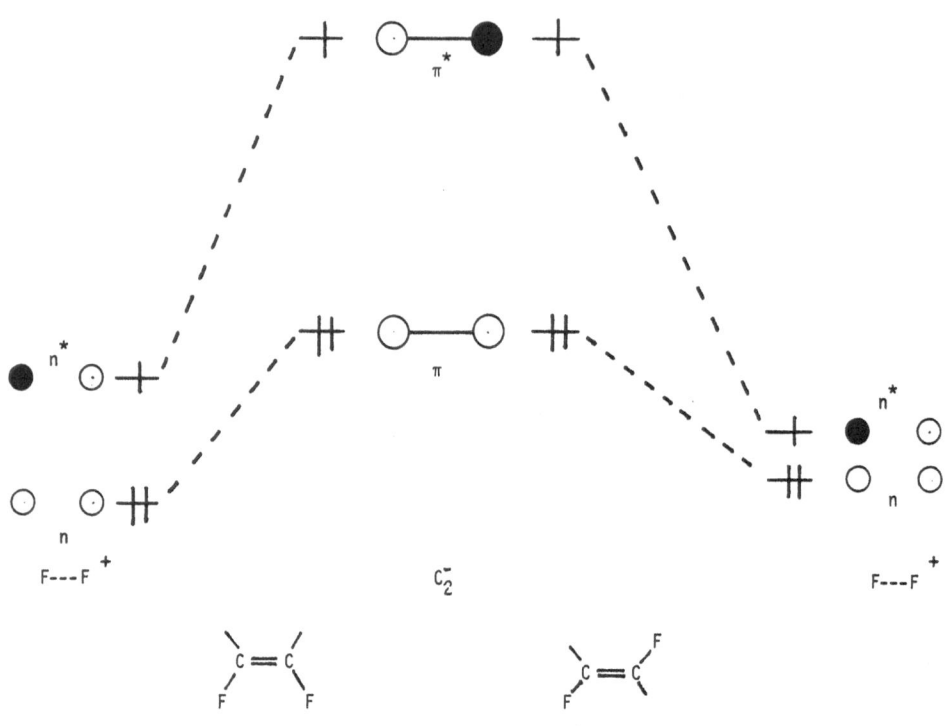

Figure 3. Pi bond diagrams of cis and trans 1,2 difluoroethylene. Note that core-ligand bond is more "ionic" in the trans isomer.

III. Implications

Cis 1,2 difluoroethylene is more stable than trans 1,2-difluoroethylene because the two highest singly occupied core orbitals have symmetry that allows them to make two two-electron multicenter bonds with two singly occupied fluorine-type ligand MO's. This is not possible in the case of the trans isomer. The characteristic property of electronegative ligands to dictate geometries which allow them to depopulate the highest lying core orbitals is nothing else but a manifestation of parity stereoselection and a reconfirmation of the often stated conclusion that D-bound are more stable than U-bound species. This phenomenon is very frequently encountered in molecular stereochemisty, as one might naturally anticipate. Indeed, many puzzling preferences of fluorinated hydrocarbons for "crowded" geometries are a reflection of the same principle, as construction of appropriate bond diagrams reveals. Thus, for example, gauche 1,2-difluoroethane is more stable than trans 1,2-difluoroethane for reasons that can be easily understood by mere inspection of the bond diagrams shown in Figure 4 and comparison with those of Figure 2.

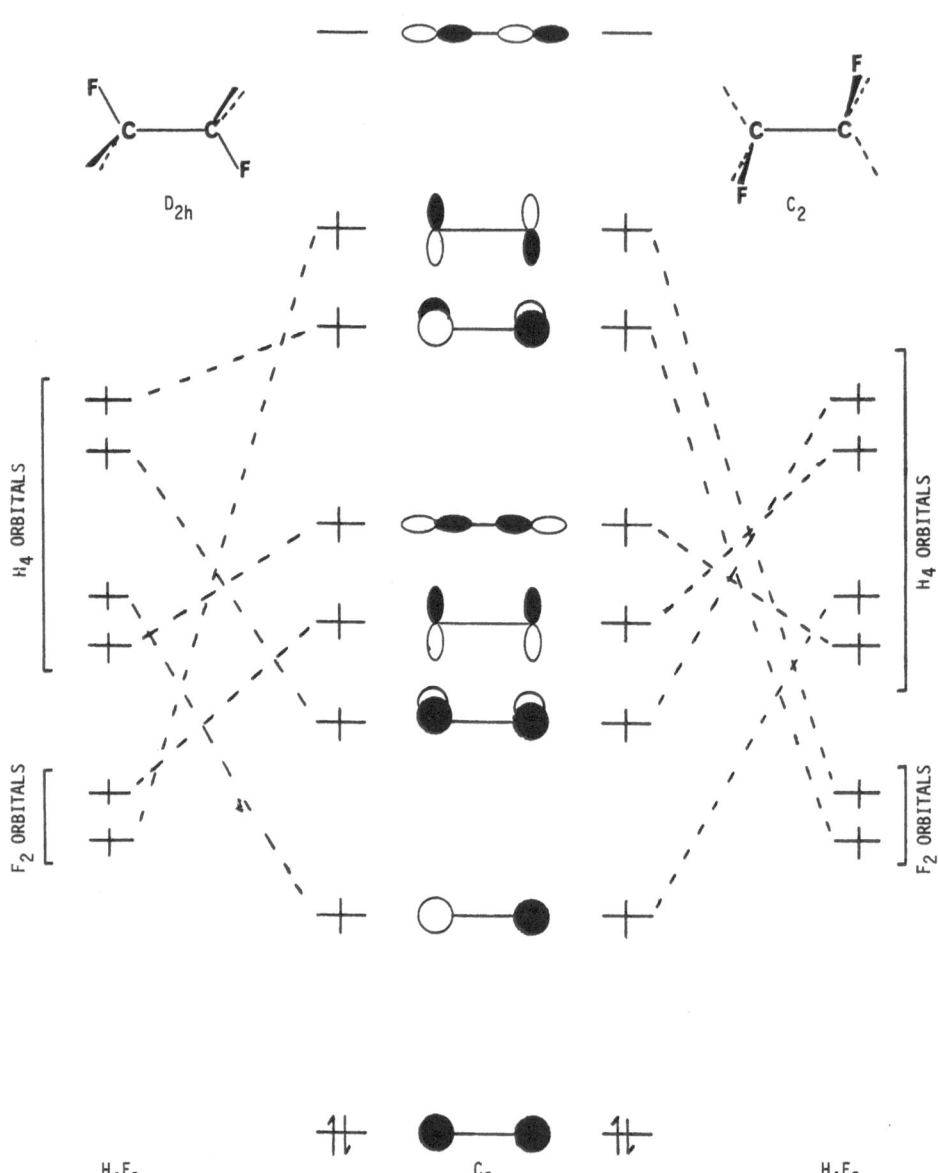

Figure 4. Compact bond diagrams of anti and gauche 1,2 difluoroethane, assuming that the F_2 and H_4 ligand MO's are segregated.

References

1. (a) Woodward, R.B.; Hoffmann, R. J. Am. Chem. Soc. 1965, 87, 395.

   (b) Longuet-Higgins, H.C.; Abrahamson, E. J. Am. Chem. Soc. 1965, 87, 2045.

   (c) Hoffmann, R.; Woodward, R.B. J. Am. Chem. Soc. 1965, 87, 2046.

2. Woodward, R.B.; Hoffmann, R. "The Conservation of Orbital Symmetry"; Verlag Chemie: Weinheim, 1970.

3. (a) Epiotis, N.D.; Larson, J.R.; Eaton, H. "Unified Valence Bond Theory of Electronic Structure" in Lecture Notes in Chemistry, Vol. 29, Springer-Verlag: New York and Berlin, 1982.

   (b) Epiotis, N.D.; Larson, J.R. Israel J. Chem. 1983, 000.

4. (a) Hoffman, R.; Olofson, R.A. J. Am. Chem. Soc. 1966, 88, 943.

   (b) Epiotis, N.D.; Cherry, W.R.; Shaik, S.; Yates, R.L.; Bernardi, F. Top. Curr. Chem. 1977, 70, 1.

   (c) Binkley, J.S.; Pople, J.A. Chem. Phys. Letters 1977, 45, 197.

   (d) Nascimento, C.C.; Brinn, I.M. Z. Naturforsch 1978, 33a, 366.

   (e) Cremer, D. Chem. Phys. Letters 1981, 81, 481.

5. (a) Wolfsberg, M.; Helmholz, L. J. Chem. Phys. 1952, 20, 837.

   (b) Hoffmann, R.; Lipscomb, W.N. J. Chem. Phys. 1962, 36, 2189.

   (c) Hoffmann, R. J. Chem. Phys. L963, 39, 1397.

6. Carlos, Jr., J.L.; Karl, Jr., R.R.; Bauer, S.H. Faraday Trans. II 1974, 70, 177.

7. See Chapter 8 of this work.

8. Almenningen, A.; Anfinsen, I.M.; Haaland, A. Acta Chem. Scand. 1970, 24, 43.

9. Bernardi, F.; Bottoni, A.; Epiotis, N.D.; Guerra, M. J. Am. Chem. Soc. 1978, 100, 7205.

Chapter 8. Structural Isomerism and the Electronic Basis for Ligand Segregation on C_2 Cores.

One of the most fascinating trends of molecular stereochemistry is the tendency of ligands to segregate into sets when attached on carbon cores. A typical example is provided by the comparison of 1,1 and cis 1,2-difluoroethylene, henceforth referred to as G and C 1,2 difluoroethylene, respectively.

| G | C |
| Isomer | Isomer |

Calculations place G substantially below C[1], with the latter being more stable than trans-1,2-difluoroethylene for reasons discussed before.[2] Both G and C isomers have an identical C_2 core. However, if we separate the ligands into two sets, one containing the two F's and the other containing the two H's, we can see that the G and C isomers differ in one important way: In the former, the F and H sets are segregated, i.e., there is only vicinal overlap of F and H AO's, while in the latter case, the two sets are brought into proximity, i.e., there is now appreciable geminal overlap of F and H AO's as they are both attached on the same carbon atom.

The energetic consequences of ligand segregation on carbon cores is a well recognized phenomenon and the occasional exceptions appear to be due to overlap repulsion of the doubly occupied orbitals of the effectively "large" ligands which is more severe in the "crowded" G geometry. Experimental and theoretical heats of formation of structural isomers which demonstrate the fact that ligands prefer segregation to aggregation have been tabulated in previous publications.[3] Two typical examples are given below with the relative energies of the isomers being computed from available thermochemical data.[4]

STRUCTURAL ISOMERS	RELATIVE ENERGIES (kcal/mol)
$CH_3CH=CHCH_3$ (trans)	+ 1.37
$(CH_3)_2CH=CH_2$	0.00
$CH_2C\ell-CH_2C\ell$	+ 0.90
$CHC\ell_2-CH_3$	0.00

Returning now to the specific case of G and C difluoroethylenes, we recognize that modern qualitative MO concepts fail to provide clear insights as to why the G isomer is so much more stable despite the greater proximity of the negatively charged fluorines in this isomer. Thus, for example, the concept of trans bond-antibond interaction tells us that the two isomers must have comparable stability because in each of them there exist two C-H bonds and two C-F antibonds which are trans-disposed.[5] Several mutually incompatible explanations based on crude models have been offered but none has emerged as an undisputable interpretation.[1,6] In this work, we use MOVB theory in order to resolve the issue. As we shall see, the "solution" offered is anything but obvious.

I. Rebonding and Rehybridization

Let us consider the transformation of A into some other isomeric structure B by means of pyramidalization, bond rotation (conformational and geometrical isomerism), atom recombination (structural isomerism), etc. When A (and B) is a polyelectronic system, this change is most often accompanied by either rebonding or rehybridization. By rebonding we mean that B is bound in an overall sense in a way which is different from that in A, e.g., A may be U- and B D-bound. By rehybridization we mean that the weights of the same resonance bond diagrams which describe A and B become readjusted as A is converted to B. A change of the former type occurs because B and A **differ in a way responsible for different orbital** symmetry interrelationships between the core and ligand orbitals and rebonding **is most frequently** a more drastic change than rehybridization. Accordingly, in comparing A and B the first task is to examine which of the two isomers is more favorably bound, i.e., if there is a great difference insofar as bond-types are concerned. If there is no significant rebonding, then our task becomes the identification of the system with the more favorable hybridization.

The difference between rebonding and rehybridization is illustrated schematic- ally in Figure 1. It is immediately evident that only the hybrid bond diagrammatic representations of the electronic structures of isomers can reveal rebonding as well as rehybridization.[7] Singular detailed bond diagrams are not as <u>explicitly</u> informative when it comes to rehybridization. For example, both isomers A and B of Figure 1b are by convention described by the same type of detailed bond diagram which, of course, must be interpreted differently in the two cases.

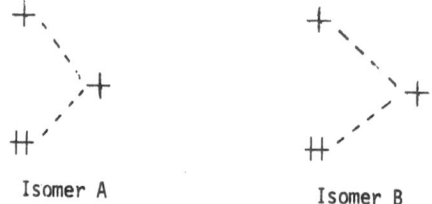

Isomer A   Isomer B

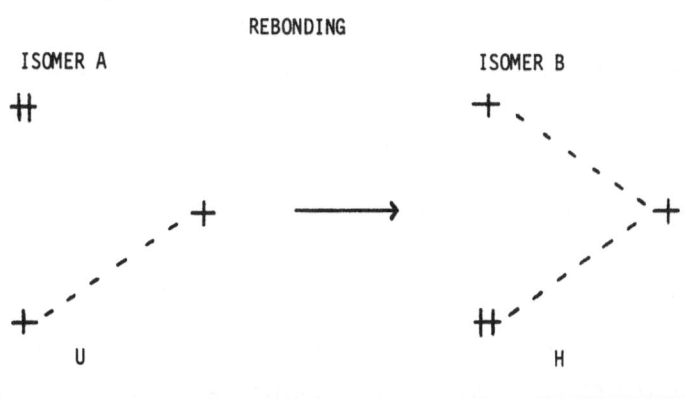

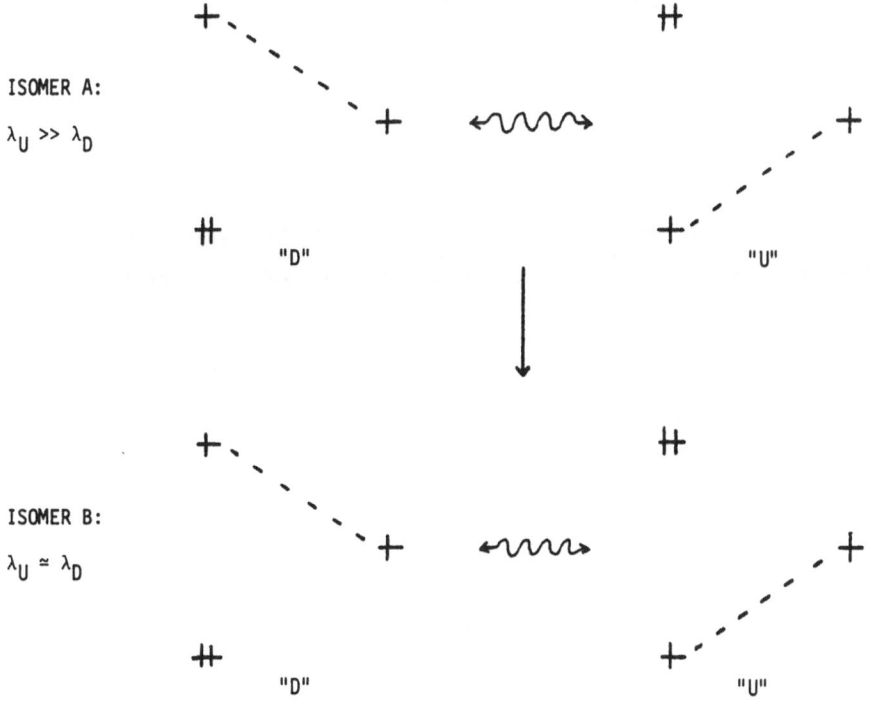

Figure 1. The difference between rebonding and rehybridization as isomer A is converted to isomer B.

With these ideas in mind, let us now consider sigma bonding in the prototypical G and C CHX = CHX isomers first through usage of compact bond diagrams. Assuming that H and X differ substantially in electronegativity (with the latter being more electronegative than the former) and that X is a univalent ligand devoid of any lone pairs, the sigma compact bond diagrams are those shown in Figure 2. We note the following:

a. The ligand MO's separate into two sets, one set being composed of two MO's localized predominantly on the H's and a second set being composed of two MO's localized mainly on the X's. We tentatively assume that ligand MO segregation is comparable in the two isomers. Later on, we shall see that this assumption is not correct and this realization will provide the key for understanding the major difference in the sigma electronic structures of the two isomers.

b. The core and ligand MO's of the two isomers can be classified as either symmetric (S) or antisymmetric (A) with respect to the σ_v symmetry planes. It is evident that because of the localization of the ligand MO's, these are geometry invariant with respect to the symmetry labels. By contrast, the symmetry classification of the core MO's is geometry dependent in the way emphasized below.

A $\dashv\vdash$ ω_5	A $\dashv\vdash$ ω_5
S $\dashv\vdash$ ω_4	S $\dashv\vdash$ ω_4
A $\dashv\vdash$ ω_3	S $\dashv\vdash$ ω_3
S $\dashv\vdash$ ω_2	A $\dashv\vdash$ ω_2
S $\dashv\!\!\vdash$ ω_1	S $\dashv\!\!\vdash$ ω_1
G Isomer	C Isomer

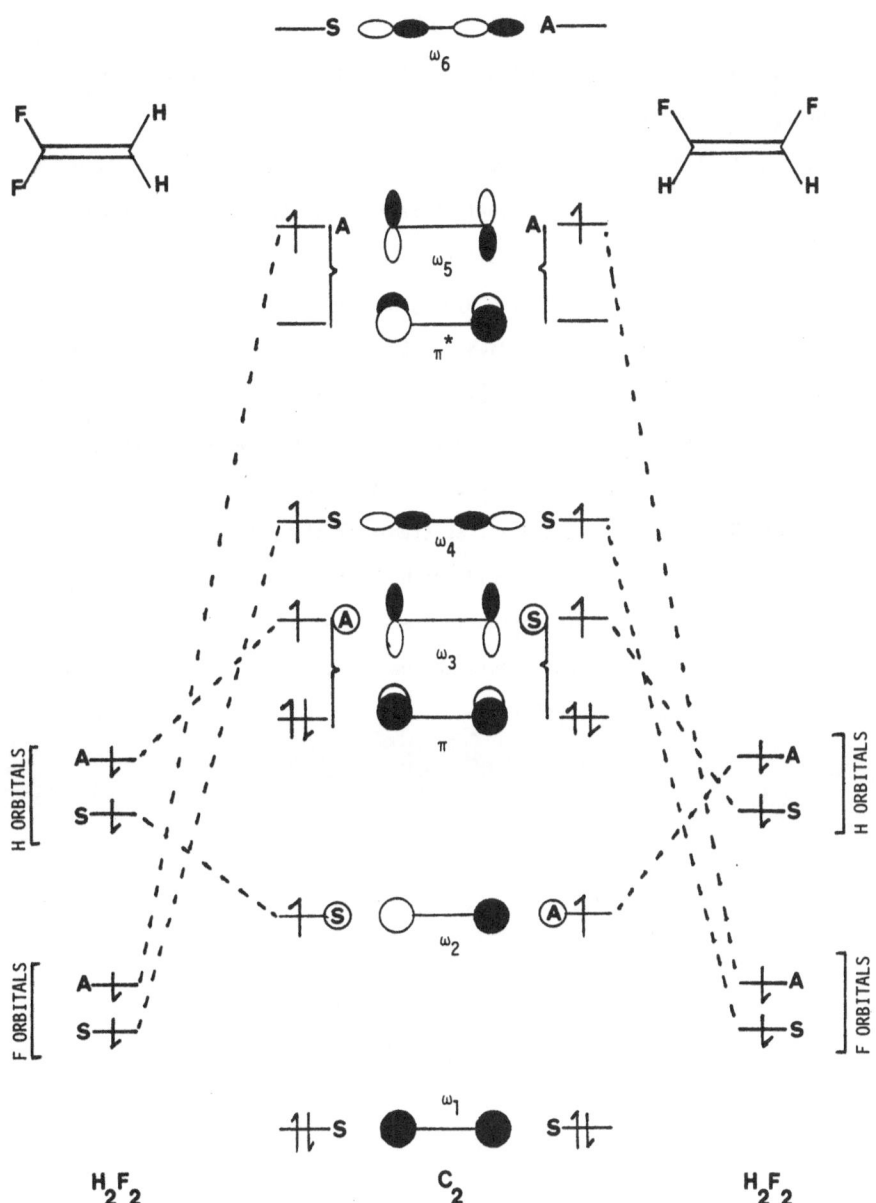

Figure 2. Compact sigma bond diagrams for 1,1- and cis-1,2- difluoroethylene, assuming that F_2 and H_2 ligand MO's are equally segregated in the two isomers. Core pi MO's are shown for calibration.

Examination of the sigma bond diagrams of Figure 2 reveals that there are
two possible conclusions one may reach depending on whether geminal H --- H nonbonded overlap, s_{HH}, is "large" or "small":

a. If s_{HH} is "large", then the two hydrogens are D-bound in C and U-bound in
G $C_2H_2F_2$. In this case, we would say that there is rebonding accompanying the
C $\longrightarrow$ G transformation in a way which favors the C isomer.

b. If s_{HH} is "small", then the two hydrogens are similarly bound in C and
G and no appreciable rebonding accompanies the C $\longrightarrow$ G transformation. In this
case, we would say that C and G should have comparable stability.
It is then evident that, if rehybridization is disregarded, the rebonding which
accompanies the C $\longrightarrow$ G transformation, whether substantial or not, would lead us
to believe that the G isomer is, at best, as stable as the C isomer and probably
less so. It follows that rebonding should not be very significant, or, at any
rate, not as important as rehybridization in the case at hand. Hence, we seek to
unravel the hybridization differences between the two isomers using the tool of
the hybrid bond diagrammatic representation.

II. The Electronic Basis of Ligand Segregation on C_2 Cores

Consider the sigma electronic structure of ethylene as represented by the sigma bond diagram of Figure 3. We may think of ethylene as the product of the partly "forbidden" union of C_2 and H_4. Replacement of two hydrogens by two X atoms or groups which have widely different electronegativity will cause a reduction of the symmetry of the ligand MO's and change the way in which the sigma bonds are made. We may think of this replacement process as a "forbiddeness"-reducing process. The question then becomes: What is the optimum placement of the H's and the X's which causes maximum reduction of the symmetry of the ligand MO's? In different language: What is the optimum geometrical arrangement of the H's and X's which most effectively removes the "forbiddenness" of the union of the core and ligand fragments?

The three possible arrangements of H's and X's are the ones which produce the three isomers trans (T), cis (C), and 1,1 (G) $C_2H_2X_2$. Let us see now what is the critical difference between the three isomers insofar as the ligand MO's are concerned. The H_2X_2 ligand MO's can be built by combining the symmetry orbitals spanning the H's, σ_H and σ_H^*, with the symmetry orbitals spanning the X's, σ_X and σ_X^*. Now, in the T and C isomers, the overlap of the two sets of symmetry orbitals is strong as each X AO can overlap with a <u>geminal</u> H AO. Thus, the final four ligand MO's tend to resemble the four ligand MO's of ethylene because strong overlap tends to counteract the effect produced by the fact that the interacting symmetry MO's are separated by a large energy gap due to the different electronegativity of H and X. By contrast, in G, the overlap of the two sets of symmetry orbitals is weak because now each X AO can only overlap with a <u>vicinal</u> H AO. Accordingly, the symmetry of the final four ligand MO's will be more effectively

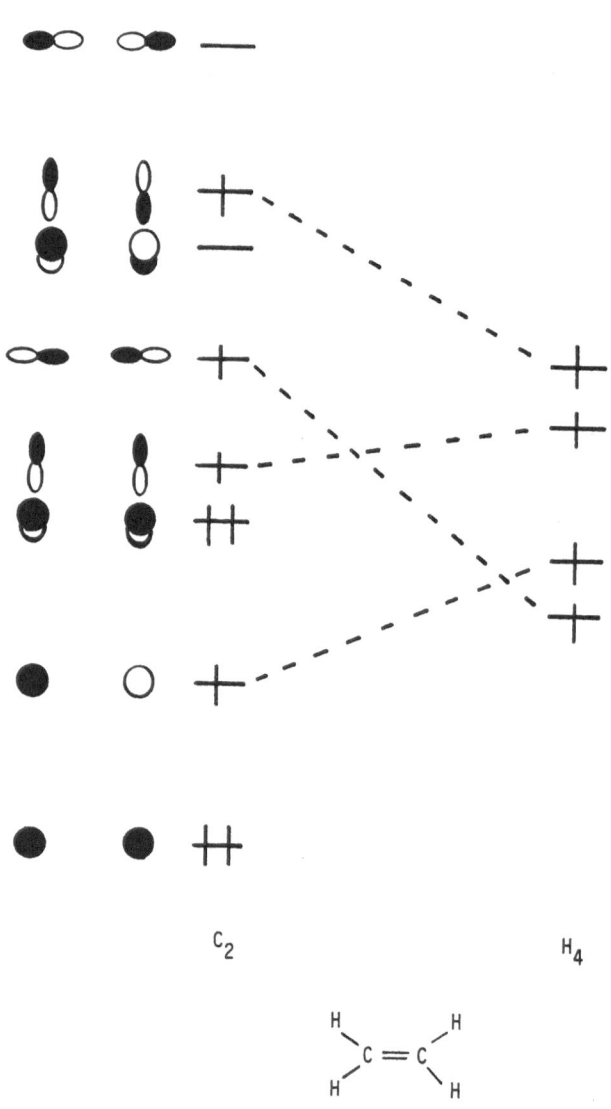

C_2

H_4

Figure 3. Compact bond diagram of ethylene.

reduced in the case of the G than in the case of either the T or C isomers. To put it crudely, G will be less "forbidden" than either T or C and, hence, more stable because of ligand segregation which maximizes ligand orbital symmetry reduction.

These considerations are excellently projected in a pictorial format by the hybrid bond diagrammatic description of the sigma bonding of G and C CHF = CHF shown in Figure 4. Because the ratio λ_D/λ_U is greater in the G isomer due to the nature of the ligand MO's, we say that the former is more D-bound than the latter and, hence, more stable. In addition, the fact that λ_D/λ_U ratio is larger in the G isomer, has the following consequences:

a. There is more charge transfer from core to fluorine ligands in the case of the G isomer.

b. Because charge transfer from core to fluorines depopulates predominantly to best "donor orbital", ω_5, which is antibonding with respect to the two carbon atoms, the C - C bond length will be shorter in the G isomer.

c. Since the two fluorines utilize primarily the ω_4 and ω_5 while the two hydrogens use mainly the ω_2 and ω_3 core MO's for binding, it follows that the $\widehat{FCF}$ angle will shrink while the $\widehat{HCH}$ angle will open up in comparison to the $\widehat{HCH}$ bond angle of ethylene. Computational and experimental results are in excellent agreement with these predictions:

1) According to computations, the F's are more negative in G than in either C or T CHF = CHF.[8]

2) The C = C bond is shorter in G than in either C or T CHF = CHF. The data (in Å) are shown below.[9]

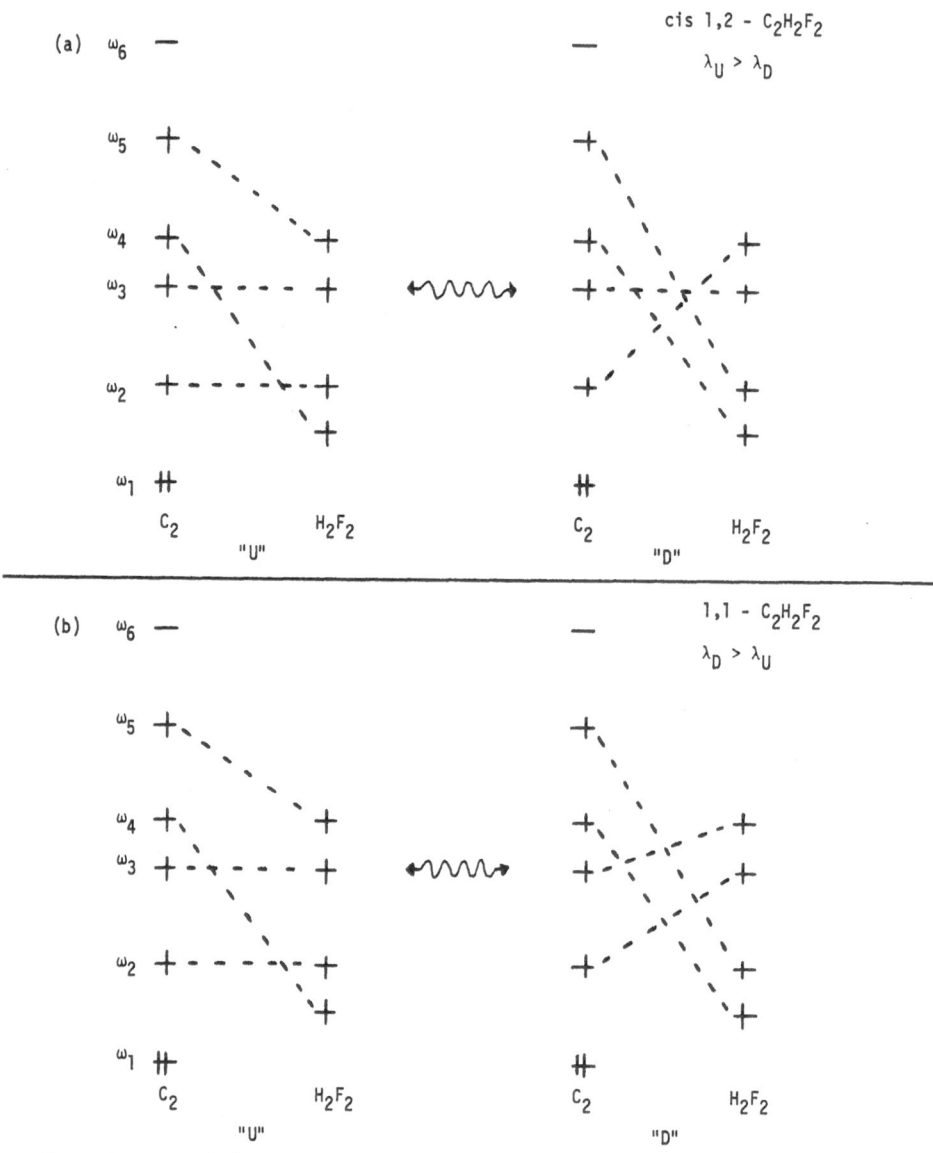

Figure 4.   a. Bond diagrammatic representation of sigma bonding in cis 1,2-difluorethylene. Because of the mixing of the H₂ and F₂ ligand MO's,"U" is the dominant contributor.   b. Bond diagrammatic representation of sigma bonding 1,1- difluoroethylene.   Because of the poor mixing of the H₂ and F₂ ligand MO's, "D" is now the dominant contributor.

F 1.315 H
109.1° C=C
F   H

F 1.331 F
C=C
H   H

F 1.329 H
C=C
H   F

H 1.339 H
117.8° C=C
H   H

3) The $\hat{FCF}$ angle of $F_2C = CH_2$ is much smaller than the $\hat{HCH}$ angle of $H_2C = CH_2$[9] and this result is reproduced by <u>ab initio</u> computations which artificially do not include the fluorine lone pairs.[10]

In the analysis presented above, we have neglected the fluorine lone pairs, i.e., we have treated CHF = CHF as CHX = CHX with X having greater electronegativity than H and being devoid of electron pairs. Using the same arguments as in the previous chapter, it can be shown that any sigma electon pairs of X will act in a way which depends on the electronegativity of X itself since this determines the extent of X electron pair delocalization. As X becomes increasingly electronegative, delocalization favoring the G isomer will be accentuated and G will become increasingly stabilized relative to C. By contrast, as X becomes increasingly electropositive, delocalization will be turned off and nonbonded overlap repulsion will assume a prominent role increasingly destabilizing the G form. In accord with these predictions, $F_2C = CH_2$ is computed to be very much more (>3kcal/mol) stable than CHF = CHF (cis) while $(CH_3)_2 C = CH_2$ is only 1.37 kcal/mol more stable than $CH_3CH = CHCH_3$ (trans).

How do pi electron pairs of X control the relative stability of the G and C forms? The bond diagrams of Figure 5 give an immediate answer: G has one four-electron-three-orbital bond and zero antibonds. By contrast, C has one two-electron bond and a four-electron antibond. Hence, pi effects favor G over C.

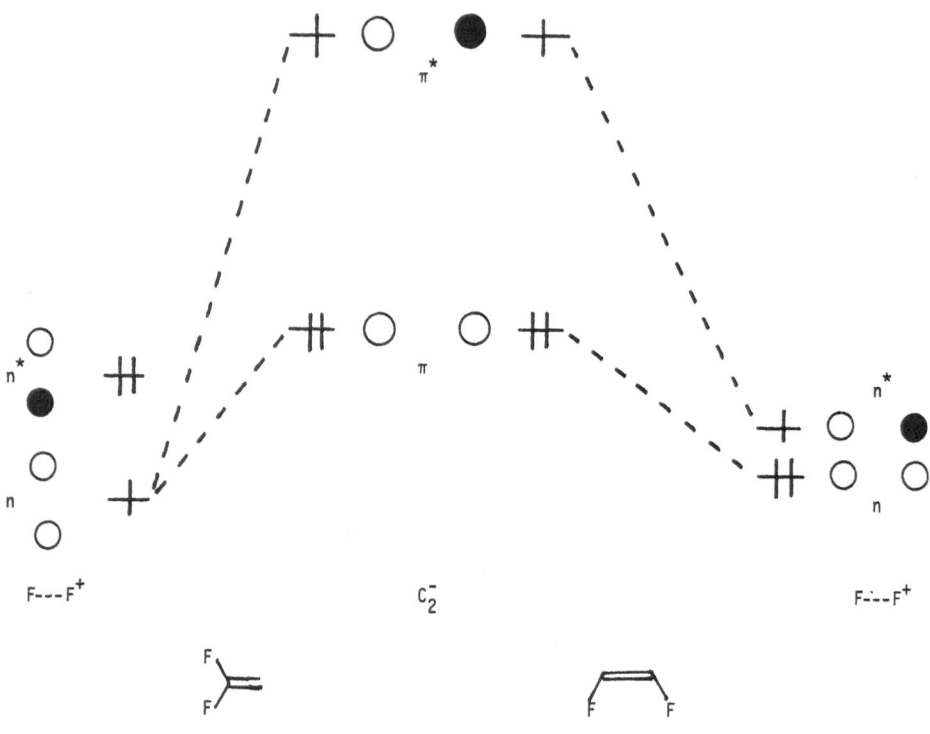

Figure 5. Pi bond diagrams of 1,1- and cis-1,2-difluoroethylene. The 1,1 isomer is free of the four-electron antibond present in the cis-1,2 isomer and it has a delocalized four-electron-three-orbital rather than a single two-electron core-ligand bond.

III. Computational Tests

As we have seen, the relative energies of trans-1,2, cis-1,2, and 1,1 $C_2H_2F_2$ are dictated by orbital symmetry. Since parity stereoselection is "contained" within HMO and EHMO theory, we expect that computations of $C_2H_2H'_2$, where H' is an electronegative pseudo atom having the VOIE of fluorine, will clearly reproduce the relative energies of the three isomers found by experiment and high level computations, in a qualitative sense. The data shown in Table 1 leave no doubt that orbital symmetry is responsible for sigma bonding which gets progressively better as we go from trans to cis to 1,1 difluoroethylene. Since parity stereo- selection is rooted in the determinantal nature of the electronic wavefunction and can never be annihilated but only attenuated by substituent effects, it is gratifying to see that the same stereochemical preferences are manifested when H' is an electropositive pseudo atom. So, here is a problem which can be easily understood at the level of EHMOVB theory and which, when dealt with in the confines of the equivalent EHMO theory, becomes a cause of controversy simply because of the conceptual intractability of MO theory, in general.

Table 1. Extended Hückel Energies (eV) of the Three Isomers of $C_2H_2H'_2$.[a]

VOIE H'[b] (eV)	Cis	Trans	Geminal
- 3.6	-221.8106	-221.5298	-221.9644
-13.6	-228.1522	-228.1522	-228.1522
-23.6	-248.9464	-248.5918	-249.3958
-33.6	-373.6960	-373.5950	-373.8248

[a] Standard geometric parameters were used throughout.

[b] The off-diagonal matrix elements were kept at the values they have for the VOIE of H' being -13.6 eV.

References

1. Epiotis, N.D.; Larson, J.R.; Yates, R.L.; Cherry, W.R.; Shaik, S.; Bernardi, F. J. Am. Chem. Soc. 1977, 99, 7460.

2. See the previous chapter.

3. Epiotis, N.D.; Cherry; W.R.; Shaik, S.; Yates, R.L.; Bernardi, F. Top. Curr. Chem. 1977, 70, 1.

4. Cox, J.P.; Pilcher, G. "Thermochemistry of Organic and Organometallic Compounds"; Academic Press: New York, 1970.

5. (a) See Ref. 3.

   (b) Brunck, T.K.; Weinhold, F. J. Am. Chem. Soc. 1979, 101, 1700.

6. (a) Kollman, P. J. Am. Chem. Soc. 1974, 96, 4362.

   (b) Greenberg, A.; Sprouse, S.D.; Liebman, J., to be published.

7. Epiotis, N.D.; Larson, J.R.; Eaton, H. "Unified Valence Bond Theory of Electronic Structure" in Lecture Notes in Chemistry, Vol. 29; Springer-Verlag: New York and Berlin, 1982.

8. This fact is borne out by practially all kinds of calculations: Epiotis, N.D. unpublished results. See also Ref. 1.

9. Callomon, J.H.; Hirota, E.; Kuchitsu, K.; Lafferty, W.J.; Maki, A.G.; Pote, C.S. in Landolt-Bornstein "Numerical Data and Function Relationships in Science and Technology" Vol. 7., New Series, "Structure Data on Free Polyatomic Molecules"; K. H. Hellwege, Ed.; Springer-Verlag: West Berlin, 1976.

10. For such a "computational experiment" see Ref. 6a.

Chapter 9. The Saga of "Hypervalent" Molecules.

In a previous work, we have outlined the Molecular Orbital-Valence Bond (MOVB) theory of chemical bonding, based on the core(C)-ligand(L) dissection.[1] In previous chapters, we have applied this brand of theory to a variety of structural problems with the aim to demonstrate its conceptual and formal advantages over previous and current qualitative theoretical approaches to bonding. We now observe that the Core-Ligand (C-L) dissection allows a classification of molecules which contain an even number of valence electrons into four major types depending upon the presence or absence of electron pairs or holes in the core and/or ligand fragments in the <u>perfect pairing</u> (R) Configuration Wavefunction (CW) representing the entire system. The four categories of molecules are the following:

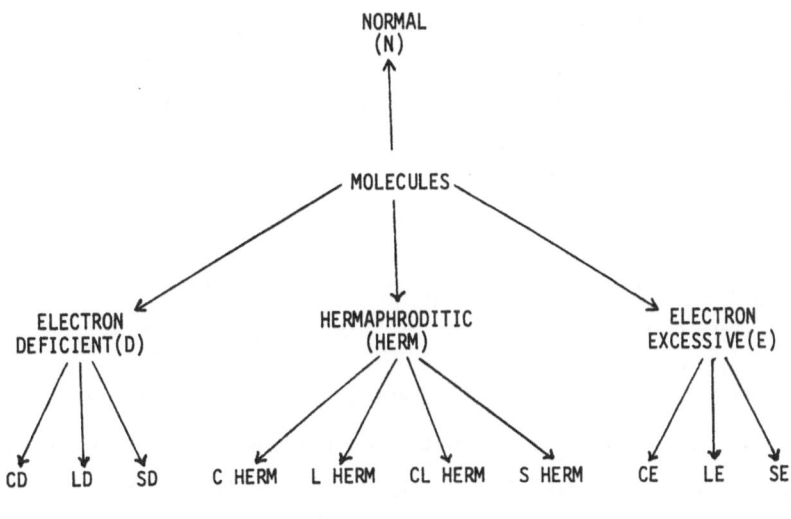

C = Core, L = Ligand, S = System

A CD molecule has one or more vacant orbitals (holes) in the core, an LD molecule has one or more holes in the ligands, and an SD molecule has holes in both the core and ligands. A CE molecule has one (or more) doubly occupied orbital (electron pair) in the core, an LE molecule has one (or more) electron pair in the ligands, and an SE molecule has electron pairs in both core and ligands. A C-HERM molecule contains holes and pairs in the core, an L-HERM molecule has holes and pairs in the ligands, a CL-HERM molecule contains core holes and ligand pairs, or vice versa, and an S-HERM molecule has holes and pairs in both core and ligands. Finally, an N molecule has no pairs or holes in the entire system. Needless to say, this classification is model dependent, i.e., it is inspired by the C-L dissection. We emphasize that the counting of holes and electron pairs is always made by reference to the R CW. An illustrative example of a CE system is given below, with Γ_m being the orbital symmetry label.

$$\Gamma_c \; + \cdots \cdots + \; \Gamma_c \qquad\qquad \Gamma_c \; + \qquad\quad + \; \Gamma_c$$

$$\Gamma_b \; + \cdots \cdots + \; \Gamma_b \qquad\qquad \Gamma_b \; + \qquad\quad + \; \Gamma_b$$

$$\Gamma_a \; \text{⧺} \qquad\qquad\qquad\qquad \Gamma_a \; \text{⧺}$$

| C | L | | C | L |

Bond Diagram   R CW

It is evident that construction of a bond diagram, a direct projector of the R CW, immediately identifies a molecule as one of the eleven possible types defined above.

Up until this point, we have dealt mostly with CE, CD, C-HERM, and N systems. An example of a CE system is H_2O, an example of a CD system is BeH_2, an example of a C-HERM system is C_2H_2, and an example of an N system is CH_4.[1] We now turn our attention to other types of molecules noting that the term "electron excessive", or, "hypervalent" is used by chemists in order to single out LE and SE molecules and the term "electron deficient" is used in order to denote CD, LD, and SD molecules. As a result, the impression is created that such molecular species are exceptional. Indeed, molecules of these types have been treated as entities subject to bonding rules which are different from those "for conventional" molecules! These differences are only apparent and they simply disappear at the level of a unified theory of electronic structure such as MOVB theory. Using this theoretical apparatus, we expect to understand the stereochemistry of "hypervalent" molecules using exactly the same concepts that we applied to the stereochemistry of "conventional" molecules. Because of the nature of the problem itself as well as because it constitutes a linguistic improvement, we shall henceforth adopt the convention of writing detailed bond diagrams so that the electron distribution is identical to the one predicted by the resonance bond diagram which we surmise that it is the major contributor of the MOVB wavefunction. For example, if Ξ_2 is the major contributor of the four-electron - three-orbital wavefunction in the hybrid bond diagrammatic representations, the detailed MOVB diagram will be written as indicated below.

HYBRID REPRESENTATION

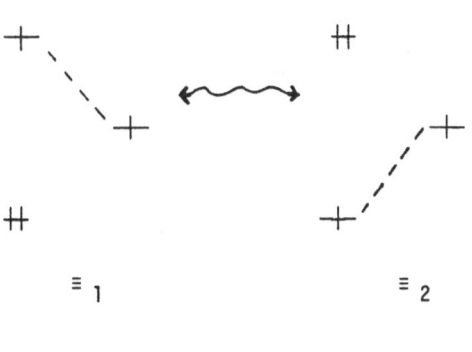

$$\lambda_2 > \lambda_1$$

DETAILED MOVB DIAGRAM

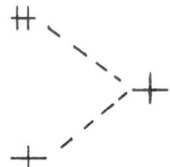

In this way, the detailed MOVB diagram projects very explicitly the <u>lowest</u> <u>energy</u> perfect pairing CW of the system in question.

In this chapter, we focus our attention primarily on SE molecules. However, instead of examining each known SE species in detail, we confine our attention to only one species, namely, SX_4 (X=H,F) in order to demonstrate with clarity how primitive our understanding of SE "hypervalent" molecules, in general, has been in the past. In order to project clearly the conceptual advantages of MOVB over MO theory, we must walk down the same alley as other invetigators have done, only now substituting their MO theoretical vehicle by corresponding MOVB theory. Since the most thorough qualitative analysis of SX_4 has been that of Chen and Hoffmann[2a] and since their treatise was based on Extended Hückel MO (EHMO) theory,[2b-2d] we first use EHMOVB theory, a brand of MOVB theory which is exactly equivalent to EHMO theory, in order to analyze the electronic structure of SX_4. Subsequently, we revert to MOVB theory and examine how the overview is modified as a result of transition to a higher level of theory.

I. The Model Problem

In theoretical studies of hypervalent SE AX_n molecules, it has been routinely assumed that AH_n adequately models AF_n regardless of the nature of A. To put it crudely, theoreticians have implicity assumed that understanding SH_4 means understanding SF_4, SeF_4, and TeF_4. These assumptions can now be proclaimed risky simply because, in a previous chapter, we have shown that F differs from H in terms of nonbonded AO overlap, As differs from N in terms of indirectly generating smaller nonbonded AO overlap, and that this nonbonded AO overlap is a primary determinant of molecular stereochemisty in A_2X_4 systems with 14 valence electrons. Thus, we attributed the gauche preference of N_2H_4 to large geminal H---H overlap, the diminished gauche preference of N_2F_4 to reduced geminal F---F overlap, and the trans preference of As_2R_4 to the very small geminal R---R overlap. It follows then that SH_4 is a poor model of SF_4 and SF_4 a poor model of SeF_4. Furthermore, since the gross three dimensional shape of AF_4 molecules, where A belongs to the VII B family (S, Se, Te, Po), is roughly invariant of A, there is the possibility of the following devilish coincidence: $\underline{SH_4, SF_4, SeF_4 \text{ and all other related}}$ $\underline{\text{molecules have the same gross stereochemistry but for entirely different reasons}}$. The experimentally determined geometries of SF_4 and SeF_4 are shown below.[3]

186.9°

F ———— S ———— F (1.646Å)

F   F (1.545Å)

101.6°

190.8°

F ————Se———— F (1.771Å)

F   F (1.682Å)

100.6°

With these ideas in mind, we now try to solve two different stereochemical problems at the EHMOVB level:

a. The stereochemisty of SH_4, where we assume that nonbonded H1s---H1s overlap is substantial (see Table 1).

b. The stereochemisty of SF_4, where we treat F as a monovalent electronegative ligand by neglecting the lone pairs and we assume that nonbonded F2p---F2p overlap is zero (see Table 1).

In doing so, we demand that the theory provides straightforward answers to the following two key questions:

a. Which of the four different geometries of SX_4 shown below is the lowest energy one, <u>assuming that all S-X bond lengths are identical</u>?

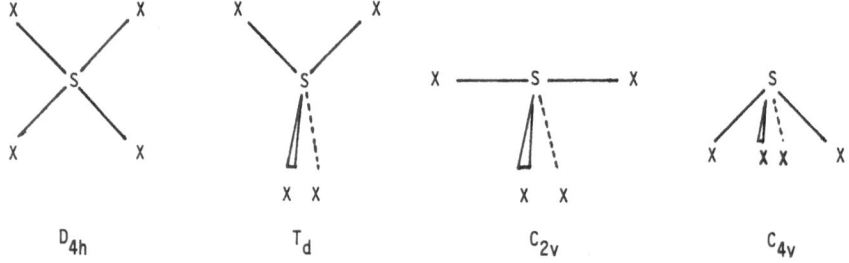

b. What, if any, is the motivation for linear and angular deformation of the C_{2v} structure?

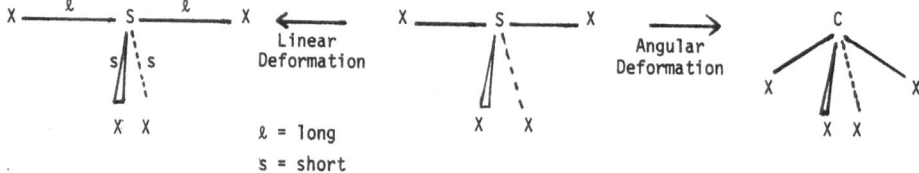

In all following discussions, we use the core-ligand dissection according to which an AX_n molecule is viewed as a core, A, and a ligand, X_n, fragment, joined by core-ligand bonds and antibonds.

Table 1. Geminal Nonbonded AO Overlap in Planar SX_4 (X=H,F).

A.   X = H

	H1s
H1s	.24436

B.   X = F

	F2s	$F2p_x$	$F2p_y$	$F2p_z$
F2s	.00873			
$F2p_x$		.00241		
$F2p_y$			.00241	
$F2p_z$	.01265			.01812

II. EHMOVB Theory of SH$_4$

An MOVB bond diagram immediately conveys the following important information:

a. The excitation requirement for bond formation.

b. The number of bonds and antibonds.

c. The strength of each bond.

On this basis, comparison of different MOVB diagrams corresponding to different geometries of one and the same molecule leads to the direct prediction of the optimum geometry of the molecule itself. The bond diagrams for D_{4h}, T_d, C_{2v}, and C_{4v} SH$_4$ are shown in Figures 1 and 2 and the self-evident conclusions can be stated as follows:

a. D_{4h} requires a very large (3s $\longrightarrow$ 3p) core excitation for the formation of three core-ligand bonds.

b. T_d has three two-electron core-ligand bonds and one four-electron core-ligand antibond but requires no core excitation.

c. C_{2v} has three core-ligand bonds and a <u>partial</u> four-electron core-ligand antibond, as revealed by the resonance bond diagrammatic representation of this system shown in Figure 3. The hallmark of the C_{2v} structure is prevention of substantial core-ligand overlap repulsion by accomodation of the ligand electron pair in a ligand orbital, σ_4, which overlaps only weakly with the core orbital which houses the core electron pair, 3s, at the expense of modest excitation energy. This is the reason why Ξ_1 is the dominant contributor to the MOVB wavefunction of C_{2v} SH$_4$.

d. C_{4v} has three core-ligand bonds, zero core-ligand antibonds, and requires no core excitation.

Neglecting, for the time being, the effect of nonbonded ligand interaction, we conclude that the preferred geometry will be the C_{4v} one.

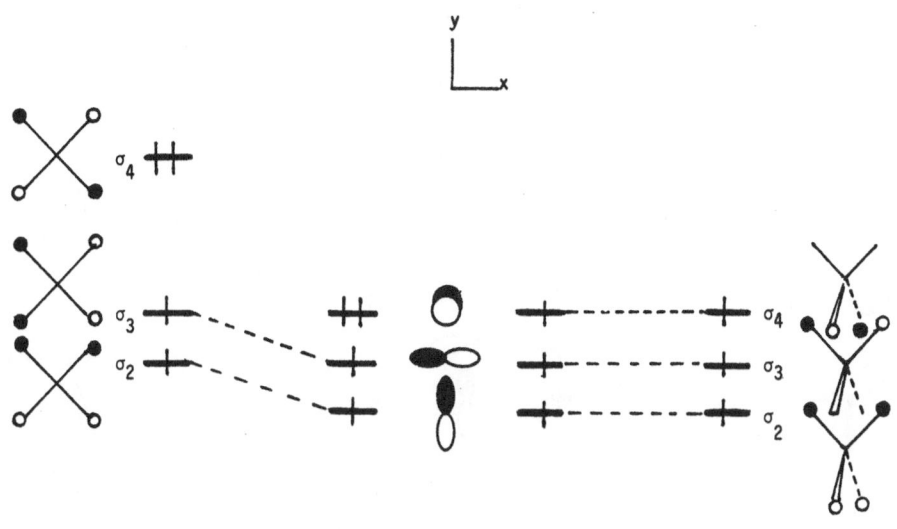

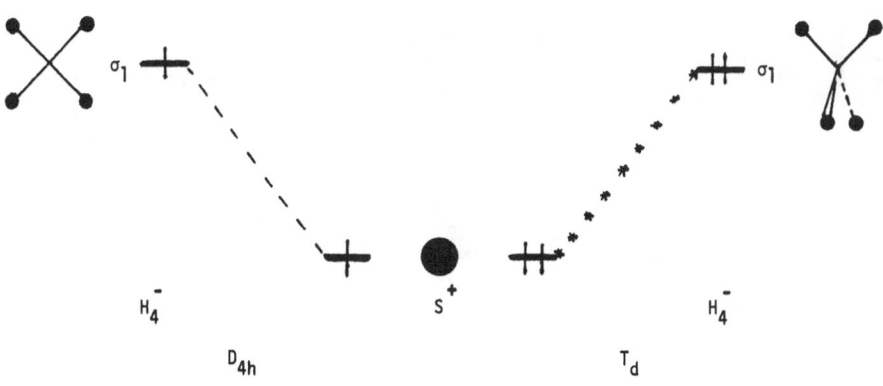

Figure 1. Detailed bond diagrams for planar (D_{4h}) and tetrahedral (T_d) SH_4.

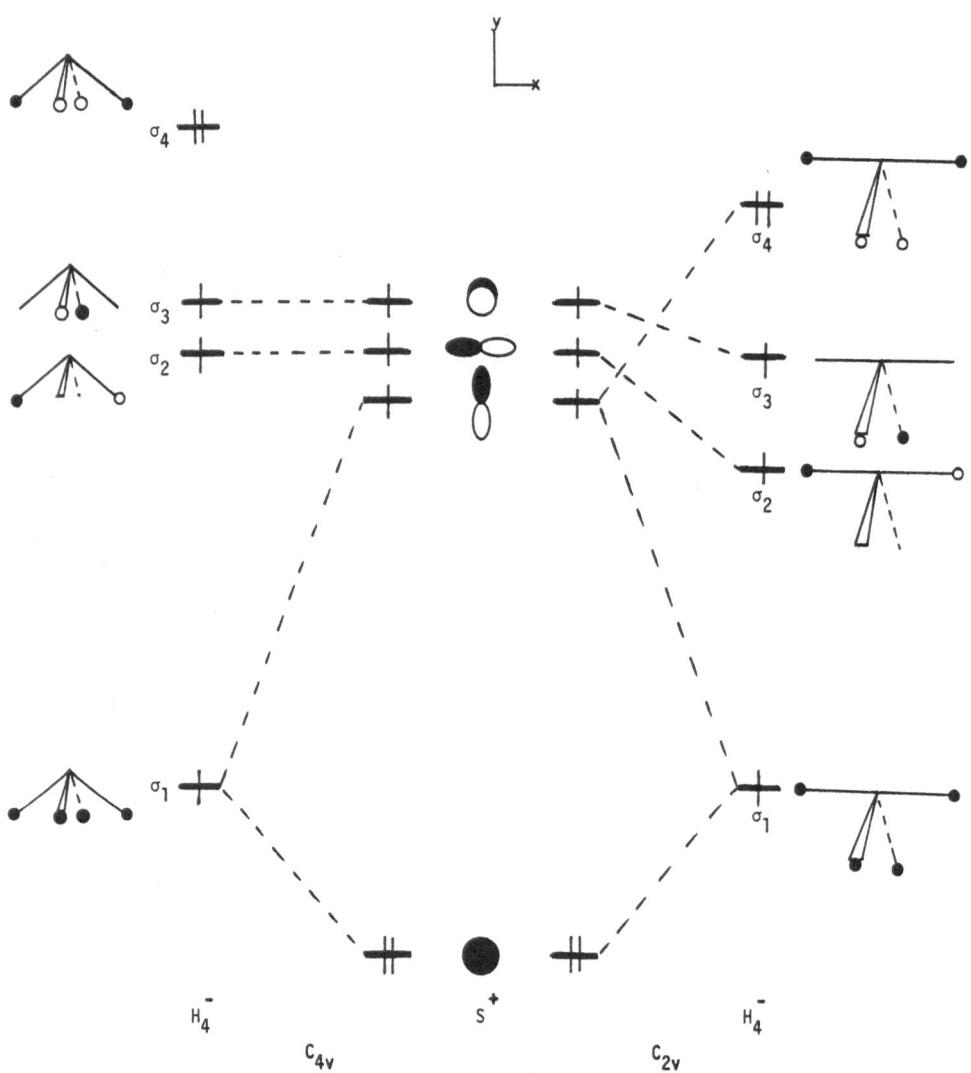

Figure 2. Detailed bond diagrams for pyramidal (C_{4v}) and C_{2v} SH_4. In the C_{2v} form, it is assumed that the nonzero $3s$-σ_4 overlap integral is negligible.

C_{2v} SH_4

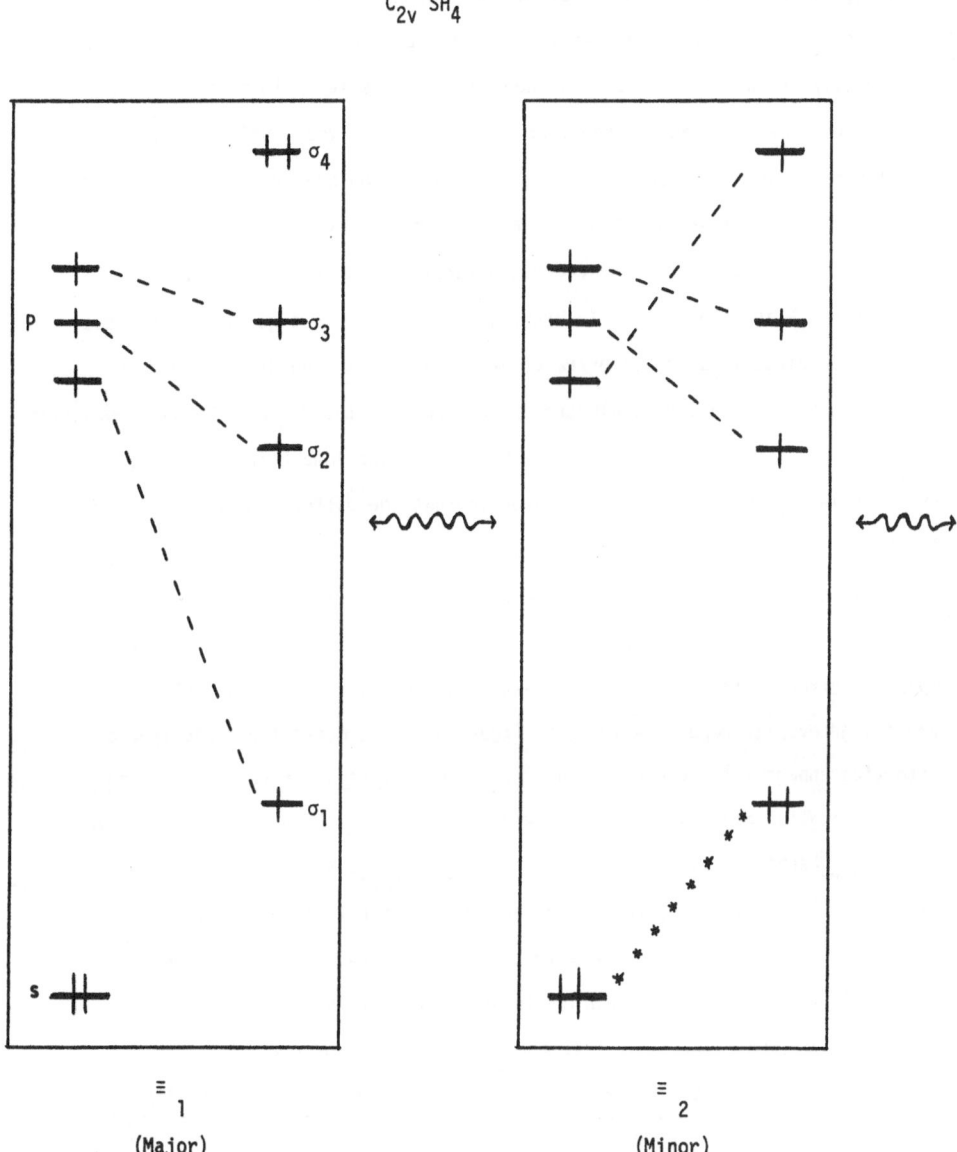

Figure 3. Resonance (or, hybrid) bond diagrammatic representation of C_{2v} SH_4. Asterisks indicate a four-electron antibond. In $\equiv_1$, stronger core-ligand bonding is obtained at the expense of excitation energy. The reverse occurs in $\equiv_2$.

Next, let us consider the effect of the substantial H1s--H1s nonbonded overlap. We note that in all four geometries there exists a ligand electron pair and that the stability of this pair depends on nonbonded AO overlap. In a more general sense, we can say that each of the four geometries involves a different degree of natural Ligand Nonbonded Repulsion (LNR). The question now arises: Which of the four different geometries can best sustain a distortion which takes the nonbonded ligands away from each other and reduces natural LNR without significantly impairing core-ligand bonding? In attempting to answer this question, we discover that C_{2v} is unique in a way illustrated by comparing C_{2v} and C_{4v}. In both geometries, there is a ligand electron pair occupying the σ_4 MO. However, reduction of the energy of the ligand orbital which contains the lone pair is much more favorable in C_{2v} than in C_{4v}. The reason is that the distortion necessary for stabilizing the σ_4 orbital entails the weakening of only one core-ligand bond in C_{2v} while it tends to destroy all three core-ligand bonds in C_{4v}. This point is illustrated in Figure 4. In addition, linear distortion in C_{2v} has another very important effect: It diminishes the $3s-\sigma_1$ overlap integral thus alleviating core-ligand overlap repulsion in $\equiv_2$ in Figure 3. This means that this type of distortion enhances the contribution of $\equiv_2$, i.e., <u>it causes electron deexcitation. This deexcitation mechanism exists only because appreciable nonbonded AO overlap gives rise to</u> substantial energy splitting of the ligand MO's. It should be noted that σ_1 as well as σ_4 belong to the same C_{2v} irreducible representation so that point group theoretical arguments by themselves would fail to reveal the important fact that $3s-\sigma_4$ overlap is near zero but $3s-\sigma_1$ overlap is large though the overlapping orbitals have congruent symmetry in both cases. Hence, it is predicted that C_{2v} will undergo linear distortion in order to stabilize the σ_4 electron pair as well as angular distortion in order to minimize core-ligand

σ_4: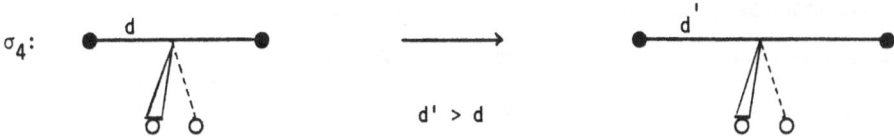

$$d' > d$$

Linear distortion lowers energy of σ_4 and electron pair it contains in addition to promoting ligand deexcitation. It also weakens the $2p_x - \sigma_2$ bond (Figure 2).

σ_4: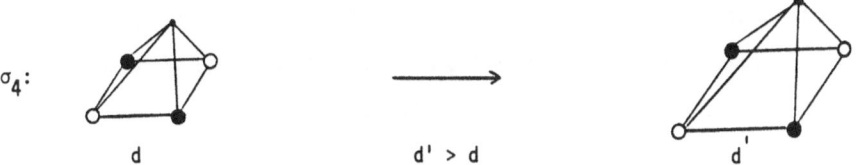

$$d' > d$$

Distortion lowers energy of σ_4 and electron pair it contains. However, it also weakens all bonds (Figure 2).

Figure 4. The energetic consequences of C_{2v} and C_{4v} distortions.

overlap repulsion due to the interaction of the core and ligand lone-pairs. The angular distortion expresses the desire of the molecule to achieve a C_{4v} geometry in which a core-ligand antibond ceases to exist. The final distorted C_{2v} geometry is the best compromise geometry which allows for ligand electron pair stabilization and avoidance of significant core-ligand overlap repulsion.

At this point, we realize that, because chemical stereoselection in ground state molecules, such as SH_4, is the final resolution of "conflicts of interest" and because EHMOVB theory is empirical and, thus, parametrization dependent, EHMOVB computations ought to be analyzed with extreme care. For example, in the case at hand, the choice of the Slater exponents, ζ, plays a pivotal role because a "small" ζ for H1s generates strong core-ligand interaction while a "large" ζ has the opposite effect. As core-ligand overlap interaction increases as a result of sequential modification of the EHMOVB parameters and, in particular, the Slater exponents, the geometries which do not involve core-ligand antibonds will tend to be favored over the ones that they do. Since the "perfect" parameterization is undefined, EHMOVB computations can produce different sets of results depending on the parameter input. Thus, instead of seeking the "best" parameterization and becoming entangled in empirical gamesmanship, we are content to say that "perfect" EHMOVB theory will most likely predict a C_{4v} geometry for SH_4 with the possibility existing that the C_{2v} structure will become optimum if nonbonded ligand AO overlap is large. In other words, there is a plausible EHMOVB scenario according to which SH_4 is C_{2v} in order to best accomodate the ligand electron pair.

III. EHMOVB Theory of SF$_4$

In SF$_4$, the reduced nonbonded overlap of the fluorine 2p AO's changes the problem to the extent that one can now assume that nonbonded ligand AO overlap plays only a minor role and that it can be neglected. As a result, and assuming further that each fluorine can be treated as a monovalent ligand devoid of electron pairs, the bond diagrams for the various geometries of SF$_4$ will be the same as those for SH$_4$ with one major difference. The ligand MO's will be degenerate. The absence of ligand nonbonded interaction now guarantees that the sole determinant of molecular stereochemistry will be core-ligand overlap interaction. It follows that the C$_{4v}$ geometry will attain lower energy than the C$_{2v}$ geometry and that there will be no longer any great motivation for linear distortion of C$_{2v}$ since the potential energetic superiority of C$_{2v}$ SH$_4$ at the EHMOVB level was traced to a more effective stabilization of the ligand lone pair via reduction of nonbonded AO overlap and this approaches zero in SF$_4$ itself. In saying this, it is important to remember that the bonds of C$_{2v}$ SX$_4$ are made in such a way so that there will always be a difference between "axial" and "equatorial" bonds on pure symmetry grounds. It is the sign and magnitude of this difference (large in SH$_4$ and small in SF$_4$) which is at the core of our argument here.

We now reconstruct the bond diagrams of T$_d$, C$_{2v}$, and C$_{4v}$ SF$_4$ under the assumption that the ligand MO's are degenerate in order to demonstrate the following:

a. T$_d$ is an "antiaromatic"-like, C$_{2v}$ a "nonaromatic"-like and C$_{4v}$ an "aromatic"-like structure.

b. The electron distribution in C$_{2v}$ is such so that the axial fluorines are electron richer than the equatorial ones.

The bond diagrams shown in Figure 5 have a common denominator: Each is made up of two two-electron-two-orbital (2e-2o) subsystems and one six-electron-four-orbital

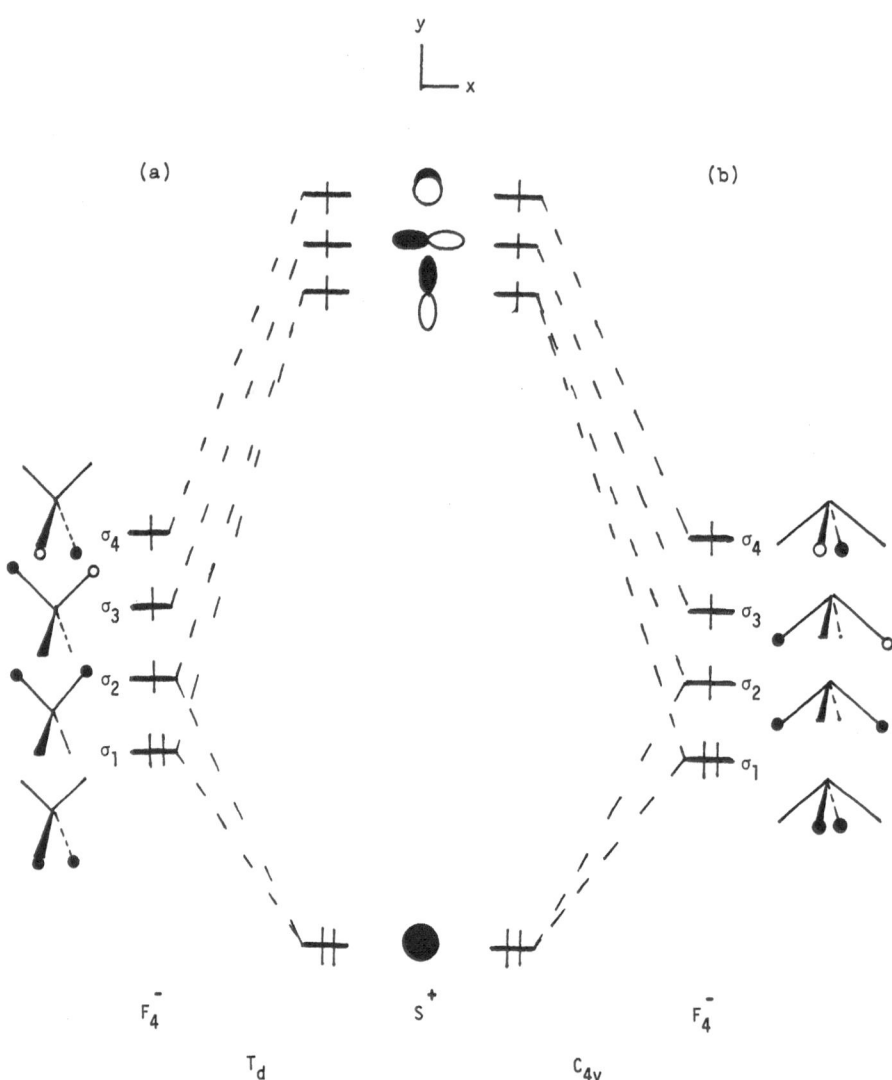

Figure 5. Detailed bond diagrams of T_d, C_{4v}, and C_{2v} SF$_4$ in which it is assumed that nonbonded AO overlap is zero. Also, F lone pairs are neglected.

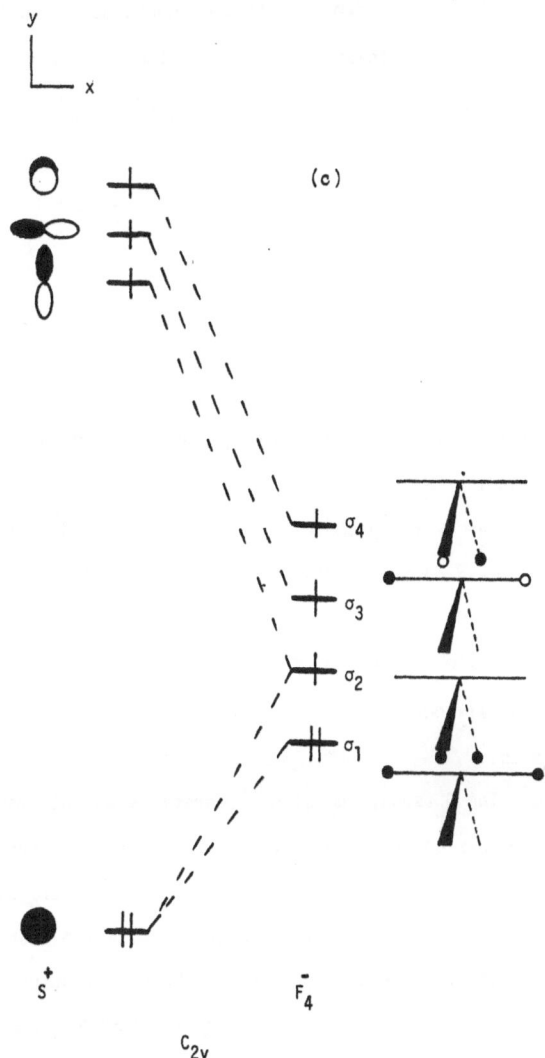

Figure 5. Detailed bond diagrams of T_d, C_{4v}, and C_{2v} SF_4 in which it is assumed that nonbonded AO overlap is zero. Also, F lone pairs are neglected.

(6e-4o) subsystem. According to the Independent Bond Model (IBM)[1], the total wave-function of each species, Ψ, can be written as a product of three subsystem wave-functions, Ω_A ($3p_x$, $\sigma_3/2$ subsystem), Ω_B ($3p_z$, $\sigma_4/2$ subsystem), and Ω_C ($3s$, $3p_y$, σ_1, $\sigma_2/6$ subsystem):

$$\Psi = \Omega_A \cdot \Omega_B \cdot \Omega_C$$

and

$$E = E_A + E_B + E_C$$

The critical difference between T_d and C_{2v} and between C_{2v} and C_{4v} lies in the corresponding Ω_C subsystems. We recognize the following:

a. The Ω_C subsystem of T_d involves cyclic Möbius CW interaction because the four orbitals define a Möbius AO cycle containing six electrons. This is shown in Figure 6a.

b. The Ω_C subsystem of C_{4v} involves cyclic Hückel CW interaction because the four orbitals define a Hückel AO cycle containing six electrons. This is shown in Figure 6b.

c. The Ω_C subsystem of C_{2v} represents an intermediate situation because the four orbitals do not form a cyclic array. This is shown in Figure 6c.

Thus, we have been able to reduce the problem of SF_4 stereoselection to the funda-mental problem of parity stereoselection with the unshakable conclusion that, in the absence of nonbonded AO overlap, the C_{4v} structure is the uncontested "aromatic"-like global energy minimum of SF_4, at the EHMOVB (=EHMO) level of theory.

What are the electronic distributions of C_{2v} and C_{4v} SF_4? The compact diagrams shown in Figures 5b and 5c make evident the following:

a. In C_{2v}, the ligand lone pair is mainly contained within an axial MO. Hence, the axial fluorines will be more negatively charged than the equatorial ones.

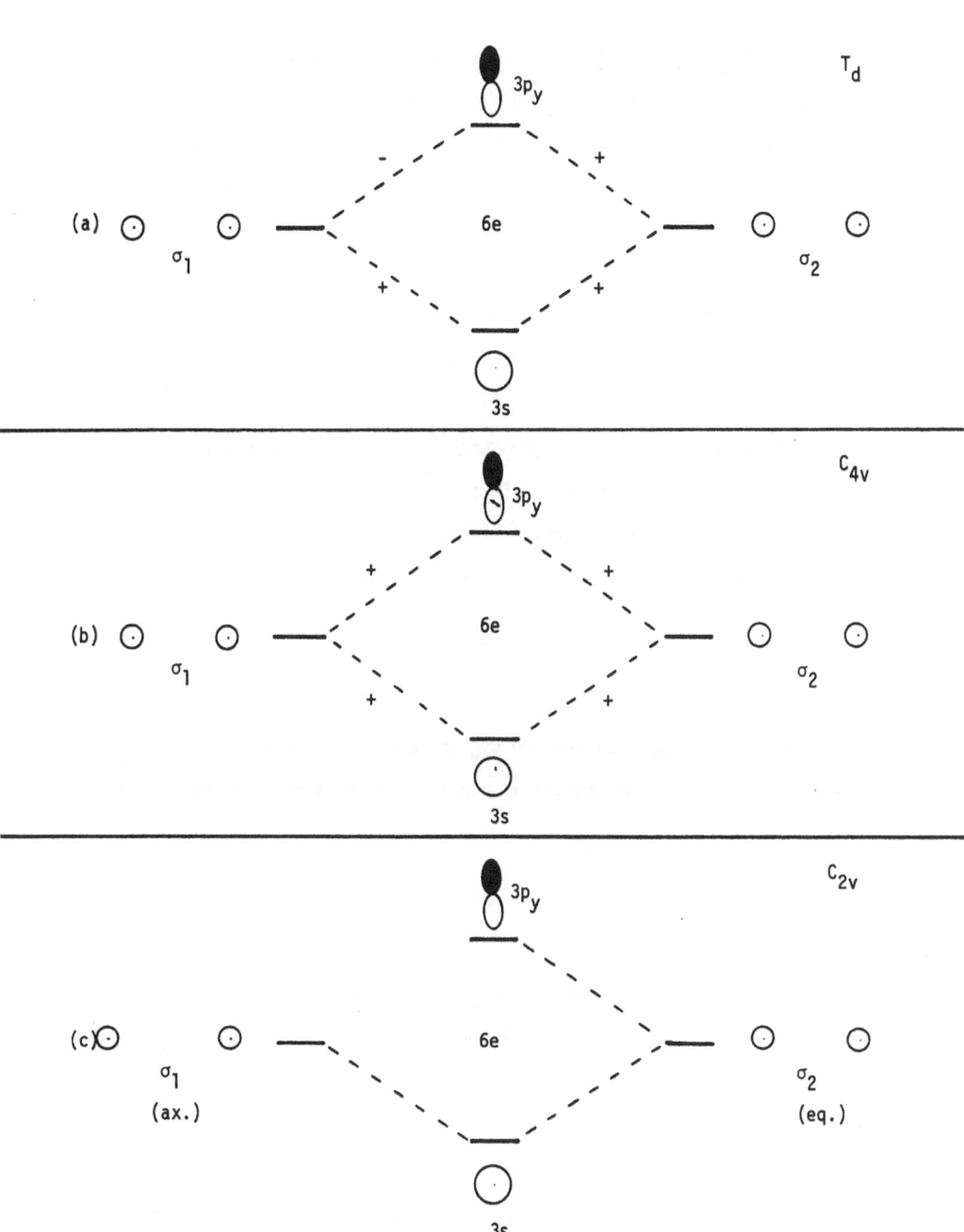

Figure 6. The mode of six-electron-four-orbital interaction in T_d, C_{4v}, and C_{2v} SF_4.

b. In C_{4v}, all fluorines will have an equal negative charge by symmetry.

We conclude that SF_4 will be C_{4v} at the level of EHMOVB (=EHMO) theory because of fundamental laws of parity stereoselection. This is in distinct contradiction to the results of experimental studies which show that SF_4 has a distorted C_{2v} structure!

The prediction of EHMOVB theory that, if nonbonded AO overlap becomes diminished, the C_{4v} will become more favorable than the C_{2v} geometry has been subjected to a computational test. By using the same parameterization, we found the following:

a. SH_4 prefers a C_{2v} over a C_{4v} geometry because of reduced natural LNR traceable to the ligand lone pair, even if all S-H bond lengths are taken to be the same. Furthermore, linear distortion of the C_{2v} further reduces the energy of this species. The computed relative energies (in eV) are given below.

	C_{2v}	$C_{2v}(S-H_{ax}>S-H_{eq})$	C_{4v}	T_d
Rel. E	0.00	-1.97	+9.00	+14.40

b. Replacement of H by F causes the C_{4v} to become more stable than the C_{2v} form because F...F nonbonded AO overlap is about twenty-fold smaller than H...H nonbonded AO overlap (See Table 1).

	C_{2v}	$C_{2v}(S-H_{ax}>S-H_{eq})$	C_{4v}	T_d
Rel. E	0.00	-3.13	-1.91	+6.91

These results leave little doubt that the proper rationalization of the shape of SF_4 and all SE molecules is "outside" EHMO theory. The reason why the C_{2v} distortion tendency is greater in SF_4 than in SH_4 is the presence of ligand lone pairs

which generate core-ligand antibonds with the core S3s electron pair. The reader
is reminded that in our analysis we treated fluorine as a monovalent ligand by
neglecting the F lone pairs since these introduce extra core-ligand overlap
repulsion in all geometries under consideration.

IV. MOVB Theory of SX_4 (X=H,F)

What happens when we replace EHMOVB by MOVB theory? EHMOVB theory neglects electron-electron interaction. As a result, minimization of "classical" coulomb repulsion becomes an important consideration only at the level of MOVB theory. Since, the electronegative fluorines are negatively charged as a result of the operation of "nonclassical" overlap effects "contained" in EHMOVB theory, we can say that SX_4 will attempt to minimize nonbonded coulomb repulsion by distortion and that, for reasons discussed before, only the C_{2v} structure provides an acceptable mechanism for doing so. Thus, both the electron distribution and the distortion potential of the various geometries of SF_4 are correctly predicted in a qualitative sense by EHMOVB theory but the motivation for distortion differs at the levels of EHMOVB and MOVB theory as spelled out below:

Molecule	Theory	Primary Motivation for Linear C_{2v} Distortion[*]
SH_4	EHMOVB	Minimization of nonbonded overlap repulsion.
SF_4	EHMOVB	None, assuming nonbonded AO overlap to be zero.
SH_4	MOVB	Minimization of nonbonded overlap repulsion.
SF_4	MOVB	Minimization of coulomb repulsion

[*] The effect of the F lone pairs has been neglected.

These considerations lead to the conclusion that, if complete surfaces are computed, SH_4 will be either C_{4v} or C_{2v} but SF_4 C_{4v} at the level of EHMO theory (assuming that there exists a "perfect" parametrization). By contrast, at the level of MOVB theory, there is now motivation for both SH_4 and SF_4 being C_{2v}. This forecast is made on the grounds that only C_{2v} affords an energetically inexpensive mechanism of distortion which relieves nonbonded repulsion of both "nonclassical" and "classical" character.

Thus, we argue that the fact that SeF_4 is distorted C_{2v}, coupled with the realization that the fluorines are too far apart for any appreciable nonbonded interaction, means that the stereochemistry of AF_4 (A = S, Se) "hypervalent" molecules is controlled by an interplay of "nonclassical" bonding effects, which allow ready distortions only in the C_{2v} geometry, and "classical" coulomb nonbonded repulsions which can be minimized without adversely affecting core-ligand bonding in the same geometry. In addition, we conclude that SH_4 is not at all a model of SF_4 at the EHMOVB level. Furthermore, since the electronegativity difference between S and H is much smaller than that between S and F (and Se and F), relief of couTomb nonbonded repulsion is probably not very important in SH_4. Hence, the reason why SH_4 is C_{2v} (if this is indeed the case) is different from the reason why SF_4 is C_{2v} and, as a result, SH_4 is not at all a model of SF_4 at the MOVB level.

MOVB theory is equivalent to SCF-MO-CI theory. SCF-MO theory is a constrained form of MOVB theory which is, nonetheless, acceptable for many problems of interest. Hence, the results of SCF-MO computations are reasonable tests of the MOVB theoretical analysis outlined above. The most significant finding is that SCF-MO computations have trouble deciding whether SH_4 or SF_4 has a C_{2v} or a C_{4v} structure, a testimony of the "aromatic" stabilization of C_{4v} which allows it to compete with the C_{2v} structure which affords relief of nonbonded repulsion. Thus, different basis sets yield different results for SH_4[4] and the same thing is true of SF_4.[5]

What is the "design lesson" learned from this analysis? If the core and ligand atoms have widely different electronegativity, minimization of nonbonded coulomb repulsion is best achieved by the C_{2v} geometry. By contrast, if the electronegativity difference is small, we expect core-ligand overlap to be the dominant factor causing the molecule to adopt an "aromatic" C_{4v} geometry. On this basis, we surmise that the unstable SH_4 and OF_4 species may very well be found to have C_{4v} geometry by "perfect" SCF-MO-CI or VB theory.

V. The Electronic Structure of SE "Hypervalent" Molecules

SH_4 is a poor model of SF_4 because the large nonbonded AO overlap in the former is absent in the latter. Because F---F nonbonded AO overlap is very small in SF_4 and near zero in SeF_4, we can say that the C_{2v} geometry of all AF_4 molecules (A = S, Se, Te) and their derivatives is the result of the interplay of nonbonded "classical" coulomb repulsion and "nonclassical" overlap core-ligand bonding and not the result of nonbonded "nonclassical" overlap repulsion. The existing analogies between the prototypical geometries of AF_4 (A = S, Se, Te) and those of AX_3 and AX_5 (A = Cl, Br, I) are depicted in Figure 7. The construction and inter-pretation of the bond diagrams for the protypical geometries of AX_3 and AX_5 species is an excellent exercise in MOVB theory which is left to the reader.

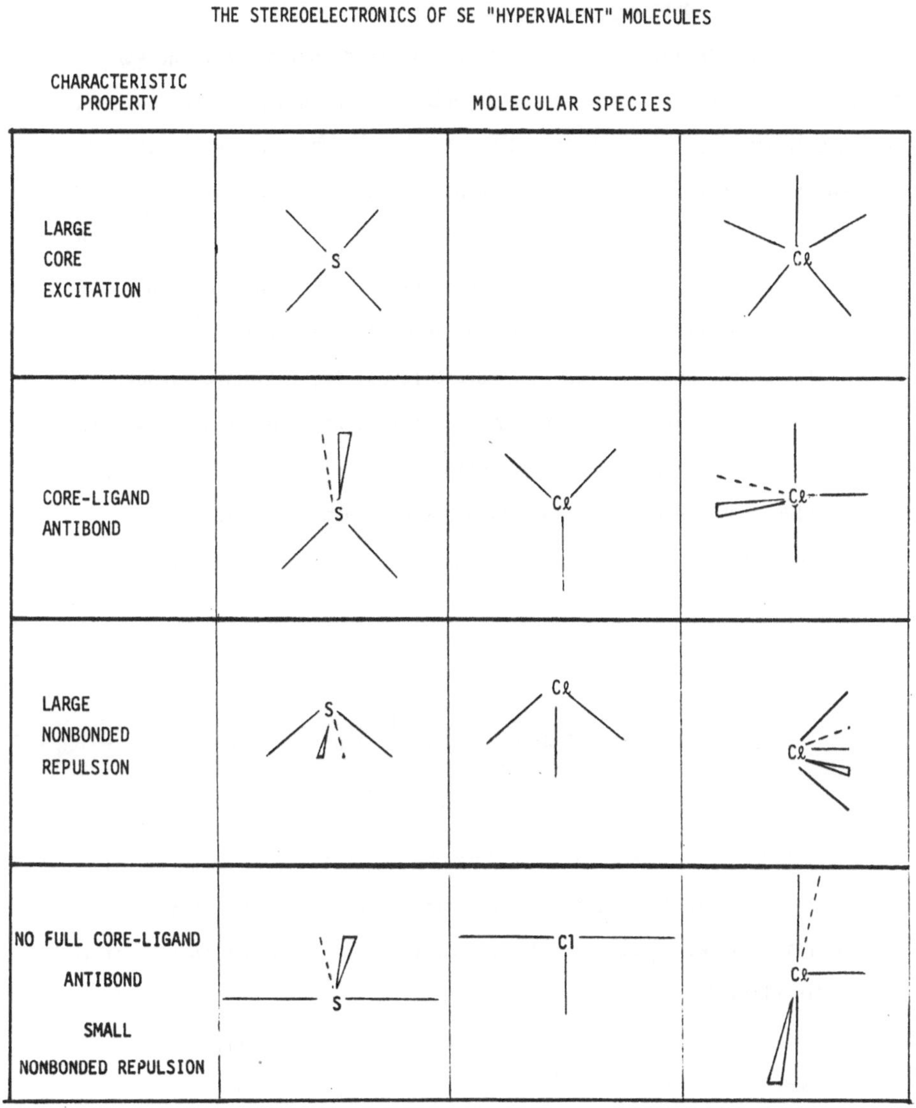

THE STEREOELECTRONICS OF SE "HYPERVALENT" MOLECULES

CHARACTERISTIC
PROPERTY   MOLECULAR SPECIES

Figure 7. Common denominators of the electronic structure of SX_4, $C\ell X_3$, and $C\ell X_5$ SE "hyper-valent" molecules.

VI. The Structures of Inert Gas Fluorides

The way in which an inert gas, such as Xe, is bound to two, four and six fluorine atoms is a topic of considerable importance which has provoked many arguments and disagreements. In order to understand exactly why XeF_2, XeF_4, and XeF_6 have the geometries that they appear to have on the basis of experimental data, it is instructive to consider bonding in the "ideal" $D_{\alpha h}$, D_{4h}, and O_h geometries for XeF_2, XeF_4, and XeF_6, respectively, with an eye towards observing common denominators (if any) and differences with respect to other "hypervalent" molecules. To this extent, the bond diagrams shown in Figure 8 reveal the following trends:

a. The extent of charge transfer from the central atom to the ligands and, hence, the approximate charge distribution within each molecule, can be predicted by simply writing down the R CW of the entire system, most likely the leading term of the complete wavefunction, and assigning charges accordingly. In this way, we predict:

$$Xe^{+1} (F_2)^{-1} \qquad Xe^{+2} (F_4)^{-2} \qquad Xe^{+3} (F_6)^{-3}$$

Indeed, the number of electrons transferred from Xe to F_n is approximately, 1 in XeF_2, 2 in XeF_4, and 3 in XeF_6 according to nonempirical SCF-MO computations.[6]

b. The ratio of two-electron bonds to four-electron antibonds in each of the three xenon fluorides is:

		XeF_2	XeF_4	XeF_6
Bonds/Antibonds	:	1	2	3
XeF Bond Length (Å)	:	1.98	1.94	1.89

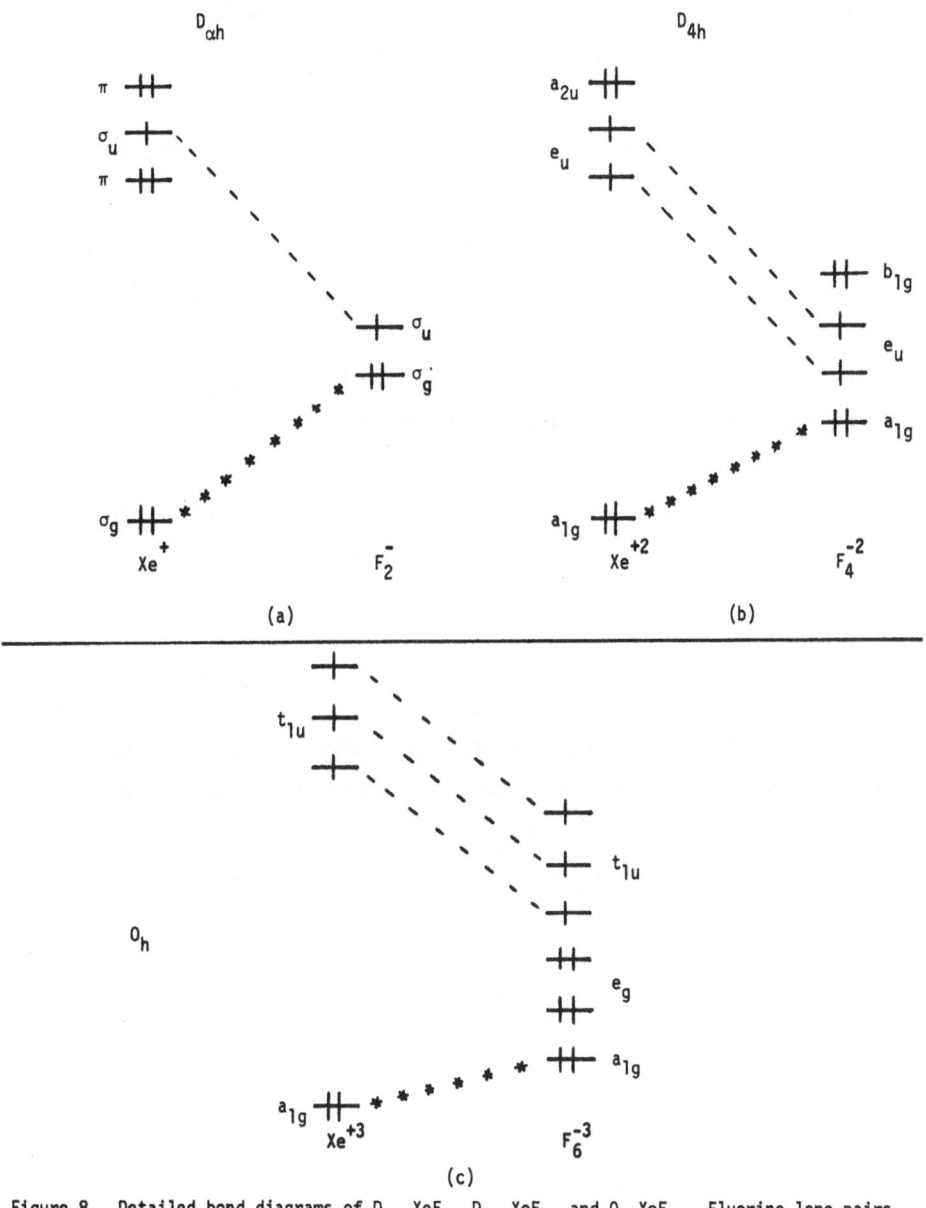

Figure 8. Detailed bond diagrams of $D_{\alpha h}$ XeF_2, D_{4h} XeF_4, and O_h XeF_6. Fluorine lone pairs are neglected.

Again, this is made immediately evident by mere inspection of the bond diagrams. This means that the Xe-F bond must decrease in length along the series $XeF_2 \longrightarrow XeF_4 \longrightarrow XeF_6$, as found by experiment.[3]

   c. All three species have a common feature: <u>One core-ligand four-electron antibond</u>. Hence, we conclude that all are analogous to the "antiaromatic"-like T_d AX_4 (A = S, Se, Te), D_{3h} AX_3(A = Cl, Br, I), and D_{5h} AX_5 (A = Cl, Br, I). <u>It follows that XeF_2, XeF_4, and XeF_6 will seek some other geometry in which the core-ligand antibond is eliminated without reduction of the already existing core-ligand bonds</u>. In this light, let us now consider the extreme XeF_2 and XeF_6 systems and ask the following question: Can we identify geometries which permit the lone pairs to avoid each other in each of the two systems? It is immediately apparent that the experiment will fail in the case of XeF_2 because the only conceivable alternative geometry is a C_{2v} geometry which, in fact, not only fails to protect the 5s core electron pair from the σ_1 electron pair but it also generates additional lone pair overlap repulsion in the way made evident by the bond diagram shown in Figure 9. By contrast, in the case of XeF_6, we <u>can</u> find a geometry which will effectively accomplish the same thing that a C_{2v} geometry accomplishes for SF_4, a C_{4v} for IF_5, etc. One such three dimensional arrangement of the atoms of XeF_6 is shown below with the corresponding bond diagram presented in Figure 10. <u>The characteristic feature of this form of XeF_6 is the partial insulation of the core and the ligand lone pairs</u>. The analogy between the SF_4 and the XeF_6 problem is obvious. SF_4

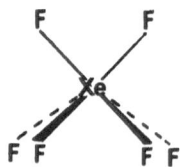

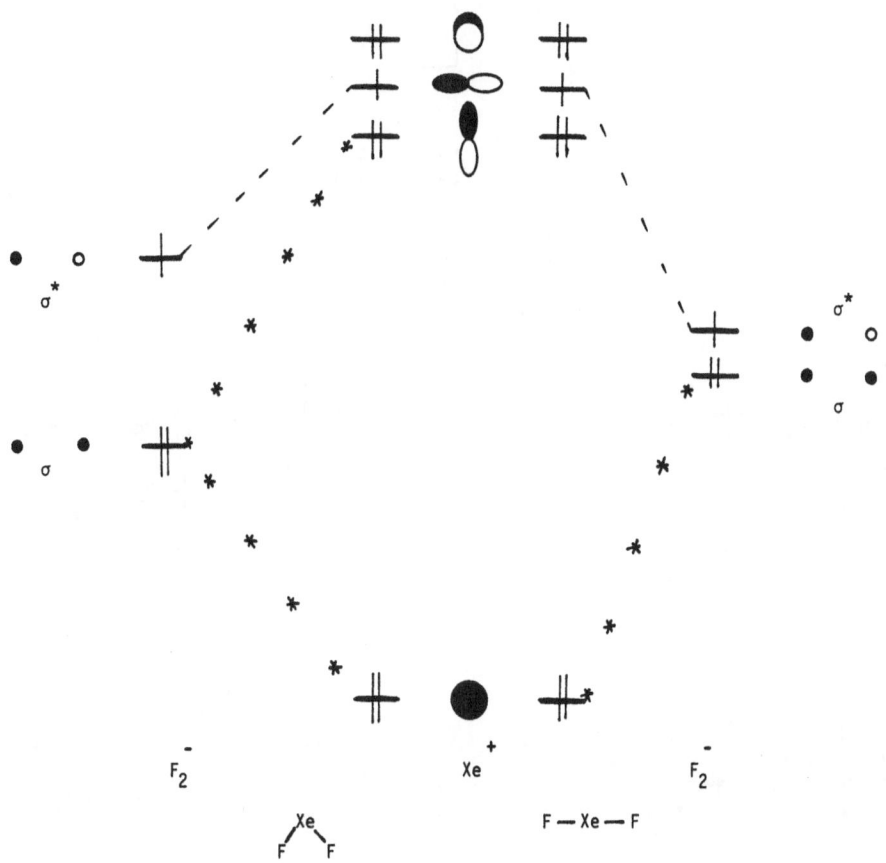

Figure 9. Detailed bond diagrams of bent and linear XeF$_2$.

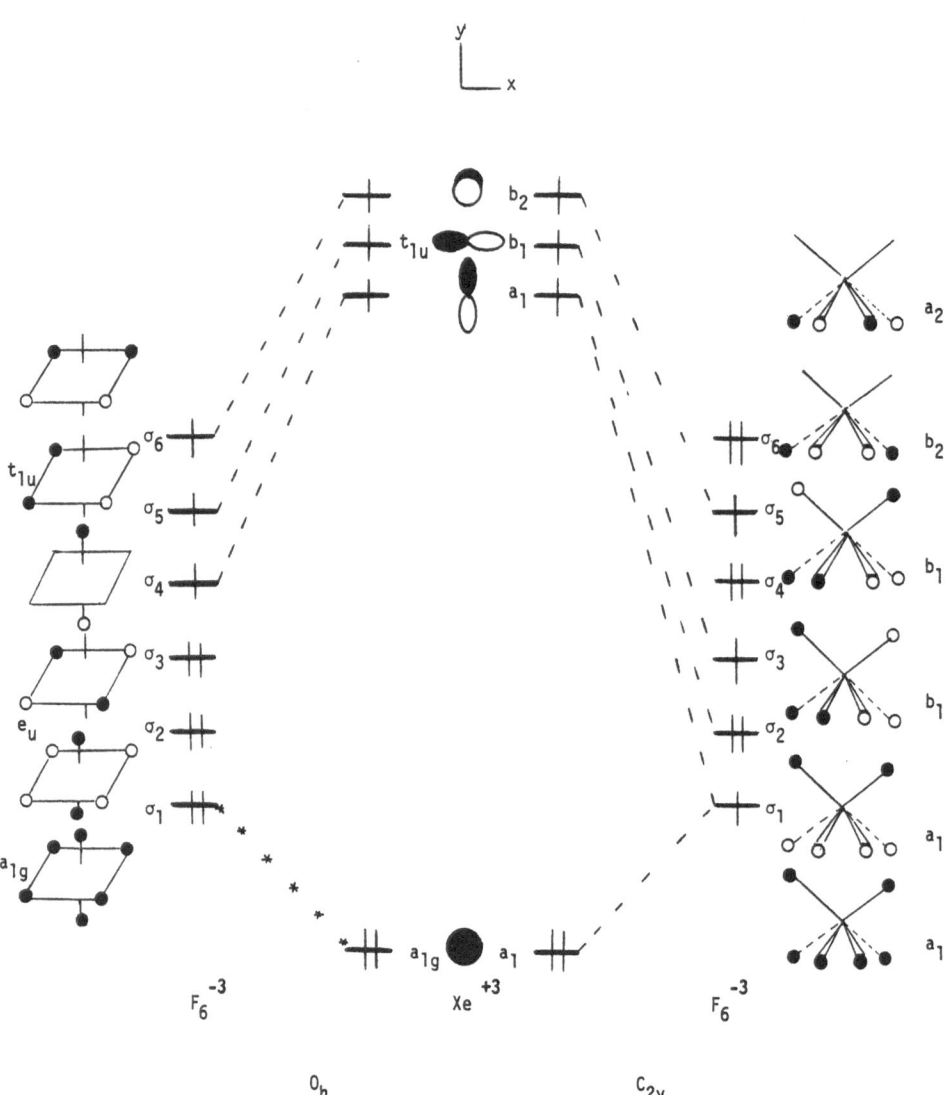

Figure 10. Detailed bond diagrams of O_h and C_{2v} XeF_6. It is assumed that $4s$-σ_2 and $4p_x$-σ_4 overlaps in the C_{2v} form, while nonzero, are negligible.

changes from T_d to C_{2v} in order to partly "hide" one ligand electron pair from one core electon pair. The same thing occurs in the $O_h \longrightarrow C_{2v}$ transformation of XeF_6. Note that although the 3s AO of S and one ligand MO of SF_4 (σ_4 in SH_4 in Figure 2) have the same symmetry label, they overlap only minimally and the same is true of the 5s AO of Xe and one ligand MO of XeF_6 (σ_2 in XeF_6 in Figure 10). This is what provides effective "insulation" of core and ligand electron pairs and constitutes the basis of stereoselection in all SE "hypervalent" molecules.

VII. Why are First Row "Hypervalent" Molecules Unstable?

If we denote an atom of the nth column and mth row of the Periodic Table by A_{nm}, then we recognize that stable "hypervalent" molecules having A_{n1} cores are non-existent. On the other hand, as m changes from 1 to 2, to 3, etc., i.e., as we move down a column of the Periodic Table, the likelihood of identifying stable "hyper valent" species increases. Thus, for example, OF_4 is unstable but SF_4 does exist. What is the origin of this trend?

Two things happen as we sweep down a column of the Periodic Table: The atoms become increasingly electropositive and their "covalent radii" increase. As a result, electron transfer from core to ligands required for the formation of the maximum number of core-ligand multicenter bonds becomes more favorable and ligand nonbonded repulsion decreases. Thus, for example, formation of the maximum number of core-ligand bonds in C_{2v} SF_4 requires a transfer of one electron from S to (F_4) according to the bond diagram of Figure 2. Since S is much more electropositive than O, an $S^+(F_4)^-$ electronic configuration is much more favorable than an $O^+(F_4)^-$ one. Furthermore, natural LNR in SF_4 is much smaller than in OF_4 simply because the longer S-F bonds place the ligands further away from each other in SF_4.

There is yet a third important factor which differentiates first row from second, third, etc., row atoms: The former do not have low lying unoccupied d orbitals available for bonding while the latter do so. On the basis of the theory of Vacant Orbital Participation (VOP), outlined in Chapter 2, it is a simple matter to understand how core d orbitals participate in the bonding of "hypervalent" molecules. Specifically, introduction of core d orbitals of appropriate symmetry will create out of the ligand "lone pairs" two-electron multicenter bonds and, in addition, it will cause hole hybridization of the existing multicenter bonds. Since the latter effect is much weaker than the former, we can neglect the hole hybridization and concentrate on the bond creation aspect of d orbital participation.

On these grounds, it is a simple matter to demonstrate that, in practially all "hypervalent" molecules containing ligand "lone pairs", one can always find core d orbitals having the same symmetry as the ligand "lone pairs" (which, in combination, can yield two-electron mutlicenter bonds), in any geometry one chooses to examine. Hence, core d orbital participation, even if strong, will not be the primary factor determining the relative energies of the various geometries of a "hypervalent" molecule.

We depart this section with the following message: If d orbital participation is assumed to be insignificant, "hypervalent" molecules are actually hypovalent molecules, e.g., SF_4 has three rather than four core-ligand bonds. On the other hand, if d orbital participation is strong "hypervalent" molecules are worthy of their title, e.g., SF_4 has four core-ligand bonds.

References

1. Epiotis, N.D.; Larson, J.R.; Eaton, H. "Unified Valence Bond Theory of Electronic Structure" in Lecture Notes in Chemistry, Vol. 29; Springer-Verlag: New York and Berlin, 1982.

2. (a) Chen, M.M.L.; Hoffmann, R. J. Am. Chem. Soc. 1976, 98, 1647.

    (b) Wolfsberg, M; Helmholtz, L. J. Chem. Phys. 1952, 20, 837.

    (c) Hoffmann, R.; Lipscomb, W.N. J. Chem. Phys. 1962, 36, 2189.

    (d) Hoffmann, R. J. Chem. Phys. 1963, 39, 1397.

3. Callomon, J.H.; Hirota, E.; Kuchitsu, K.; Lafferty, W.J.; Maki, A.G.; Pote, C.S. in Landolt-Bornstein "Numerical Data and Function Relationships in Science and Technology", Vol. 7, New Series, "Structure Data on Free poly-atomic Molecules", K. H. Hellwege, Ed.; Springer-Verlag: West Berlin, 1976.

4. (a) Schwenzer, G.M.; Schaefer, III, H.F. J. Am. Chem. Soc. 1975, 97, 1393. These computations predict a C_{2v} structure.

    (b) Gleiter, R.; Veillard, A. Chem. Phys. Letters 1976, 37, 33. These computations predict a C_{4v} structure.

5. Radom, L.; Schaefer, III, H.F. Aust. J. Chem. 1975, 28, 2069. These computations predict a C_{4v} structure with one type of basis set and C_{2v} with another.

6. Basch, H.; Moskowitz, J.W.; Hollister, C.; Hankins, D. J. Chem. Phys. 1971, 55, 1700.

Chapter 10. The Molecular Orbital-Valence Bond Theory of Inorganic Chemistry.

The purpose of this paper is to rephrase the fundamental concepts of inorganic chemistry in the language of MOVB theory[1] as a prelude to a reexamination of the electronic structure of inorganic molecules. For illustrative purposes we reformulate the concepts of the coordinate bond,[2] the Dewar-Chatt-Duncanson (DCD) model,[3] and the concept of high and low spin complexes[4] and we apply MOVB theory to the problem of the ground stereochemistry of prototypical inorganic complexes, the thermodynamics of olefin coordination in complexes, and the problem of the "trans effect" in the equilibrium geometries of inorganic complexes.[5] In order to exemplify our approach, we make pairwise comparisons of states and geometries using MOVB bond diagrams which are constructed by adhering to the following conventions:

a. Some of the applications involve specific complexes for which there is a large body of structural data. Some others have as target prototype molecules, e.g., ML_2L_2', so that a general concept is better brought to focus. In dealing with such prototype systems, we **assume** that M is a d^n metal with zero charge and L is a neutral two-electron sigma donor, unless otherwise specified.

b. The convention spelled out above is necessary for making an unambiguous assignment of <u>formal</u> charges of metal and ligand fragments. These are always made by reference to the perfect pairing CW projected by the MOVB bond diagram. It should be kept in mind that these <u>formal</u> charges will increasingly tend to be the <u>real</u> charges as the perfect pairing CW tends to make an increasingly greater contribution to the MOVB wavefunctions. Because of the weak overlap binding ability of metals, this limit is not often reached. Nonetheless, formal charges are informative as they alert us of the bond-forming potential of a metal and a set of ligands and they will be consistently displayed at the bottom of each bond diagram.

I. The Coordinate Bond

Most chemists are used to represent the electronic structure of ground state LiF by the Lewis formula, $Li^+ F^-$, realizing that this is only an approximate representation which implies that the ground state wavefunction of LiF is a resonance hybrid of the type shown below, with the principal contributor being the structure $Li^+ F^-$.

$$\overset{+}{Li} \; \overset{..}{\underset{..}{:F:}} \longleftrightarrow Li \cdot \cdot \overset{..}{\underset{..}{F:}} \longleftrightarrow \overset{-}{Li:} \; \overset{+}{\underset{..}{F:}}$$

$$\Phi_1 \qquad\qquad \Phi_2 \qquad\qquad \Phi_3$$

Major
Contributor

This fact can be made explicitly clear by the VB bond diagram show below where we have omitted the three fluorine lone pairs.

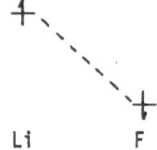

Li   F

This VB bond diagram is interpreted as being the representation of the optimal VB wavefunction which contains all VB structures that can be generated by permuting the two electrons between the Li2s and F2p AO's. In other words, the bond diagram is the representative of the wavefunction shown above, where the Φ_i's are termed Configuration Wavefunctions (CW's).

Similarly, most inorganic chemists represent a coordinate bond between a metal (M) and a ligand (L) as follows:

$$M \longleftarrow :L$$

The Lewis formula shown above is again taken to imply a resonance hybrid of the type shown below.

```
-2  +2         -   +
M:  L  ⟷  M·  ·L  ⟷  M :L
                     Major
                     Contributor
```

The above representation is nothing else but a representation of a highly polar, or, "ionic" two-electron bond which can be represented by a VB bond diagram which is entirely analogous to that of LiF.

In MOVB theory with core-ligand dissection, any molecule can be represented by a bond diagram which shows explicitly how the core and ligand fragments are bound by multicenter bonds of different degrees of "covalency", or, "ionicity". The Lewis formulae, the ordinary Single Determinant (SD) MO descriptions, and the MOVB representations of typical coordinate bonds are compared in Figure 1.[1] The MOVB diagrams show explicitly the number and character of the various bonds which connect core and ligands within a given system and they describe properly bond dissociation. This latter process cannot be described correctly by SD MO theory and this implies that SD MO theories can run into difficulties when applied to inorganic problems, in general, for two reasons:

a. Chemical transformations, whether organic, inorganic, or otherwise, which involve bond breaking and bond making cannot be dealt with by SD MO theory.

b. Many inorganic complexes have weak core-ligand "covalent" bonds which cannot be described adequately by SD MO theory since they are essentially "partly-broken" bonds.

It follows that only the MOVB bond diagrams provide a formally correct representation of bonds, in general, and, thus, only MOVB diagrams can serve as the basis for unifying the conceptual frameworks of different subdisciplines of chemistry.

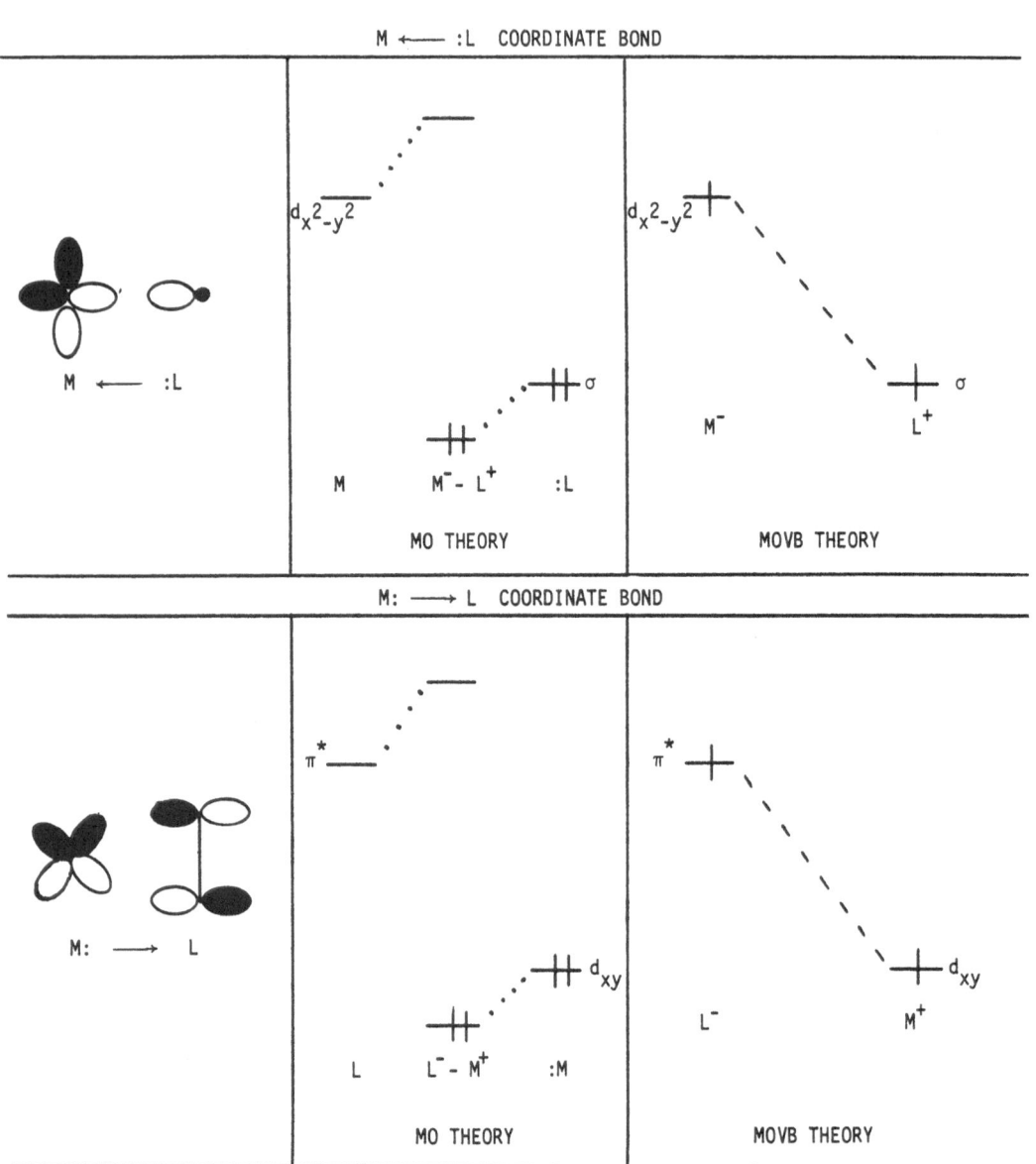

Figure 1. Pictorial MO and MOVB descriptions of coordinate bonds. Note that, at the level of MOVB theory, formal charges are assigned by reference to the perfect pairing CW projected by the bond diagram.

With this in mind, we can say that the concept of the coordinate (multicenter) bond exists in MOVB theory, now as a theoretically well defined concept. In Figure 2 we show the bond diagrammatic representation of $PtCl_4^{-2}$ as well as the corresponding MO interaction diagram one is apt to encounter in the inorganic literature. Although both MOVB and SD MO theory depart from the same fragment symmetry orbitals, the formalisms, the conceptual frameworks, and the reliability of each of the two methods are entirely different. The MOVB diagram and associated concepts already "contains" hybridization, which can be conceptualized at the level of SD-MO theory only if usage of Perturbation Theory (PT) is justified and then only at the level of high order PT, and coulomb correlation, not accounted for by SD MO theory. Though much progress has been made by application of approximate SD MO theory to inorganic problems,[6] MOVB theory has formal and conceptual advantages which promise to reveal facets of chemical bonding which have hitherto remained unrecognized. "Surprising" conclusions of MOVB theory have already been discussed in previous chapters.

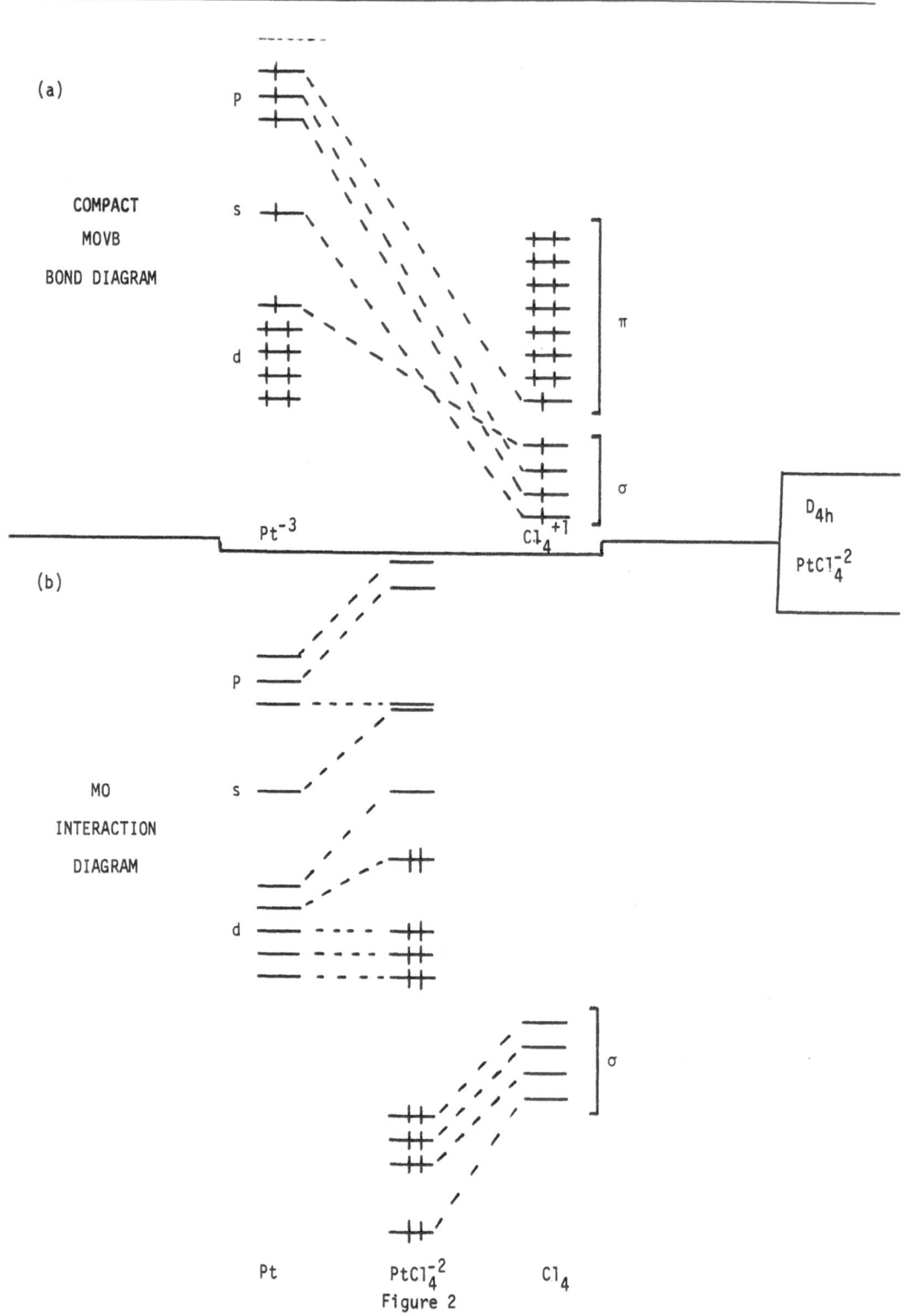

Figure 2

II. Bonding and Back-Bonding

The Dewar-Chatt-Duncanson (DCD) model describes the formation of two
coordinate bonds resulting from the overlap of two metal and two ligand orbitals.
In other words, the DCD model of bonding of organometallic complexes is nothing
else but a simple model of metal-ligand double multicenter bonding. Now, since
two multicenter bonds can be formed in three distinct ways, i.e., they can be U,
H, or D bonds,[1] it turns out that the DCD model describes two coordinate bonds of
one flavor, namely D. Since we have seen that organic chemistry is virtually with
no exception the chemistry of U and H bonding,[7] it is clear that the DCD model is
a restrictive model. In addition, since a metal and a set of ligands are most
frequently bound by more than two multicenter bonds, the DCD model in itself is
not adequate for dealing with any problem of chemical bonding in an a priori sense
unless one can pinpoint exactly which two bonds play the most important role in
the chemical process under investigation. Finally, the associated concepts of
forward and backward donation are of little use when we face the problem of
state identification, i.e., the electron allocation responsible for metal-ligand
binding. These points are illustrated by reference to the example of metal-pi
ethylene ($M-\pi C_2H_4$) bonding which is said to be accomplished through dative bonding
involving the doubly occupied ethylenic π MO and a vacant (d, s, or, p) AO of the
metal and back bonding involving a filled orbital of the metal and the vacant
ethylenic π^* MO. The bond diagrams of Figure 3 contrast D-bonding in $M-\pi C_2H_4$ with
U-bonding in C_2H_8 (cyclobutane). The MOVB diagrams of two different states, which
are similar to the extent that they both involve forward and backward donation
but different insofar as the allocation of electrons in the four orbitals is concerned, are shown in Figure 4. The concepts of dative and back bonding cannot
help us predict which has higher and which has lower energy. The DCD model as

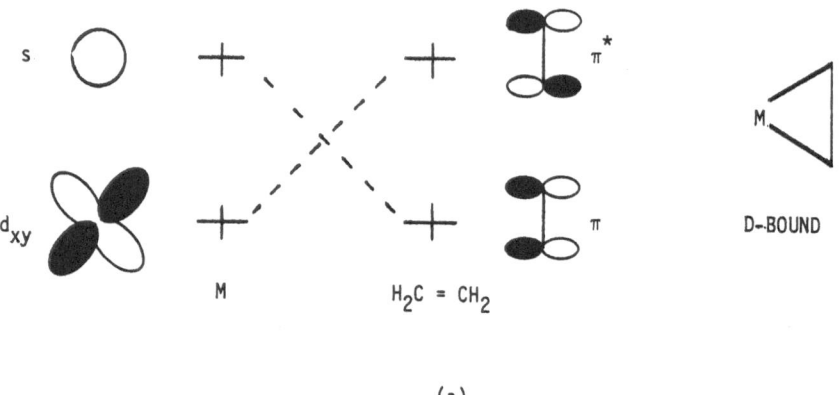

(a)

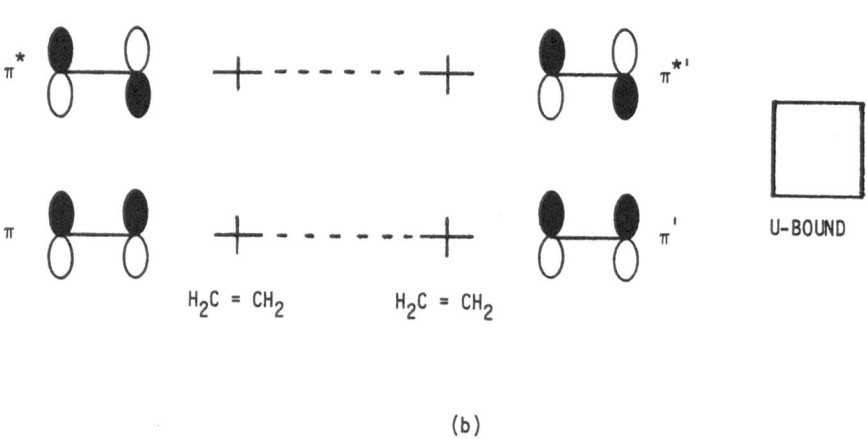

(b)

Figure 3. a. A D-bound organometallic system.
b. A U-bound organic molecule.

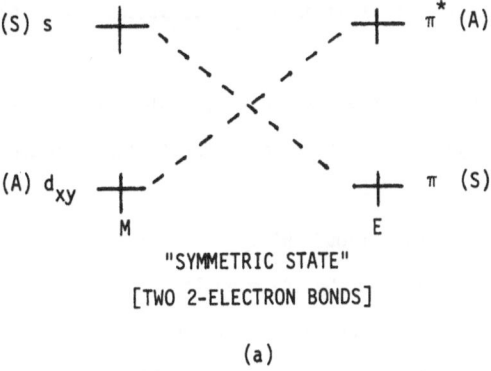

"SYMMETRIC STATE"
[TWO 2-ELECTRON BONDS]

(a)

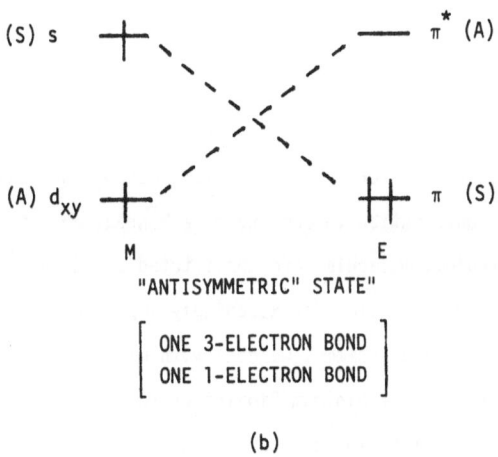

"ANTISYMMETRIC" STATE"

$$\left[\begin{array}{c} \text{ONE 3-ELECTRON BOND} \\ \text{ONE 1-ELECTRON BOND} \end{array}\right]$$

(b)

Figure 4. Two different electronic states of the D-bound organometallic system of Figure 3a. Note how "different states" means "different number and types of bonds" in MOVB theory.

originally formulated is actually applicable only to states involving electron pair bonds.

The MOVB CW's which are required for the description of M - πC_2H_4 (see Figure 3a) are shown in Figure 5. In this case, metal-ligand bonding arises from two sources:

    a. Spin pairing within the various CW's.

    b. CW interaction brought about by the monoelectronic Charge Transfer (CT) interaction matrix element, d_{ij}, which connects two CW's which differ by one occupied spin orbital and which define ionic delocalization, as illustrated below.

$$
\begin{array}{ccc}
X_b - \quad - X_c & & X_b - \quad + X_c \\[2pt]
X_a \,\text{\tiny{++}}\quad \text{\tiny{++}}\, X_d & \xleftrightarrow{\;d_{73}\;} & X_a + \quad \text{\tiny{++}}\, X_d \\[6pt]
\Phi_7 & \mathbf{d}_{73}\,\alpha\,\beta_{ac} & \Phi_3
\end{array}
$$

The d_{ij}'s depend on interfragmental AO resonance integrals, β_{mn}'s, and the magnitudes of the latter depend on the nature of the atoms involved in a way that has been discussed in a previous chapter. As these atoms become increasingly stronger overlap binders, the CT contribution to bonding is enhanced, and vice versa. In most organic chemical problem, molecules are constituted of strong overlap binders and the CT mechanism of bond formation is exceedingly important. By contrast, in most inorganic or organometallic systems, we deal with weak interfragmental bonding. As a result, the low lying CW's of minimum "ionic" character, e.g., Φ_1 and Φ_7, mix only weakly with the higher lying "ionic" charge transfer CW's, e.g., Φ_3 - Φ_6 and Φ_9, Φ_{10}, and the total wavefunction increasingly resembles one of two CW's: The one having the electrons occupying the lowest energy orbitals of the fragments without suffering severe interelectronic repulsion (e.g., Φ_7) or the perfect pairing (R) CW (e.g., Φ_1). Empirical theories which neglect "classical" coulomb interaction (e.g., HMO theory) and semiempirical or nonempirical SD MO theories, which

VB CW COMPOSITION OF MOVB CW's

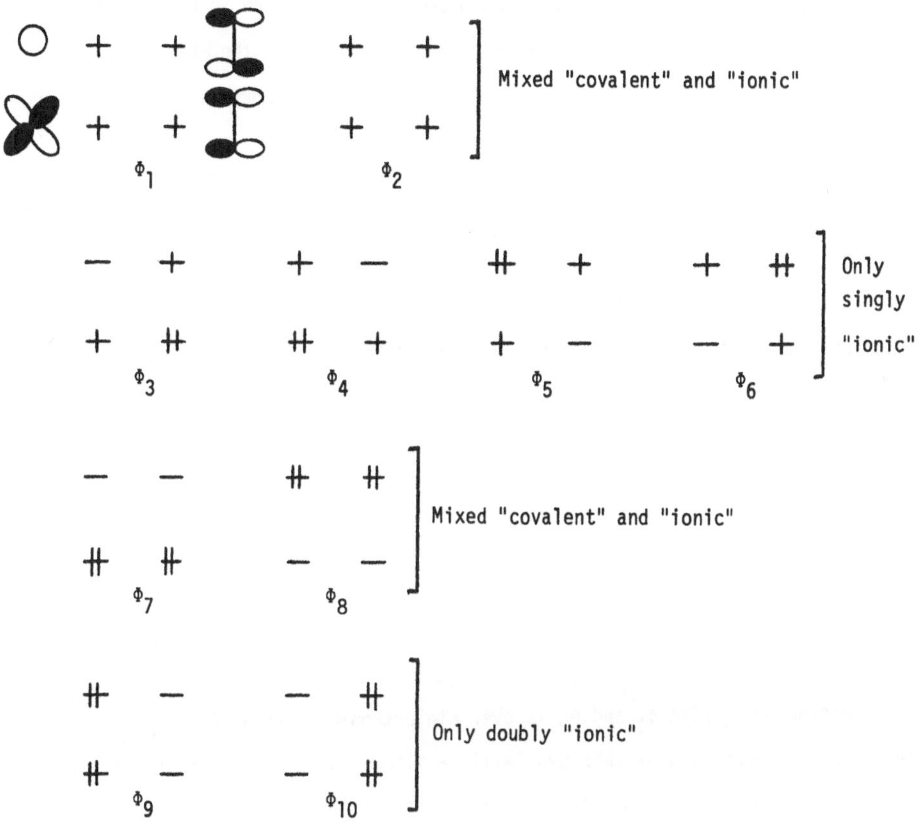

Figure 5. The MOVB CW's necessary for the treatment of the D-bound organometallic system of Figure 3a and their VB CW composition.

incorrectly describe coulomb correlation because of the constraint approximation, cannot properly describe weak bonds or bond dissociation and tend to exaggerate the importance of the CT bonding mechanism. Thus, for example, the MOVB wavefunction of the $M-\pi C_2H_2$ complex at the level of SD MO theory looks like:

$$\Psi \propto \Phi_1 \longleftrightarrow \Phi_3 \longleftrightarrow \Phi_4 \longleftrightarrow \Phi_7$$

By contrast, the correct MOVB wavefunction looks like:

$$\Psi \propto \Phi_7$$

or

$$\Psi \propto \Phi_1 \longleftrightarrow \Phi_7 \longleftrightarrow \Phi_2$$

The interaction of Φ_7 with Φ_1 and Φ_2 is over bielectronic correlation matrix elements (W_{ij}) and bonding of this type will be discussed in a subsequent chapter. For the time being, we are content to point out that, since the DCD model is often used in conjunction with SD MO theoretical analyses, it is important that we remember well the formal deficiencies of SD MO theory so that we can discern which concepts derived by application of the DCD model are "right" and which are "wrong".

III. High versus Low Spin Complexes

Consider the system ML_2 which contains four electrons in the four AO's shown
in Figure 6. Treating the ML_2 system as a composite of M (core) and L_2 (ligand),
we can write bond diagrammatic representations of three low lying zero order states
as shown in Figure 6. For calibration, the corresponding MO descriptions for
strong core-ligand interaction are also given in Figure 6. It is immediately
obvious that the three states involve different types of core and ligand excitations
as well as different types of bonds, with both excitation energy and core-ligand bond
strength increasing in the order C > B > A. Accordingly, depending upon the overlap
binding ability of the constituent atoms, i.e., depending upon the magnitudes of
the interfragmental AO resonance integrals, β_{ML}, we distinguish two diametrically
opposite situations:

a. $|\beta_{ML}|$ is small. In this case, the ML_2 system will opt for minimal excitation
and the relative energies of the three states will be: C > B > A (A most stable).

b. $|\beta_{ML}|$ is large. In this case, the ML_2 system will opt for strong bond
formation at the expense of excitation energy. As a result, the relative energies
of the three states will now become: A > B > C (C most stable).
Here, we interject that we always compare the global minima of the three states,
with that of A expected to occur at long and with that of C expected to be found
at short core-ligand distance. Since the A state is dissociative because it involves
one four-electron antibond,[8] it follows that there will exist choices of M and L for
which the bound "high spin" triplet B will lie below the bound "low spin" singlet C
and other choices for which exactly the opposite will be true. The way in which a
change of the overlap binding ability of atoms brings about such a state switchover
has been discussed and exemplified in great detail in a previous chapter. For
the time being, the reader is simply reminded that MOVB theory makes possible a

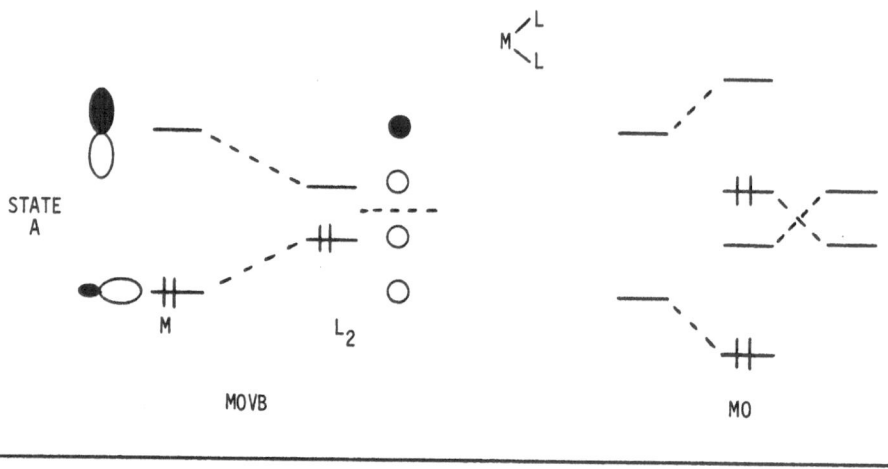

STATE A

MOVB

MO

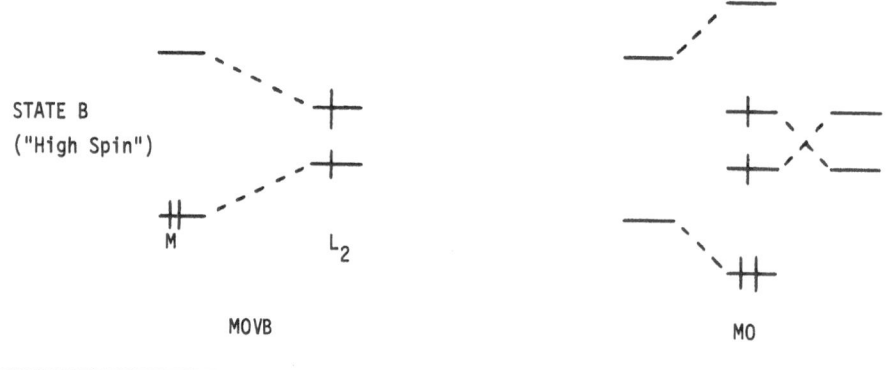

STATE B
("High Spin")

MOVB

MO

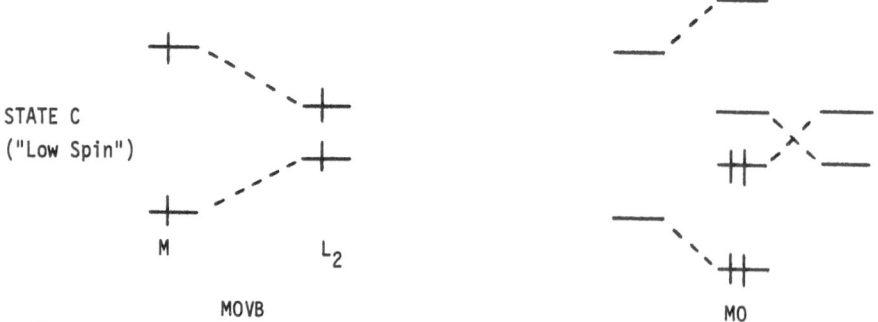

STATE C
("Low Spin")

MOVB

MO

Figure 6. The MOVB and corresponding single determinant MO descriptions of three low lying states of ML_2.

qualitative description of the state manifolds of molecules and that state switchovers effected by different combinations of atoms are entirely predictable, always
in a qualitative, trendwise fashion. In other words, while the organic chemist
thinks of most ground state organic molecules as closed shell species with open
shell entities being exceptional, and whereas the inroganic chemist has invented
equivalent but different-sounding terms for describing different states of inorganic
systems, i.e., "low spin" (e.g., singlet C) and "high spin" (e.g., triplet B), MOVB
theory transgresses interdisciplinary barriers and forms the basis for a general
understanding of why ground state molecules have the electronic configuration they
do and the rational design of systems likely to exist in one or another spin state.

The prototypical examples of low and high spin octahedral complexes are
$Fe(CN)_6^{-3}$ and $Fe(H_2O)_6^{+3}$, respectively. If we treat every ligand, whether CN^- or
H_2O, as a two-electron sigma donor, we obtain the bond diagrams for FeX_6^{+3} shown in
Figure 7. The difference between the low and high spin complex is simple: In going
from the former to the latter, two core d electrons are relocated in such a way so
that two doubly occupied AO's of the core become singly occupied while two two-
electron core-ligand bonds are replaced by two <u>weaker</u> three-electron core-ligand
bonds. In exchange for the bond strength reduction, the two d-electrons are now permitted to descend to the low lying ligand orbitals and, in addition, the system profits from better overall coulomb correlation, an attribute of all "high spin" relative to "low spin" complexes.[9] The relocation of the two electrons is tantamount to
a deexcitation due to core $\rightarrow$ ligand charge transfer (CT) defined by reference to
the perfect pairing CW's of the low and high spin forms as illustrated in Figure 7.
Thus, the difference between low and high spin FeX_6^{+3} is exactly analogous to the
difference between state C and state B of ML_2 (Figure 6).

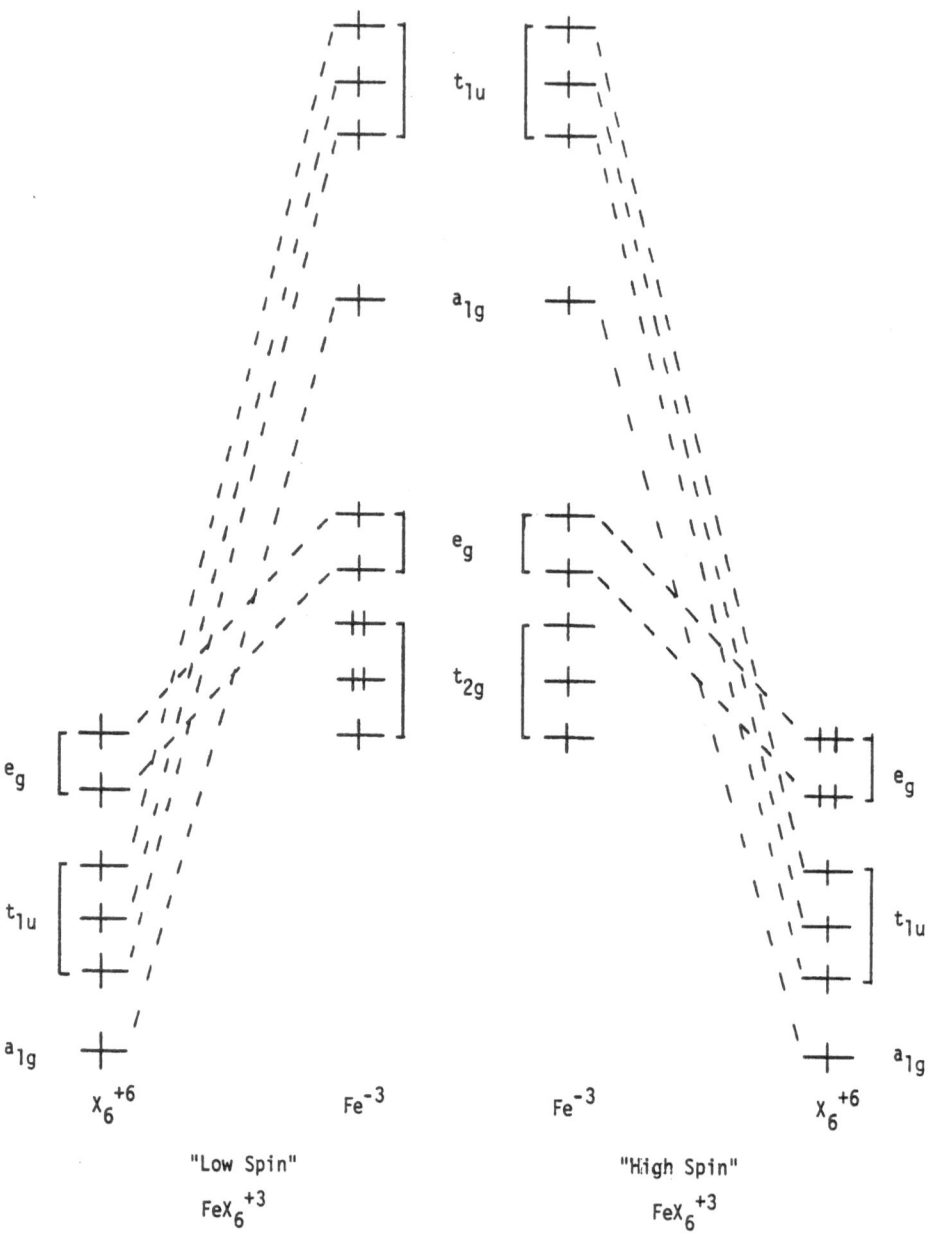

Figure 7. Sigma bond diagrams of "high spin" and "low spin" FeX$_6^{+3}$ complex. Charges are <u>formally</u> assigned by reference to the perfect pairing CW projected by the bond diagram and by assuming that X is neutral. The geometry of both species is presumed to be octahedral.

Next, we recognize that, in addition to sigma orbitals and sigma electrons, ligands possess pi orbitals and pi electrons and that, in the case of $(CN^-)_6$ there exist doubly occupied as well as <u>low lying</u> pi unoccupied MO's while, in the case of $(H_2O)_6$, there exist only doubly occupied and <u>high lying</u> unoccupied pi MO's. The filled pi MO's create core-ligand four-electron antibonds and, on this basis, there is no fundamental difference between the two types of ligands. On the other hand, the fact that only $(CN^-)_6$ has low lying occupied MO's means that only $(CN^-)_6$ can create three additional core-ligand bonds via the overlap of the three vacant orbitals of $(CN^-)_6$ with the three occupied t_{2g} d orbitals of the core. Because of this, the approximate bond diagrams for high and low spin $Fe(H_2O)_6^{+3}$ are those shown in Figure 7 and the proper bond diagrams for $Fe(CN)_6^{-3}$ those shown in Figure 8. The bonding changes that accompany the conversion of a low to high spin complex in the two cases are the following:

A. $Fe(H_2O)_6^{+3}$.

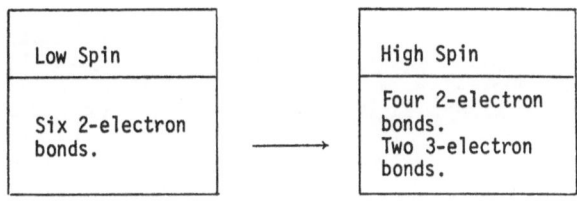

Low Spin	High Spin
Six 2-electron bonds.	Four 2-electron bonds. Two 3-electron bonds.

B. $Fe(CN)_6^{-3}$.

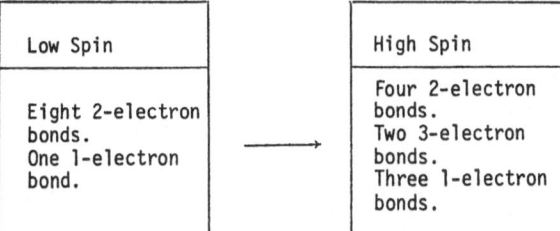

Low Spin	High Spin
Eight 2-electron bonds. One 1-electron bond.	Four 2-electron bonds. Two 3-electron bonds. Three 1-electron bonds.

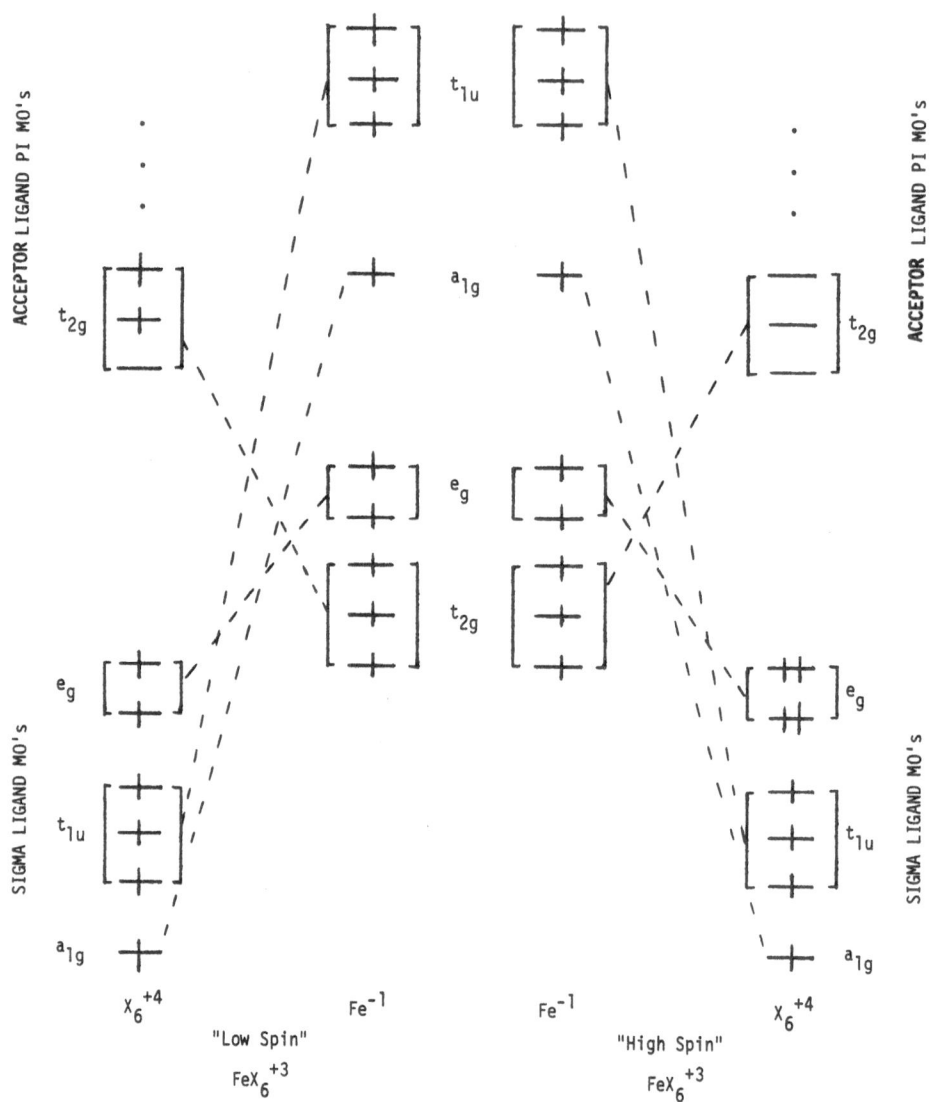

Figure 8. Bond diagrams of "high spin" and "low spin" FeX_6^{+3} which include all sigma donor (nonbonding) and pi acceptor (antibonding) orbitals of the ligands. Additional doubly occupied pi bonding or nonbonding orbitals of the ligands have been disregarded since they can only act as to perturb the existing bonds. Note how inclusion of the ligand pi antibonding orbitals reduces the metal <u>formal</u> negative charge.

Clearly, the price of destroying four two-electron bonds and replacing them by two weaker one-electron and two weaker three-electron bonds in $Fe(CN^-)_6^{-3}$ is greater than simply destroying two two-electron bonds and replacing them by two three-electron bonds in $Fe(H_2O)^{+3}$. As a result, rebonding of the former type is resisted while rebonding of the latter types becomes energetically favorable since the partial loss of core-ligand bonding is counteracted by deexcitation via core $\rightarrow$ ligand CT and improved coulomb correlation.

Finally, one can easily show that, while the "high" and "low spin" forms of one and the same species in a fixed geometry are related in a "ground versus excited state" sense, the optimal geometries of the two states need not be (and, most often, are not) identical. An often cited paradigm is the fact that NiX_4 has a D_{4h} "low spin" form but a T_d "high spin" form with either of them becoming the ground state depending on the nature of $X(d^8$ Nickel). The reader will find it amusing to construct the MOVB bond diagrams of singlet D_{4h} and triplet T_d $NiCl_4^{-2}$, identify the different bonds of each species, and deduce why the two have comparable energies.

IV. <u>Applications</u>

The purpose of the previous section was to project the fact that some of the most important ideas of chemical bonding which inorganic chemists are familiar with are nothing else but special elements of a complete theory of chemical bonding such as the MOVB theory. Thus, a coordinate bond is an electron pair bond, the DCD model describes D-type double bonding, and a "high" and a "low" spin complex are only two of many possible low lying valence states of a complex. Recognizing the obvious, namely, that the partial concepts we have spoken of above have been extremely useful for rationalizing and, sometimes, predicting experimental results, we now proceed to demonstrate that many fundamental problems of inorganic chemistry can be easily "solved" through application of MOVB theory, a theory which has no limitations in terms of essential formal "correctness", conceptual clarity, and range of applicabiltiy. In presenting illustrative applications of the theory, we recognize that in constructing bond diagrams for inorganic complexes, both sigma and pi ligand electrons must be taken into consideration. However, in many comparisons, the number and strength of core-ligand bonds formed via the utilization of the ligand pi electrons is roughly invariant. Whenever this is the case in the examples discussed in this section, the bond diagrams are simplified to the extent that only the sigma-type bonds linking core and ligands are shown in them.

A. Ground State Inorganic Stereochemistry

Consider the basic stereochemical problem represented by the chemical equation shown below.

We ask the question: Which of the two forms of the ML_2E complex, where M is a d^{10} metal, L is an arbitrary ligand, and E stands for ethylene, is more stable: The planar (P) one or the staggered (S) form? The answer can be immediately obtained by the construction and interpretation of the corresponding bond diagrams. These are shown in Figure 9 and they plainly make clear that both P and S contain four multicenter bonds. Thus, neither the concept of the coordinate bond nor the DCD model can help us decide which of the two is more stable. However, mere inspection of the two bond diagrams reveals that the fundamental difference between the two geometries amounts to a symmetry label "reversal", i.e., one ligand MO of S (σ^*) has b_1 while the corresponding ligand MO of P (σ^*) has b_2 symmetry. Since the core contains a b_1 doubly occupied d-orbital and since the direction of primary CT is from core to ligands, the S form suffers from overlap repulsion between a lone pair and a bond pair while the P form is protected by symmetry from it! Clearly, the planar geometry will be the minimum energy geometry of the ML_2E complex, as found by experiment.[10] The same analysis, leading to the same conclusions, can be given for ML_2O_2.[11]

A second stereochemical problem related to the one discussed above is the conformational "equilibrium" shown below:

M: A d^8 metal

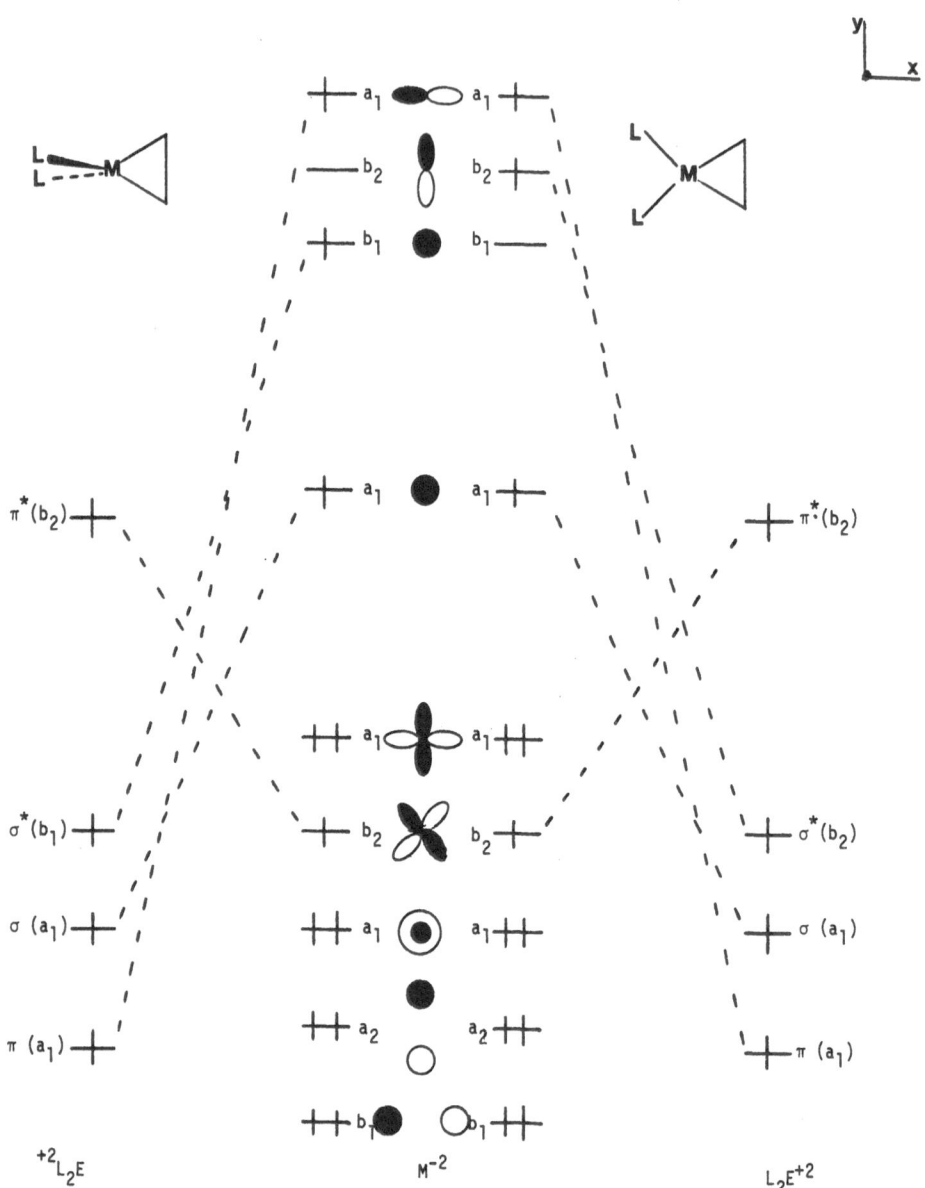

Figure 9. Compact bond diagrams of planar (eclipsed) and perpendicular (staggered) L_2ME, where L is a two-electron ligand, E = olefin, and M is d^{10} species. Formal charges are assigned by reference to the perfect pairing CW and by assuming that M and L are neutral.

The bond diagram shown in Figure 10 demonstrate that this problem is entirely analogous to the previous one, i.e., the P and S geometries differ by one symmetry label reversal (π^* is b_2 in P and b_1 in S) so that primary CT in the $2p_y - \sigma_2$ bond of P does not cause overlap repulsion with the d_{xz} lone pair while primary CT in the same bond of S does cause overlap repulsion with the d_{xy} lone pair. Once again, we predict that the P should be more stable than the S structure, assuming always that nonbonded ligand repulsion will play a secondary role. This latter assumption appears to be satisfactory in the previous ML_2E but not in the present ML_3E species. Indeed, experiment shows that organometallic complexes of the latter type exist in an S geometry.[12]

We end this section with a side comment: Both of the problems discussed here have been treated before by Hoffmann and his coworkers[10c] using the EHMO method. Both of these problems turn out to be "within" EHMO theory. Hence, by translating MOVB into EHMOVB (=EHMO) theory, we can show that our own conclusions are essentially equivalent to those of Hoffmann, et al. In the next section, we take up a problem which lies "outside" EHMO theory.

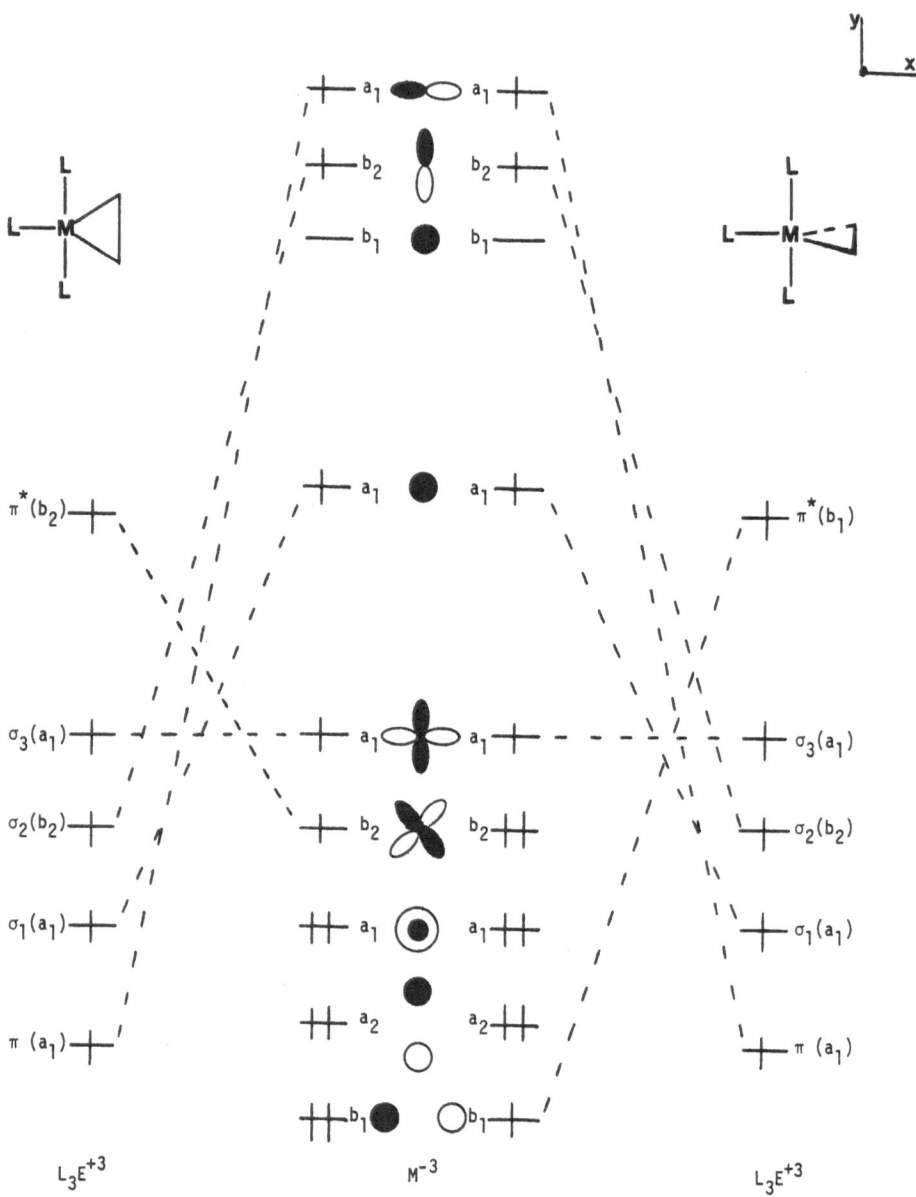

Figure 10. Compact bond diagrams of planar and perpendicular L_3ME. The definitions of L and E are as in Figure 9 but M is a d^8 species.

B. Substituent Effect on Complex Stability

Consider the equilibrium shown below

$$L_4M \; + \; \underset{X}{\overset{X}{\rangle}}\!\!=\!\!\underset{X}{\overset{X}{\langle}} \; \underset{\longleftarrow}{\overset{K}{\rightleftharpoons}} \; L_4M$$

M: A d^8 Metal

We ask the question: How does the equilibrium constant K depend on the nature of X? We compare three reaction systems, namely, one with X = H, one with X = F, and one with X = CN. Furthermore, we assume that, to a first approximation the L_4M - C_2X_4 binding can be described by the DCD model, i.e., that this L_4M - C_2X_4 complex can be treated as a "four electrons in four orbitals" problem because the nonbonded distance between L and C is such as to "allow" the ligand orbitals to separate into two sets, one spanning L_4 and one $\dot{C}_2X_4$.

Let us now examine in detail and in a step-by-step manner the requirement for metal-olefin bonding starting with ground singlet d^8 metal and ground singlet ethylene:

a. The metal is excited to the lowest triplet state of appropriate symmetry. In this case, this is the $d's'$ triplet state.

b. The planar ethylene is excited to the lowest triplet state of appropriate symmetry. In this case, it is the $\pi\pi^*$ triplet state.

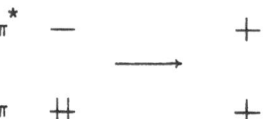

c. The planar triplet ethylene relaxes to the cis-bent geometry required for bonding to the metal. This cis triplet ethylene can be represented by the bond diagram of Figure 11. Cis bending is favorable because it allows the highest singly occupied ω_7 core orbital to define a bond with the σ_2 ligand orbital, a process which allows the corresponding electron pair to achieve low energy (by "dropping" into the σ_2 ligand MO) while localizing the two odd electrons in the low lying ω_2 and ω_4 core orbitals. The hybrid representation of cis triplet ethylene is shown below and it makes clear that the way triplet ethylene is going to bind to the metal is partly through utilization of the low lying singly occupied olefin MO's ω_2 and ω_4 (brought into play by contributor Ξ_1) which have two important properties:

1) High carbon 2s AO character. Because of the greater overlap binding ability of 2s relative to 2p, this will tend to create strong metal-olefin overlap binding.

2) Low one-electron energies. This will tend to promote CT from metal to olefin which will additionally strengthen the metal-olefin bonds due to the con- tribution of CT CW's becoming increasingly important.

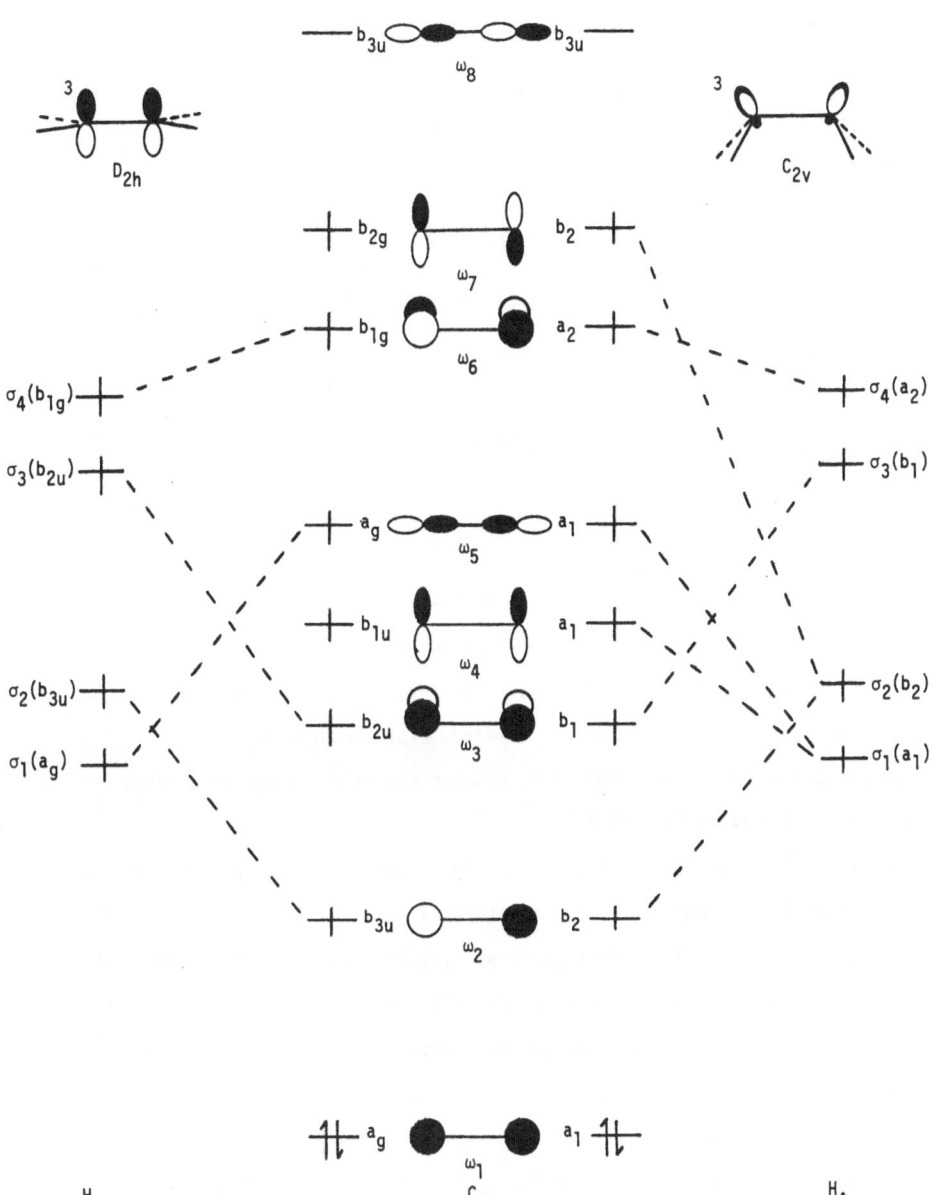

Figure 11. Detailed bond diagrams of planar (D_{2h}) and cis-bent (C_{2v}) triplet ethylene.

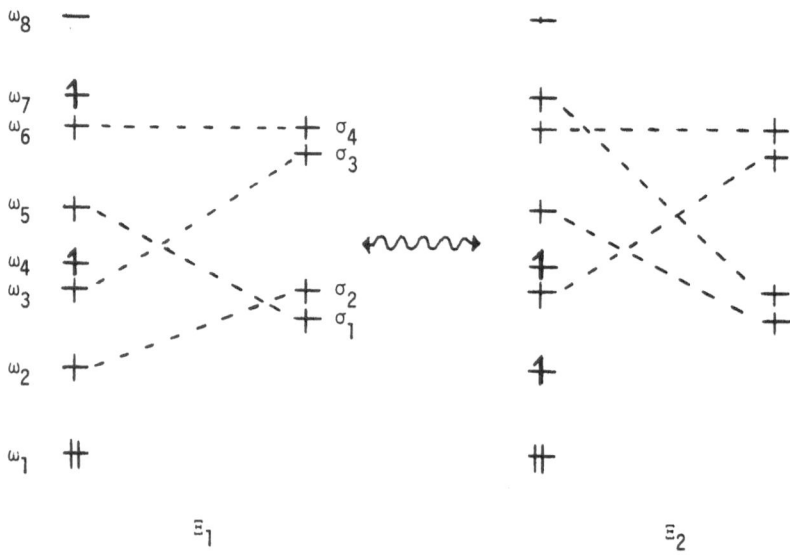

$$\Xi_1 \qquad\qquad\qquad \Xi_2$$

We conclude then that the thermodynamic stability of the metal-ethylene complex relative to the ground metal and ground ethylene is a function of the triplet excitation of planar ethylene, the relaxation energy of the planar triplet ethylene, and the hybridization of the relaxed triplet ethylene.

How do all of these determinants of complex stability change as ethylene is replaced by perfluoroethylene (TFE)?

a. Because TFE has a weaker pi bond than ethylene,[13] its triplet excitation energy is most likely lower than that of ethylene.

b. Because of the electronegativity of fluorine, the Ξ_2 contributor of the resonance hybrid of relaxed cis triplet TFE will become more important, the carbon AO's utilized for sigma bonding more electronegative and stronger overlap binders because of a greater percentage s character, and the metal-TFE bonds stronger than those linking metal and ethylene.

Finally, replacement of TFE by tetracyanoethylene (TCNE) will adversely affect the metal-olefin bonding because TCNE does not have the abnormally weak pi bond of TFE. TFE will tend to utilize carbon 2s AO's more than TCNE because the sigma electronegativity of fluorine is greater than that of CN. Thus, only the greater pi electron acceptor ability of TCNE will favor M-TCNE over M-TFE.

Hence, we conclude that TCNE will bind more strongly than TFE to a metal only if the metal (or, the second fragment, in general) is a modest or powerful electron donor which accentuates the charge transfer mechanism of enhancing the stability of the complex.

In order to understand better how MOVB differs from HMOVB (equivalent to HMO) theory, let us consider the problem of adiabatic $M-CX_2=CX_2$ (X=H,F) binding in which no geometry relaxation and, hence, rehybridization, is permitted to occur. How do predictions regarding substituent effects differ at the two levels of theory?

The form of the MOVB wavefunction for the complex $M-CX_2=CX_2$ is the one shown below.

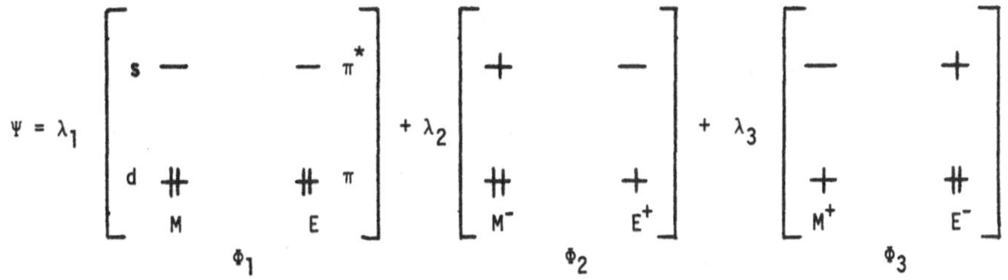

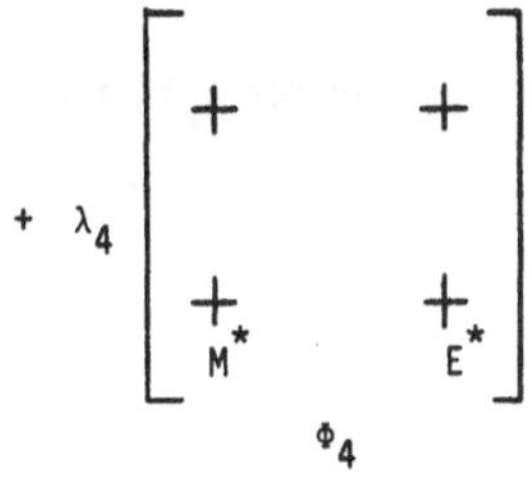

M = Metal

E = Olefin

$\lambda_3 > \lambda_2$

If we were to carry out a Hückel MOVB (HMOVB) computation, we would find that $\lambda_1 > \lambda_3 > \lambda_4$ because of the integral approximations at this level of theory.[1] Accordingly, we would say that, since $\lambda_3 > \lambda_4$, the stability of the complex depends on the donor ability of L_4M and the acceptor ability of C_2X_4. Thus, we would predict that as the electron affinity of C_2X_4 increases, or, the energy of the Lowest Unoccupied MO (LUMO) of C_2X_4 is depressed, K will increase. Since the HOMO's and the LUMO's of C_2H_4 and C_2F_4 have very similar energies, we would predict $K_F \approx K_H$.[14] By contrast, if we were to carry out a proper MOVB computation, we would find that $\lambda_1 > \lambda_3 \approx \lambda_4$ since now interelectronic repulsion effects are "contained" in the theory. Accordingly, we would say that, since $\lambda_3 \approx \lambda_4$, the stability of the complex depends greatly on the $\pi\pi^*$ triplet excitation energy C_2X_4. Since this is related to the C - C pi bond dissociation energy, we would predict $K_F > K_H$ because the weakest C - C pi bond is decidely that of C_2F_4 and the strongest that of C_2H_4. Why this is so will be discussed in Chapter 19.

It is well known that C_2F_4 and most fluoro olefins form more stable complexes with metals than C_2H_4 and its (nonfluoro) derivatives.[15] For example, Vaska has shown that K_{eq} of the reaction shown below is 100 times larger for C_2F_4 than for C_2H_4:[16]

$$Ir\ I(CO)(PPh_3)_2 + Olefin \xrightleftharpoons{K} Ir(olefin)\ I(CO)(PPh_3)_2$$

These experimental results can in no way be accomodated within the framework of HMO-DCD theory. In fact, they have often been presented as interesting oddities. By contrast, MOVB theory allows for a natural explanation of the (unjustifiably) "unexpected" great coordination ability of C_2F_4. Furthermore, it has been known for some time that fluorines in fluorinated olefin complexes are generally bent back to a greater extent than other substituents. The MOVB analysis presented above makes clear exactly why this is so. Noting that in Φ_1, Φ_3, and Φ_4 there are two, one, and zero net pi bonding electrons within the olefin fragment (if overlap is neglected) and recalling that Φ_4 makes a predominant contribution only when the olefin bears fluorine substituents (because of sigma-pi hybridization that weakens the C - C pi bond discussed in Chapter 6), it is clear that the greater bending of fluoro olefins is brought about by the perfect pairing CW which essentially has two high spin open shell fragments (metal and ligands) coupled into an overall singlet. This known trend cannot be reproduced by EHMO theoretical calculations for reasons explained before.[17]

C. The Trans Effect

The term "trans effect" is a code word for a number of presumably related phenomena which allegedly reflect the influence of a group on a ligand located in a trans position within an inorganic complex.[5] In applying MOVB theory to this problem, we are now cognizant of the fact that many of what people have long thought to be related problems are indeed very different ones.[15] Thus, we now seek to explain the following observations related to square planar $d^8MA_2X_2$ complexes, where M is the metal and A and X two ligands of different actual or effective group electro-negativity (with A being more electropositive and X more electronegative), which we believe that they define a "homogeneous" trans effect problem:

a. In the absence of severe ligand nonbonded repulsion, the cis isomer tends to be more stable (in an enthalpic sense) than the trans isomer.[18]

b. The M-X bond lengths differ in the two geometric isomers with the one in the cis being longer than the one in the trans form.

$$
\begin{array}{ccc}
PEt_3 & Cl & PPhEt_2 \\
| & | & | \\
Cl - Pt - Cl & Et_3P - Pt - Cl & H - Pt - Cl \\
|\quad 2.32 \text{ Å} & |\quad 2.42 \text{ Å} & |\quad 2.42 \text{ Å} \\
PEt_3 & Et_3P & PPhEt_2
\end{array}
$$

Traditionally, inorganic chemists have thought of this structural difference as reflecting a universal trans influence of a variable substituent Z on a fixed substituent Y. Thus, in the case of the above example, it is said that Et_3P has a stronger trans effect than Cl with respect to the M-Cl bond length. The conse-quences of this type of thinking are evident when we realize that the unsymmetrical $PtHCl(PPhEt_2)_2$ system has also been utilized in order to further classify ligands according to their trans influence with the conclusion being that H is as good a trans effector as Et_3P. However, we now recognize that all three Pt complexes shown above involve entirely different types of bonding and that the bonding of the symmetrical cis and trans species is substantially different from that of the unsymmetrical species. Since the cis-trans comparison is the most convenient one because of maintainance of symmetry, we say that the trans effect problem in geometric isomers is tantamount to the problem of geometric isomerism itself and the trans effect is closely related to the enthalpic preference for the cis geometry exhibited by such d^8 metal complexes.

    c. The s character of the M-A bonds, as measured by the J (M-A) nuclear spin coupling constant, is greater in the cis isomer. A typical example is given below.[5b]

	trans-$[PtCl_2(Bu_3P)_2]$	cis$[PtCl_2(Bu_3P)_2]$
J(Pt-P) (in c/s)	2380 ± 4	2508 ± 6

    d. There is ample physical evidence showing that in the $PtCl_2(PR_3)_2$, platinum-phosphorus bonds are stronger in the cis isomer and that the converse is true of platinum-chlorine bonds.[19]

All these observations can be explained by simply drawing the bond diagrams for the cis and trans isomer of planar MA_2X_2 and noting the difference in the bonding of the two isomers ordained by symmetry. For illustrative purposes, we use the case of $PtCl_2(PR_3)_2$ noting the following:

    1) The metal has eight d electrons.

    2) Both Cl and PR_3 ligands have doubly occupied orbitals of pi symmetry. However, because of their much higher energy, only the pi doubly occupied orbitals of Cl need be considered.

    3) Only PR_3 has low lying vacant sigma-antibonding orbitals of pi symmetry.

    4) Doubly occupied and unoccupied ligand orbitals which have sigma symmetry

can be neglected, to a first approximation, by assuming that they act as polarizers
of already existing, so to speak, sigma bonds. It can actually be shown that inclusion of them in the bond diagram would only strengthen the conclusions reached
in their absence.

The bond diagrams for cis and trans $PtCl_2(PR_3)_2$ are shown in Figure 12. Using
standard MOVB concepts, we interpret them as follows:

a. In the absence of severe nonbonded ligand repulsion due to the
interaction of doubly filled orbitals, the cis isomer will have lower energy
than the trans isomer for two main reasons:

1) It has three core-ligand pi bonds whereas the trans has only two.

2) It is D-bound whereas the trans form is U-bound through sigma core-
ligand overlap.

b. In the cis geometry, the metal s AO is primarily used for bonding the
more electropoistive P. As a result, the Pt-P bonds have greater metal s
character than the Pt-Cl bonds. In the trans geometry, the trend is reversed
as the metal s AO is now used primarily to bind the Cl ligands.

c. The trans $\longrightarrow$ cis transformation generates greater core-ligand
overlap repulsion because of symmetry reduction, the situation being entirely
analogous to the one encountered in the case of the trans $\longrightarrow$ gauche conversion
of N_2H_4. Hence, in the hypothetical absence of pi effects, we predict that
Pt-Cl and Pt-P bonds will be longer in the cis geometry. Indeed, Pt-Cl
bonds are longer but Pt-P bonds are shorter in the cis isomers. The
reason for the last trend is that the PR_3 ligands are bound to the metal
by two pi bonds in cis but only one is trans and this pi effect counter-
acts the opposing sigma effect.

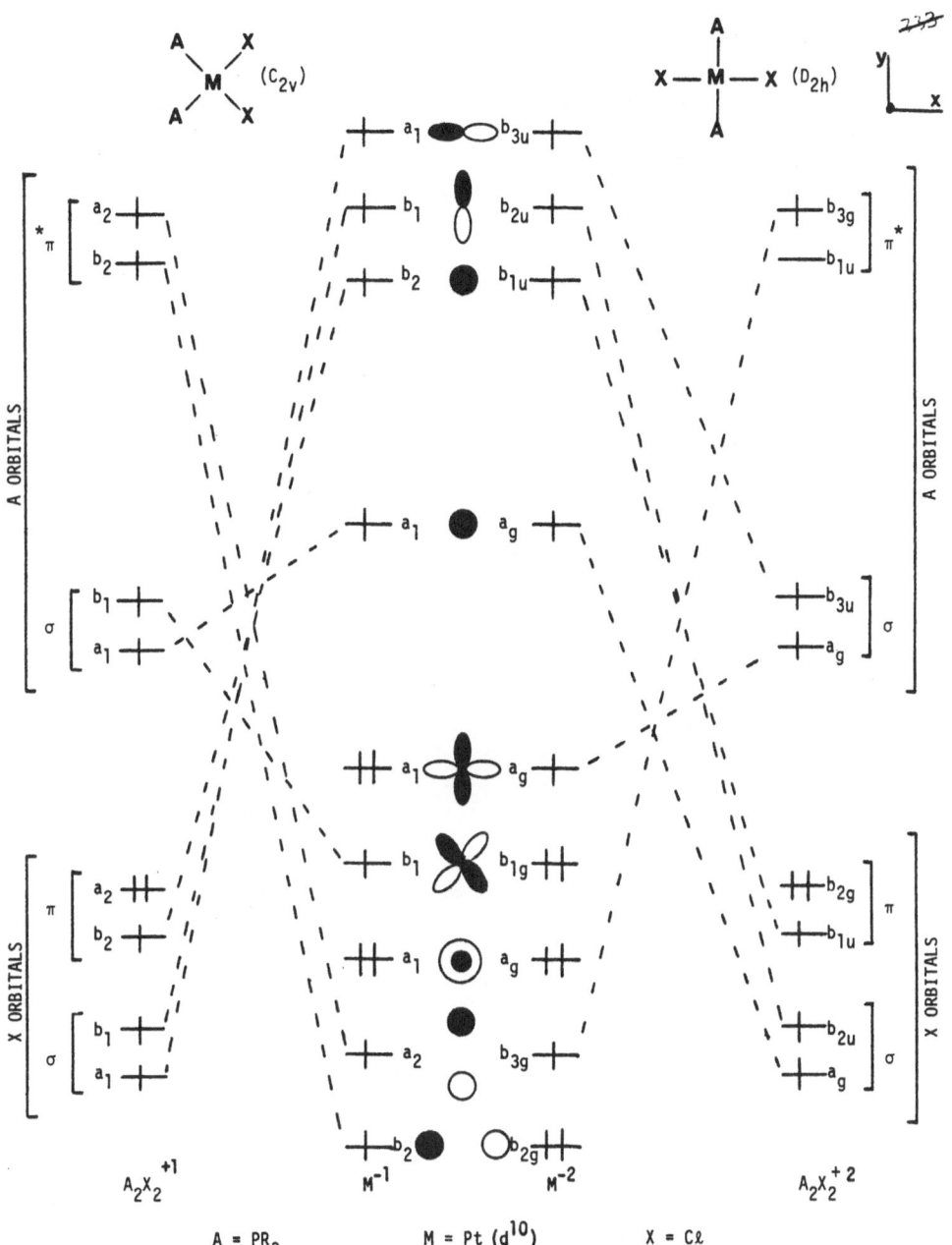

Figure 12. Compact bond diagrams of cis and trans planar PtCl$_2$(PR$_3$)$_2$.

For assisting the reader to best understand exactly how the two geometric isomers differ in sigma bonding, we have constructed the bond diagrams shown in Fiugre 13 and we have used bold dashed lines to indicate the two core-ligand bonds which are formed differently in the two species as a result of orbital symmetry constraints.. The reader will have no difficulty seeing that <u>sigma</u> bonding in cis is better than in trans MA_2X_2 ($M=d^8$ metal; A,X=two-electron donor ligands) for exactly the same reasons that it is better in cis than in trans $C_2H_2F_2$!

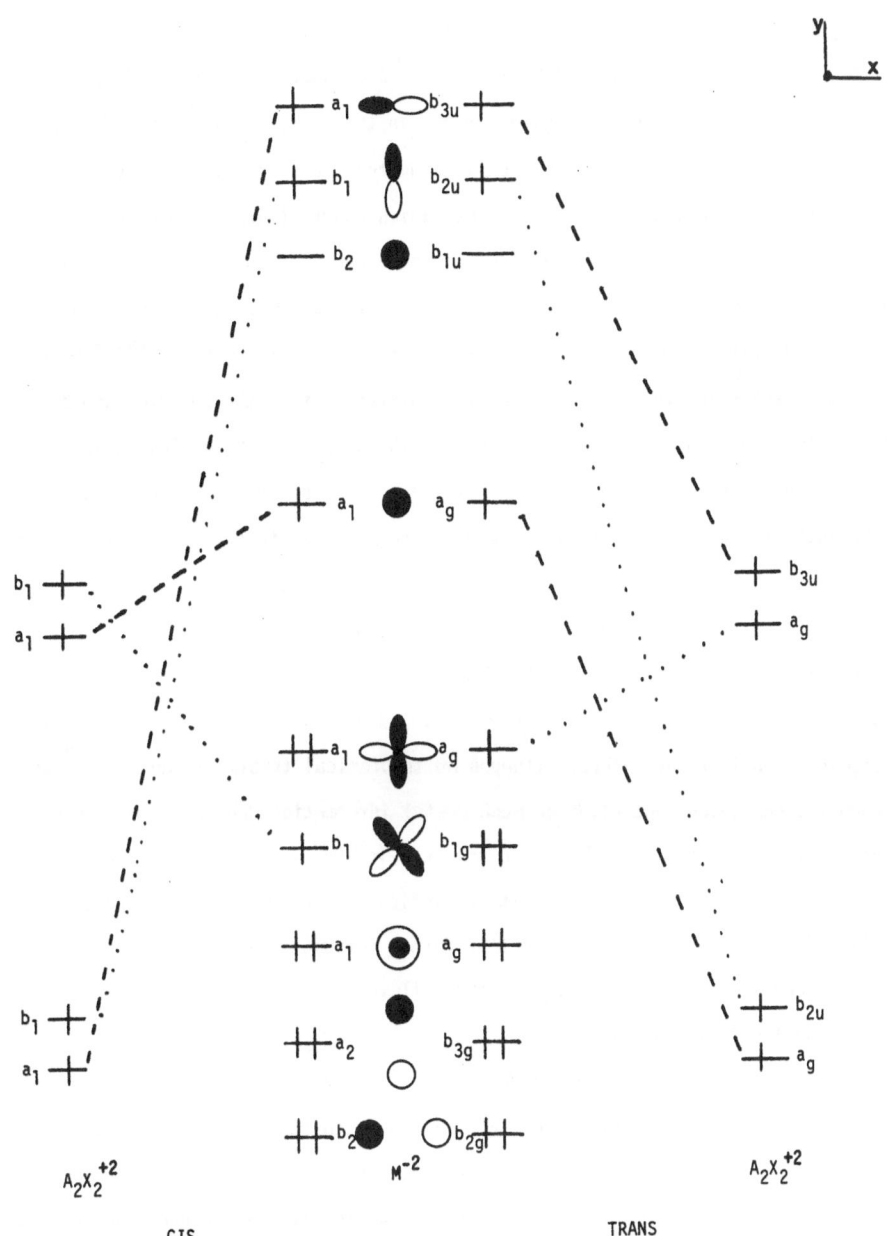

Figure 13. The difference in sigma bonding between cis and trans planar $PtCl_2(PR_3)_2$ (M = Pt, X = Cl, A = PR_3).

Conclusion

Some years ago, McWeeny, Mason, and Towl[20] published an ingenius MOVB theoretical justification of the DCD model. Even further in the past, Syrkin[21] provided an ingenious hybridization argument for explaining the trans effect. We have now seen how MOVB theory forecasts the problems into which EHMOVB (=EHMO) theory can run into because of overdelocalization, i.e., we have seen why the McWeeny-Mason-Towl model is actually better than the EHMO variant of the DCD model, and we have provided an interpretation of the "thermodynamic" trans effect (based on the concept of parity stereoselection) which is not very different from the one proposed by Syrkin. Thus, we can say that the seeds of MOVB theory have been planted in inorganic chemistry a long time ago. That these have not grown is a consequence of the lack of proper formalism which would have made the theory applicable to all chemical problems and would have cast aside any doubts about its operational significance. In this treatise, I have uniformly used the same concepts to deal with all types of problems and the MOVB theoretical applications to inorganic and organometallic systems described here are nothing but links of an unending chain of chemical applications. Thus, although MO theoretical treatments do exist[22] and MO computations abound and although much useful information can be and has been obtained by using these methods, our confidence in MOVB theory is based on the fact that it is capable of making difficult predictions where current MO theoretical models fail because of eith conceptual or formal deficiencies. We say then that VB theory provides an excellent chance for building "strong" bridges between organic and inorganic chemistry.[23]

We end by asking the following question: Is there a fundamental difference between $A'X_n$ where A' is a main group atom and $A''X_n$ where A'' is a transition metal? The answer is straightforward: Because the valence AO's of A' are stacked in the order $ns \rightarrow np$, $A'X_n$ molecules are "forbidden" species if X --- X nonbonded overlap is substantial. By contrast, the fact that the valence AO's of A" increase in

energy in the order nd → (n+1)s → (n+1)p entitles these molecules to be

formally "allowed" species, where the terms "forbidden" and "allowed" refer to the

way in which the symmetries of the core and ligand MO's match. Since the nature

of bonding within AX_n determines its stereochemistry, we can say that many differences

between main group and transition element molecules which have yet to be discovered

will be primarily due to the difference in core-ligand orbital matching as illus-

trated below, where the arrows indicate the way in which the orbital nodal planes

increase:

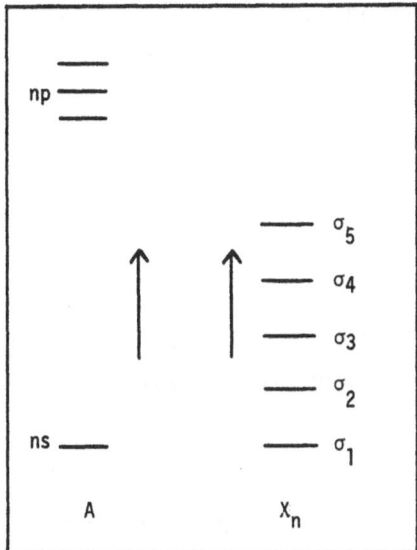 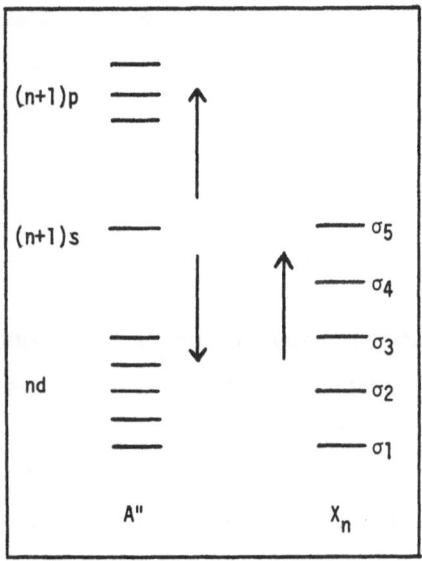

Low core AO (e.g., ns) symmetry   High core AO (e.g., ns) symmetry

matches that of low ligand MO (e.g.,   matches that of low ligand MO (e.g.,

σ_1), etc.: Parallel Overlap. σ_1), etc.: Diagonal Overlap.

References

1. Epiotis, N.D.; Larson, J.R.; Eaton, H. "Unified Valence Bond Theory of Electronic Structure" in Lecture Notes in Chemistry, Vol. 29; Springer-Verlag: New York and Berlin, 1982.

2. (a) Basolo, F.; Johnson, R.G. "Coordination Chemistry"; W. A. Bemjamin, Inc.: New York, 1964.

   (b) Cotton, F.A.; Wilkinson, G. "Advanced Inorganic Chemistry", 2nd Ed.; Interscience: New York, 1966.

3. (a) Dewar, M.J.S. Bull. Soc. Chim. France 1951, 18, C71.

   (b) Chatt, J.; Duncanson, L.A. J. Chem. Soc. 1953, 29, 39.

4. Huheey, J.E. "Inorganic Chemistry", 2nd Ed.; Harper and Row: New York, 1978.

5. (a) Chernyaev, I.I. Ann. Inst. Platine (U.S.S.R.) 1929, 4, 243.

   (b) Venanzi, L.M. Chem. in Britain, 1968, 5, 162.

6. For important contributions to the theory of the electronic structure of transition metal complexes using the EHMO method see:
   (a) Elian, M.; Hoffmann, R. Inorg. Chem. 1975, 14, 1058 and subsequent papers of the Hoffmann school.

   (b) Burdett, J.K. Chem. Soc. Rev. 1978, 7, 507.

7. Methane, the prototypical molecules of organic chemistry, is the "result" of the "forbidden" union of C and H_4.

8. In this analysis, we assume that an energy minimum due to coulomb correlation (van der Waals attraction) is absent and that state A is purely repulsive.

9. This is due to the fact that the percentage of "ionic" CW's, which involve strong interelectronic repulsion, decreases as the number of open shell electrons having the same spin increases. For explanations, see the chapter dealing with the electronic structure of excited states.

10. (a) Itel, S.D.; Ibers, J.A. Adv. Organomet. Chem. 1976, 14, 33.

    (b) Nelson, J.H.; Jonassen, H.B. Coordination Chem. Rev. 1971, 6, 27.

    (c) An extensive list of primary references can be found in: Albright, T.A.; Hoffmann, R.; Thibeault, J.C.; Thorn, D.L. J Am. Chem. Soc. 1979, 101, 3801.

11. (a) Vaska, L. Accounts Chem. Res. 1976, 9, 175.

    (b) Valentine, J.S. Chem. Rev. 1973, 73, 235.

    (c) Choy, Y.J.; O'Connor, C.J. Coordination Chem. Rev. 1972, 9, 145.

12. Black, M.; Mais, R.H.B.; Owston, P.G. Acta Cryst. 1969, B25, 1753.

13. The pi C-C and strengths of $CH_2=CH_2$ and $CF_2=CF_2$ are discussed in:

    (a) Benson, S.W. J Chem. Ed. 1965, 42, 502 (C_2H_4).

    (b) Wu, E.C.; Rodgers, A.S. J. Am. Chem. Soc. 1976, 98, 6112 (C_2F_4).

14. The lowest ionization potentials of $CH_2=CH_2$ and $CF_2=CF_2$ are almost identical. The electron affinity of $CH_2=CH_2$ is nearly the same as that of $CF_2=CFC\ell$, a close relative of $CF_2=CF_2$, as the vertical electron attachment energies differ by only 0.24 eV: Burrow, P.D.: Modelli, A.; Chiu, N.S.; Jordan, K.D. Chem. Phys. Letters, in press. Hence, it is safe to assume that $CH_2=CH_2$ and $CF_2=CF_2$ do not differ significantly in either their pi donor or pi acceptor ability.

15. For experimental evidence in support of the generalization that many fluoro-olefins form more stable complexes with metals than olefins with other substituents and that, in particular, C_2F_4 binds more strongly to metals than C_2H_4, see:

    (a) Cramer, R. J. Am. Chem. Soc. 1967, 89, 4621.

15. (continued)

(b) Cramer, R.; Kline, J.B.; Roberts, J.D. J. Am. Chem. Soc. 1969, 91, 2519.

(c) Fields, R.; Germain, M.M.; Haszeldine, R.N.; Wiggins, P.W. J. Chem. Soc. A 1970, 1969.

(d) Guggenberger, L.J.; Cramer, R. J. Am. Chem. Soc. 1972, 94, 3779.

(e) Stone, F.G.A. Pure Appl. Chem. 1972, 30, 551.

16. Vaska, L. Accounts Chem. Res. 1968, 1, 335.

17. For example, see Chapter 6.

18. The trans isomer of square planar $Pt(II)A_2X_2$ complexes is favored by entropy and the cis by enthalpy: Chatt, J.; Wilkins, R.G. J. Chem. Soc. 1952, 273, ibid. 1956, 525.

19. Pidcock, A.; Richards, E.R.; Venanzi, L.M. J. Chem. Soc. A 1966, 1707.

20. McWeeny, R.; Mason, R.; Towl, A.D.C. Discuss. Faraday Soc. 1969, 47, 20.

21. Syrkin, Y.K. Bull. Acad. Sci. U.S.S.R., Classe Sci. Chim. 1948, 69.

22. For a review of MO treatments of the problems dealt with in in chapter, see: Mingos, D.M.P. Adv. Organomet. Chem. 1977, 15, 1.

23. For a different qualitative VB treatment of inorganic molecules, including "hypervalent" molecules, the reader is referred to the interesting work of Harcourt: Harcourt, R.D. "Qualitative Valence-Bond Description of Electron-Rich Molecules", Lecture Notes in Chemistry, Vol. 30; Springer-Verlag: New York and Berlin, 1982.

Chapter 11. How to build Bridges by Molecular Orbital-Valence Bond Theory:
The Structures of A_2X_4 Molecules.

A_2X_4 molecules with ten valence electrons, assuming that the X ligands are
monovalent groups, have a choice of adopting one of the following two "extreme"
geometries: A planar, D_{2h}, or, a perpendicular, D_{2d}, conformation. From experi-

mental studies, it is well known that, depending on the nature of A and X, some
prefer one and others prefer the other geometry. Heuristic models such as Valence
Shell Electron Pair Repulsion (VESPR),[1] hyperconjugation,[2] etc.[3] are of no use in
tackling problems of this type. The traditional approach has been to construct
Mulliken-Walsh (MW) MO correlation diagrams[4] and attempt to find an answer in the
way in which the one-electron energies of the occupied MO's change as one conformer
is converted to the other. In performing this task, one inexorably finds that some
MO's "go up" and some "go down" in energy with the net result being uncertain in
most problems of interest. Of course, familiarity with available experimental facts
always allows one to claim that one or another MO is responsible for the overall
energy trend. In modern times, MW MO correlation diagrams are actually constructed
through explicit computation in which case one is hard pressed to provide a justification for the cataloguing of occupied MO energy changes the moment he could very
easily plot the total energy and dispense with the problem in this way. On the
other hand, if one rejects the explicit calculation, he is faced with the formidable
problem of generating symmetry adapted MO's in a reference geometry and mixing them
under the different symmetry constraints imposed along the conformational

isomerization pathway using either a variational or a high order perturbation procedure. It is no wonder then that the most fundamental problems of molecular stereochemistry are still enshrouded by a mantle of uncertainty and that an alternative approach is needed for a clear understanding of the factors which determine the stereochemical "decisions" of molecules. That is to say, what is needed is a qualitative theory of bonding which can be implemented on the "back of an envelope", so to speak, without compromising severely its formal correctness and without involving crude approximations, and which is capable of generating a "two-sentence" answer to a problem in a way which inspires further experimental or theoretical exploration of other related problems. MOVB theory[5] has been "designed" to meet these standards and in this paper we exemplify its analytical and predictive power by applying it to the problem of the stereochemistry of A_2X_4 molecules.

MOVB theory with core-ligand dissection tells us that how favorable a given geometry is depends on two factors:

a. The energy expenditure for preparing the core and ligand fragments for bonding by means of core and ligand excitation.

b. The strength of the ensuing core-ligand bonds.

Unfortunately, geometries which optimize both excitation and bond energies are rare because the nodal structures of orbitals dictate that excitation can be averted only at the cost of bond strength reduction. In other words, molecules containing main group atoms can be thought of as the result of a "forbidden" union of core and ligands in a reference geometry and removal of "forbiddenness" can only occur by symmetry reduction which entails reduction of core-ligand bonding. For example, linear water has strong O-H bonds but it involves high O excitation while bent water has weaker O-H bonds but it involves lower O excitation. This intellectural picture should be contrasted to the one emerging from the MW MO correlation diagram approach in which the transition from linear to bent H_2O is seen as accompanied by one MO rising and two MO's declining in energy, with a fourth remaining relatively unaffected. This tempts one to reach the incorrect conclusion that bent is more stable than linear H_2O because there is more AO mixing in the former. This often used argument is tantamount to a nonunderstanding of rehybridization and the blame for this must be placed squarely on the shoulder of MO theory as a conceptual tool and, in particular, to its inability to project clearly that rehybridization sacrifices bonding for deexcitation or it enhances bonding at the expense of excitation, at least in most problems of interest!

The Stereochemistry of B_2H_4 and Related Molecules

The compact bond diagrams of planar and perpendicular B_2H_4 as shown in Figure 1. It is immediately obvious that the D_{2h} form is the maximum overlap form and the D_{2d} the minimum excitation form. That is to say, the difference between the two geometries are the following two:

a. The ω_4 - σ_3 and ω_7 - σ_4 overlap integrals in D_{2h} are larger than the ω_4 - σ_3 and ω_3 - σ_4 (or, the ω_4 - σ_4 and ω_3 - σ_3) overlap integrals in D_{2d} and, thus, the planar geometry is characterized by two multicenter bonds which are stronger than the corresponding ones in the perpendicular geometry.

b. The core configuration required for bonding in the D_{2h} is $\omega_1^2\omega_2^1\omega_4^1\omega_5^1\omega_7^1$ while that required for bonding in the D_{2d} is $\omega_1^2\omega_2^1\omega_3^1\omega_4^1\omega_5^1$.
Thus, a transition from the perpendicular to the planar conformation is accompanied by large core excitation. This excitation is of the $\pi \rightarrow \pi^*$ type and since the energy gap separating the ω_3 and ω_7 B_2 core MO's is large, the excitation factor is expected to dominate the spatial overlap factor and, as a result, the D_{2d} form is expected to have lower energy than the D_{2h} form although the latter will have stronger core-ligand bonds than the former.

The bond diagrams of trans (C_{2h}) and gauche (C_2) hydrogen peroxide (H_2O_2) are shown in Figure 2. It is immediately evident that the C_2 preference of H_2O_2 and the D_{2d} preference of B_2H_4 are due to the fact that the $C_2 \rightarrow C_{2h}$ transformation in H_2O_2 necessitates a large $\pi \rightarrow \pi^*$ O_2 core excitation and the $D_{2d} \rightarrow D_{2h}$ conversion of B_2H_4 requires exactly the same: A large $\pi \rightarrow \pi^*$ B_2 core excitation. We say then that MOVB theory reveals a common denominator in two apparently unrelated problems. Experiment shows that H_2O_2 exists preferentially in a C_2 geometry[6] and computations forecast that D_{2d} has lower energy than D_{2h} B_2H_4.[7]

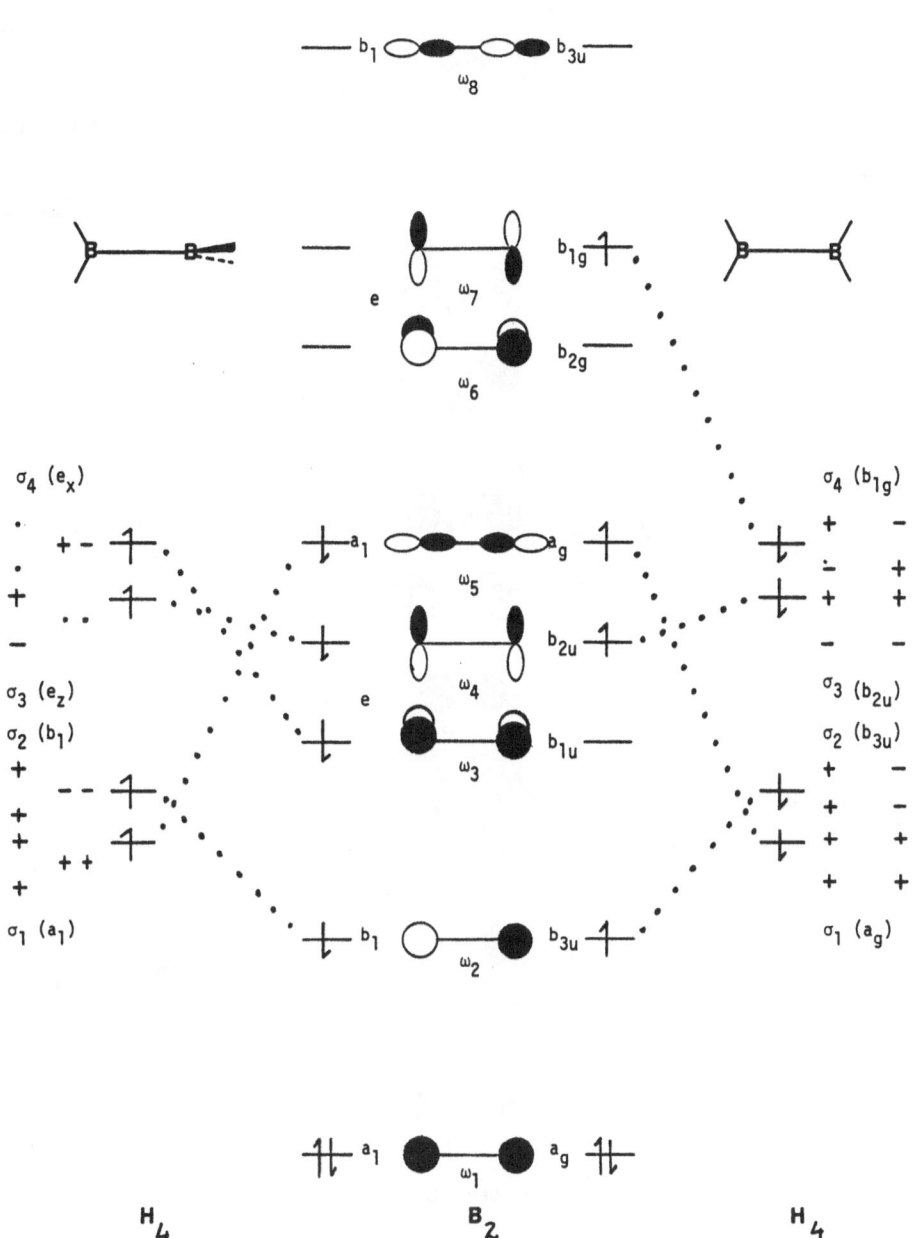

Figure 1. Compact bond diagrams of Planar (D_{2h}) and perpendicular (D_{2d}) B_2H_4.

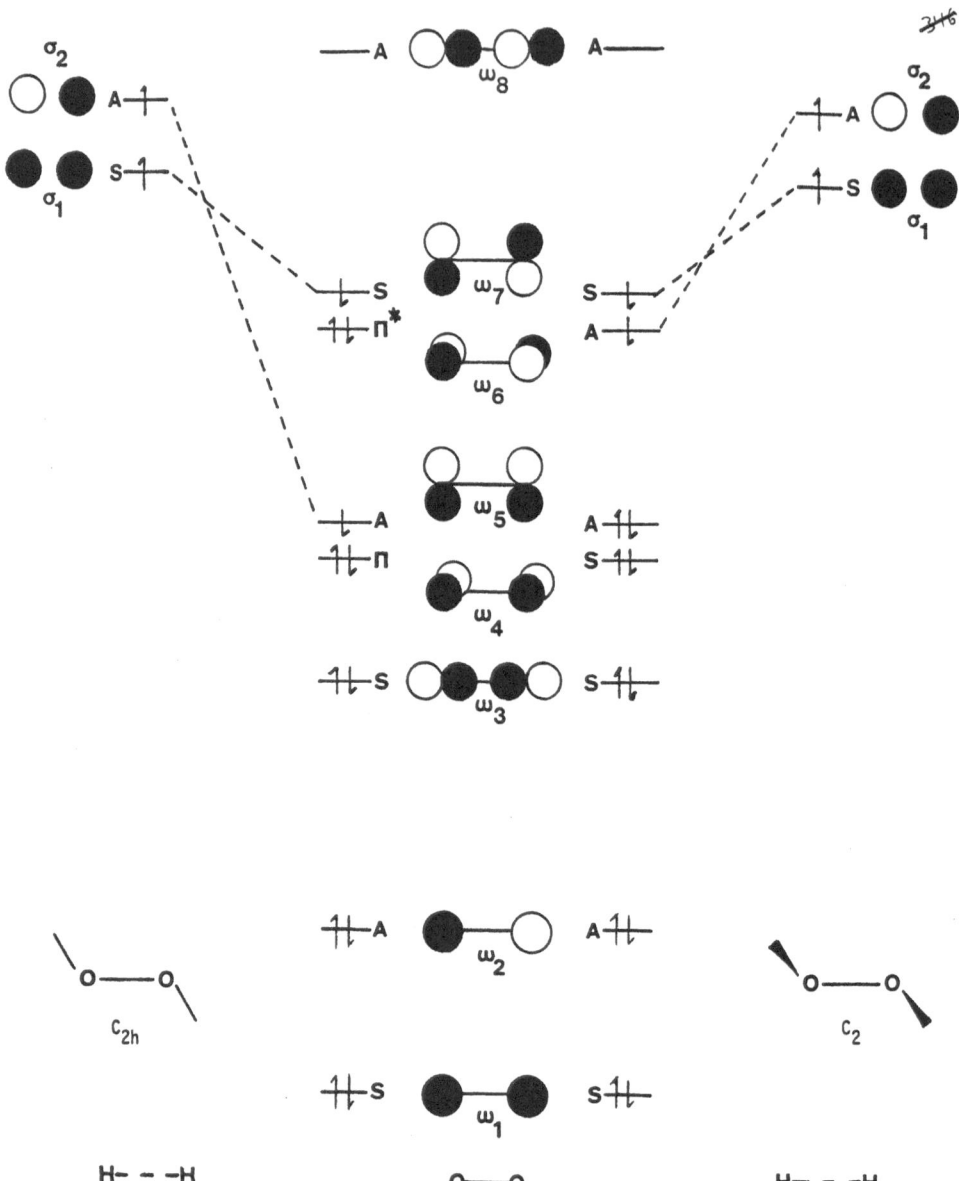

Figure 2. Compact bond diagrams of trans planar (C_{2h}) and gauche (C_2) H_2O_2.

Let us now replace the hydrogens by fluorines in B_2H_4 and inquire as to the stereochemical consequences of this type of substitution. These are projected by the two bond diagrams shown in Figure 3. It is now apparent that the existing holes in the B_2 core of B_2H_4 have disappeared[8] and we can no longer speak of differential core excitation in the D_{2h} and D_{2d} geometries of B_2F_4. The reason for this is that the two B_2 core holes combined with two doubly occupied ligand orbitals of appropriate symmetry to give rise to two additional core-ligand multicenter bonds in each of the D_{2h} and D_{2d} forms. All other doubly occupied ligand MO's can be neglected because they act as perturbers of existing core-ligand bonds and, thus, they play a secondary role in determining molecular stereochemistry. With the effect of differential excitation wiped out, we now predict that the planar form, which involves stronger core-ligand spatial overlap, will attain lower energy than the perpendicular form. Indeed, experimental investigations have demonstrated that B_2F_4 is planar.[9]

Here now, are some major chemical implications of the analysis presented above:

a. A ten-electron A_2H_4 system prefers naturally the D_{2d} conformation. However, replacement of at least two hydrogens by two substituents each of which contains at least an electron pair is expected to annihilate the differential excitation aspect of the D_{2h} and D_{2d} geometries and render the planar geometry more favorable because this affords maximum core-ligand spatial overlap. This means that many of the well known structures of organic molecules are as we know them because of the all too often disregarded, yet profoundly important, overlap factor. For example, the prototypical 1,3 butadiene system which exists in a planar transoid conformation is nothing but a $C_2H_2X_2$ system with ten sigma valence electrons and with each X having a pi-type electron pair (X = $\overset{\ominus}{CH_2}$).

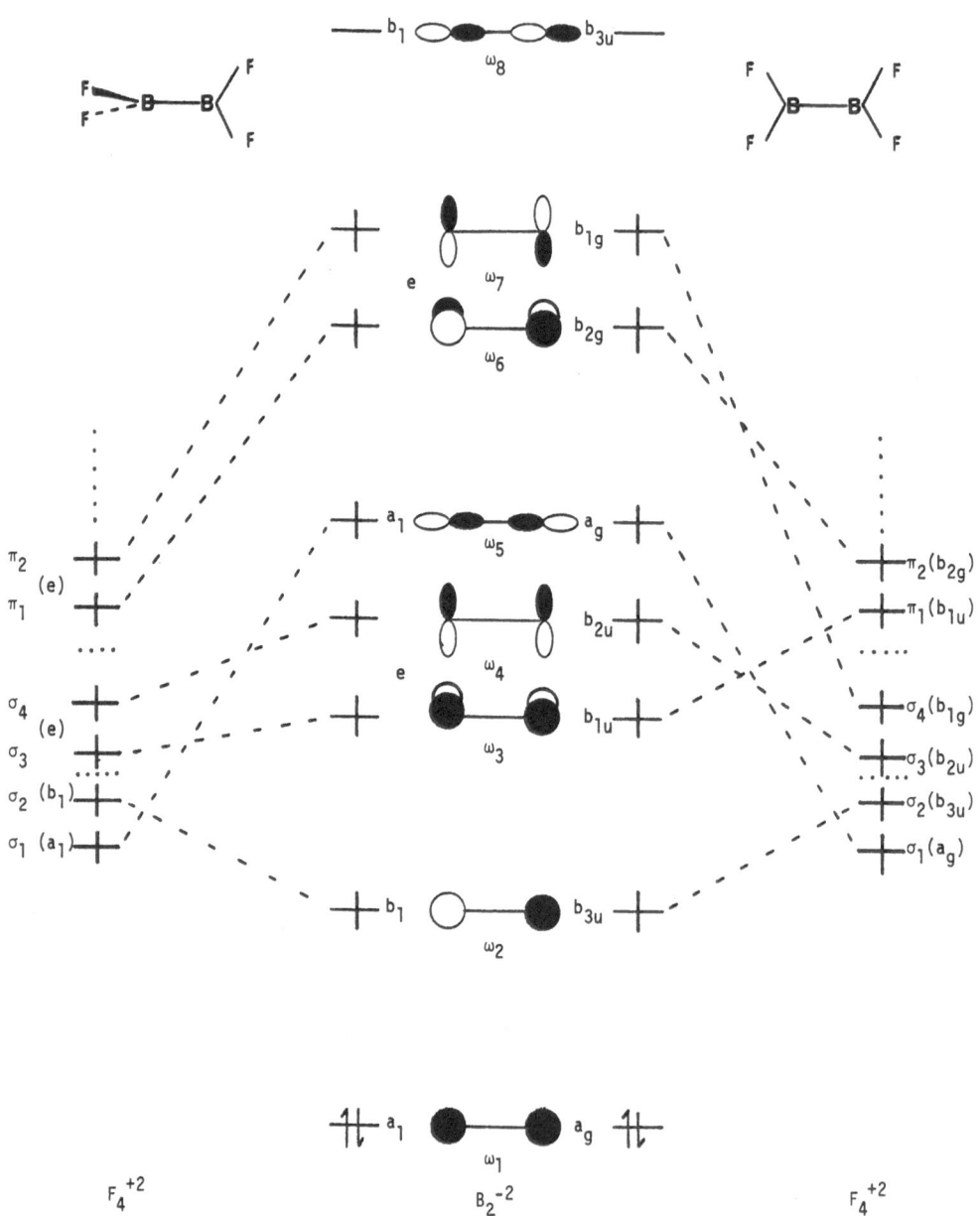

Figure 3. Compact bond diagrams of D_{2h} and D_{2d} B_2F_4. Only the two lone pair MO's of the F_4 ligand fragment which have symmetry appropriate for generating core-ligand bonds are shown. It is assumed that ω_8 cannot participate in bonding because it has very high energy.

The same thing is true of acrolein (CH_2=CH-CH=O) and glyoxal (O=CH-CH=O). Molecules of this type as well as the 2-monosubstituted and the 2,3-disubstituted parents are thus expected to have a planar (trans) conformation in order to take advantage of strong core (C_2) -ligand ($CH_2^{\ominus}HH^{\ominus}CH_2$, $O^{\ominus}HH^{\ominus}CH_2$, $O^{\ominus}HH^{\ominus}O$, etc.) binding. Experimental results are mostly in agreement with these expectations. From an examination of the literature, Devaquet, et al. concluded that "The most stable conformer of butadiene, acrolein, and glyoxal is planar s-trans. With a few exceptions this rule also holds for the mono- and di-substituted parents of these three basic dienes." Some data is summarized in Table 1.

At this point, we open a parenthesis in order to emphasize the self evident: Substituents with lone pairs have an excellent chance of planarizing A_2X_4 by creating two new core-ligand bonds. For this to happen the perfect pairing (R) CW implied by the bond diagrams of Figure 3 must make a dominant contribution to the total wavefunction. In turn, the R CW requires relocation of two electrons from ligands to core. For example, in B_2F_4 the R CW necessary for strongly predisposing B_2F_4 towards a D_{2h} geometry is $(B_2)^{-2}(F_4)^{+2}$. It follows that whenever such a two-electron transfer is energetically unfavorable, as is the case when the core and the ligands are neutral, (e.g., B_2 plus F_4 in B_2F_4) the correct description of the molecular system will involve less than six full core-ligand bonds and the predisposition twoards planarity will be small. In addition, one must take into consideration the "classical" coulomb and overlap repulsion of the nonbonded ligands which acts as to disfavor the D_{2h} geometry. As a result, the energetic advantage of D_{2h} over D_{2d} depends on the nature of A and X and most often is small.

Table 1 identifies a number of A_2X_4 molecules which adopt a D_{2d} geometry despite the fact that the X groups carry lone pairs. Particularly noteworthy is the trend shown below.

$$B_2F_4 \longrightarrow B_2Cl_4 \longrightarrow B_2Br_4$$
$$D_{2h} \qquad\qquad D_{2d} \qquad\qquad D_{2d}$$

What is the reason behind the occurrence of this trend? A_2X_4 can be considered as an A_2 core and an X_4 ligand fragment bound by a maximum of six bonds. We can represent the bonding in these molecules by the simple formula given below, where C stands for core and L for ligands:

$$\overset{-2 \qquad +2}{C \equiv\equiv\equiv L} \qquad\qquad \begin{array}{l} C = B_2 \\ L = X_4 \end{array}$$

We now recall that as C and L become weak overlap binders there will be an increasing tendency for adoption of a geometry which is compatible with maximal deexcitation. As we have already seen, the overlap binding ability of atoms declines as we move down a column of the Periodic Table. Hence, the above trend makes perfectly good sense reflecting an increasing unwillingness of the molecules to sustain ligand $\longrightarrow$ core charge transfer, a form of interfragmental excitation, much like electron promotion within a fragment is a form of intrafragmental excitation.

b. As we have discussed, the D_{2d} preference of 10-electon A_2H_4 systems and the C_2 preference of 14-electron A_2H_2 systems are both due to the same factor: Lower core excitation energy in the preferred geometry. We have also seen that the

Table 1. Gas Phase Structures of A_2X_4 and AX_2Y_2 Molecules.

X	X_2B-BX_2	$X_2\overset{+}{C}-\overset{+}{C}X_2$	$\overset{+2}{X_2}N-\overset{+2}{N}X_2$	$\overset{-}{O}X\overset{+}{C}-\overset{+}{C}X\overset{-}{O}$ $(O{=}C-C{=}O)$	$H_2\overset{-}{C}X\overset{+}{C}-\overset{+}{C}X\overset{-}{C}H_2$ $(H_2C{=}C-C{=}CH_2)$
H	D_{2d} (calc.)[1]			Planar(trans)[7]	Planar(trans)[10]
O^-	—	D_{2d} (calc.)[5]	D_{2h}[6]		
F	D_{2h}[2]				Planar(trans)[11] (calc.)
Cl	D_{2d}[3]			Planar(trans) + Gauche[8]	Planar(trans)[12]
Br	D_{2d}[4]			Planar(trans) + Gauche[9]	

Table 1.   References

1. Pepperberg, I.M.; Halgren, T.A.; Lipscomb, W.N. Inorg. Chem. 1977, 16, 363.

2. Danielson, D.D.; Patton, J.V.; Hedberg, K. J. Am. Chem. Soc. 1977, 99, 6484.

3. Ryan, R.R.; Hedberg, K. J. Chem. Phys. 1969, 50, 4986.

4. Danielson, D.D; Hedberg, K. J. Am. Chem. Soc. 1979, 101, 3199. See also:
    Odom, J.D.; Saunders, J.E.; Durig, J.R. J. Chem. Phys. 1972, 56, 1643.

5. Clark, T.; Schleyer, P.v.R J. Comput. Chem. 1981, 2, 20.

6. Bibort, C.H.; Ewing, G.E. J. Chem. Phys. 1974, 61, 1248. Snyder, R.G.;
    Hisatsune, I.C. J. Mol. Spectrosc. 1957, 1, 139.

7. Currie, G.N.; Ramsay, D.A. Com. J. Phys. 1971, 49, 317.

8. Hagen, K.; Hedberg, K. J. Am. Chem. Soc. 1973, 95, 1003.

9. Hagen, K.; Hedberg, K. J. Am. Chem. Soc. 1973, 95, 4796.

10. (a) Aston, J.G.; Ssasz G.J.; Wooley, H.W.; Brickwedde, F.G. J. Chem. Phys.
         1946, 14, 67.

    (b) Almenningen, A; Bastiansen, O.; Traetteberg, M. Acta Chem. Scand. 1958,
         12, 1221.

    (c) Batuer, M.; Ouischchenko, A.; Matreera, A,; Azonova, N.I. Dokl. Chem.
         (Engl. Transl.) 1960, 543.

    (d) Mavais, D.J.; Sheppard, N.; Stoicheff, B.P. Tatrahedron 1962, 17, 163.

11. Devaquet, A.J.P.; Townsend, R.E.; Hehre, W.J. J. Am. Chem. Soc. 1976, 98, 4068.

12. Ssasz, G.J.; Sheppard, N. Trans. Faraday Soc. 1953, 49, 358.

differential core excitation effect in the case of A_2H_4 system can be eliminated by
replacing the H's by substituents having lone electron pairs which can combine with
the core holes to define additional core-ligand multicenter bonds so that the core-
ligand bonds are cumulatively stronger in the D_{2h} geometry. In an analogous sense,
the differential core excitation effect in A_2H_2 systems can also be eliminated by
replacing the H's by substituents having low lying vacant orbitals which can combine
with the core lone pairs to define additional core-ligand bonds, with the bonds
being cumulatively stronger in the C_{2h} geometry. Since $-\overline{A}-\overline{A}-$ linkages are ubiquitous
in molecules of biological importance, one wonders to what extent nature uses
effective ligand modification in order to turn the "swivel" having an A-A arm in
order to move the attached fragments closer, or, further apart from each other for
the purpose of catalysis, control, etc.

  c. Substituents with low lying vacant orbitals will predispose the A_2X_4
systems towards a D_{2d} geometry and substituents with lone pairs will predispose the
$A_2 X_2$ systems towards a C_2 geometry because they cannot annihilate the differential
core excitation effect by giving rise to a complete set of core-ligand multicenter
bonds. All these prediction have never been tested in any systematic way and work
in this direction would be most welcome.

  d. Is there any way to convince the skeptical reader who is suspicious of new
ways of thinking that the analysis given in this paper is the "truth" and that
stereochemical diversity in nature is the result of the all-important conflict
between optimal fragment excitation and optimal fragment bonding? The ten-valence-
electron system B_2H_4 is perpendicular because the D_{2d} geometry is compatible with a
deexcited B_2 core. Similarly, the fourteen-valence-electron system H_2O_2 is gauche
because again the C_2 geometry is compatible with a deexcited O_2 core. What happens

in the case of the twelve valence electron C_2H_4? As demonstrated by the bond diagrams in Figure 4, there is now no core deexcitation accompanying the D_{2h} to D_{2d} conversion, only loss of core-ligand spatial overlap! Hence, ethylene represents a molecule which is planar because deplanarization in no way affects the fragment excitation energies and only causes weakening of core-ligand bonds. The three singlet "diradical" states of "perpendicular" ethylene in MOVB bond diagrammatic notation are shown in Figure 5.

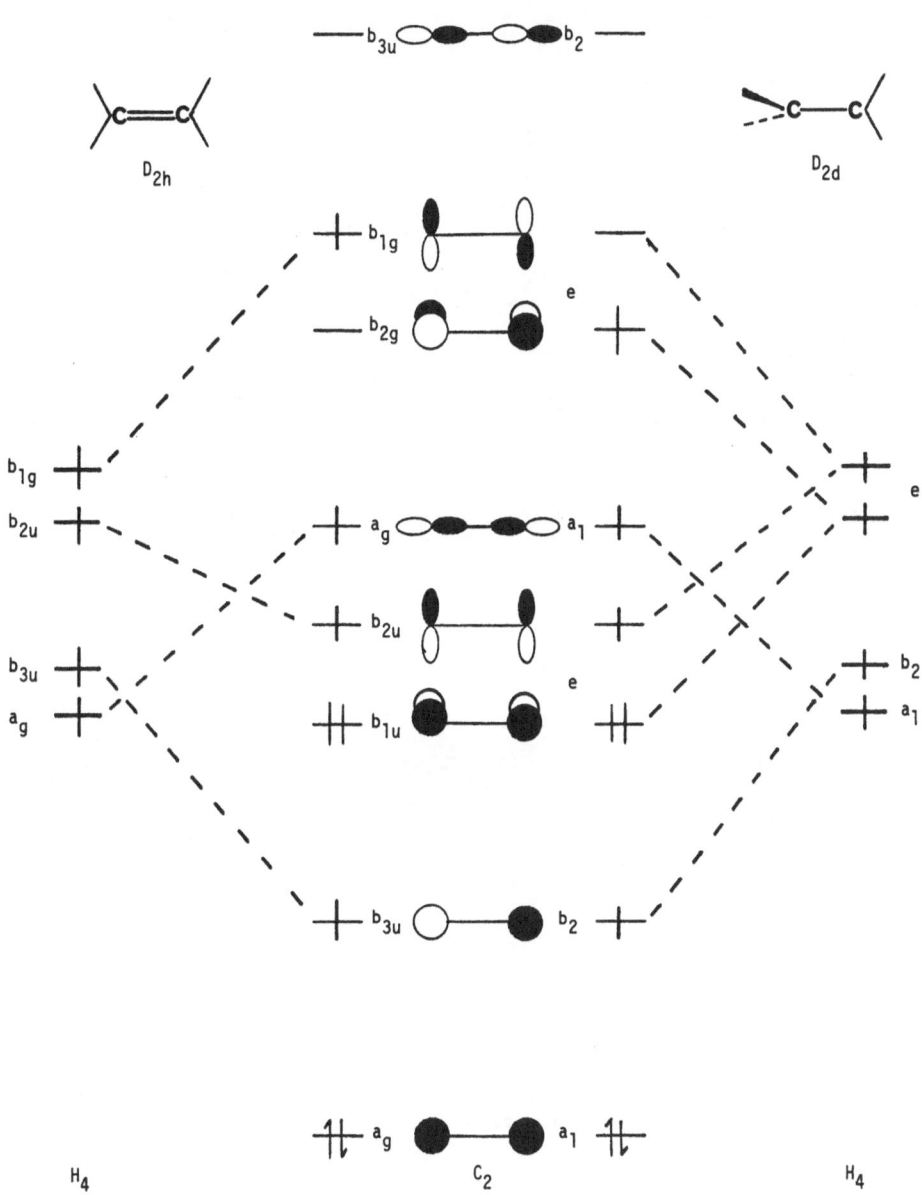

Figure 4. Compact EHMOVB bond diagrams of planar (D_{2h}) and perpendicular (D_{2d}) C_2H_4. The correct representation of D_{2d} C_2H_4 at the MOVB level is obtained by a linear combination of bond diagrams as shown in Figure 5.

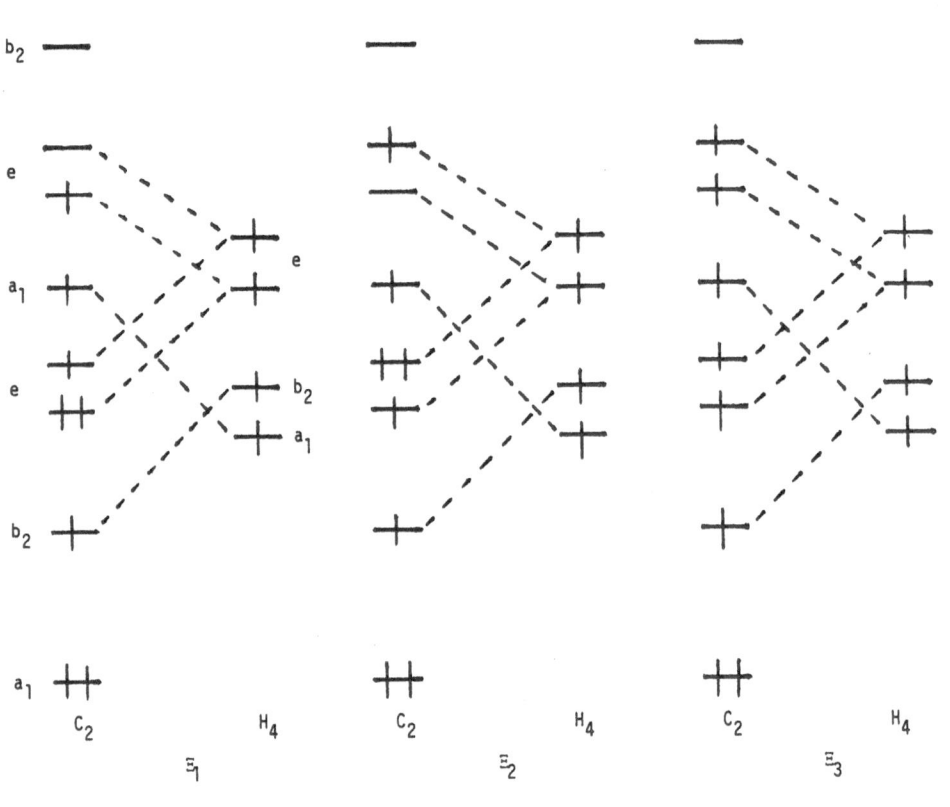

"DIRADICAL" STATES

$$\Psi_1 \propto \Xi_1 - \Xi_2$$

$$\Psi_2 \propto \Xi_1 + \Xi_2$$

$$\Psi_3 = \Xi_3$$

Figure 5. Bond diagrammatic representation of the three "diradical" states of D_{2d} C_2H_4.

References

1. (a) Sidgwick, N.V. "The Electronic Theory of Valency"; Oxford University Press: London, 1927.

   (b) Gillespie, R.J. "Molecular Geometry"; van Nostrand Reinhold: London, 1972.

2. (a) Wheland, G.W. J. Chem. Phys. 1934, 2, 474.

   (b) Pauling, L.: Springall, H.S.; Palmer, K.J. J. Am. Chem. Soc. 1939, 61, 927.

   (c) Mulliken, R.S. J. Chem. Phys. 1939, 7, 339.

   (d) Mulliken, R.S.; Rieke, C.A.; Brown, W.G. J. Am. Chem. Soc. 1941, 63, 41.

   (e) Dewar, M.J.S. "Hyperconjugation"; Ronald Press Co.: New York, 1962.

3. (a) Bartell, L.S. J. Chem. Ed. 1968, 45, 754.

   (b) Pearson, R.G. "Symmetry Rules for Chemical Reactions"; Wiley and Sons, Inc.: New York, 1976.

4. (a) Walsh, A.D. J. Chem. Soc. 1953, 2260, 2266, 2288, 2296, 2301.

   (b) Walsh, A.D. Progress in Stereochemistry 1954, 1.

   (c) Mulliken, R.S. J. Am. Chem. Soc. 1955, 77, 887.

5. Epiotis, N.D.; Larson, J.R.; Eaton, H. "Unified Valence Bond Theory of Electronic Structure" in Lecture Notes in Chemistry, Vol. 29; Springer-Verlag: New York and Berlin, 1982.

Chapter 12. Why Benzene prefers to substitute and an Olefin likes to add?

Hückel MO (HMO) theory,[1,2] predicts that pi benzene is more stable than three pi ethylenes. This conclusion seems to be compatible with the fact that the heat of hydrogenation of benzene is much less than the heat of hydrogenation of three cyclohexenes[3] and the known unwillingness of benzene to undergo addition reactions, opting for "aromatic" substitution instead. These data have prompted an on-going preoccupation with "resonance energies", "aromaticity", and the like. In this chapter, we suggest that, while the experimental facts are indisputable, the concepts which chemists have devised over more than a century are most likely erroneous and that benzene is pi destabilized but operationally "aromatic".

Figure 1 shows the relative energies of the following four species:
CT: 1,3,5 cyclohexatriene, in which there is no pi bond interaction and no sigma bonds holding the fragments together. This hypothetical species is modelled by three acetylenes.

BZ: D_{6h} benzene.

CTH_6: The hydrogenation product of CT. This hypothetical species is modelled by three ethylenes.

CY: D_{6h} cyclohexane modelled by the lowest energy conformational isomer of cyclohexane.

Figure 1 also contains a set of definitions of energy differences which will be used consistently in this work. ΔE and $\Delta E'$ are resonance energies which have sigma and pi components, if one makes the usual assumption that pi and sigma bonds are formed independently of each other and H_1 and H_2 are the heats of hydrogenation of CT and BZ, respectively. The following things become immediately evident:

a. The fact that BZ is more stable than CT in no way reveals whether the six pi electron interaction favors the CT $\rightarrow$ BZ conversion or whether it acts in an opposite direction. Chemists brought up with HMO theory believe that, indeed,

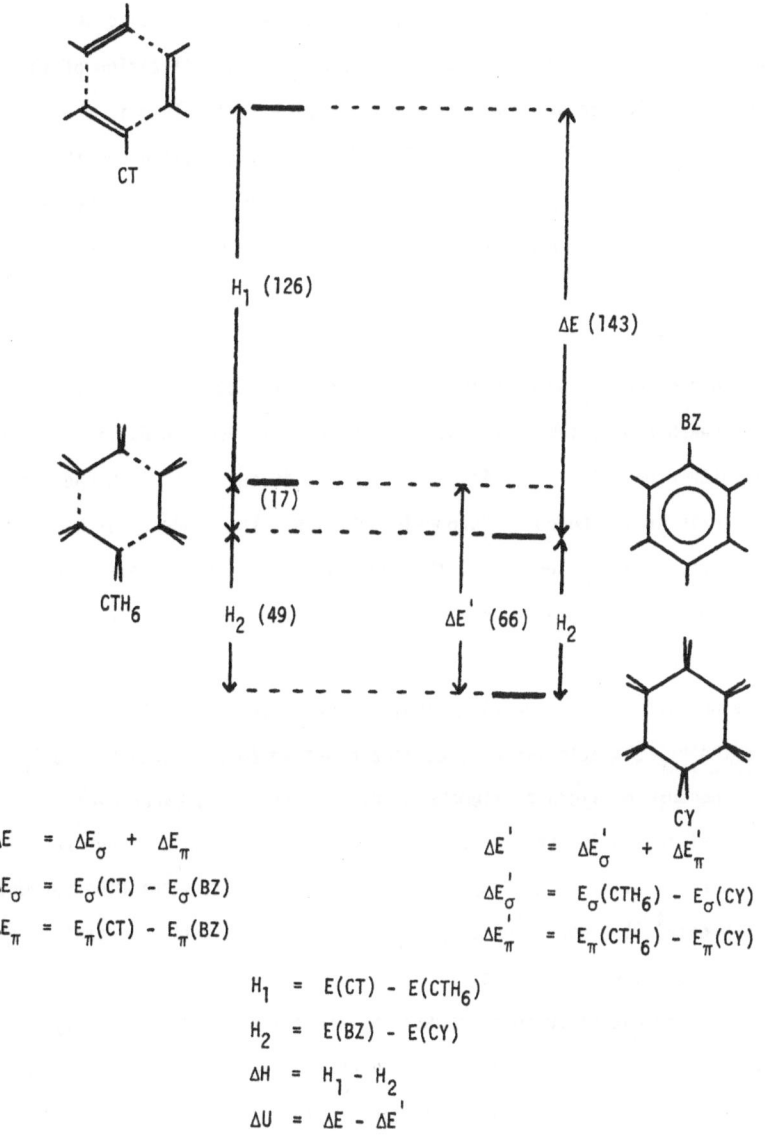

$$\Delta E = \Delta E_\sigma + \Delta E_\pi$$

$$\Delta E_\sigma = E_\sigma(CT) - E_\sigma(BZ)$$

$$\Delta E_\pi = E_\pi(CT) - E_\pi(BZ)$$

$$\Delta E' = \Delta E'_\sigma + \Delta E'_\pi$$

$$\Delta E'_\sigma = E_\sigma(CTH_6) - E_\sigma(CY)$$

$$\Delta E'_\pi = E_\pi(CTH_6) - E_\pi(CY)$$

$$H_1 = E(CT) - E(CTH_6)$$

$$H_2 = E(BZ) - E(CY)$$

$$\Delta H = H_1 - H_2$$

$$\Delta U = \Delta E - \Delta E'$$

Figure 1. The relative energies of CT, CTH_6, CY and BZ as inferred from thermo-chemical data and definitions used in the text.

BZ is _pi stabilized_ relative to CT but the experimental evidence is also consistent with BZ being _pi destabilized_ relative to CT with the sigma stabilization of the former relative to the latter overriding the adverse pi effect. This is the explanation we shall advance in this work. That is to say, we will argue that, while at the level of HMO theory ΔE_π is positive, it becomes actually negative at the higher level of theory which includes the effect of electron-electron interaction.

   b. The fact that the heat of hydrogenation of CT is greater in absolute magnitude than that of BZ is the uncontestable basis of the concept of benzene "aromaticity". Furthermore, the fact that H_1 is more positive than H_2 is compatible with ΔE_π being either negative or positive. Hence, it is the sign of ΔH and not that of ΔE_π which is at the heart of "aromaticity", i.e., the tendency of benzenoid molecules to undergo substitution rather than addition reactions. Chemists have always been fond of pi effects only because there has been a simple enough theory (pi HMO theory) which permitted the calculation of ΔE_π and disdainful of sigma effects because no comparable theoretical tool has been available.

   c. Since the key characteristic of BZ is a negative ΔH and since BZ has D_{6h} symmetry either because of sigma or sigma plus pi stabilization, experimental observations pertaining to the static geometry of BZ tell us nothing about why BZ has the geometry it does. For example, the fact that there is "pi ring current" according to NMR data[4] is a consequence of D_{6h} symmetry which tells us nothing as to whether ΔE_π is positive or negative.

   There is an important difference between ΔE_π and ΔH: The first quantity is a mere energy difference while the second is a difference of two energy differences. It is exactly this fact that makes the prediction of the sign of ΔH easy while rendering the prediction of ΔE_π a tricky problem for anyone who has forgotten what is "in" and what lies "out" of HMO theory. The reason for this is that, in going from CT to BZ, there is sigma bond formation accompanied by pi delocalization. Now, such delocalization lowers the one-electron energy of a system, because of the interaction of "covalent" and "ionic" CW's, <u>at the expense of strong inter-electronic repulsion,</u> the characteristic attribute of "ionic" CW's. Hence, in comparing CT and BZ, one must consider one-electron and two-electron effects while recognizing that HMO theory describes <u>overdelocalization</u> because of the neglect of electron-electron interaction, i.e., it misrepresents one-electron effects by totally neglecting two-electron effects. By contrast, addition of $3H_2$ to CT as well as addition of $3H_2$ to BZ require intrafragmental excitations which cause similar readjustments of the relative weights of the "ionic" and "covalent" CW's. Hence, in comparing H_1 and H_2, one may neglect interelectronic repulsion effects. We then conclude that the problem of benzene "aromaticity" can be considered solved if one can explain why ΔH is positive, assuming that interelectronic repulsion effects are unimportant, and predict whether ΔE_π is indeed positive, as predicted by HMO theory, or actually negative as a result of bielectronic destabilization dominating mono-electronic stabilization of the cyclic pi system of benzene. We now show that ΔH has the same sign as ΔE and that ΔE_π does not have necessarily the same sign as ΔE. Furthermore, since $\Delta E' + H_1 = \Delta E + H_2$, the correct interpretation of benzene "aromaticity" is tantamount to a proper rationalization of the sign of either ΔH or ΔU and that ΔE_π actually leads to the conclusion that benzene is pi destabilized. Thus, the fact that benzene resists addition, while an olefin readily undergoes such a reaction, can be explained in two different, yet equivalent, ways with the ΔH explanation being the more physically meaningful one as it directly links the pi electronic structure of two reactants to their differing ability to form two "new" sigma bonds to replace one "old" pi bond.

Why there is nothing Magic about Six Pi Electron Delocalization in Benzene

In general, H_i reflects two factors:

a. The energy required for preparation of the reacants for bonding, i.e., the requisite reactant excitation energies.

b. The energy payback due to bond formation.

Since we make C - H bonds in both CT and BZ, the AO resonance integrals describing bond formation are kept constant. Thus, the only difference between CT and BZ with respect to reaction with H_6 is due to the differing excitation energies. The bond diagrams which show explicitly how CTH_6 is formed by adding CT and H_6 and how CY arises from the union of BZ and H_6, are shown in Figure 2. Both CTH_6 and CY can be regarded as the products of the "forbidden" unions of CT plus $3H_2$ and BZ plus $3H_2$, respectively. It is immediately evident that there is a greater excitation requirement in the case of BZ than in the case of CT. Hence, ΔH is predicted to be positive. Since ΔH is a measure of relative reaction exothermicity, we say that CT is a hyperdesmic and BZ is a hypodesmic molecule, in a relative sense. We now defend this conclusion in some detail.

In constructing the bond diagrams, we have made three assumptions. First, both CTH_6 and CY can be regarded as the products of the "forbidden" unions of CT plus $3H_2$

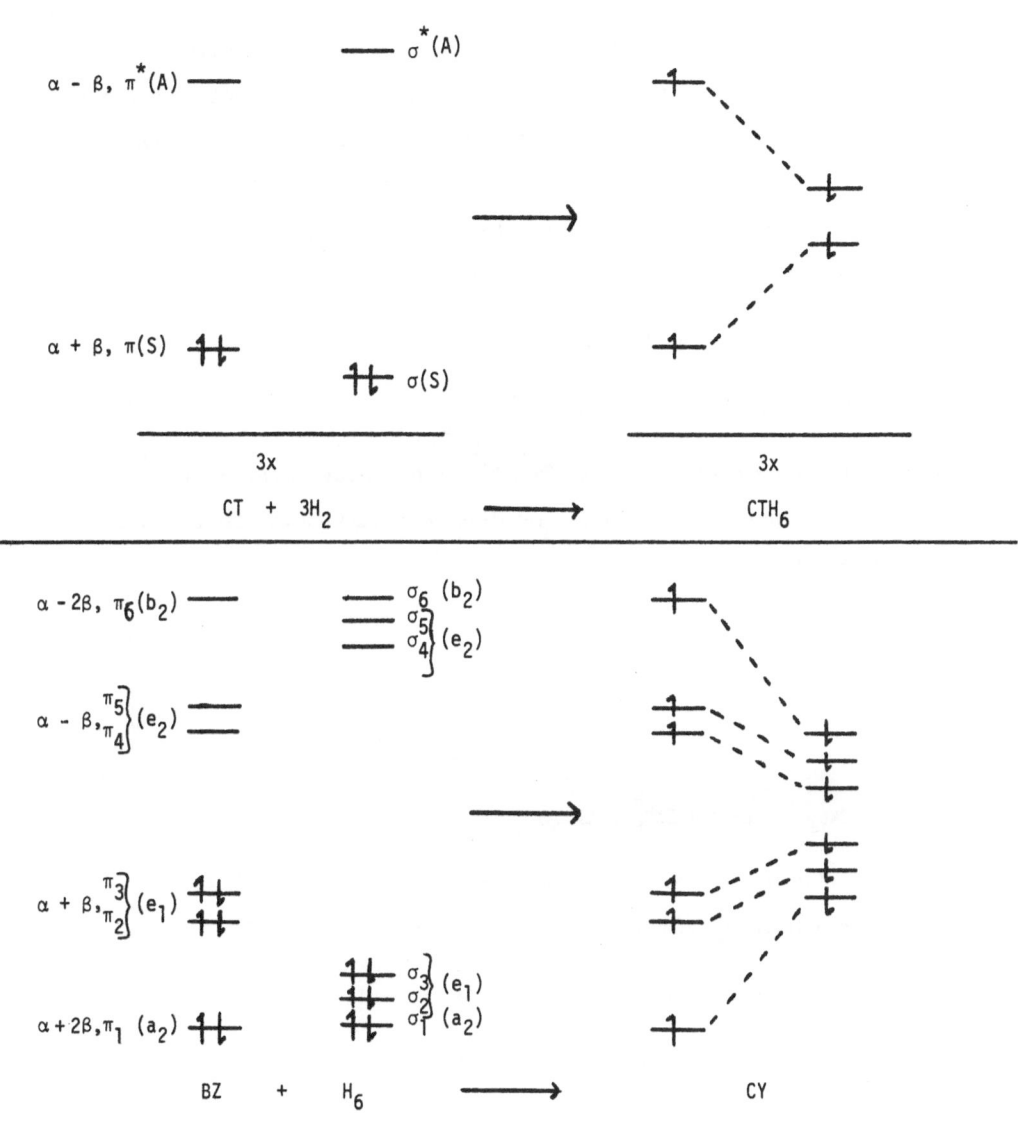

Figure 2. Bond diagrammatic representation of the reaction CT + $3H_2 \rightarrow CTH_6$ and BZ + $H_6 \rightarrow$ CY. For simplicity, it is assumed that nonbonded hydrogen AO overlap in the products is zero. If this assumption is not made, there is an even greater energy requirement for excitation for bond formation in the BZ + $3H_2$ system.

and BZ plus $3H_2$, respectively. We used as our orbital basis the pi Hückel MO's (HMO's) of CT (equivalent to three pi acetylenes) and BZ, the sigma HMO's of H_6 and we neglected overlap. Second, we assumed that the dominant Configuration Wavefunction (CW) of the total wavefunction of each composite system is the perfect pairing (R) CW of the total system so that the total excitation energy EXC_T in each case is defined, using CT as an example, as follows:

$$3\pi^2\sigma^2 \longrightarrow 3\pi^1\pi^{*1}\sigma^1\sigma^{*1}$$

$$EXC_T = (Q_\pi^* - Q_\pi) + (Q_H^* - Q_H) + \Delta J \tag{1}$$

where Q_π is the one-electron energy of $3\pi^2$, Q_H the one-electron energy of $3\sigma^2$, etc., and ΔJ is the change of interelectronic repulsion accompanying excitation. Thus, we can now define:

$$EXC_\pi^0 = Q_\pi^* - Q_\pi \tag{2}$$

$$EXC_H^0 = Q_H^* - Q_H \tag{3}$$

and

$$EXC_T \simeq EXC_\pi^0 + EXC_H^0 + \Delta J \tag{4}$$

Using the Pairing Theorem and assuming that the HMO coulomb integral, α, is zero, EXC_T can be expressed in terms of HMO energies, ε_i and ε_i', as follows:

$$EXC_\pi^0 \simeq -2\sum_i^{occ} \varepsilon_i \text{ where } \varepsilon_i = <\pi_i|\hat{H}|\pi_i> \tag{5}$$

$$EXC_H^0 \simeq -2\sum_i^{occ} \varepsilon_i' \text{ where } \varepsilon_i' = <\sigma_i|\hat{H}|\sigma_i> \tag{6}$$

$$EXC_T = -2\sum_i^{occ} \varepsilon_i - 2\sum_i^{occ} \varepsilon_i' + \Delta J \tag{7}$$

Finally, we did assume for simplicity that the excitation energy of the H_6 fragment is geometry invariant. As a result, we obtain:

$$\Delta EXC_T = EXC_T^0(CT) - EXC_T^0(BZ) + \Delta\Delta J \tag{8}$$

$$= EXC_\pi^0(CT) - EXC_\pi^0(BZ) \quad (\text{by assuming } \Delta\Delta J = 0)$$

$$= -2\sum_i^{occ} \epsilon_i(CT) + 2\sum_i^{occ} \epsilon_i(BZ)$$

$$= +2\beta$$

Equation (8) simply says that CT will be thermodynamically more reactive towards $3H_2$ than BZ, i.e., that ΔH will be positive.

$$\Delta H \quad \alpha - \Delta EXC_T \tag{9}$$

Next, let us compute the energy difference between pi CT and pi BZ, ΔE_π. We have:

$$\Delta E_\pi = Q_\pi(CT) - Q_\pi(BZ) + \Delta J' \tag{10}$$

We can tentatively assume that $\Delta J' \simeq 0$ and express ΔE_π in terms of HMO energies as:

$$\Delta E_\pi = 2\sum_i^{occ} \epsilon_i(CT) - 2\sum_i^{occ} \epsilon_i(BZ) \tag{11}$$

$$= -2\beta$$

The question now arises: Is $\Delta J' \simeq 0$ a good approximation? The answer is a resounding NO! The reason why $\Delta J'$ cannot be neglected and the reason why HMO theory fails to predict barriers in "allowed" pericyclic reactions can be easily understood by inquiring as to the error committed as a transition is made from MOVB (equivalent to SCF-MO-CI) to HMOVB (equivalent to HMO) theory. We immediately recognize that, if we view each of CT and BZ as a composite of three HC = CH fragments (F_1, F_2, F_3), there exists a set of purely "ionic" charge transfer

MOVB CW's which contain VB (not MOVB) CW's that can only contribute to the VB wave-function of BZ but not to that of CT, e.g., Φ_1 and Φ_2 shown below. "Excusing" inter-electronic repulsion at the level of HMO theory makes Φ_1 and Φ_2 unduly important and causes BZ to be more stable than CT.

Φ_1:
$$\pi^* \quad - \quad + \quad -$$
$$\pi \quad + \quad +\!\!+ \quad +\!\!+$$
$$F_1^+ \quad F_2^- \quad F_3$$

Φ_2:
$$\pi^* \quad +\!\!+ \quad - \quad -$$
$$\pi \quad +\!\!+ \quad - \quad +\!\!+$$
$$F_1^{-2} \quad F_2^{+2} \quad F_3$$

A related example drawn from our previous work is the constitution of the VB wave-function of pi cyclobutadiene with noninteracting pi bonds (CB*) and Möbius pi cyclobutadiene (MCB), two hypothetical systems analogous to CT and BZ. According to VB theory, and in complete analogy to MOVB theory, the advantage of MCB over CB* lies in four unique "ionic" VB CW's which contribute to the wavefunction of the former but not the latter system. Once again, neglect of interelectronic repulsion at the level of HMO theory causes these unique "ionic" CW's to achieve undue prominence and stabilize the "allowed" or "aromatic" MCB complex relative to the "nonaromatic" CB* more than it deserves".

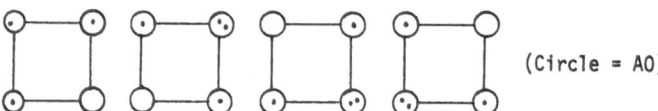 (Circle = AO)

According to MOVB theory, the culprit "ionic" CW's responsible for undue stabilization of MCB relative to CB* at the HMO level are the charge transfer CW's shown below.

$$\pi^* \quad - \quad + \quad + \quad - \quad +\!\!+ \quad + \quad + \quad +\!\!+$$
$$\pi \quad + \quad +\!\!+ \quad +\!\!+ \quad + \quad + \quad - \quad - \quad +$$
$$F_1^+ \quad F_2^- \quad F_1^- \quad F_2^+ \quad F_1^{-*} \quad F_2^{+*} \quad F_1^{+*} \quad F_2^{-*}$$

It is then evident that our assumption that $\Delta J' = 0$ is not justified, i.e., we can not neglect interelectronic repulsion because we deal with interbond excitations (charge transfer) in the case of BZ which can only occur at the expense of severe interelectronic repulsions and which, if neglected, would lead one to believe that BZ is more stable than CT while, if included, they can lead to exactly the opposite conclusion, i.e., that ΔE_π is negative. That pi CT is more stable than pi BZ has been implicityly suggested about a half century ago when Eyring used the Diatomics In Molecules method[5] to show that $3A_2$ is more stable than hexagonal A_6 provided that A is a strong overlap binder.[6] A_6 can become more stable than $3A_2$ only when A is a weak overlap binder (e.g., Li) in which case the greater stability of hexagonal A_6 is due to the fact that such atoms bind to a large extent via "classical" coulomb interactions in the absence of interatomic charge transfer. These ideas have been confirmed by <u>ab initio</u> computations.[7]

We conclude the following:

a. The resonance energy of benzene can be thought of as $\Delta E_\sigma + \Delta E_\pi$ or as $\Delta E' + \Delta H$. The $\Delta E_\sigma + \Delta E_\pi$ energy decomposition tells us that the resonance stabilization of benzene is due to a positive ΔE_σ which overpowers a negative ΔE_π. The $\Delta E' + \Delta H$ energy partition tells us that the same effect is due to the fact that pi benzene is hypodesmic relative to three noninteracting pi ethylenes, i.e., ΔH is positive and, of course, because $\Delta E'$ is positive. The advantage of the $\Delta E' + \Delta H$ partition is that it <u>effectively</u> renders unnecessary the consideration of bielectronic effects. In either language, one can say that, if sigma effects were unimportant, benzene would have D_{3h} and not D_{6h} symmetry.

Since

$$\Delta EXC_T \simeq -\Delta Q + \Delta\Delta J$$

and

$$\Delta E_\pi \simeq \Delta Q + \Delta J'$$

where

$$\Delta Q =. Q_\pi(CT) - Q_\pi(BZ)$$

and because only $\Delta\Delta J$ but not $\Delta J'$ can be assumed to be negligible, ΔQ is a reliable measure of ΔEXC_T but not of ΔE_π.

    c. The pi delocalization energy of benzene according to HMO theory is $-\Delta Q$. This means that the pi delocalization energy of benzene is not a measure of how much more stable is pi benzene relative to three pi ethylenes but rather how much energy is gained by attaching H_6 to CT rather than to BZ! At this point, note a devilish coincidence: At the level of HMO theory, ΔQ is taken to be equal to ΔE_π which is then said to be a measure of the <u>resonance energy</u>, ΔE (see Figure 1):

$$\Delta E \; \alpha \; \Delta E_\pi = +\Delta Q$$

However, according to MOVB theory, we have:

$$\Delta E \; \alpha \; \Delta H \; \alpha - \Delta EXC_T \; \alpha + \Delta Q$$

This means that all resonance energy (ΔE) <u>trends</u> predicted using HMO delocalization energies ($-\Delta Q$'s) are correct for the wrong reasons. Since both MOVB and HMO theories predict that the resonance energy will increase as the <u>absolute magnitude</u> of ΔQ increases, both succeed albeit for different reasons!

    d. How certain are we that benzene is pi destabilized and that its geometry is due to sigma bonding constraints? According to the Wolfsberg-Helmholz approximation of AO resonance integrals, the ability of an orbital to generate bonding via overlap interaction is related to its ionization energy. Thus, in attempting to

place pi BZ within a series of systems with which it can be compared, one realizes
that it is much closer to hexagonal H_6 (an energy maximum due to overlap repulsion
of the component H_2 fragments) than to hexagonal Li_6 (an energy minimum due to
"classical" coulomb interaction) because the C2p orbital has an ionization energy
which is closer to that of the H1s than to the Li2s orbital. I think that pi BZ
is closer to the nonmetallic H_6 rather than the metallic Li_6, for, after all, organic
chemistry is said to be the chemistry of the "covalent" bond, and that it is the
resistance of the sigma bonds to distortion that prevents the molecule from adopting
a D_{3h} geometry. The important role of sigma effects can be readily appreciated
once it is realized that square antiaromatic cyclobutadiene is only slightly higher
in energy than rectangular cyclobutadiene simply because the cost for "antiaromaticity
avoidance" is sigma bond stretching.[8]

e. While benzene is pi destabilized relative to the separated components,
borazine is probably stabilized because the polarity of the B-N pi bonds "justifies"
the incursion of the "ionic" VB CW's which stabilize the cyclic relative to the
noninteracting system.

f. There has been a schizophrenic attitude in the chemical community: Most
chemists have embraced HMO theory because it apparently rationalizes the reactivity
properties of benzene. The same chemists have repressed the fact that HMO theory
incorrectly predicts no barriers in "allowed" reactions. This "nightmare" can now
be dispelled: There are barriers in "allowed" reaction for (almost) the same reason
that benzene is pi destabilized and HMO theory is wrong on both counts because it
neglects interelectronic repulsion. Benzene bypasses addition while linear polyenes
readily enter into such reactions simply because ΔH is positive.

References

1. Hückel, E. $\underline{Z}$. Physik. 1930, 60, 423; ibid. 1931, 70, 204; ibid. 1932, 76, 628. Hückel, E. $\underline{Z}$. Electrochem. 1937, 43, 752.

2. For pedagogical presentation and application of Hückel MO theory, see, inter alia:
   (a) Streitwieser, Jr., A. "Molecular Orbital Theory for Organic Chemists"; John Wiley and Sons, Inc.: New York, 1961.
   (b) Heilbronner, E.; Bock, H. "Das HMO-Modell und Seine Anwendung"; Verlag Chemie, Gmbh: Weinheim, 1968.

3. From thermochemical data, the heat of hydrogenation of benzene (to form cyclohexane) is 49 kcal/mol and that of cyclohexene (to form cyclohexane) 29 kcal/mol. Thus, the heat of hydrogenation of three cyclohexenes is much larger than that of benzene.

4. (a) Pople, J.A. $\underline{J}$. Chem. Phys. 1956, 24, 1111.
   (b) Pople, J.A.; Untch, K.G. $\underline{J}$. $\underline{Am}$. $\underline{Chem}$. $\underline{Soc}$. 1966, 88, 4811.

5. Glasstone, S.; Laidler, K.J.; Eyring, H. "The Theory of Rate Processes"; McGraw-Hill: New York, 1941, and references therein.

6. According to DIM theory with neglect of non-neighbor overlap, $3A_2$ have a total energy of $Q + 3T$ while hexagonal A_6 an energy of $Q' + 2.61T$, where Q is the coulomb and T the exchange integral of VB theory. For neutral fragments, $Q \simeq Q'$, which means that $3A_2$ lie below A_6 even if non-neighbor overlap is neglected. For a pedagogical discussion of pi ethylene (A_2) and pi benzene (A_6) see: Sandorfy, C. "Electronic Spectra and Quantum Chemistry"; Prentice-Hall: Englewood Cliffs, NJ, 1964.

7. Dixon, D.A.; Stevens, R.M.; Herschbach, D.R. <u>Faraday</u> Discussion Chem. <u>Soc.</u> <u>1977</u>,
    62, 110. This paper is of central importance as it clearly projects the
    differing binding mechanisms of nonmetallic and metallic atoms, something having
    profound chemical implications.

8. For computational results, see: Borden, W.T.; Davidson, E.R.; Hart, P. <u>J</u>. <u>Am</u>.
    <u>Chem</u>. <u>Soc.</u> <u>1978</u>, 100, 388.

Chapter 13. Why "Effective" Bonds exist when "Real" Bonds are Absent: The
Electronic Structure of the (1.1.1.) Propellane.

In a recent communication[1], Wiberg and Walker reported the synthesis of
the presumably "superstrained" (1.1.1.) propellane which, according to previous
SCF-MO computations by Newton and Schulman[2], has no bond linking the bridgehead
carbons, but, nonetheless, it is comparatively very stable[1]. I now show that MOVB
theory[3,4] easily accounts for these "strange" observations by comparing the
electronic structures of the following three molecules.

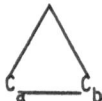      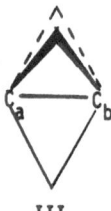

I   II   III

Each molecule can be viewed as a central diatomic core, C_a- C_b, attached to a
set of ligands. For example, I is thought of as a composite of a C_a - C_b core
and a ligand fragment made up of four univalent hydrogens and one divalent
methylene, etc. Each H is assumed to bind via its 1s AO and each CH_2 via one
sp and one 2p carbon AO. Furthermore, we note that III differs from I and II
to the extent that it has only three "inside" methylene ligands while the latter
two molecules have "outside" hydrogen and "inside" methylene ligands. As we
shall see, it is this difference that it is ultimately responsible for the fact
that C_a - C_b is a bond in I and II but only a nonbond in III.

The compact bond diagrams for I, II, and III are shown in Figures 1, 2,
and 3[5]. In MOVB theory, a compact bond diagram is the principal resonance bond
diagram of the complete resonance bond diagrammatic representation of a given
molecule and as such it constitutes a pictorial way for showing exactly the one
arrangement of the electrons into bonding and nonbonding pairs (with respect to
the two fragments) which is primarily (but not exclusively) responsible for the

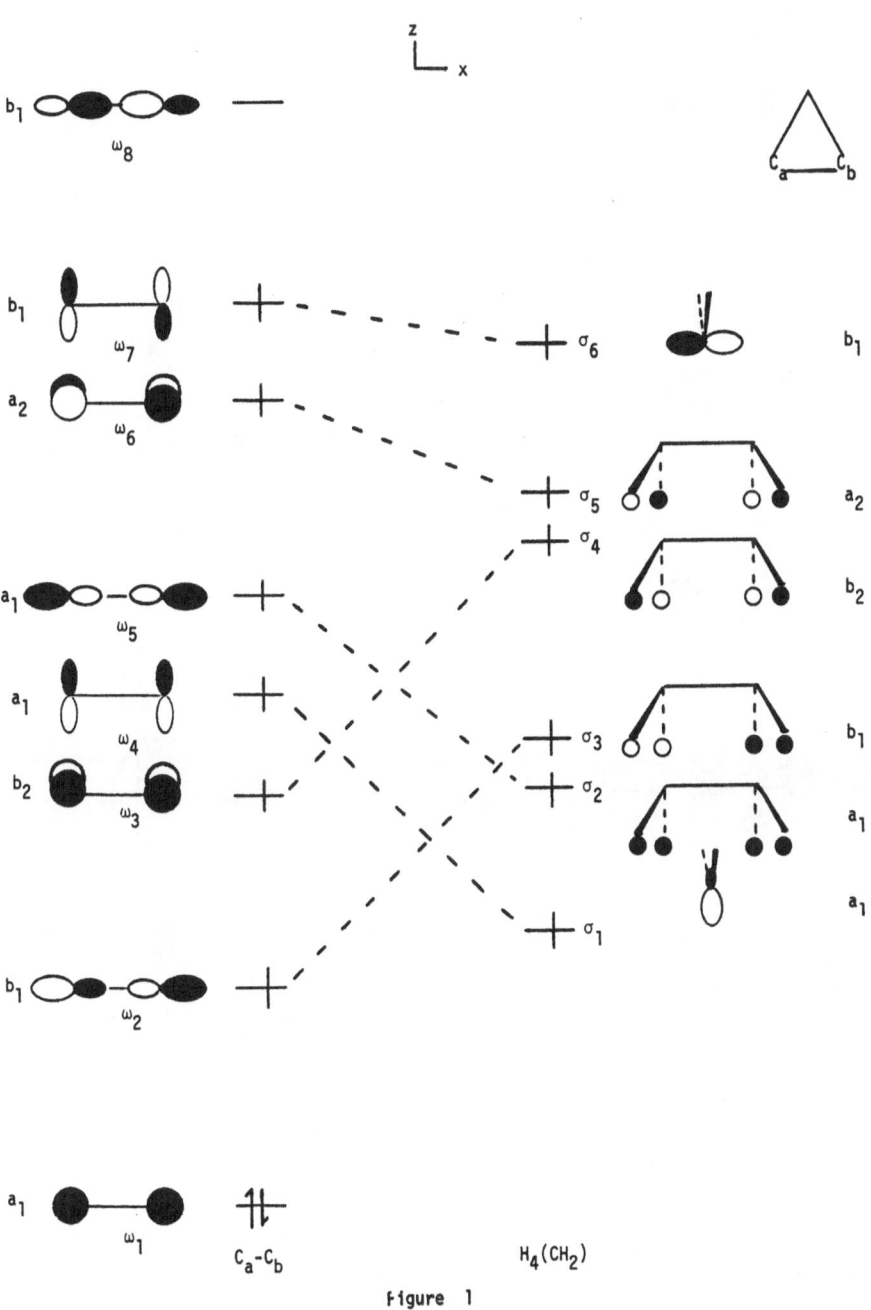

Figure 1

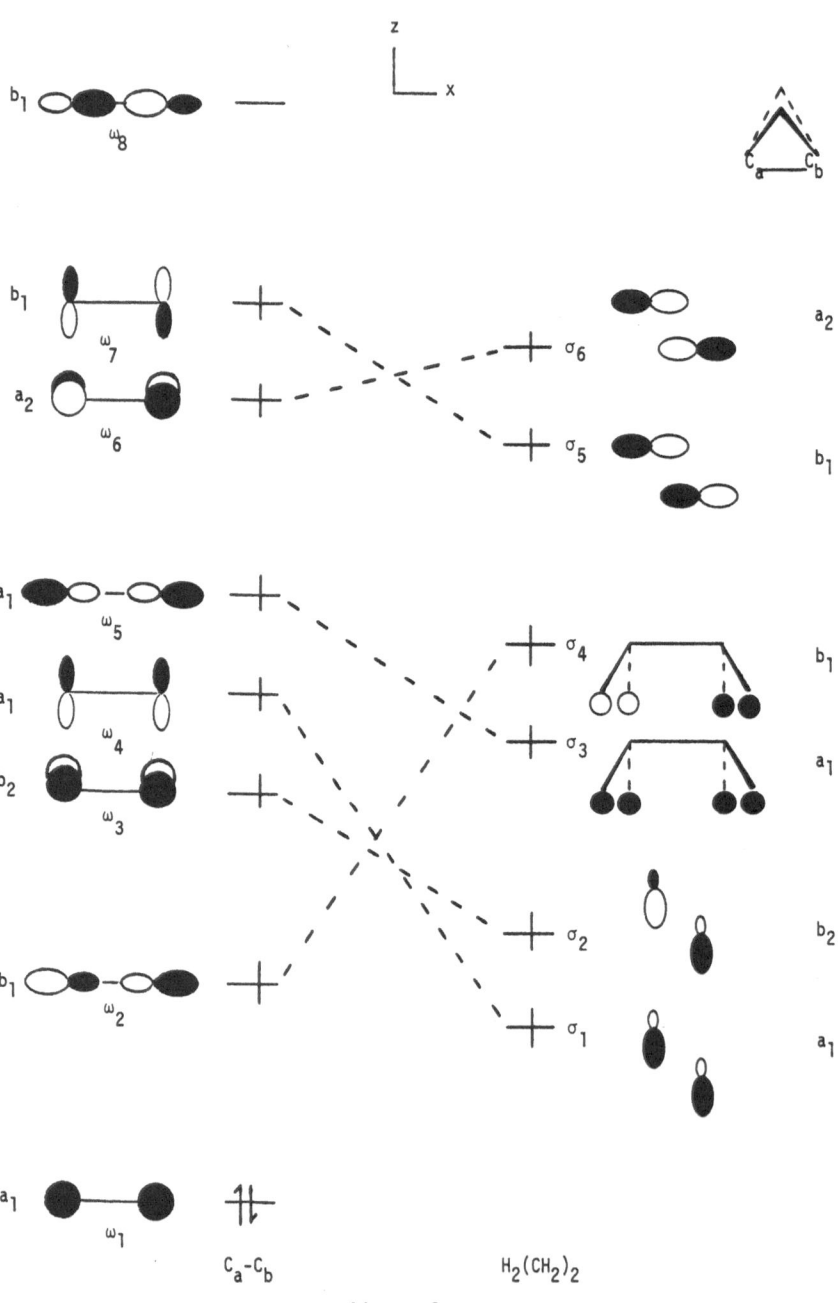

Figure 2

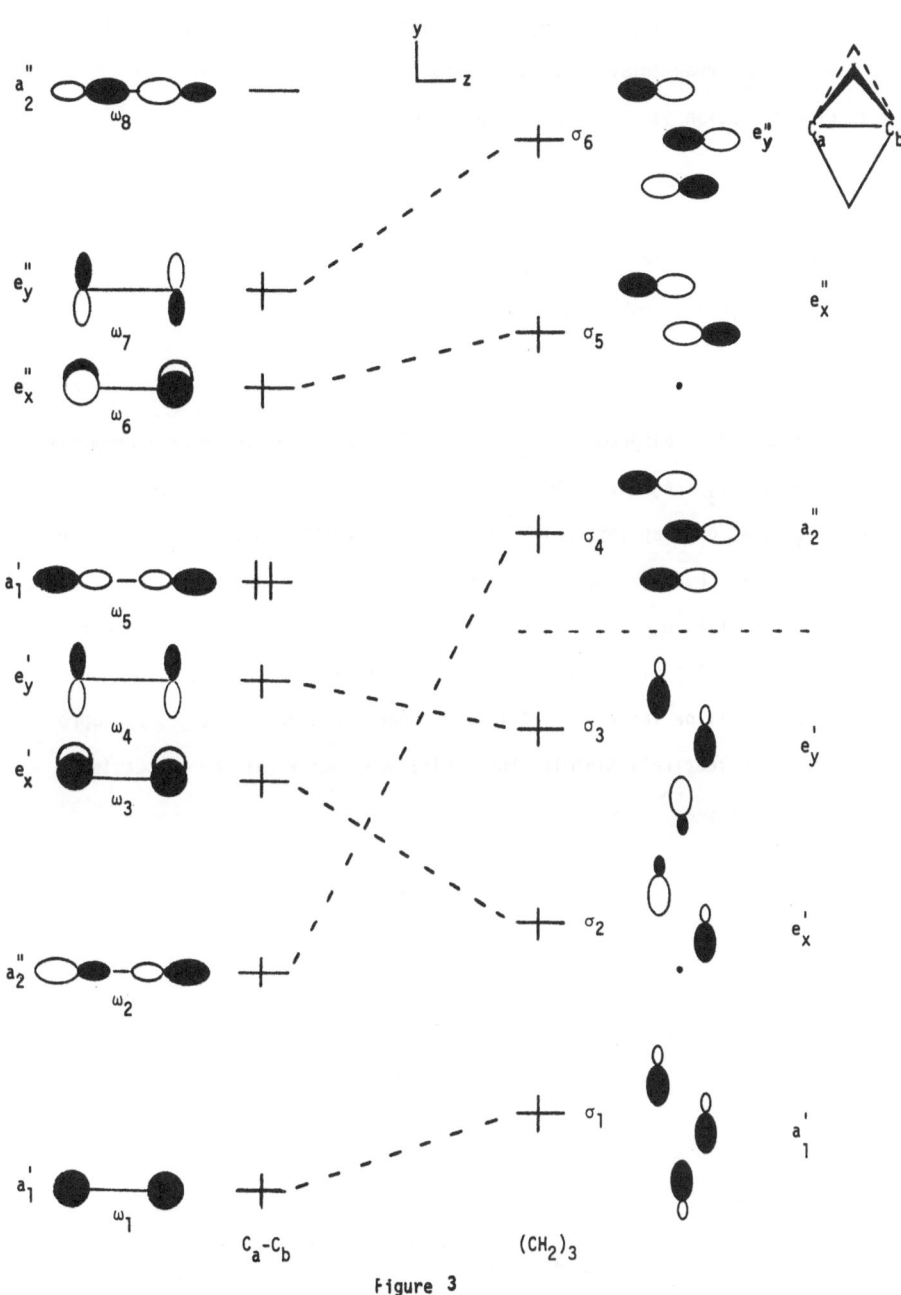

Figure 3

Figure 1. Compact bond diagram of D_{3h} cyclopropane viewed as a composite of C_a - C_b and $H_4(CH_2)$ fragments. Note deexcited (ω_1 doubly occupied) core. Orbital symmetry classification is with respect to the C_{2v} point group.

Figure 2. Compact bond diagram of C_{2v} (1.1.0.) bicyclobutane viewed as a composite of C_a - C_b and $H_2(CH_2)_2$ fragments. Note deexcited core.

Figure 3. Compact bond diagram of D_{3h} (1.1.1.) propellane viewed as a composite of C_a - C_b and $(CH_2)_3$ fragments. Note excited (ω_5 doubly occupied) core. In this case, the σ_1 - ω_5 overlap integral is small. As a result, the system opts for a strong core-ligand bond formed by the strong overlap of ω_1 and σ_1 at the expense of modest core excitation energy,instead of the formation of a weak core-ligand bond through the overlap of ω_5 and σ_1 in the absence of core excitation. The C_a - C_b distance will be larger in (1.1.1.) propellane than in, e.g., cyclopropane in order to effectively minimize the excitation energy cost by decreasing the ω_1 - ω_5 energy gap.

existence of the molecule itself. The three compact bond diagrams shown in Figures 1 - 3 "speak for themselves" rendering lengthy discussions of their meaning unnecessary. The key difference between I and II, on one hand, and III, on the other, is that , while in the former two species six strong core-ligand bonds can be formed by keeping ω_1 (a C_a - C_b "bonding" MO) doubly occupied, core excitation is required in III for the formation of six strong core-ligand bonds because the ω_5 - σ_1 overlap integral is small so that the (1.1.1.) propellane opts for holding two electrons in the C_a - C_b "nonbonding" MO, ω_5, in order to make a strong ω_1 - σ_1 core-ligand bond. Hence, in I and II, there is a real C_a - C_b bond while, in III, there is a real C_a - C_b nonbond. Does this mean that it will be easy to convert III to the corresponding 1,3 diradical by appropriate deformation, or, equivalently, does this mean that there is no effective C_a - C_b bond in III? The answer is a resounding NO!!!

Because of the spatial extension of the sp carbon AO of methylene, there are one, two, and three T - bonds in I, II, and III, respectively, where by T - bond we imply the core-ligand bond generated by the overlap of carbon sp AO's with the appropriate core MO as shown explicitly in the bond diagrams. An example is given below.

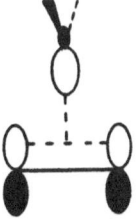

T - Overlap

A deformation which increases the C_a - C_b distance is expected to weaken significantly a T - bond and this is the reason why none of the three molecules is
willing to be converted to the corresponding diradical by such a C_a - C_b stretch.
However, note that, although I, II, and III have one, two, and three T - bonds,
respectively, C_a - C_b stretch will not be three times more adverse in III compared
to I and one and one-half times more adverse in III compared to II for the simple
reason that the cores of the three molecules have different configurations so that
the increase of the C_a - C_b distance will affect I and II cores much more adversely
than the III core because of the double occupancy of ω_1 in I and II as compared to
the double occupancy of ω_5 in III. The final conclusion is then that the deformation of interest will raise the energy of all three systems despite the fact
that I and II have a C_a - C_b bond while III has only a C_a - C_b nonbond.

The above analysis provides a simple rationalization of the following facts:

a.   I and II have a "strained" core bond while III has a "strained" core nonbond
according to SCF-MO computations.[2]

b.   As the $-CH_2-$ groups are replaced by $-CH_2-CH_2-$ groups so that the T - bonds
are replaced by core-ligand bonds which are little affected by an increase of the
C_a - C_b distance, the energy required for converting a propellane to the corresponding diradical decreases. Examples taken from the computational work of
Wiberg and Walker[1] are shown below.

$$\Delta H \approx 5$$

$$\Delta H \approx 30$$

$$\Delta H \approx 65$$

c.   The large positive enthalpy associated with the conversion of III to the corresponding diradical is the result of "side bond" (e.g., C_a - C) weakening.

A message: As we have discussed and illustrated before , the enthalpy of the process A $\longrightarrow$ B reflects two things: Change in total fragment excitation and change of the total fragment overlap. Whenever predictions are made by considering only one of these two factors, these will be either accidentally correct or simply incorrect. In anticipating that one can make a diradical out of (1.1.1.) pro-pellane because there is no real core C_a - C_b bond according to computations, one commits the error of not considering the energy change brought about by the re-adjustment of all, rather than one arbitrarily chosen, bonds as well as the change of the excitation energies of the core and ligand fragments. An example of the error one makes by focusing attention on a single overlap population element (i.e., one bond) rather than the total overlap population (i.e., all bonds) has been given, in an MO frame, in an older work.[6] Examples of the error one commits by considering only total overlap populations have been given in ref. 1.[7]

In conclusion, we suggest that the original computations of Newton and Schulman are correct in suggesting no C_a - C_b bonding in the (1.1.1.) propellane and that the stability of this species and its unwilligness to open up to a di-radical, two important conclusions of Wiberg and Walker, are due to the "side bonds" that "clamp" C_a and C_b together,despite the absence of a real C_a - C_b bond,in order to preserve their own integrity. We say that the energy penalty involved in the conversion of the (1.1.1.) propellane to the corresponding diradical is the result of an effective C_a - C_b bond which is present in the absence of a real C_a - C_b bond.

References

1. Wiberg, K.B.; Walker, F.H. _J. Am. Chem. Soc._ 1982, 104, 5239.

2. Newton, M.D.; Schulman, J.M. _J. Am. Chem. Soc._ 1972, 94, 773.

3. Epiotis, N.D.; Larson, J.R.; Eaton, H. "Unified Valence Bond Theory of Electronic Structure", in Lecture Notes in Chemistry, Vol. 29; Springer-Verlag: New York and Berlin, 1982.

4. Parts of the MOVB theory of chemistry and illustrative applications have been presented in conferences:

   (a) NATO Advanced Study Institute on Topic in Theoretical Organic Chemistry, Gargnano, Italy, 1979.

   (b) International Symposium on Stereoelectronic Effects in Organic Chemistry, St. Andrews, Scotland, 1980.

   (c) Symposium on Theoretical Aspects of Fluorine Chemistry, 183rd American Chemical Society Meeting, Los Vegas, USA, 1982.

   (d) International Symposium on Theoretical Organic Chemistry, Dubrovnik, Yugoslavia, 1982.

   For applications, see also:

   (a) Epiotis, N.D. _Pure Appl. Chem._ 1983, 000.

   (b) Epiotis, N.D.; Larson, J.R. _Israel J. Chem._ 1983, 000.

5. The compact bond diagrams have been constructed assuming zero nonbonded interactions. The energy ranking of the ligand MO's is only approximate and the core and ligand orbital stacks are shown without implying that their relative placement is as depicted.

6. Epiotis, N.D.; Cherry, W.R.; Shaik, S.; Yates, R.L.; Bernardi, F. _Topics Curr. Chem._ 1977, 70, 49-54.

7. See p. 266-297 and especially the admonitions on p. 282-284 of Ref. 1.

Chapter 14. The Detailed Electronic Structure of Carbocyclic Molecules and the Concept of Superaromaticity.

The literature on the electronic structure of carbon cyclic compounds such as cyclopropane, cyclohexane, benzene, etc., is so enormous that citation of some works but not others is mandatory and, thus, unfair.[1] The essential point is that many aspects of the bonding of these molecules have been successfully dealt with but a fundamental understanding of the electronic structure of these molecules still eludes us. Why do I say so?

H_6 is more stable as $3H_2$ than as hexagonal H_6.[2] This means that electron pair bonds repel each other regardless of whether 4N or 4N + 2 electrons are involved, in contrast to commonly held beliefs fostered by Hückel MO theory. Indeed, in a previous chapter we advanced the argument that benzene is pi destabilized and explained why Hückel MO theory fails to show this pi destabilization. Now, imagine six carbons with one electron per valence AO coming together to form a six membered benzene ring of D_{6h} symmetry in conjunction with a D_{6h} cyclic array of six hydrogens with one electron per valence AO. Assuming that the 2s AO of one carbon overlaps only with the 2s AO of its neighbor, that the $2p_x$ AO of one carbon overlaps only with the $2p_x$ AO of its neighbor, etc., we realize that one cyclic C_6 sigma AO array (containing six electrons) will tie up the cyclic H_6 AO array (also containing six electrons) in forming the six C - H bonds, leaving three other C_6 AO arrays each containing six electrons. Hence, by analogy to the H_6 system, benzene must not exist, i.e., it must be unstable with respect to 3 HC≡CH. The fact that benzene not only exists but it does so in a D_{6h} geometry tells us that the <u>interaction</u> of the AO rings defined above is of critical importance. That is to say, the sigma C - C bonds of benzene are not "nuts and bolts" standard C - C bonds involving sp^2 carbon AO's but, rather, constituents of a sigma framework which has unique stability characteristics that remain to be discovered.

The purpose of this chapter is to show that the exploitation of MOVB theory is restricted only by the imagination of the practicioner by applying this methodology to cyclopropane, a cyclic carbon system which, if properly understood, sheds light

on the electronic structure of all carbocyclic systems and yields answers to

questions such as those posed indirectly above.

Theory

In dealing with cyclopropane, we define a C_3 core and three H_2 ligands with each H_2 being attached to one and the same carbon. By inscribing an equilateral triangle in a circle, we now define the following orbital basis:

a. Three carbon 2s AO's which form the S ring.

b. Three carbon 2p AO's of pi symmetry which define the Π ring.

c. Three carbon 2p AO's which are aligned with the radius of the circle in which the triangle is inscribed and which define the R ring.

d. Three carbon 2p AO's which are aligned perpendicular to the radius of the circle and which define the P ring.

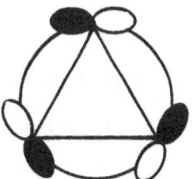

e. Three MO's of sigma symmetry, each spanning one H_2 fragment being an in-phase combination of two H1s AO's. These define the L_σ ring.

f. Three MO's of pi symmetry, each spanning one H_2 fragment being an out-of-phase combination of two H1s AO's. These define the L_π ring.

Sigma-type H_2 MO: Pi-type H_2 MO:

We now proceed as follows:

a. We write the MO's of each ring, as indicated in Figure 1. This can be easily done by anyone with mild exposure to Hückel MO theory.

b. We recognize that the various rings overlap with each other in the way indicated below, assuming for the moment no symmetry constraints.

   1) Pi Overlap.

$$L_\pi - \Pi$$

   2) Sigma Overlap.

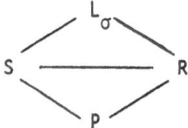

c. We start with three carbons, each having one electron per valence AO, and three triplet H_2's, each having one electron per valence MO, and we seek to form three core bonds (i.e., three C - C bonds) and six core-ligand bonds (i.e., six bonds linking C_3 and $3H_2$). Of the latter six, three will be sigma- and three pi-type core-ligand bonds. To this extent, we apportion the electrons in the way indicated in Figure 1, noting the following:

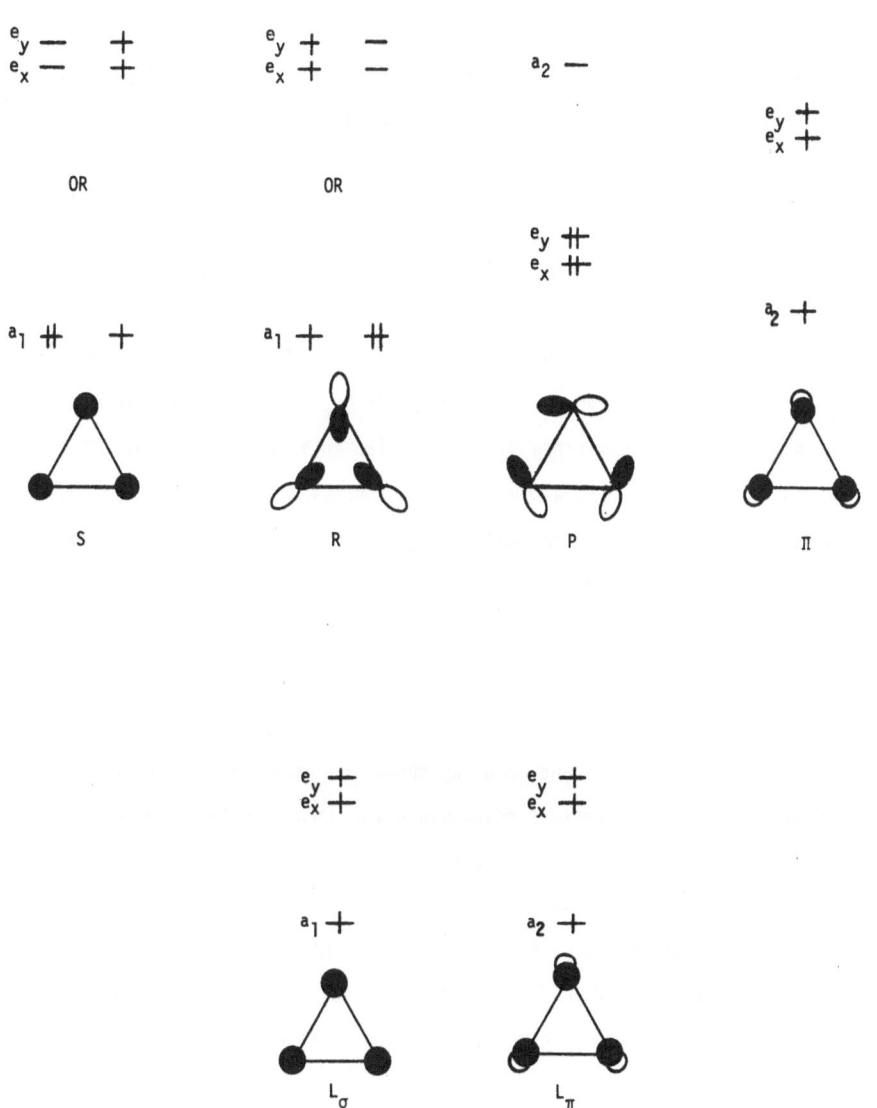

Figure 1. The MO's of the various cyclic AO arrays of cyclopropane. Note that all except the P ring are Hückel arrays. The P ring is a Möbius array. The symmetry classification of the MO's is with respect to the D_3 point group. Electron allotment produces six C-H and three C-C bonds.

1) Three C - C bonds are primarily due to the double occupation of the bonding orbitals of P and R - S. Thus, the three C - C bonds constitute an "aromatic" system in the conventional sense because P is a Möbius and S - R a Hückel AO ring.

2) Three pi core-ligand bonds are formed by pairing L_π and Π.

3) Three sigma core-ligand bonds are formed by pairing L_σ and R - S. The notation R - S implies that R and S are hybridized.

The MOVB bond diagram which can now be constructed using as ingredients all rings with the specified electron occupancy would be an accurate description of the bonding only if the various rings did not overlap in a cyclic fashion. However, this is not the case at hand. As a result, we must examine the sigma and pi sub-systems in detail if we are to understand why C_3H_6 is bound and we must not be fooled by the fact that four electrons occupy the bonding MO's of a Möbius AO ring and two electrons occupy the bonding MO's of a Hückel AO ring thus conferring C - C bond "aromaticity" to the system at hand.

According to the Independent Bond Model (IBM) there are three pi subsystems responsible for three pi core-ligand bonds as shown in Figure 2a. On the other hand, there are also three subsystems responsible for core as well as sigma core-ligand bonding as shown in Figure 2b. Because of cyclic overlap, we must determine the signs of the MO overlap integrals in order to find out whether this cyclic interaction causes stabilization or destabilization. These are as shown in Figure 2b. It is immediately apparent that the sigma rings form three Möbius "aromatic" MO (not AO) arrays because each MO array is occupied by four electrons. This

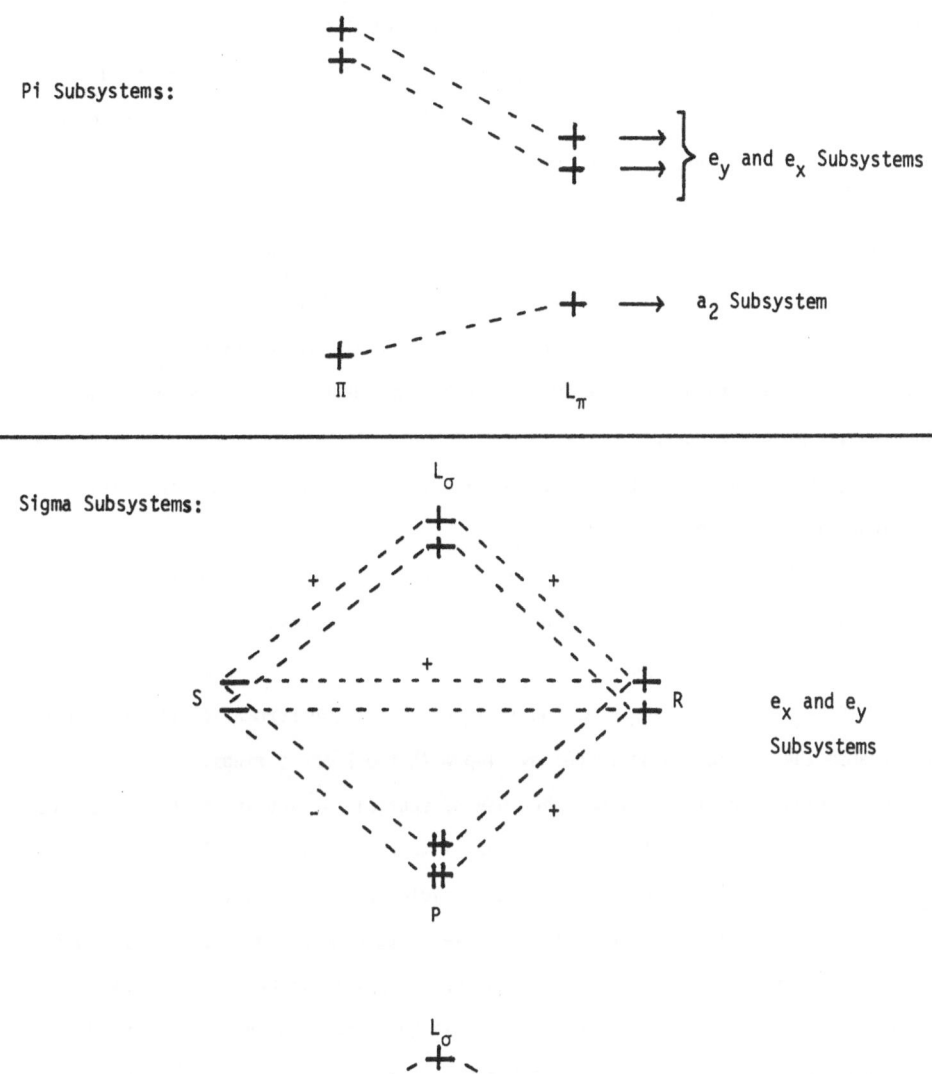

Figure 2. Pi and sigma subsystems of C_2H_6. The pi subsystems define C-H pi bonding and the sigma subsystems C-H <u>and</u> C-C bonding. The signs of the MO overlap integrals within each sigma subsystem are such that the MO's form Möbius arrays each of which accommodates four electrons.

means that the correct MOVB diagram which takes into account the cyclic MO overlap of the sigma rings is a representative of an MOVB wavefunction which has resulted from the Hückel interaction of the MOVB Configuration Wavefunctions of appropriate symmetry. At this point, we coin the term "superaromatic" and we mean this to indicate that a system is stabilized by MO "aromatic" interaction of fragments which in themselves are already stabilized by AO "aromatic" interaction. We conclude that, on top of a conventional "aromatic" AO stabilization of the three C - C bonds there is "aromatic" MO stabilization of the entire sigma frame of C_3H_6 rendering it a "superaromatic" system. This is the reason why cyclopropane and other cyclic molecules exist.

Using the procedure delineated above, it is easy to generate the following additional important conclusions:

a. As the number of carbons of C_nH_{2n} and C_nH_n increases, the overlap of AO's within a ring as well as the overlap of ring MO's changes so that the AO and MO "aromaticity" components of "superaromaticity" change. A maximum seems to be reached in D_{6h} C_6H_{12} and C_6H_6. At the limit of $n = \alpha$, "superaromaticity" is turned off because the R ring can no longer overlap with the P and S rings.

b. Cyclopropane is "strained" because AO spatial overlap in the P ring causes weak Möbius AO "aromatic" stabilization and sigma ring spatial overlap, in general, causes weak Möbius MO "aromaticity". Hence, cyclopropane is weakly sigma "superaromatic". By contrast, planar cyclohexane seems to provide optimum spatial overlap for AO and MO "aromatic" stabilization and it appears to be strongly sigma "superaromatic". The reason why cyclohexane is "puckered" and not planar is the same reason why N_2H_4 is gauche and not trans. By the same token, $C_3H_3^{\oplus}$ exists because of weak sigma superaromaticity and C_6H_6 is D_{6h} because of the strong sigma "superaromaticity" and not because of pi "aromaticity".

c. Each sigma subsystem in cyclopropane defines one core and one core-ligand bond. Because of the fact that the e MO's of P are bonding while those of the other sigma rings are antibonding, two of the three C - C bonds are generated by the peripheral overlap of the carbon 2p AO's. The third C - C bond is generated by the radial overlap of the carbon 2p AO's in combination with the cyclic overlap of the carbon 2s AO's (hybridization). To put it crudely, the three C - C bonds of cyclopropane are described by the lower two MO's of the Möbius AO array of P and the lowest MO of the Hückel AO array of R - S. This conclusion is identical to the one reached by MO analysis of the bonding of cyclopropane.

References

1. For review, see: Liebman, J.F.; Greenberg, A. <u>Chem</u>. <u>Revs</u>. <u>1976</u>, 76, 311.

2. Dixon, D.A.; Stevens, R.M.; Herschbach, D.R. <u>Faraday</u> Discussion <u>Chem</u>. <u>Soc</u>. <u>1977</u>, 62, 110.

Chapter 15. The Explicit Theory of "Real" Electrocyclizations of Closed and Open Shell Molecules.

The purpose of this chapter is threefold:

a. To develop the explicit theory of pericyclic reactions by reference to the electrocyclization reaction in a way that makes new analogies possible.

b. To show that the stereoselection rules for model and "real" closed shell electrocyclizations are identical.

c. To show that the stereoselection rules for model radical electrocyclizations are opposite to those for "real" radical electrocyclizations because these reactions are not simple "n electrons in n pi AO's" reactions but, rather, sigma-pi transfer reactions of the type that current qualitative theory cannot deal with in a formally satisfactory and conceptually intelligible way. The MOVB theory of stereoselection of "real" radical electrocyclic reactions destroys the impasse which has been reached in the field of radical chemistry with regards to the stereoselectivity of radical reactions, in general.

I. Many-Fragment MOVB Theory

Consider the well known and fundamental problem of the construction of the pi Hückel MO's of benzene starting from an arbitrary set of fragment pi MO's. One can envision three ways of solving this problem:

a. By dissecting benzene into two fragments A and B, as shown in Figure 1a.

b. By dissecting benzene into three fragments, A, B, and C, as shown in Figure 1b. It is now evident that, unlike in (a), the fragment MO's can interact in a cyclic (in fact, multicyclic) manner.

c. By dissecting benzene again into three fragments, A, B, and C, as shown in Figure 1c. Now, unlike in (b), the fragment MO's cannot interact in a cyclic manner.

Obviously, there are other dissection modes. However, restricting our attention to the three fragment choices specified above, we recognize that mode I and III are less cumbersome than II simply because advantage is taken of symmetry so that the number of nonzero MO interaction matrix elements is kept at a minimum. Nonetheless, any of the three fragment MO basis sets is, of course, adequate for solving the assigned problem.

Let us now consider the following problem: Is Hückel AO or Möbius AO benzene more stable at the level of Hückel MO (HMO) theory? This is a problem of parity stereoselection and, in this case, the stereochemical information is conveyed by the signs of the interaction matrix elements if the fragment orbitals overlap in a cyclic manner, or, by the mode of orbital interaction if the fragment orbitals overlap in a noncyclic manner. Thus, dissection II can lead to a solution of the problem only if the signs of the interaction matrix elements are determined correctly while dissections I or III can lead to an answer without the necessity of determining the parity of interaction matrix elements. Translating into the language of Hückel

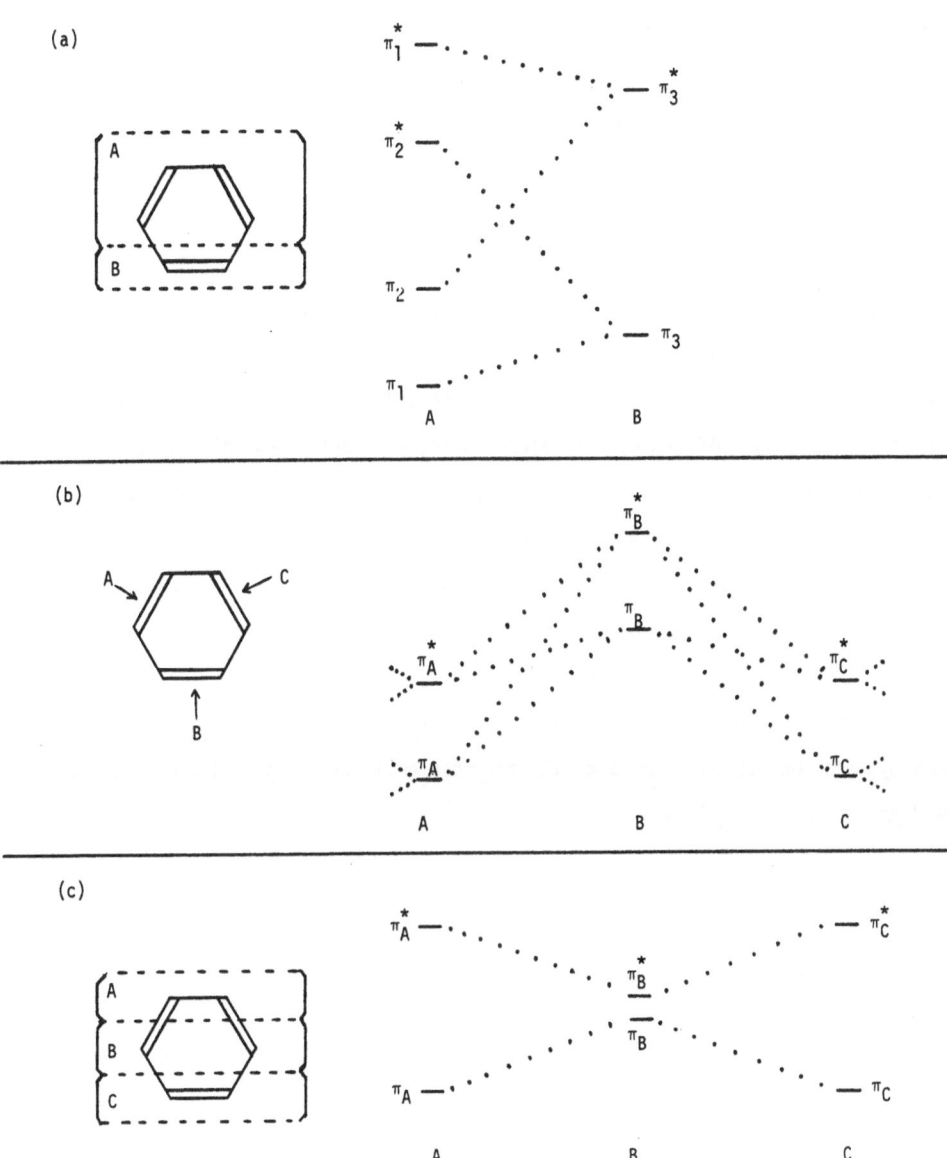

Figure 1. Three different dissections of a cyclic six-AO system. In the dissections shown in (a) and (c) there is no cyclic MO overlap and there is no necessity of determining the signs of the interaction matrix elements.

MOVB theory,[1] we say that, according to dissection II, parity stereoselection is expressed via the signs of the interaction matrix elements over Configuration Wavefunctions (CW's), while, according to dissection I or III, it is expressed via the excitation energies of the CW's. Obviously, the easy way to solve the problem is via dissection I or III, although dissection II will also give the same answer. Dissection I is called the core-ligand (two-fragment) dissection and it is the cornerstone of MOVB theory as practiced in the previous works. Dissection II requires utilization of VB-type concenpts and this renders it undesirable for dealing with large systems. We now recognize that dissection III is entirely compatible with the MOVB concepts developed using the core-ligand dissection simply because the total state energy is independent of the signs of the interaction matrix elements because the basis orbitals cannot interact in a cyclic manner. In general, we can always define a many-fragment MOVB theory provided that the dissection is such so as to enforce noncyclic fragment orbital (<u>not</u> <u>CW</u>) overlap. The advantage of this brand of theory is that it allows us to examine large systems in a very convenient way. This attribute is demonstrated in this work by reference to the problem of the stereoselectivity of thermal electrocyclizations.

II. The Concept of the Transfer Reaction

Consider the simple S_N2 reaction shown below

$$I:^- + H_3C - I \longrightarrow I - CH_3 + I:^-$$

In symbolic language, this can be represented as follows:

$$B_1:^- + A - B_2 \longrightarrow B_1 - A + B_2:^- \quad \text{with } B_1 = B_2$$

We say that the above is a <u>transfer reaction</u> in which the fragment A originally bound to B_2 is transferred from B_2 to B_1. An S_E2 reaction can be symbolized likewise:

$$B_1^+ + A - B_2 \longrightarrow B_1 - A + B_2^+$$

Again, this can be thought of as a transfer reaction. Finally, the hypothetical reaction shown below can be formalized as a pi transfer reaction by employing a dissection which enforces intrafragment orbital orthogonality so that there is no cyclic orbital overlap. The chosen fragmentation mode and the symbolic representation are also shown below.

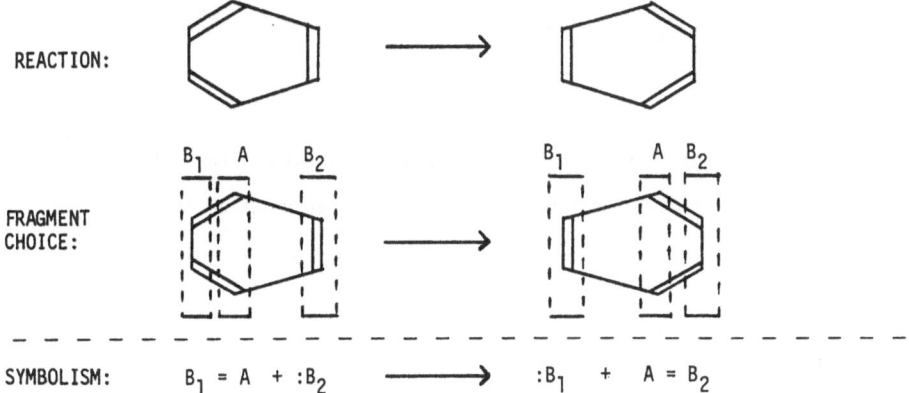

REACTION:

FRAGMENT CHOICE:

SYMBOLISM: $B_1 = A + :B_2 \longrightarrow :B_1 + A = B_2$

It follows that the familiar Diels-Alder reaction of 1,3-butadiene and ethylene is an example of a transfer reaction in which the A fragment is transferred from the B to the C fragment.

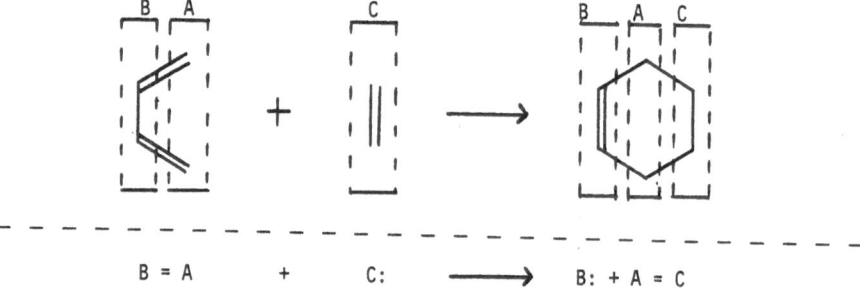

$$B = A \quad + \quad C: \quad \longrightarrow \quad B: + A = C$$

The bond diagrams for the prototypical S_N2, S_E2, and Diels-Alder transition states shown in Figure 2 make evident the fact that the Diels-Alder reaction is a combined S_N2 - S_E2 reaction.

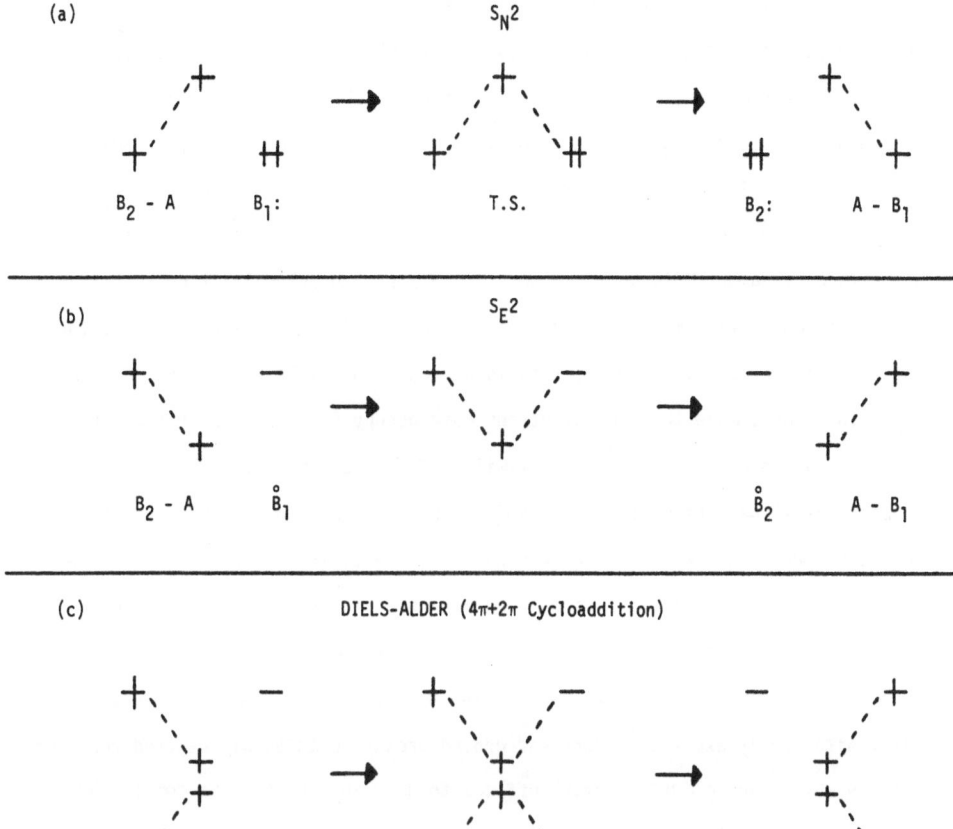

Figure 2. Bond diagrammatic representations of the prototypical S_N2, S_E2, and Diels-Alder reactions.

What is the difference between a 4s + 2s and a 4s + 2a Diels-Alder reaction complex according to the three-fragment MOVB theory which envisions this cyclo-addition as a transfer reaction? The answer can be given directly by construction of the appropriate MOVB diagrams and identifying the differences in bonding between the two complexes. The diagrams shown in Figure 3 speak for themselves: In the 4s + 2s union, four electrons are confined to the lower three and two electrons to the upper three fragment MO's and through spin pairing and delocalization they define always two multicenter bonds joining the central fragment with either of the two outside fragments. By contrast, this is no longer possible in the case of the 4s + 2a union. Now, more than two electrons must occupy the three upper fragment MO's for bonding to occur. In a manner familiar from the application of MOVB theory with core-ligand dissection to chemical problems, we can say that the 4s + 2s species is D-bound (low CW excitation energy requirement) while the 4s + 2a species is U-bound (higher CW excitation energy requirement). Furthermore, the MOVB bond diagrams make plain that, in the case of the 4s + 2s union, reactants are smoothly converted to products while, in the case of the 4s + 2a union, ground reactants correlate with doubly excited product and ground product with doubly excited reactants and this is why a single MOVB diagram suffices to describe the 4s + 2s complex but two (differing in orbital occupancy by two electrons) are required for describing the 4s + 2a complex.

We will now show that closed shell electrocyclization ractions are actual sigma-pi transfer reactions very much analogous to the Diels-Alder transfer reaction and that the conrotation and disrotation transition state complexes are bound elec-tronically in a manner analogous to the antarafacial and suprafacial (or, vice versa) transition states of 4N + 2 electron cycloadditions.

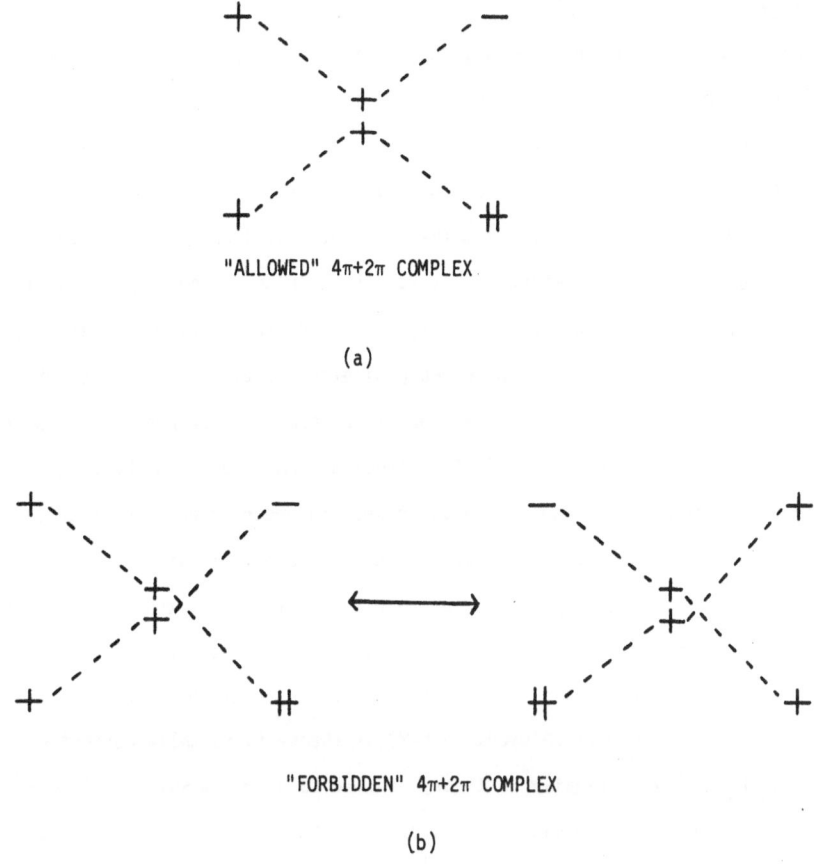

"ALLOWED" 4π+2π COMPLEX

(a)

"FORBIDDEN" 4π+2π COMPLEX

(b)

Figure 3. Bond diagrammatic representations of the "allowed" and "forbidden" 4π + 2π cycloaddition complexes.

III. Implicit versus Explicit Theory

The Woodward-Hoffmann[2] and the Longuett-Higgins-Abrahamson[3] treatises of
pericyclic reactions, in general, and electrocyclizations, in particular, are based
on the construction of simple MO correlation diagrams,[4] starting with a minimum
basis of reactant MO's (usually pi MO's) and following their evolution to product
MO's by assuming that an element of symmetry is maintained along the reaction
coordinate. This is a very ingenious idea, for the ensuing arguments are symmetry
arguments which cannot be defeated. However, in solving the problem in this way,
one cannot follow the precise electronic reorganization required for the conversion
of reactants to products. That is to say, one loses track of how bonds are broken
and made and how different stereochemical arrangements make such bond breaking and
bond making worse or better in an energetic sense. In order to be able to deal with
the complete problem of reaction stereoselection in an explicit rather than implicit
manner one must have at his disposal a clear and formally correct theory of chemical
bonding which is applicable to any chemical problem without restrictions. Since
monodeterminantal MO theory fails to describe properly bond dissociation, MO
approaches of this type cannot be used. SCF-MO-CI theory is formally correct but
conceptually intractable. Finally VB theory applied to a system which is larger
than "four electrons in four orbitals" is too cumbersome to serve as a qualitative
theory of chemical bonding. These facts are behind the following noteworthy trend:
All theoretical treatises of pericyclic reactions which have followed the Woodward-
Hoffmann publication of 1965 are _implicit_ treatises simply because there has been
no theoretical vehicle to take one beyond this stage.[5] MOVB theory is the qualitative
theory deliberately constructed to meet these apparently insurmountable problems of
the conventional brands of theory. This is now illustrated by reference to the
electrocyclizations of the prototypical allyl cation, allyl anion, 1,3-butadiene,

and allyl radical, all treated as complete systems, i.e., by taking cognizance of the role of all pi as well as sigma electrons and projecting it through construction of MOVB bond diagrams. However, before we discuss the actual construction and implications of these diagrams, let us first define the nature of the problem we have undertaken to solve.

IV. The MOVB Viewpoint for Isomerizations

Consider the isomerization of trans CHX = CHX effected by rotation about the
C = C bond. Chemists are used to thinking of this process as a rigid rotation about
a C = C double bond which results in the complete breaking of the C - C pi bond at
the dihedral angle of 90°. In other words, chemists view CHX = CHX as a composite
system of two CHX units linked by a sigma and a pi bond.

Alternatively, one may consider CHX = CHX as a composite of a rigid core, C_2, and a
set of ligands, H_2X_2, which change positions in space as the isomerization proceeds.
This is the viewpoint of MOVB theory with core-ligand dissection. The species having
a 90° dihedral angle is now energetically unfavorable relative to the planar cis and
trans isomers because of loss of core-ligand spatial overlap.

Adopting now the MOVB theoretical viewpoint, we recognize that the electrocyclization of 1,3-butadiene to cyclobutene may occur by either con- or dis-rotatory
motion of the hydrogen pairs attached on the "outer" carbons, C_1 and C_4, while the
rest of the molecule remains "frozen". In other words, ring closure is not effected
by rotation about carbon-carbon bonds, as most chemists like to think when dealing
with this problem, but by keeping the orbitals of the C = CH - CH = C core fixed
and simply relocating the outer ligands so that part of sigma core-ligand overlap is

progressively replaced by <u>pi</u> core-ligand overlap. In this way, we transfer the H_4 ligand fragment from the sigma to the pi frame of the planar C = CH - CH = C core fragment. In short, an electrocyclization is a transfer reaction much like the Diels-Alder reaction (or any 4N + 2 electron cycloaddition). Hence, we expect to obtain MOVB diagrams for con- and dis-rotation reaction complexes analogous to those for 4s + 2s and 4s + 2a (or, <u>vice</u> <u>versa</u>) Diels-Alder reaction complexes.

V. The Explicit MOVB Theory of Electrocyclizations

We illustrate the MOVB theory of electrocyclization by reference to the proto-typcial reaction shown below.

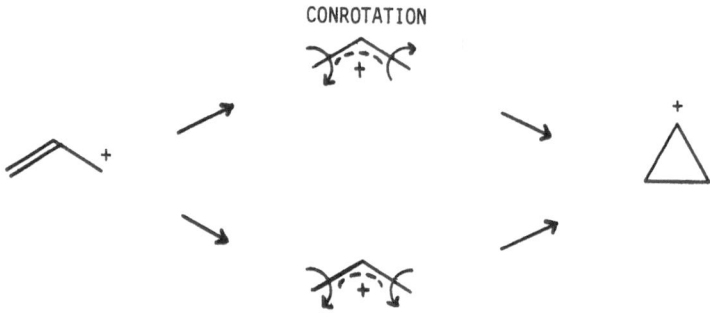

CONROTATION

DISROTATION

In attempting to construct MOVB diagrams for the reactant and product, we convene that we view each of them as a composite of a fixed sigma C_3 framework, a fixed pi C_3 framework, and a set of H_5 ligands, with the stereochemistry of the reaction being dependent on the preferred motion of the four "outer" hydrogen ligands, i.e., the two H's attached on C_a and the two H's attached on C_c. The compact MOVB representations of the allyl cation and the cyclopropyl cation are shown in Figure 4. Realizing that, in both cases, we have a total of 17 valence AO's and 16 electrons, we recognize the following facts:

a. In the case of the allyl cation, there are three pi C_3 MO's with only one being doubly occupied. In additon, there are five C - H bonds which, according to MOVB theory, correspond to five core-ligand sigma bonds due to the overlap of five core sigma MO's with five ligand MO's, each containing a single electron in the perfect pairing (R) CW which corresponds to the bond diagram as written. This leaves out two doubly occupied core sigma MO's primarily responsible for the sigma C - C bonds, and two corresponding core sigma unoccupied MO's of very high energy.

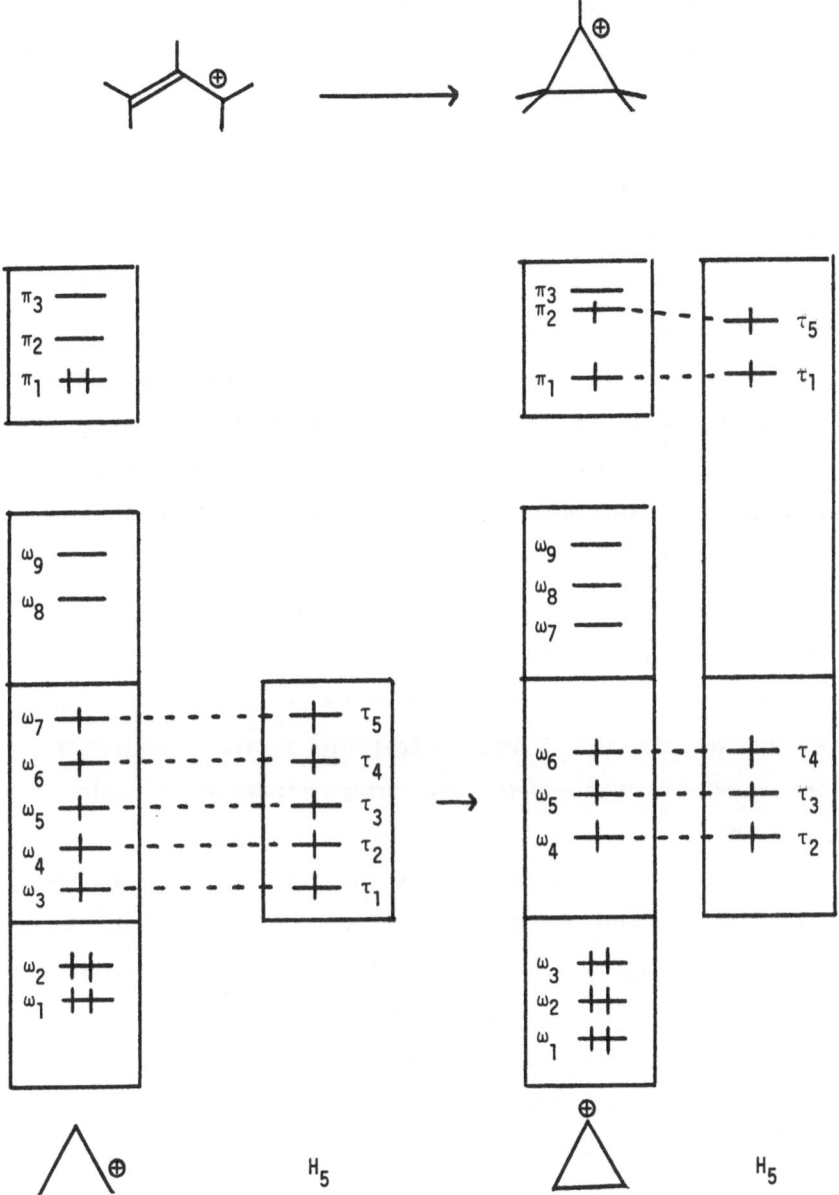

Figure 4. Sigma and pi bonding in reactant allyl cation and product cyclopropyl cation. Note that both reactants and products (in their ground state) belong to the totally symmetric representation because an even number of electrons occupies each subsystem.

b. The consequences of transforming allyl to cyclopropyl cation are straightforward: One sigma bond connecting two carbons is generated by a single electron
deexcitation of the sigma core which produces an additional occupied core sigma MO
and an additional vacant core sigma MO of high energy. At the same time, a coupled
single electron excitation of the pi core becomes responsible for the formation of
two new core-ligand pi bonds in conjunction with the two ligand MO's formerly
responsible for two core-ligand sigma bonds. The electronic reorganization which
accompanies the electrocyclization of allyl cation can now be fully understood by
comparing the MOVB bond diagrams shown in Figure 4. These make clear that what we
deal with is a destruction of two sigma bonds in exchange for sigma core deexcitation to produce two pi core-ligand bonds at the price of core pi excitation.
This is a prototypical transfer reaction and the MOVB diagrams of Figure 4 are
tantamount to a clear and theoretically correct qualitative description of such a
reaction. The argument can be best understood by focusing our attention on the
principal actors of the drama, namely, orbitals ω_3, ω_7, τ_1, τ_5, and π_1, π_2, and π_3
having the shapes shown in Figure 5 and constructing the partial MOVB diagrams shown
in Figure 6. Note that the original two sigma bonds have pi pseudo-symmetry and
that they are converted to pi core-ligand bonds through rotation of the in-plane
outer ligands by 90°.

At this stage, it is extremely important that the reader understands exactly
how we have made the transition from the complete to the partial bond diagram, e.g.,
why we have neglected the ω_4, ω_5, and ω_6 core and τ_2, τ_3, and τ_4 ligand orbitals
and the electrons occupying them. The reason is that one core orbital in con-
junction with one ligand orbital will essentially define the sigma carbon-hydrogen
bond emanating from the "middle" carbon. The other two sigma-core ligand bonds
are invariant to rotation, so to speak. This can be understood by reference to
the pseudorotation ("tumbling") of the $\overset{\circ}{A}H_2$ fragment illustrated below.

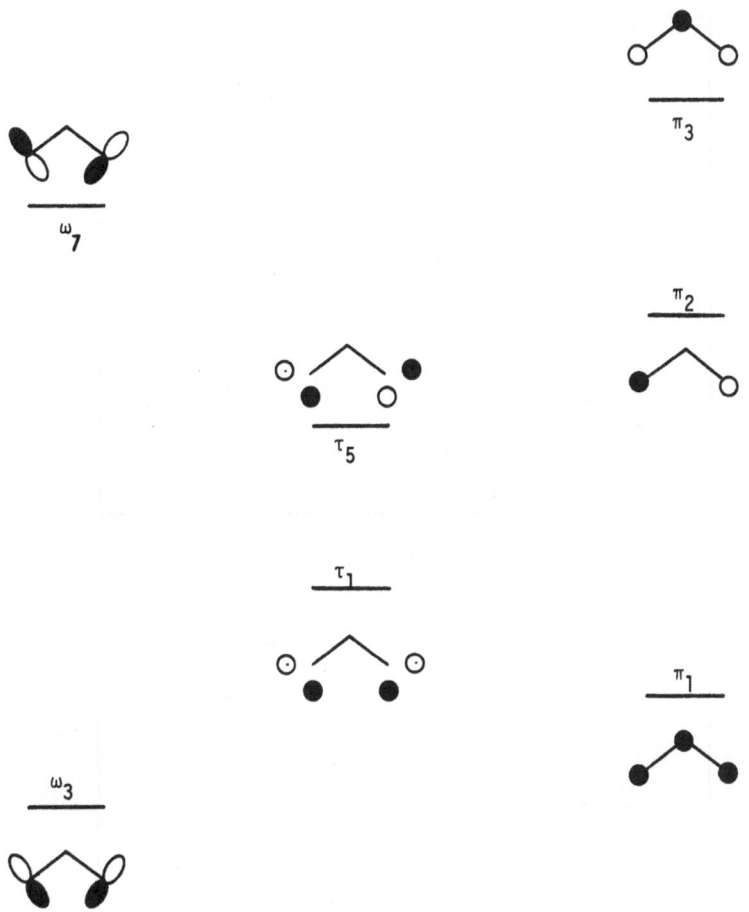

Figure 5. The critical core (C_3) and ligand (H_5) orbitals which are intimately involved in the sigma $\rightarrow$ pi transfer of the outer four hydrogen ligands.

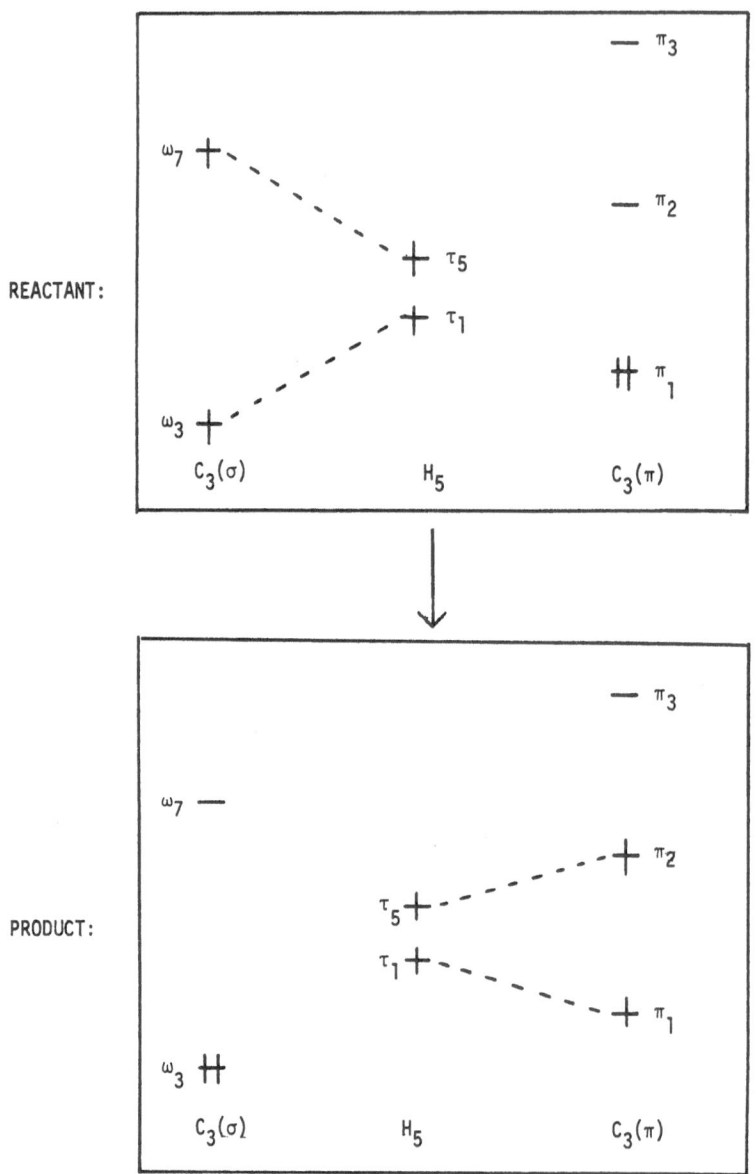

Figure 6. Simplified bond diagrammatic representation of the electronic reorgani-
zation accompaying the ring closure of $CH_2\!\!=\!\!CH\!-\!\overset{\oplus}{C}H_2$, i.e., how "old" bonds are
broken and "new" bonds are formed according to MOVB theory.

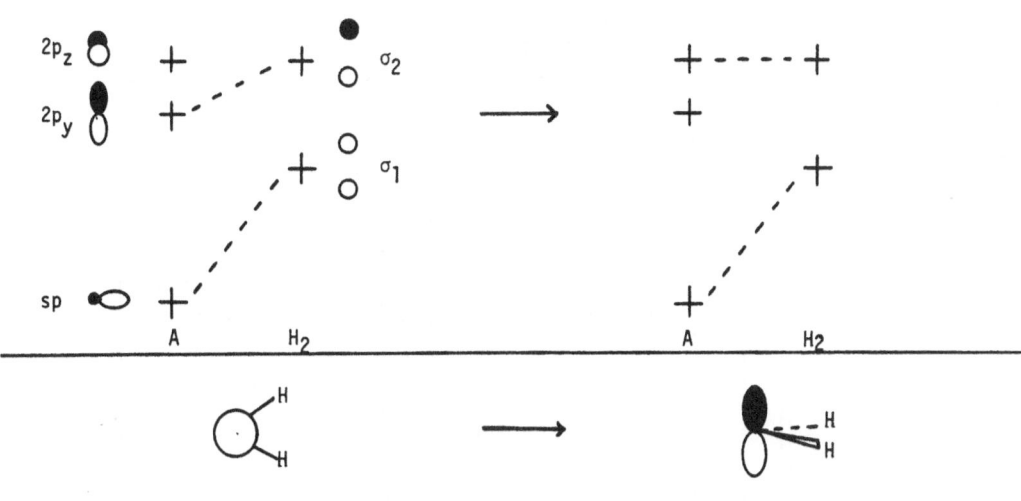

The above process is a true (not pseudo) pi to pi transfer with the sp_A - σ_1 sigma bond retained throughout. Two of the core-ligand bonds which have been neglected in constructing the partial bond diagram are sigma bonds of exactly this type.

Finally, it is important to note that the order symmetric (e.g., ω_3) below antisymmetric (e.g., ω_7) carbon core orbital will always be imposed as the terminal carbons of the core are brought into proximity and the corresponding AO's begin to interact strongly in a "through space" sense, although the original order in the open unperturbed form may or may not have been the same. That is to say, depending on the system at hand, one may encounter either of the two situations depicted below:

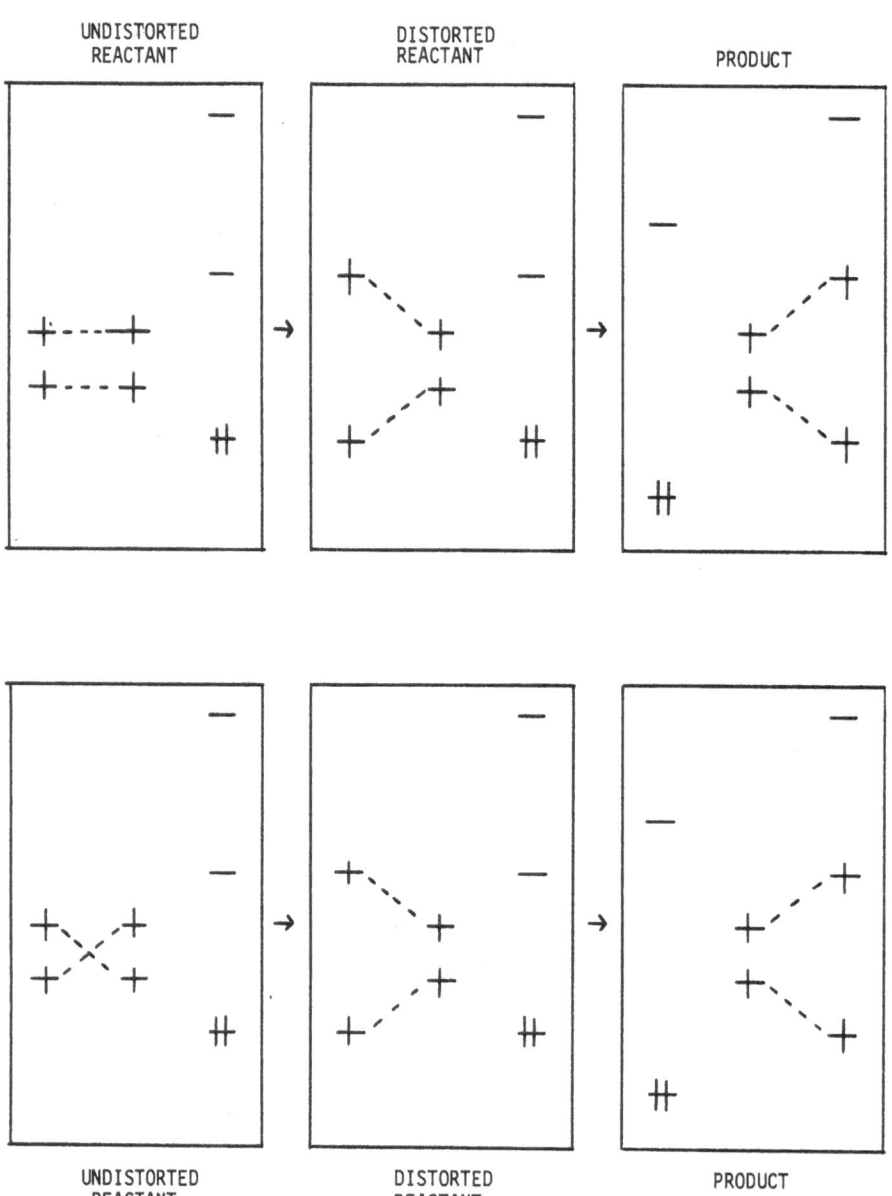

UNDISTORTED REACTANT   DISTORTED REACTANT   PRODUCT

Stereoselection is imposed in the second stage that involves conversion of perturbed reactants to products. The energy order of the core and ligand orbitals can be determined, to a first approximation, through Extended Hückel MO (EHMO) computations. In the system of interest, $C_3H_5^+$, the symmetric ω_3 lies underneath the antisymmetric ω_7 even at the unperturbed reactant level.

(a) CONROTATION COMPLEX

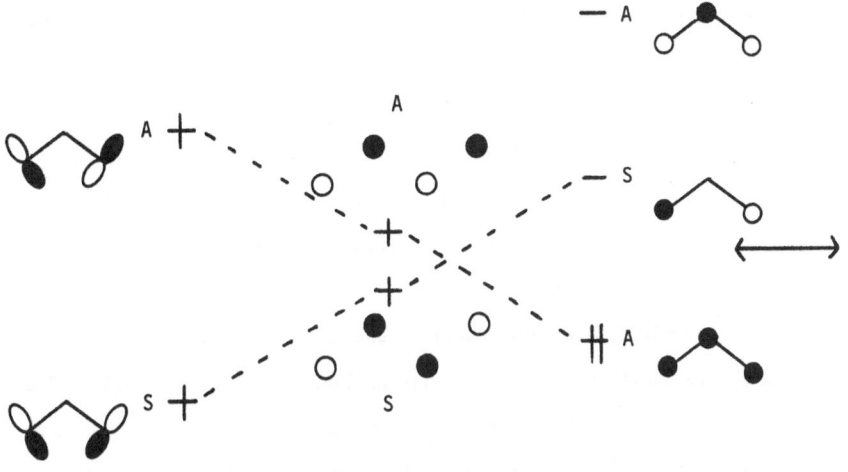

(b) DISROTATION COMPLEX

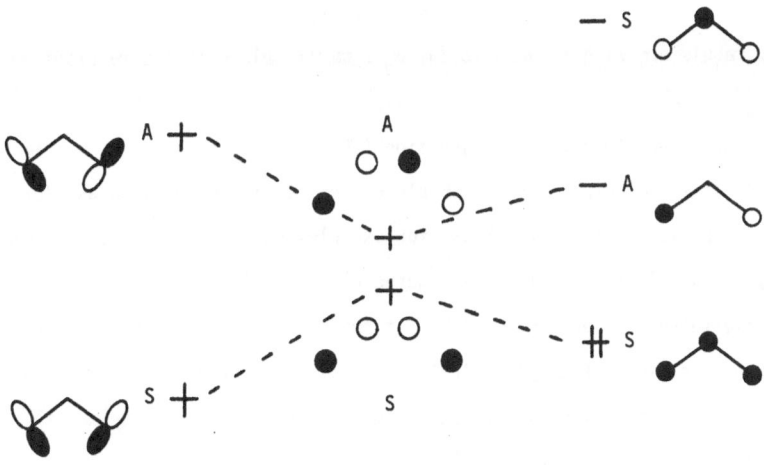

Figure 7. Conrotation and disrotation $CH_2 = CH - \overset{\oplus}{C}H_2$ complexes. In (a), reactants are converted to excited products, while, in (b), there is smooth conversion of reactants to products. Also compare Figure 7 with Figure 3.

The MOVB diagrams for the conrotation and disrotation transition states are shown in Figure 7 and it is evident that the difference between an axis of symmetry (conrotation) and a plane of symmetry (disrotation) is expressed only in the pi MO's of the C_3 core. It is easy to see that the former is entirely analogous to that for 4s + 2s and the latter is exactly analogous to that for 4s + 2a cycloadditions (compare Figures 7 and 3). Hence, allyl cation closes to cyclopropyl cation in a disrotatory fashion for the same reason that 1,3-butadiene adds to ethylene in a suprafacial manner.

The three-fragment MOVB theory described and applied above to the case of the allyl cation electrocyclization can be easily applied to any electrocyclic ring closure of a closed shell system with the conclusion always being that the stereoselectivity of the reaction is dictated by the symmetry labels of the pi HOMO and pi LUMO of the core pi system in the following way:

	HOMO	LUMO	RING CLOSURE
	S	A	Disrotatory
Symmetry type	A	S	Conrotatory

The symmetry labels are with respect to the σ_v symmetry plane of the reactant polyene system.

As we shall see, Frontier Configuration (FC) theory tells us that stereoselectivity in 4N + 3 electron systems should be qualitatively the same as stereoselectivity in 4N electon systems. Thus, for example, we expect that 1,3-butadiene, allyl anion, and allyl radical will all undergo electrocyclization in a conrotatory fashion. While there is every indication that this is the case in the first two systems, the ring closure of allyl radical-type systems is predicted by computations to be near disrotatory with indicative (but not definitive) experimental results pointing in the same direction.[6] What is wrong?

FC theory is a truncated form of VB theory which can be applied in a qualitative sense to model systems. It predicts that Möbius will have lower energy than Hückel trigonal H_3^- and that, similarly, Möbius H_3 will also be more stable than Hückel H_3. It is then evident that, since the conclusions of FC theory for a model system are unassailable, there is a fundamental difference between "real" and model systems and that this difference surfaces and becomes critical in odd electron systems. Thus, we expect that if we preform an MOVB theoretical treatment of the electrocyclization of the allyl radical, following exactly the same reasoning as in the case of the allyl cation electrocyclization, a fundamental difference between even and odd electron systems will be revealed. Indeed, this is the case as illustrated by the MOVB diagrams of reactants and products (Figure 8) as well as the MOVB diagrams for the conrotation and disrotation transition state complexes (Figure 9) of the allyl radical electrocyclization. In stark contrast to the allyl cation electrocyclization case, excitation of the core pi system required for product formation generates a pi system with an odd number of singly occupied orbitals in each of the two possible low energy R CW's so that the symmetry of the total wavefunction changes as we go from reactants to products, whether the process occurs by disrotation or conrotation. However, the important things is that only one of the two modes of ring closure, namely, disrotation, is consistent with four electrons being confined in the lower and three electrons in the higher energy orbitals of the fragments throughout the transformation. We say then that, with the exception of the symmetry aspects of reactants and products, suprafacial $4\pi + 2\pi$ cycloaddition, allyl cation disrotation, and allyl radical disrotation have one and the same key feature: They keep more than one-half of the electrons critically involved in the reaction within the lower energy set of orbitals of the fragments during the entire course of the reaction!

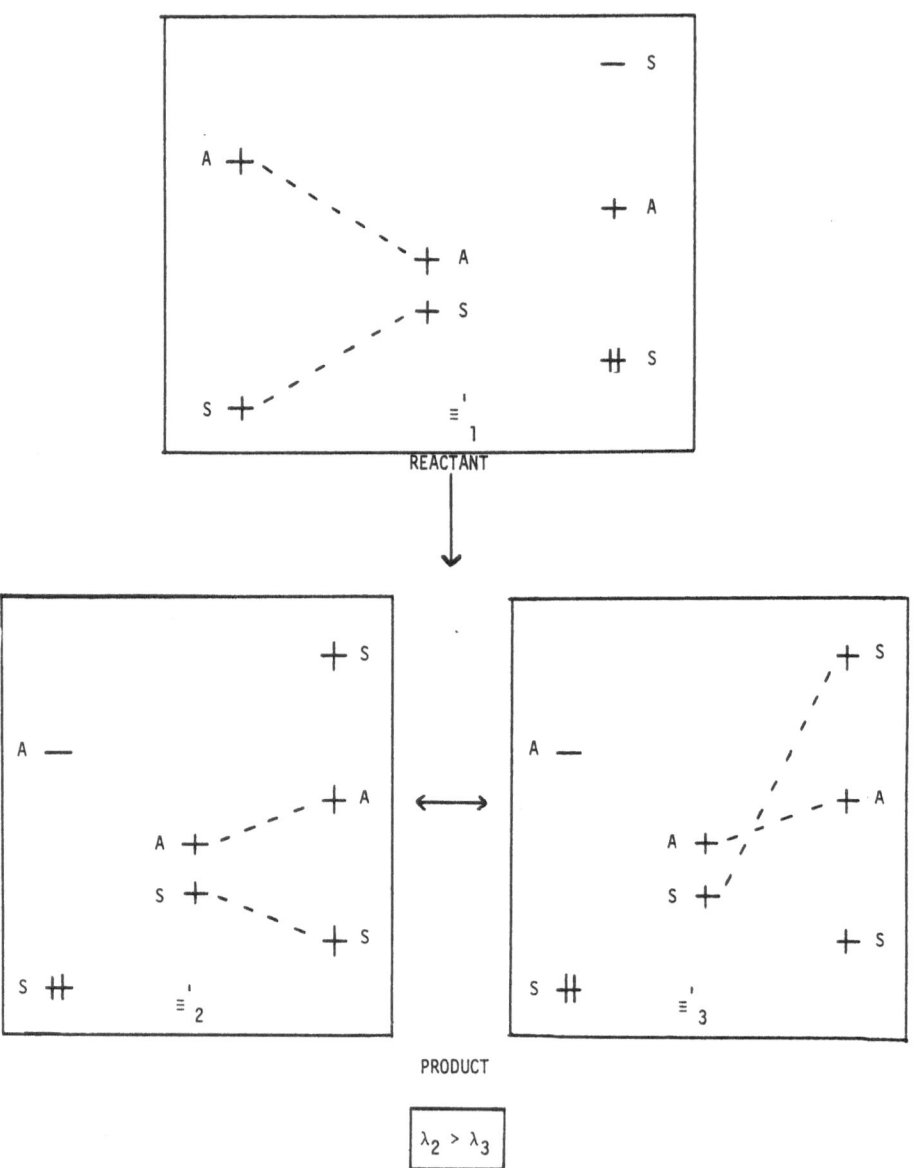

Figure 8. Sigma and pi bonding in reactant allyl radical and product cyclopropyl radical. Note that reactants and products have different symmetry. $\lambda_2 > \lambda_3$ because the nodal structures of π_1 and π_3 ensure that the $\tau_1 - \pi_1$ is significantly greater than the $\tau_1 - \pi_3$ overlap integral.

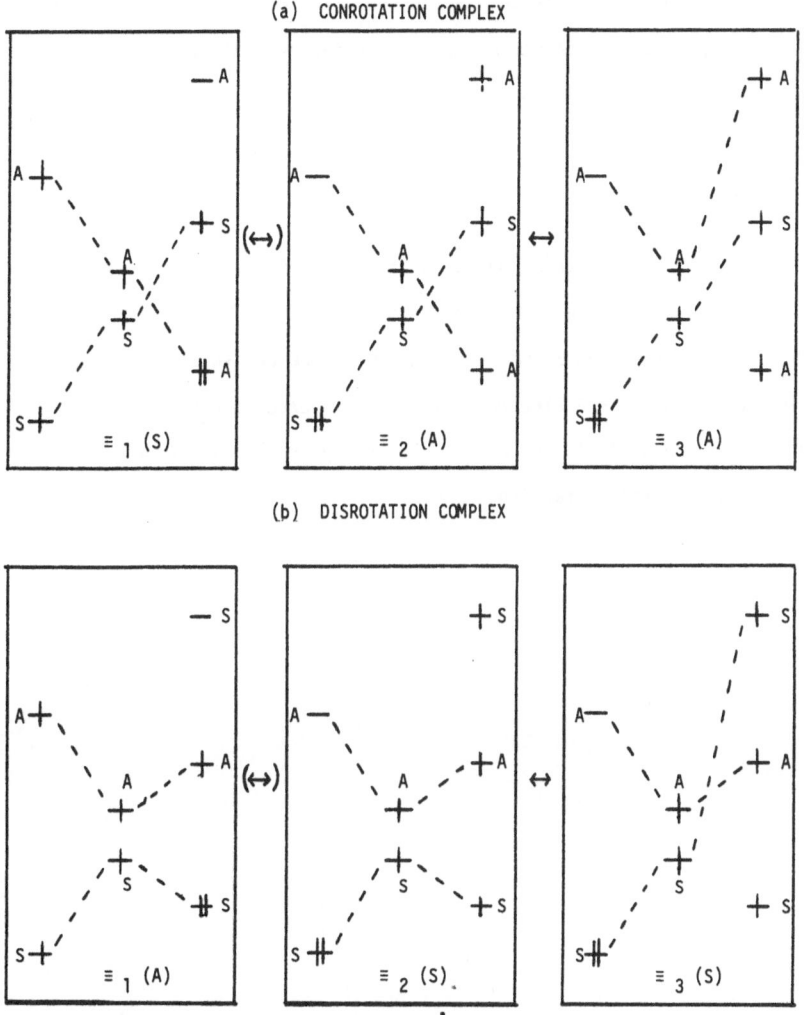

Figure 9. Conrotation and disrotation $CH_2 = CH - \overset{\bullet}{C}H_2$ complexes. In both cases, only distortion can cause interaction of all resonance bond diagrams $\equiv_1$, $\equiv_2$, and $\equiv_3$. In the absence of distortion, disrotation is favored over conrotation because it allows the exclusive accommodation of four electrons in the three lowest energy orbitals and three electrons in the three highest energy orbitals in $\equiv_1$ and $\equiv_2$. By contrast, core-ligand bonds can only be formed by delocalizing all seven electrons in the higher lying orbitals in the case of conrotation . Compare the funneling of the "4" and "2" electron sets in Figures 3 and 7 with that of the "4" and "3" sets in this Figure.

This analysis leads to a stereoselection rule for odd electron systems which is, in a sense, opposite to the one stated before for even electron systems:

	SOMO	LUMO	RING CLOSURE
	A	S	Disrotatory
Symmetry type			
	S	A	Conrotatory

Since 4N electron polyenes have A HOMO's and S LUMO's, 4N + 2 electron polyenes have S HOMO's and A LUMO's, 4N - 1 electron polyenes have A SOMO's and S LUMO's and 4N + 1 electron polyenes have S SOMO's and A LUMO's, we arrive at the following conclusion: 4N + 2 and 4N + 3 electron polyenes will undergo electrocyclization in a disrotatory fashion while 4N and 4N + 1 electron polyenes will undergo electrocyclization in a conrotatory fashion.

References

1. Epiotis, N.D.; Larson, J.R.; Eaton, H. "Unified Valence Bond Theory of Electronic Structure" in Lecture Notes in Chemistry, Vol. 29; Springer-Verlag: New York and Berlin, 1982.

2. Woodward, R.B.; Hoffmann,R. J. Am. Chem. Soc. 1965, 87, 395.

3. (a) Longuet-Higgins, H.E.; Abrahamson, E. J. Am. Chem. Soc. 1965, 87, 2045.

   (b) Hoffmann, R.; Woodward, R.B. J. Am. Chem. Soc. 1965, 87, 2046.

4. These are constructed using the philosophy of the United Atom method.

5. For review, see: Epiotis, N.D.; Shaik, S.; Zander, W. "Rearrangements: A Theoretical Approach" in "Rearrangements in Ground and Excited States", de Mayo, P., Ed., Vol. 2; Academic Press: New York, 1980.

6. Marvell, E.N. "Thermal Electrocyclic Reactions"; Academic Press: New York, 1980.

PART TWO

BEYOND MONODETERMINANTAL MO THEORY

INTRODUCTION

In the first part of this work, we applied MOVB theory to problems which, in principle, are "within" either EHMO (which does not include electron-electron interaction effects) or SCF-MO (which imperfectly accounts for interelectronic repulsion) theory in order to demonstrate the conceptual potency of the approach. In this second part, we focus our attention to problems which are "soluble" only at the SCF-MO-CI level, in order to demonstrate that the fundamental VB and MOVB theoretical concepts developed earlier are sufficient for dealing with any kind of problem. In the treatment of regular atomic arrays with one AO per atom, VB theory is simple enough so that it can replace MOVB theory as a qualitative tool. In this spirit, we begin with a discussion of the concept of the Frontier Configuration, which is the foundation of qualitative VB theory, and we develop the VB theory of molecular spin selection and weak binding (e.g., F_2), two problems which are treatable only at the level of SCF-MO-CI theory. Subsequently, we show that a general understanding of anticooperativity can be achieved by VB or MOVB theory and we discuss the problem of sigma-pi hybridization which is treatable only at the level of SCF-MO-CI theory. Finally, we use MOVB theory in order to explore the stereochemical consequences of weak binding (which are evident only at the level of SCF-MO-CI theory) and in order to make qualitative predictions of "correlation effects" in MO theory.

Chapter 16. Frontier Configurations and a New Classification of Annulenes.

Organic "diradicals" are important for the synthetic chemist who wants to exploit them as precursors of target molecules, for the mechanistic chemist who seeks to unravel reaction pathways, for the quantitative theoretician who is anxious to test different computational schemes on such molecules because of the formalistic intricacies involved, and for the qualitative theoretician who seeks to understand how these species are bound. Specialists of the latter two types most often adhere to MO theory and they discuss the electronic properties of "diradicals" in the following way: They depart from Hückel MO theory and point out why neglect of interelectronic repulsion renders it inapplicable to problems involving "diradicals". Then, the discussion shifts to the SCF-MO level and various formal drawbacks and resulting pitfalls are recognized. Finally, one is forced to examine the problem at the SCF-MO-CI level, something which guarantees that the potential audience of the paper is exponentially reduced and that the ensuing discussion is rendered cumbersome and lengthy. In a recent work,[1] we advanced the argument that qualitative Valence Bond theory has the formal correctness and conceptual clarity which can allow one to dispense with problems which are hard to deal with within the MO theoretical framework in the space of a paragraph or two. In particular, in treating homonuclear systems involving relatively weak pi bonds, one can use the Approximate Heitler-London (AHL) theory outlined in the original monograph. Furthermore, if the problem does not involve parity distinctions, AHL theory can be substituted by the nearly half-century old Diatomics in Molecules (DIM) approach introduced by Eyring.[2-12] According to this brand of theory, the relative stability of isomers is mainly due to overlap bonding or overlap antibonding and the DIM eigenstates are excellent approximations of the corresponding SCF-MO-CI eigenstates, under the conditions stated above. Noting that cyclobutadiene(CB) and trimethylenemethane(TMM) have attracted a large amount

of attention in the last ten years and desiring to show exactly how trivial most
of the associated problems are when looked from the standpoint of qualitative VB
theory (in this case DIM theory), we have calculated the ground state energies of
these controversial organic diradicals by using procedures known since the 1930's.
The results are given in Table 1 and they constitute direct answers to many often-
discussed problems. Since the coulomb integral Q is approximately invariant,
relative energies depend on the number and signs of the exchange terms, T, each
of which describes a two-electron bond (positive sign preceding T) or a two-
electron antibond (negative sign preceding T). The following conclusions are the
same as those arrived at by ab initio SCF-MO-CI computations after considerable
controversy generated by less than adequate monodeterminantal SCF-MO calculations:

    a. Singlet CB is more stable than triplet CB.

    b. Triplet TMM is more stable than singlet TMM.

    c. Singlet TMM is less stable than allyl plus pi so it must have a perpen-
dicular conformation.

    d. The electronic structures of the various species extracted from laborious
SCF-MO-CI computations are essentially those shown in Table 1.

If we replace DIM by AHL theory so that the effect of parity on stereoselection can
be "seen", one obtains the Woodward-Hoffmann predictions that Möbius CB will have
lower energy than Hückel CB and, analogously, Möbius TMM will be more stable than
Hückel TMM. That is to say, TMM is predicted to close thermally to methylene
cyclopropane by conrotation.

    Finally, I have added to the Table the predictions of DIM theory for the
degenerate isomerization of 3 H_2 via an H_6 hexagon, a model of an "allowed"
reaction. We must conclude that barriers exist in "allowed" reactions because of
interfragmental overlap repulsion and that concepts like N-electron-N-orbital-N-
center homoconjugation are invalid.[5] It is then clear that many of the current

Table 1

Molecule	Multiciplicity	Ground State Energy	Description [a]
(square)	Singlet	Q + 2.0 T	(diagram)
(square)	Triplet	Q	(diagram)
(Y-shape)	Singlet	Q	(diagram) and (diagram)
(Y-shape)	Triplet	Q + 1.1 T	(diagram)
(linear)	Singlet or Triplet	Q + 1.0 T	(diagram)
3 (bond)	Singlet	Q + 3.0 T	3 (bond)
(hexagon)	Singlet	Q + 2.6 T	(diagram)

[a] Solid lines are bonds and dashed lines antibonds. Wriggly line is a partial bond.

problems of organic chemistry have been solved long ago. Given the preoccupation of many with HMO theory, it is, in fact, accurate to say that some "current" problems were better understood half a century ago than they are today!!

Encouraged by these results, we sought to develop a qualitative VB theory of chemical bonding based on exclusive consideration of the Frontier Configurations (FC's) of homonuclear systems comprised of strong overlap binders. The FC's are the set of equivalent HL Configuration Wavefunctions (CW's) of lowest energy and the HL CW's are the VB CW's of maximum open shell character, often referred to as "covalent" CW's. <u>FC's play in polydeterminantal VB theory a role which is analogous to that played by FO's in monodeterminantal HMO theory</u>: <u>They are the primary actors in the drama of chemical selection</u>. We now take a closer look at these critical CW's and show how these provide the basis for a "different from usual" classification of cyclic homonuclear N-electron-M-center (N/M) systems, with one AO per center, projected by the field matrix is shown in Figure 1.

Consider a 4/4 (4 electron-4 orbital) system wherein the orbitals are located at the corners of a square. There are two independent CW's having maximum open shell character, namely, the two Kekule CW's shown schematically below.

$$\Phi_1 \qquad\qquad \Phi_2$$

Next, consider a 4/4 system with the four orbitals arranged in a collinear fashion. Now, one of the maximum open shell CW's attains a much lower energy than the other.

$$\Phi_1 \qquad\qquad \Phi_2$$

Low Energy   High Energy

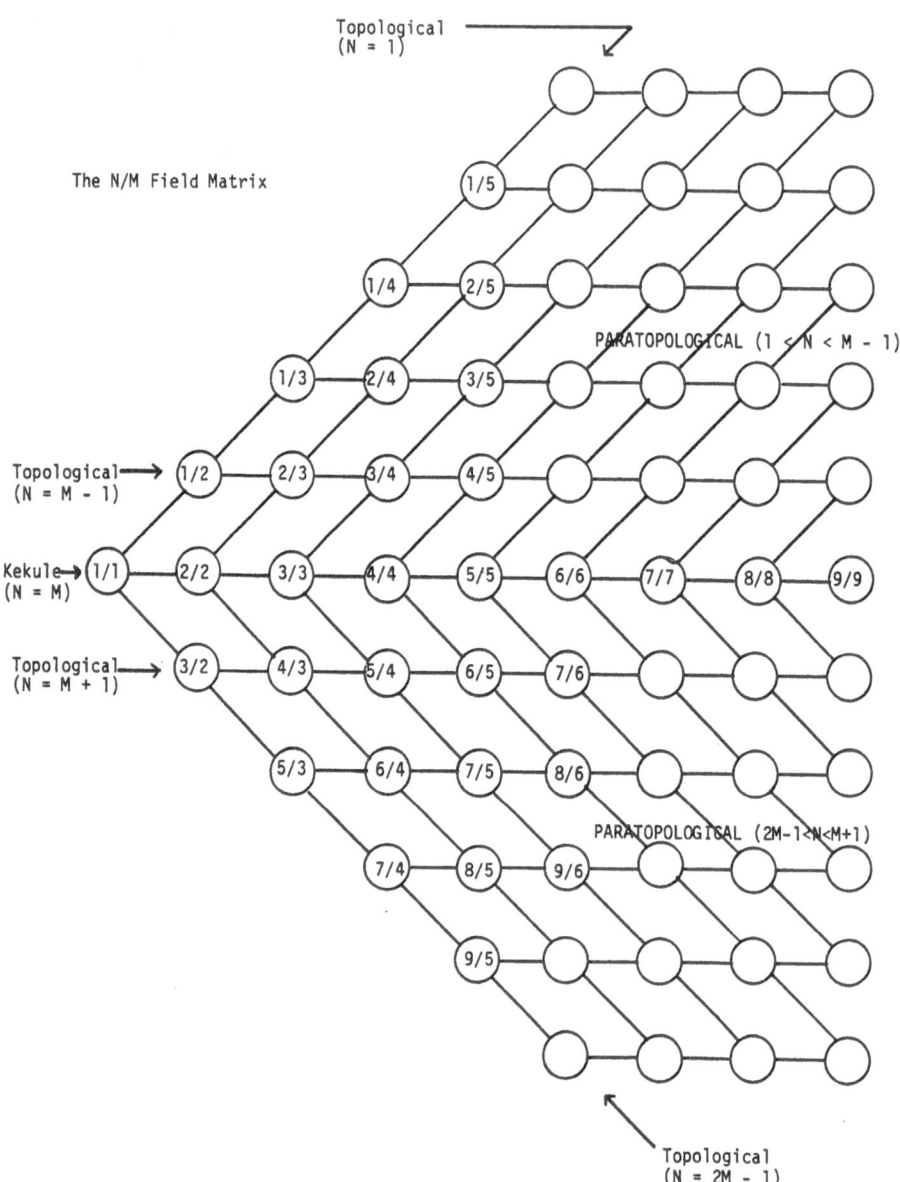

Figure 1. The N/M Field Matrix. N symbolizes the number of electrons and M the number of centers of an annulene.

Finally, consider a 6/6 system wherein the orbitals are located at the corners of a regular hexagon. Now, there are five independent CW's having maximum open shell character, with two of them forming a lower and three a higher energy set.

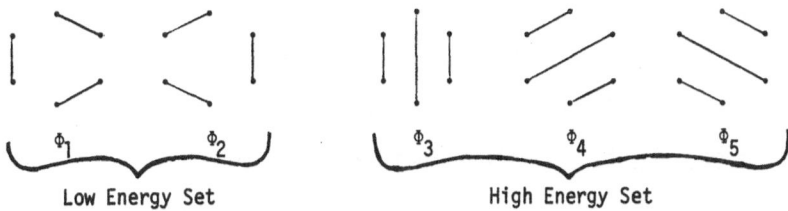

Low Energy Set　　　　　　　　　High Energy Set

We say that Φ_1 and Φ_2 in square 4/4, Φ_1 in linear 4/4, and Φ_1 and Φ_2 in regular hexagonal 6/6 are the Frontier Configurations (FC's) of the corresponding systems, i.e., the CW's with maximum open shall character (something which guarantees minimal electron-electron repulsion) which can effect maximum interatomic bonding through spin pairing. We expect that trends in chemical selection can be fully understood through approximate HL theory which departs from a basis set of FC's under the condition that we deal with homonuclear systems made up of strong overlap binders (e.g., H, F, O, N, C). That is to say, we expect that the dominant contributors to the complete "perfect" VB wavefunctions of all systems of the type specified above are the FC's and that it is these CW's which set chemical trends.

With these definitions in mind, we now turn attention to the field matrix and we note the following important trends in cyclic N/M systems:

a. In N/M systems [N are electrons and M are centers (orbitals)] in which N equals 1, M - 1, M +1, or 2M - 1, there exist M equivalent FC's if M is odd. If all the integral approximations of HMO theory are implemented and the energy of each orbital (symbolized by α in HMO theory) is set equal to zero then the energy matrix over FC's in HL theory is proportional to the energy matrix over AO's in HMO theory. Systems of this type are called topological because the HL/FC energy matrix is related to the HMO energy matrix which, in turn, is related to the topological matrix of the AO system. With the approximations stated above, the energies of the HL/FC eigenstates are proportional to the energies of the MO's of the corresponding AO system generated by HMO theory. As a result, one can differentiate Hückel FC (HFC) from Möbius FC (MFC) systems in the same way that HAO are differentiated from MAO systems. The characteristic property of topological systems is <u>strong</u> <u>cyclic</u> <u>FC</u> <u>interaction</u> brought about by the fact that each FC can interact with another one which differs from it by one occupied spin orbital with the interaction matrix element being proportional to the resonance integral of neighbor AO's, β. If M is even, then there are either M or 2M FC's with M FC's defining again an energy matrix which is related to the HMO energy matrix which, in turn, is related to the topological matrix of the AO system. In systems of this type there is once again strong cyclic FC interaction. Hence, they are also termed topological systems.

b. In triplet N/M system in which N equals M - 1 and M + 1 and M is odd, there exist M <u>equivalent</u> FC's and these systems are topological for reasons which are the same as those discussed above.

c. In singlet N/M systems in which N = M and M is even there exist only two <u>equivalent</u> FC's, commonly known as Kekule structures. By contrast, in doublet N/M systems in which N = M and M is odd there exist (M+1)/2 <u>equivalent</u> FC's. N/M systems with N = M are designated <u>Kekule</u> systems.

d. Systems with N = 0 or N = 2M are obviously described by only one CW and they will not concern us in this work.

e. In singlet and triplet N/M systems in which N assumes all integral values except 0, 1, M - 1, M, M + 1, 2M - 1, and 2M there exist either M or M/2 equivalent

FC's depending on N and M. Because the FC's segregate electron pairs in order to "insulate" electrons that have the same spin to the maximum extent possible so that exchange repulsion is minimized, these systems are no longer topological and the interacting FC's define some other topology in which elements of a cyclic array are connected in a way which is different from that encountered in (a). Systems of this type are paratopological. The characteristic property of paratopological systems is weak or zero cyclic FC interaction because each FC can at best interact with another one which differs from it by either one or two occupied spin orbitals with the interaction matrix element being proportional to β' or $2\beta s$, where β' is the resonance integral of non-neighbor AO's, β that of neighbor AO's, and s the overlap integral of neighbor AO's. Since $|\beta|>|\beta'|$, $|2\beta s|$ if $s<0.5$, and assuming that the latter condition is met in all cases of interest, paratopological are truly distinct from topological systems. A special case is triplet 2/4 in which there is only noncyclic interaction of two FC's.

Examples of prototypical N/M systems are given in Figure 2.

An N/M system may involve Hückel or Möbius AO overlap, it may be in an aggregated or nonaggregated state, it may have a "low" or "high" spin, it may be symmetrical or distorted. We can study parity stereoselection, aggregation stereoselection, spin selection, and Jahn-Teller distortion in topological, paratopological and Kekule cyclic homonuclear N/M systems, made up of strong overlap binding atoms, by using three different brands of qualitative HL theory:[1]

    a. DIM/FC theory, i.e., DIM theory with an FC basis and with the integral approximations inherent in the DIM method first proposed by Eyring.

    b. AHL/FC theory, i.e., HL theory with an FC basis and the integral approximations of EHMO theory.

    c. AHL/FC theory, i.e., HL theory with an FC basis and the integral approximations of HMO (with neglect of overlap) theory.

The simplest, proper understanding of chemical selection can be attained by
using AHL/FC theory in dealing with parity and spin selection and DIM/FC theory
is dealing with aggregation selection. Furthermore, AHL/FC may be replaced by the
simpler AHL/FC theory when dealing with off-diagonal energy matrix elements in topological systems. The forms of the energy matrix elements according to the above
three brands of theory are given in Table 2. F reflects the energy of the isolated
atoms. G reflects the "classical" coulomb interaction of the atoms. X^0 is the
overlap term which contains only even powers of AO overlap integrals. X is the
overlap term which contains even as well as odd powers of AO overlap integrals
with the latter remaining unchanged as a result of a chosen AO overlap sign inversion. L is the overlap term which changes sign as a result of a chosen AO overlap
integral sign inversion. Table 2 tells us that parity stereoselection is exerted
in three different ways:

a. Via the magnitudes of the diagonal energy terms, H_{ii}.

b. Via the magnitudes of the off-diagronal energy terms, H_{ij}.

c. Via the signs of the H_{ij}'s since there determine whether a cyclic FC
interaction is of the Hückel or Möbius type.

We defer the derivation of the equation governing parity stereoselection and the
development of predictive notions regarding aggregation stereoselection to a
separate paper to be published elsewhere and direct our attention to the problem
of spin selection viewed from the vantage point of FC theory.

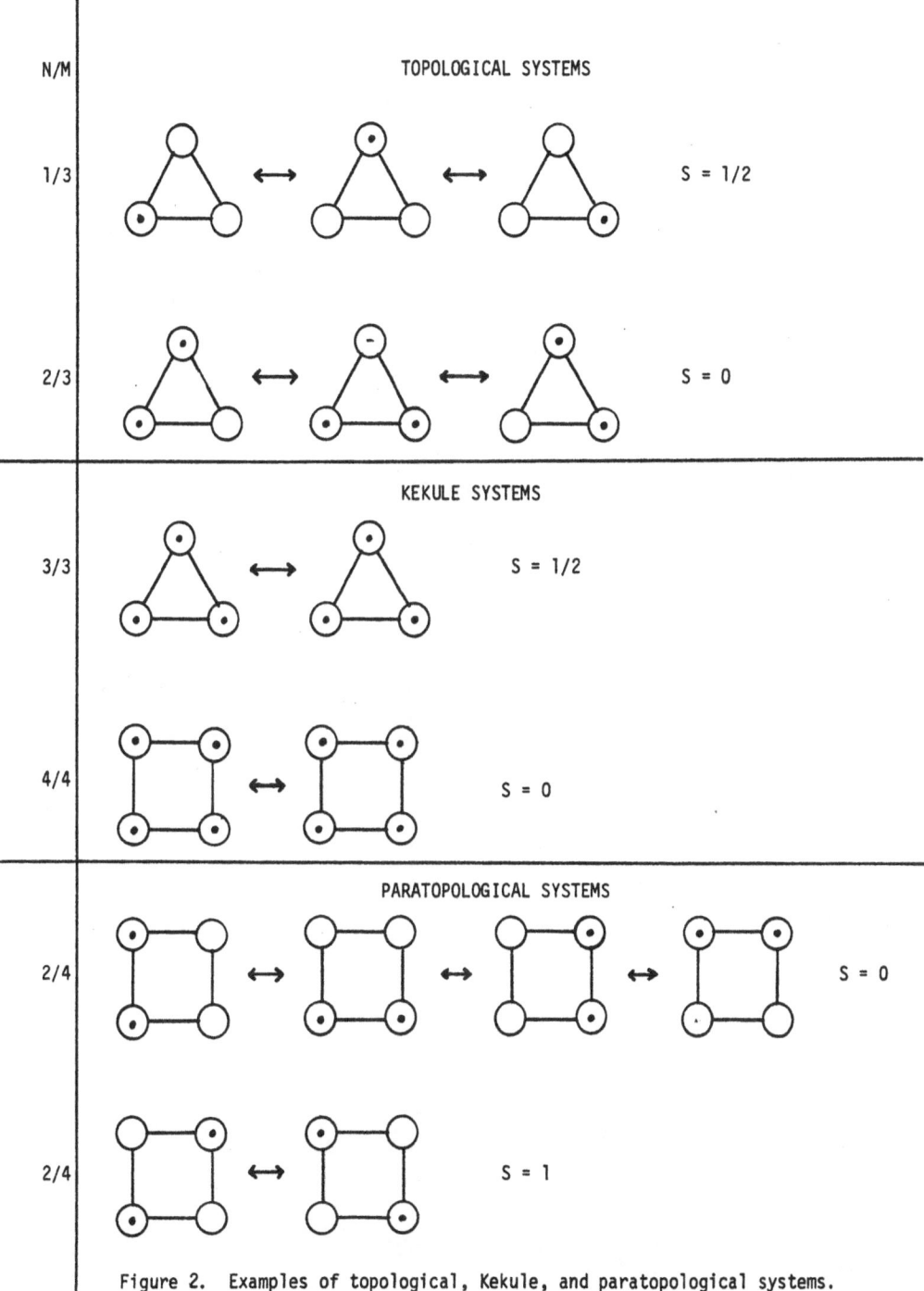

Figure 2.   Examples of topological, Kekule, and paratopological systems.

Table 2. Forms of Matrix Elements of FC-type Theories.[a]

THEORY	H_{ii}	H'_{ij}	H''_{ij}
DIM/FC [b]	$(F+G) + X^0$	$(F+G) + X^0$	
AHL/FC	$(F+G) + (X\pm L)$	$(F+G) + (X\pm L)$	$\pm(X\pm L)$
$\widetilde{AHL}$/FC	F	F	$\pm X$

[a] H'_{ij} connects two CW's having common SD's. H''_{ij} connects two CW's differing by one or two spin-orbitals.

[b] Matrix elements for Kekule systems only.

References

1. Epiotis, N.D.; Larson, J.R.; Eaton, H. "Unified Valence Bond Theory of Electronic Structure" in Lecture Notes in Chemistry, Vol. 29; Springer-Verlag: New York and Berlin, 1982.

2. (a) Eyring, H.; Kimball, G.E. J. Chem. Phys. 1933, 1, 239, 626.
   (b) Eyring, H.; Frost, A.A.; Turkevich, J. J. Chem. Phys. 1933, 1, 277.

3. Glasstone, S.; Laidler, K.J.; Eyring, H. "The Theory of Rate Processes"; McGraw-Hill: New York, 1941.

4. Pauling, L. J. Chem. Phys. 1933, 1, 280.

5. Ellison, F.O. J. Am. Chem. Soc. 1963, 85, 370. Ellison, F.O. J. Chem. Phys. 1964, 41, 2198.

6. Cashion, J.K.; Herschbach, D.R. J. Chem. Phys. 1964, 40, 2358; 41, 2199.

7. Blais, N.E.; Truhlar, D.G. J. Chem. Phys. 1973, 58, 1090.

8. Steiner, E.; Certain, P.; Kuntz, P. J. Chem. Phys. 1973, 59, 47.

9. Tully, J.C. J. Chem. Phys. 1973, 58, 1396; 59, 5122.

10. Pickup, B.E. Proc. Roy. Soc. A 1973, 333, 69.

11. Gelb, A.; Jordan, K.D.; Silbey, R. Chem. Phys. 1975, 9. 175.

12. Dixon, D.A.; Stevens, R.M.; Herschbach, D.R. Faraday Discussion of Chem. Soc. 1977, 62, 110.

Chapter 17. Frontier Configuration Theory of Spin Selection.

The spin selection problem revolves about the determination of the sign and magnitude of the energy difference between two different spin states Λ_1 and Λ_2, for a given geometry, g, and a given number of electron pairs, n. For our purposes, g can be either a Hückel AO (HAO) or Möbius AO (MAO) geometry.

$$\Delta sp(g,n) = E(\Lambda_1) - E(\Lambda_2)$$

If we restrict our attention to the lowest energy singlet (S) and triplet (T) states of M-1/M, M/M, and M+1/M species, we have:

$$\Delta sp(g,n) = E(T) - E(S)$$

The H_{ii} and H_{ij} matrix elements for singlet and triplet FC's have the forms shown in Table 1. Accordingly, $\Delta sp(g,n)$ can be expressed as follows:

$$\Delta sp(g,n) = \Delta X(g,n) + \Delta L(g,n) + \Delta I(g,n)$$

Removing the parenthesis notations for brevity, we have:

$$\Delta X = X_{ii}(T) - X_{ii}(S)$$

$$\Delta L = L_{ii}(T) - L_{ii}(S)$$

$$\Delta I = I(T) - I(S)$$

where X_{ii} is the parity invariant overlap term, L_{ii} is the overlap term responsible for parity stereoselection, and I is the stabilization energy due to the interaction of the FC's.

Now, in systems which can be represented by normal Lewis structures in which electrons are coupled to bond pairs and "lone" pairs, ΔX is always positive.

Table 1.

N/M	SINGLET		TRIPLET	
	H_{11}	H_{12}	H_{1i}	H_{1j}
M−1/M	F + G + X	$\pm\,(X\pm L)$	F + G + X	$\mp\,(X\mp L)$
M/M	F + G + X $\pm$ L	F + G + X $\pm$ L	F + G + X $\mp$ L	F + G + X $\mp$ L
M+1/M	F + G + X $\pm$ L	$\pm\,(X\pm L)$	F + G + X $\mp$ L	$\mp\,(X\mp L)$

Table 2.

N/M	Systems	ΔX	ΔL	ΔI
M−1/M	B	+	−	−
	A	+	+	+
M/M	B	+	−	~0
	A	+	+	~0
M+1/M	B	+	−	−
	A	+	+	+

A is C^+/n odd or C^-/n even. B is C^+/n even or C^-/n odd.

Furthermore, since ΔL is relatively small, being a function of a product of AO overlap integrals, we arrive at the following important conclusion: $\underline{\Delta sp\ will\ be}$ $\underline{positive\ (i.e.,\ lowest\ singlet\ below\ lowest\ triplet)\ unless\ \Delta I\ is\ large\ and}$ $\underline{negative}$. If we denote MAO systems by C^- and HAO systems by C^+, the signs of ΔX, ΔL, and ΔI predicted by AHL/FC theory are given in Table 2. We now show that only $M\pm1/M$ systems can have a large negative ΔI while M/M systems are condemned to a near zero ΔI and, as a result, triplet ground states can be expected only in the case of $M\pm1/M$ systems. We illustrate our approach by reference to the 4/3 and 4/4 systems.

According to AHL/FC theory, the lowest energy states of the singlet and triplet manifolds of a 4/3 system can be generated from three equivalent singlet and three equivalent triplet FC's as shown below.

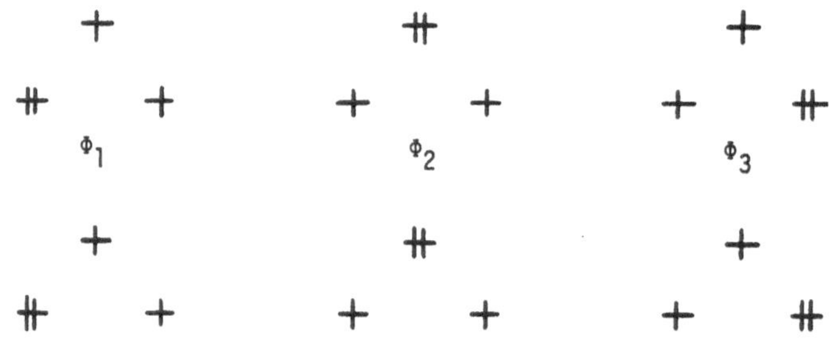

Singlet
FC's:

Common Symbol: Φ_1 Φ_2 Φ_3

Triplet

FC's:

The way in which the three FC's interact in the HAO and MAO geometries of the singlet and triplet species is shown below.

SIGNS OF FC INTERACTION MATRIX ELEMENTS

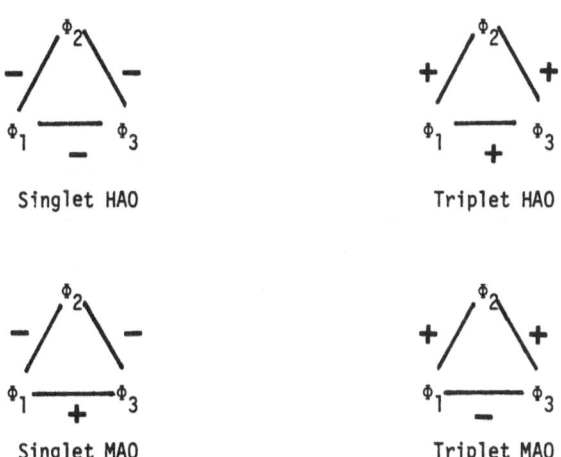

It is evident that, because of the signs of the interaction matrix elements, the FC's of singlet MAO and triplet HAO systems define Hückel CW (not AO) inter-action, while the FC's of singlet HAO and triplet MAO systems define Möbius CW interaction. With this information we can construct the interaction diagrams shown in Figure 1. It is obvious that ΔI is negative in the HAO and positive in the MAO geometry. This guarantees that the triplet MAO will lie far above the singlet MAO species. On the other hand, the HAO triplet has an excellent chance of dropping below the HAO singlet. We say then that cyclic FC interaction in the 4/3 system can cause the HAO triplet to drop below the HAO singlet.

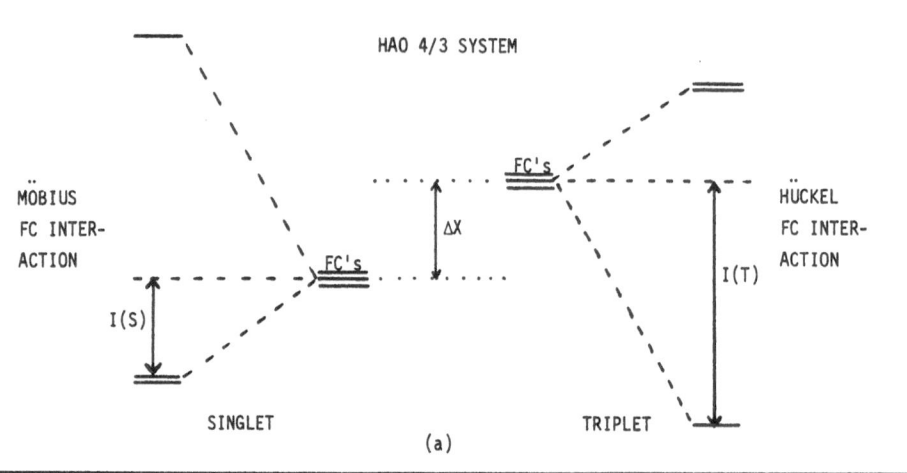

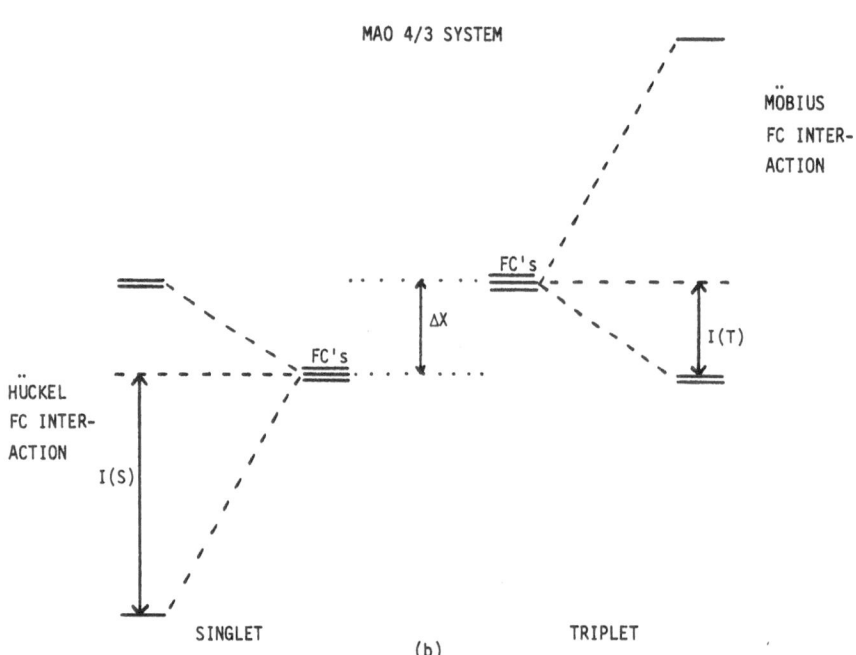

Figure 1. a. Singlet and triplet eigenstates of the HAO 4/4 system. Lowest triplet has lower energy than lowest singlet due to cyclic Hückel FC interaction in the triplet and cyclic Möbius interaction is the singlet system. b. Singlet and triplet eigenstates of the MAO 4/3 system. In contrast to (a), there is Hückel FC interaction in the singlet and Möbius FC interaction in the triplet system.

The situation becomes extremely different in a 4/4 system. Now, there are only two equivalent singlet but three equivalent triplet FC's. This mere fact could lead someone to expect that the triplet FC's being more in number than the singlet FC's will produce a triplet state having lower energy than the singlet state in either the HAO or the MAO geometry. However, this is not the case at all. For, of the three triplet FC's, only two interact as strongly so that <u>effectively</u> both the singlet and triplet FC's define <u>noncyclic</u> FC interaction in both HAO and MAO geometries. This fact is made evident by inspection of the DIM matrix elements shown below.

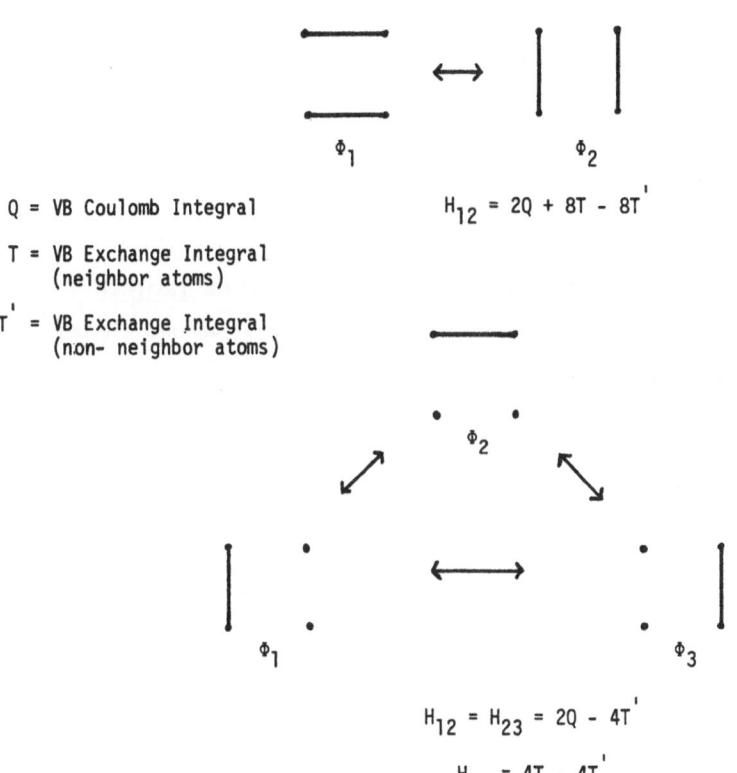

Q = VB Coulomb Integral

T = VB Exchange Integral
(neighbor atoms)

T' = VB Exchange Integral
(non- neighbor atoms)

$$\Phi_1$$

$$\Phi_2$$

$$H_{12} = 2Q + 8T - 8T'$$

$$\Phi_2$$

$$\Phi_1 \qquad \Phi_3$$

$$H_{12} = H_{23} = 2Q - 4T'$$

$$H_{13} = 4T - 4T'$$

As it can be seen, the three triplet FC's are connected by two large (H_{12}, H_{23}) and one small (H_{13}) interaction matrix elements. In fact, the latter goes to zero if AO overlap is neglected. As a result, ΔI approaches zero. The DIM interaction diagrams are shown in Figure 2. It follows that, in the 4/4 system, there is no motivation for the triplet to dip below the singlet in either the HAO or MAO geometries. The following description is then the most apt explanation of why cyclopropenyl anion (CP) has a triplet ground state while cyclobutadiene (CB) has a singlet ground state: In the HAO geometry of CP, the lowest triplet is generated by the "aromatic" interaction of the triplet FC's and the lowest singlet by the "antiaromatic" interaction of the singlet FC's. By contrast, in the HAO geometry of CB, the lowest triplet is generated by the "nonaromatic" interaction of the triplet FC's and the lowest singlet by the "nonaromatic" interaction of the singlet FC's. As a result ΔI is appreciably negative only in the former case and a triplet ground state likely only in the first few members of the "antiaromatic" M±1/M series. For, as the number of bond pairs, p, increases, spin selection in M±1/M systems decreases because H_{12} decreases in absolute magnitude (see Section V).

We can now ask the ultimate question: Is there a recipe for the rational prediction of molecules which have a triplet ground state? The answer has already been given in the form of equation (67). If we neglect ΔL, we recognize three possibilities:

a. ΔX and ΔI are both positive. As we have already seen, this guarantees that the triplet will lie above the singlet in molecules which comply with this requirement.

b. ΔX is positive but ΔI is negative. Molecules which satisfy this condition may have triplet ground states. Furthermore, these molecules belong to the class of conventional molecules, i.e., species for which classical Lewis structures can

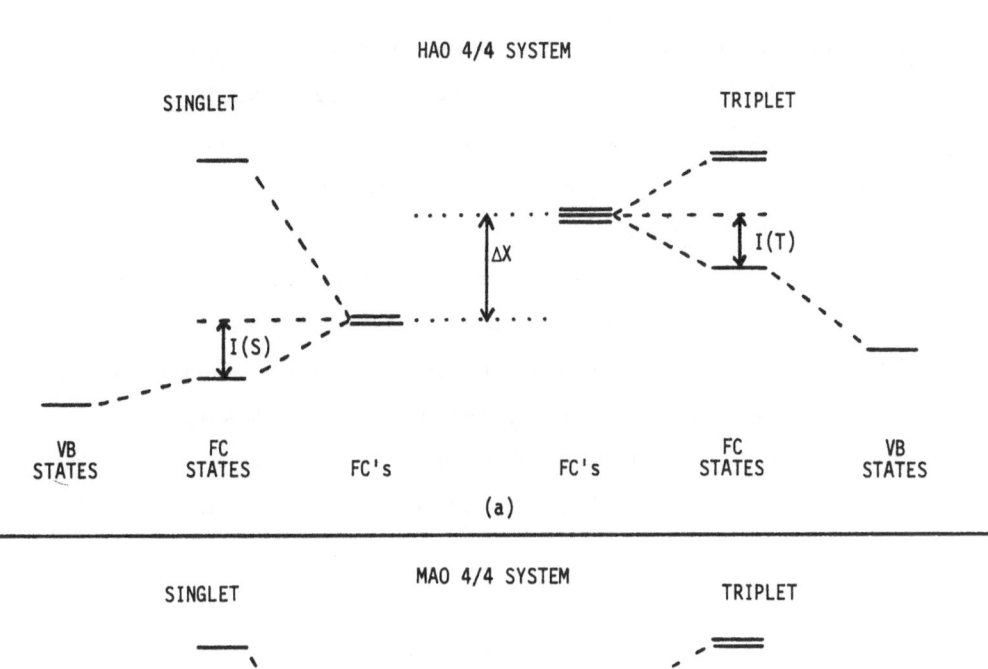

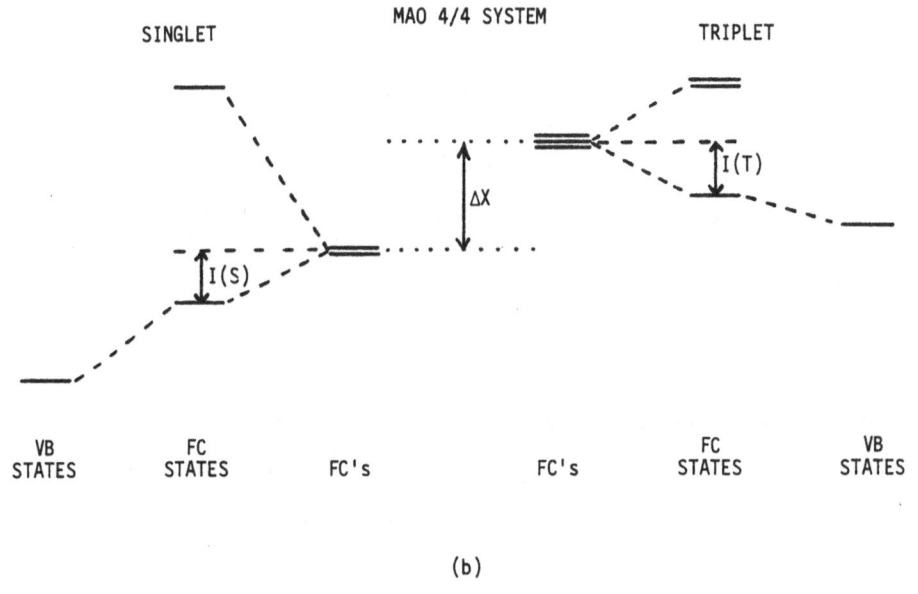

Figure 2. a. Singlet and triplet eigenstates of the HAO 4/4 system. b. Singlet and triplet eigenstates of the MAO 4/4 system. In both cases, the singlet ends up below the triplet because differential FC interaction is small and ΔX is the determining factor.

be drawn in which all electrons are paired to form bonds or "lone" pairs.

c. ΔX and ΔI are both negative. This condition cannot be met by conventional molecules because in the latter ΔX is always positive. It follows that we must look for _unconventional_ molecules. How do we do this? Again, we illustrate our approach by reference to the 4/4 system.

Consider the DIM matrix elements for the HAO 4/4 system shown diagrammatically on the left side of Figure 3. In these drawings a solid line represents a _bonding_ exchange term and a dashed line an _antibonding_ exchange term with the number on top of a line signifying the number of corresponding bonding or antibonding terms. We now observe that if we move one center of the square in the direction indicated below, we obtain a species with zero bonding or antibonding terms in the matrix elements.

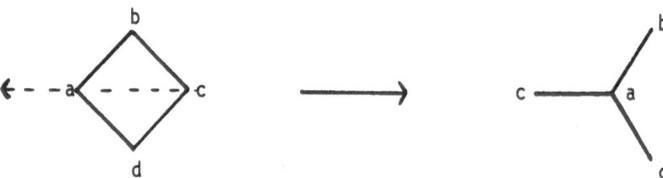

This means that the lowest eigenstate of this molecule is a doubly degenerate E state simply because $H_{12} - H_{11}S_{12} = H_{21} - H_{22}S_{21} = 0$, i.e. the two CW's cannot interact and are components of an E state. If we label the cyclic HAO 4/4 system by C^+ and the form resulting from the distortion specified above by T, we conclude that singlet T is a _destabilized_ form of the singlet C^+ system. Following the same reasoning, we can show that triplet T is the _stabilized_ form of triplet C^+

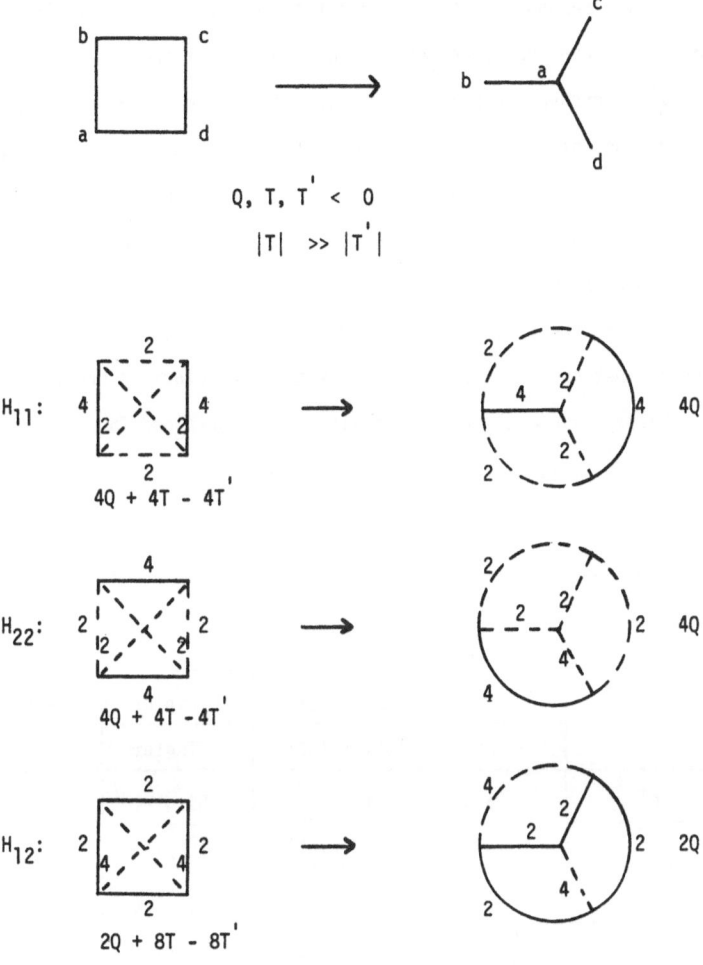

Figure 3. The change of the FC matrix elements of the singlet HAO 4/4 system upon T-distortion. Note how T distortion raises the energies of both FC's of the 4/4 system. Solid lines imply "bonding" and dashed lines "antibonding" exchange (T or T') interaction.

(see Figure 4).   The reason for this is that the T distortion causes the lowest triplet FC of T to attain much lower energy than the lowest energy triplet FC of C^+ because of relief of exchange repulsion. Thus, triplet T ends up below triplet C^+ since the interactions of the FC's are comparable in the two systems. Finally, we conclude that T is a molecule in which ΔX_{11} is zero, if non-neighbor overlap is neglected, and ΔI is negative. Hence, a molecule of this type has no choice other than having a triplet ground state.

Let us now ponder the precise implications of the analysis given above. It is easy to recognize that spin selection is a distinctly topological problem. This can be easily grasped by comparing the results we have obtained for the 4/3 system using AHL and the 4/4 system using DIM theory with neglect of non-neighbor overlap.

System	Topology	Characteristic Spin Selection Equation	Ground State	Chemical Example
4/3	C^+	$\Delta sp \;\simeq\; \lvert\Delta X\rvert - \lvert\Delta I\rvert$	Triplet	CP Anion
	C^-	$\Delta sp \;\simeq\; \lvert\Delta X\rvert + \lvert\Delta I\rvert$	Singlet	
4/4	C^+	$\Delta sp \;\simeq\; \lvert\Delta X\rvert$	Singlet	CB
	T	$\Delta sp \;\simeq\; - \lvert\Delta I\rvert$	Triplet	TMM
4/4	C^-	$\Delta sp \;\simeq\; \lvert\Delta X\rvert$	Singlet	
	T	$\Delta sp \;\simeq\; - \lvert\Delta I\rvert$	Triplet	TMM

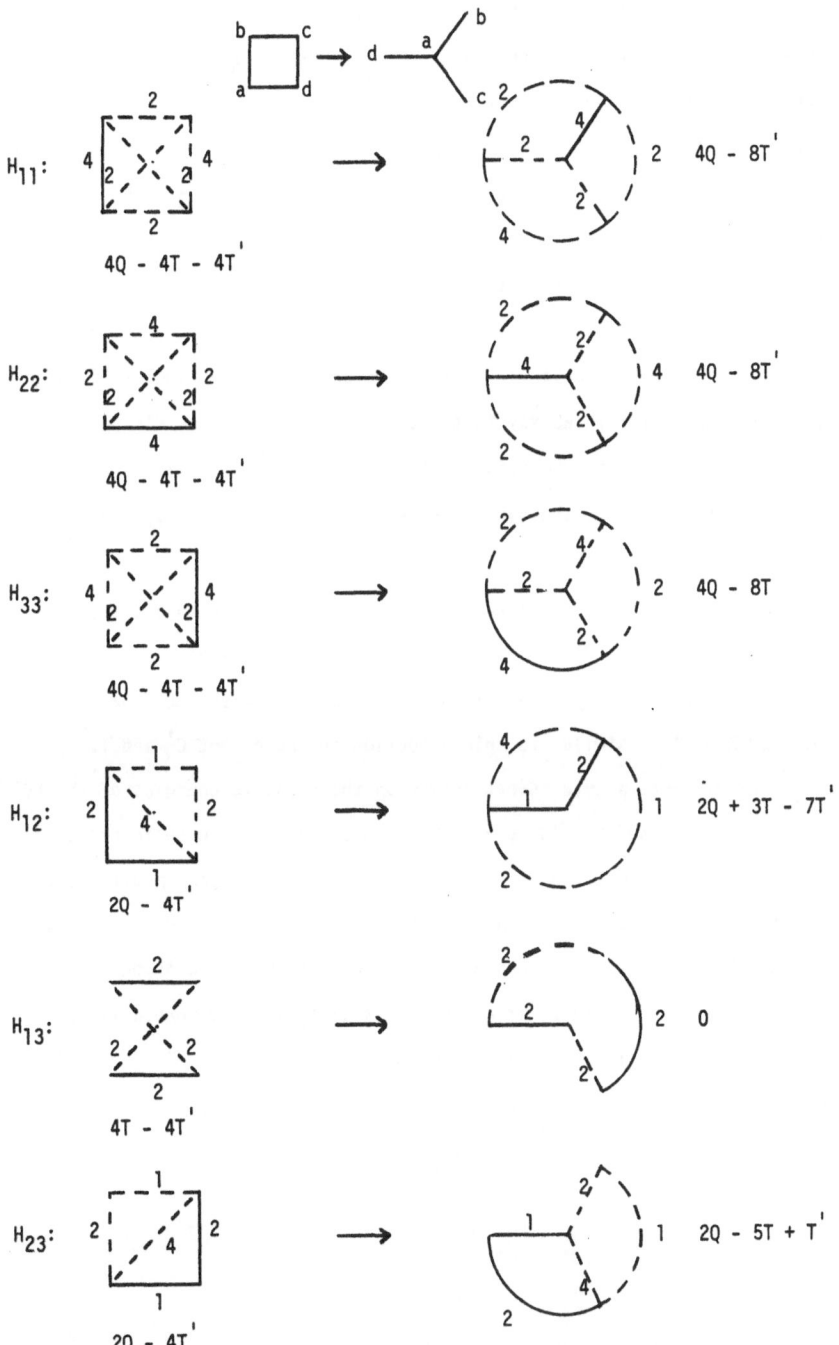

Figure 4.   The change of the FC matrix elements of the triplet HAO 4/4 system upon T-distortion.   Note how T distortion lowers the energies of two of the three FC's of the 4/4 system.

Let us now ask a different question: What will be the effect of heteroatomic replacement on spin selection? Once again, let us consider the effect of heteroatomic replacement in 4/3 and 4/4 systems.

Suppose that the 4/3 system is made up of three carbons and we replace one of them by, say, X. What will be the effect of such a modification on Δsp? It is immediately obvious that such a substitution will split the degeneracy of the FC's in the singlet as well as the triplet species. At the limit of a very electronegative or electropositive X, each of the singlet and the triplet species will be represented approximately by a single FC with the energy of the singlet FC lower than that of the triplet FC. We say that heteroatomic replacement disrupts cyclic FC interaction. As a result, the triplet state, which is the beneficiary of such an interaction, crosses above the singlet state if X is even modestly more electronegative or electropositive than C.

The situation is entirely different in the 4/4 system. In this case, the electron allocation in the FC's guarantees that the degeneracy of them will not be lifted appreciably and that spin selection in C^+ and T in which one C has been substituted by X will be similar to spin selection in the parent C^+ and T.

The effect of heteratomic substitution on the relative energies of the FC's is pictorially illustrated in Figure 5. We predict that heteroatomic replacement will tend to convert 4/3 C^+ from triplet to singlet while the same modification will not affect the spin multiplicity of 4/4 C^+ and T. In making this declaration, we recognize the obvious: If we abandon HL theory for a complete VB theory, we can find grounds for predicting that, if X becomes <u>exceedingly</u> electronegative or electropositive, the singlet will dip below the triplet in, e.g., T since CW's of the type shown below will tend to dominate the total wavefunction.

$$\begin{array}{c} \overset{\ominus}{:X} \\ \quad\quad C\!-\!C \\ \overset{\oplus}{C} \end{array}$$

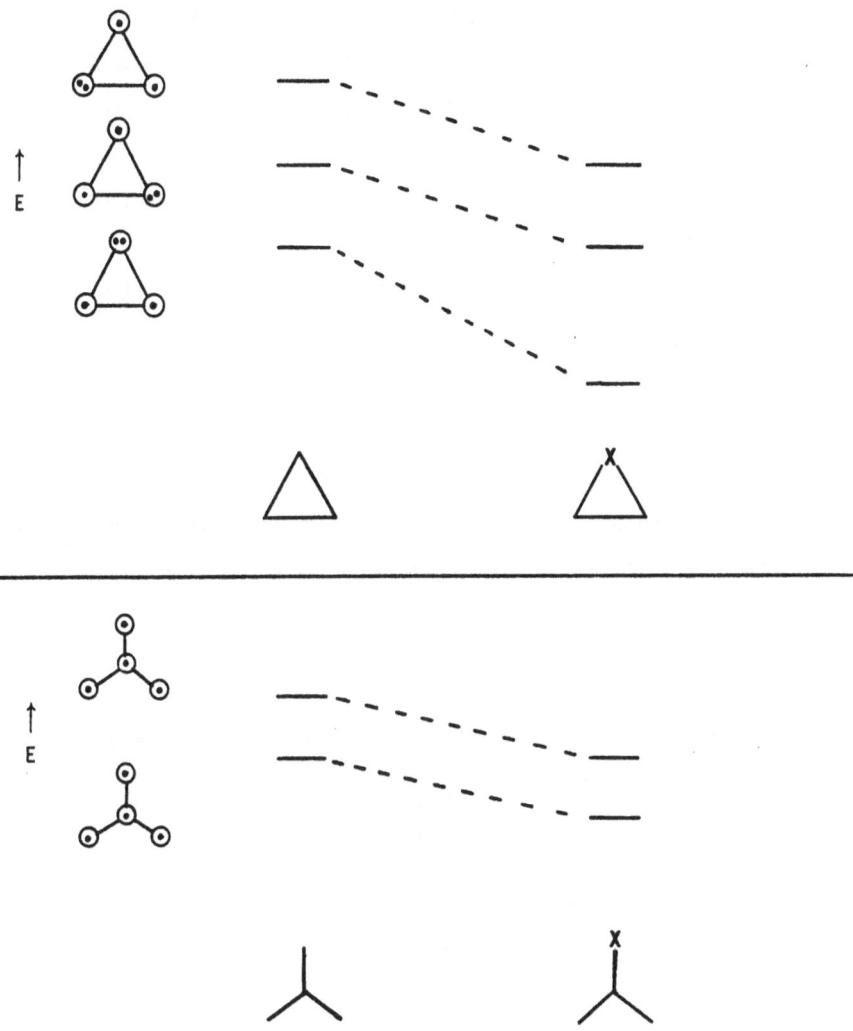

Figure 5. Effect of replacement of C by a more electronegative X on the energies of the FC's of a carbon 4/3 system and pi trimethylenemethane.

However, the important message of HL/FC theory which cannot be obscured by this realization is that <u>spin selection in topological systems will have a profoundly different response to heteroatomic substitution than spin selection in Kekule systems</u>.

The analyses presented above can be generalized to all topological and Kekule systems with our final conclusions being:

a. Topological systems with an <u>HAO geometry and even electron pairs, or, MAO geometry and odd electron pairs</u> will tend to have a triplet ground state due to cyclic Hückel triplet FC interaction. This tendency will decrease as the number of bond pairs increases.

b. Kekule systems will have a singlet ground state regardless of geometry and number of electron pairs. On the other hand, molecules which are representable by a set of electron pair bonds and "lone" pairs plus a single trimethylenemethane-type configuration, have a triplet ground state. This is illustrated below.

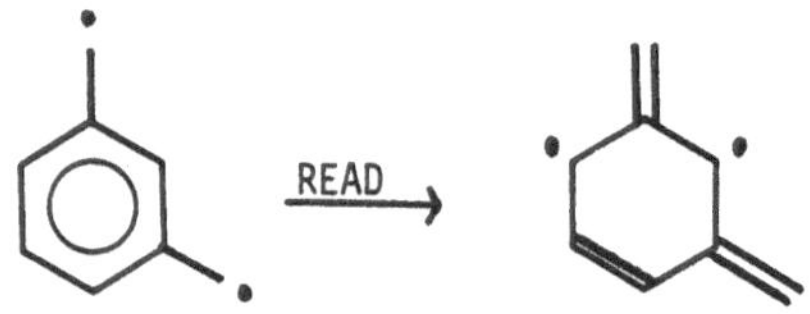

c. Substituent effects can change the spin multiplicity of topological systems from triplet to singlet. By contrast, substituent effects cannot easily convert a triplet T system to a singlet one.

What is the <u>calculational</u> evidence regarding spin selection? It is plentiful and uniformly supportive of AHL/FC theory.

a. The prototypical molecules of organic chemistry cyclopropenyl anion (CP), cyclobutadiene (CB), and trimethylenemethane (TMM), have been exhaustively studied in the last 15 years. Recent reviews summarize the appropriate data.[1] CP and TMM are computed to have triplet and CB singlet ground states.

b. The S-T energy gap decreases on going from cyclopropenyl anion to cyclo-pentadienyl cation, as predicted by HL/FC theory.

c. In a thorough PPP-CI study of "diradicals", Döhnert and Koutecky computed the S-T gaps of various species.[2] Their data reveals a remarkable trend: All species have singlet ground states with the exception of HAO/even and MAO/odd topological systems and systems resulting from a T-type distortion of a Kekule system with N even!

Are there any experimental data which suggest that AHL/FC theory is at all valid? It is normally expected that a non-Kekule molecule with nondegenerate nonbonding MO's which serve to house two electrons has a singlet ground state. By contrast, AHL/FC theory predicts that atomic replacement cannot easily change the triplet multiplicity of TMM-like molecules. In "tailor-made" experiments, two groups[3] report that the TMM-like species shown below have triplet ground states despite the fact that replacement of carbon by a more electronegative atom lifts the nonbonding MO degeneracy!

HN CH$_2$

O CH$_2$

O

↓READ   ↓READ   ↓READ

HN CH$_2$

O CH$_2$

O

CH$_2$

We end this section with a comment on how different theories produce different outlooks and create different predispositions. For example, let us consider the noncontroversial case of cyclopropenyl anion and the controversial case of cyclo-butadiene, both 4k-electron "antiaromatic" annulenes. The former molecule is computed to have a triplet ground state in D_{3h} and the latter one is computed to have a singlet ground state in D_{4h} geometry. With respect to HMO theory, the CP system can be called "normal" and the CB system "abnormal". More specifically, the former is said to comply to Hund's rule but the latter is claimed to violate it. As a result, spin selection in CB becomes a topic of controversy! By contrast, FC theory paints an entirely different picture in which the interesting thing is not that square CB turns out to be singlet but that CP turns out to be triplet! For all conventional singlets start with an advantage over the corresponding convential triplets and only under certain conditions defined by equation (67), one of them being "aromatic" FC interaction, can a triplet ground state material-ize! Hence, the intriguing system is not D_{4h} CB but, rather, D_{3h} CP insofar as spin selection is concerned. To put it crudely, a proper understanding of VB theory takes the mystery and controversy away from CB and TMM, two molecules which owe their (theoretical) notoriety to the formal and conceptual deficiencies of SD-MO theory.

We now give an example of how high-level <u>ab initio</u> computations can be used to define the limits of a qualitative theory by reference to the problem of chemical selection in paratopological systems. To this extent, we recall the information tabulated below, where by "covalent" CW's we mean the CW's of maximum open shell character, and we note that the <u>complete</u> <u>set</u> of "covalent" CW's of a paratopological system interact in a cyclic fashion by virtue of monoelectronic charge transfer,[4] much like the FC's of a topological system.

SYSTEM		FRONTIER CONFIGURATIONS
Paratopological	:	The lowest energy subset of the "covalent" CW's.
Topological	:	The complete set of "covalent" CW's.
Kekule	:	The lowest energy subset of the "covalent" CW's.

We distinguish two possibilities:

a. Paratopological systems are adequately described, <u>in a qualitative sense</u>, by the FC basis set. As a result, because of small FC interaction, parity stereo-selection is predicted to be small and spin selection such that, in even-electron systems, the singlet always lies below the triplet, regardless of the type of system (HAO or MAO) and the number of electrons ($4k$ or $4k+2$). If the FC approximation is valid, paratopological are different from topological systems insofar as chemical selection is concerned (FC Model).

b. Paratopological systems are adequately described, <u>in a qualitative sense</u>, <u>not by the FC basis set defined before</u> but, rather, by a basis set which includes all "covalent" CW's. In such an event, <u>parity stereoselection and spin selection</u> <u>are predicted to be dependent on the type of system and number of electrons in</u> <u>exactly the same way as topological system do</u>[5] under the assumption that the interaction of the FC's with the higher energy "covalent" CW's is operationally significant (Covalent Model).

In short, the critical question is the following: Is there an operationally significant difference between paratopological and topological systems? The reason why it is import to answer this question can be projected by a specific example.

One way of rationalizing the fact that square CB, a Kekule HAO system, is a ground singlet is to note that the two degenerate pi MO's can be written in a way so that they have no common AO's, in which case interelectronic repulsion no longer discriminates strongly against the singlet species at the SD-MO level so that CI "pushes" the singlet below the triplet. This is not possible in topological and paratopological HAO systems which, always following the MO theoretical argumentation, are predicted to have triplet ground states when N = 4k. For example, both D_{3h} $C_3H_3^-$ (a HAO topological system) and D_{6h} $C_6H_6^{+2}$ (a HAO paratopological system) are "four pi electron" systems expected to have triplet ground states. This MO argument leads to the same conclusions as the Covalent Model. By contrast, the FC Model predicts that D_{3h} $C_3H_3^-$ is <u>probably</u> a triplet while D_{6h} $C_6H_6^{+2}$ is definitely a singlet, since ΔI is very small in paratopological systems. What is the experimental evidence?

The spin multiplicity of the cyclohexadienyl dication has not been determined. However, the perchloro derivative was found to have a triplet ground state with appreciable odd electron delocalization to the chlorines.[6] This result taken by itself does not necessarily mean that AHL/FC theory fails in the case of paratopological systems simply because the triplet $C_6Cl_6^{+2}$ species may not be a perturbed triplet $C_6H_6^{+2}$ system but rather a molecule with <u>two odd-electron topological systems coupled to an overall triplet</u>. The first topological system is "five electrons in six C2p orbitals" and the second topological system is "eleven electrons in six Cl3p orbitals", or, "elevent electrons in six sigma Cl AO's",

as illustrated below.

For simplicity, each of the relevant six sigma-type Cl AO's is represented by a

sphere and all MO symmetry assignments are based on this convention (<u>vide infra</u>).

The MOVB bond diagram shown in Figure 6 make clear what is the common perception

and what is conceivably the species actually "seen" by Wasserman. $C_6H_6^{+2}$ is trip-

let because the high electron affinity of the dipositive core induces an electron

transfer from the highest lying doubly occupied pi or sigma b_2 MO of the ligand to

the low lying degenerate e_1 core MO's which simultaneously relieves coulomb repul-

sion by separating the positive holes by a larger distance. This electron transfer

is tantamount to a deexcitation process and it is more favorable than delocalization

in the "conventional" species because the former involves b_2(ligand) $\longrightarrow e_1$(core)

while the latter involves b_2(ligand) $\longrightarrow b_2$(core) transfer with respect to a doubly

occupied MO as the initial state and the first is a downhill while the second is an

uphill process. Of course, the more tightly bound the ligand lone pairs are, the

less the likelihood of either delocalization or electron transfer, in which case

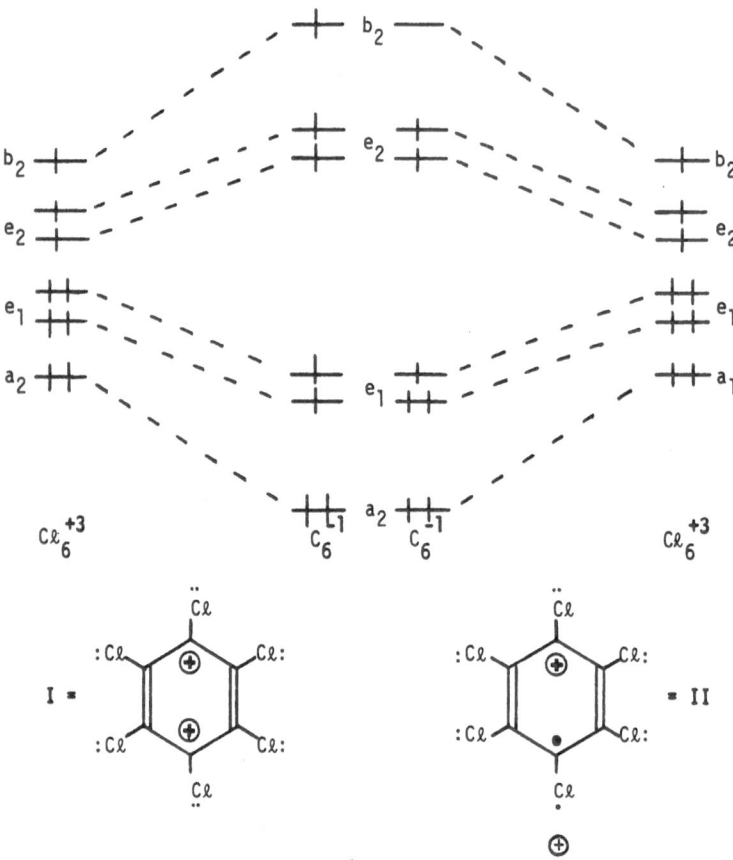

Figure 6. Two different triplet states of $C_6Cl_6^{+2}$. I is the intuitively expected paratopological diradical while II is a topological diradical. I and II are both pi radicals. MO's are classified according to the D_6 point group.

the "conventional" triplet species will have lower energy than either of the two topological triplets.

"State of the art" computations of $C_6H_6^{+2}$ would be immensely useful in testing FC theory and pinpointing whether a paratopological is operationally different from a topological system (singlet $C_6H_6^{+2}$ ground state) or merely a more "subdued" version thereof (triplet $C_6H_6^{+2}$ gound state). Furthermore, explicit calculations are essential for determining whether triplet $C_6Cl_6^{+2}$ is a perturbed triplet $C_6H_6^{+2}$ or the triplet species suggested above.

We end by drawing the attention of the reader to the fact that in 1963 Fischer and Murrell published an important paper which shows that, to our knowledge, these authors were the first to recognize the fact that spin selection can be expressed by signs in VB theory.[7] In particular, they performed HL/FC computations of the topological systems 2/3, 4/3, 4/5, 6/5, and 8/7 and they recognized that the spectrum of the energy eigenvalues of the singlet and triplet species was deter- mined by the signs of the FC overlap integrals. In addition, they realized that topological systems differ fundamentally from Kekule systems. Thus, it can be said that as early as 1963 Fischer and Murrell had reproduced through computations Hückel and Möbius function (in their case, CW) interaction, i.e., they had at hand the so called Woodward-Hoffmann rules in cryptographic code! Our own discovery of the concept of CW "aromaticity" and "antiaromaticity" was, of course, the result of previous awareness of the stereochemical implications of signs of interaction matrix elements because of our past preoccupation with qualitative MO theory and our familiarity with the Möbius-Hückel distinction, which Heilbronner and Zimmerman first brought to the attention of chemists.[8] In following chapters, we shall apply this concept to various problems which, at first sight, seem to have nothing to do with "aromaticity"! It is hoped that these applications will bring to the attention of a wider public not only the important "sign contribution" mentioned above but also some recent, and diverse, valuable contributions to VB theory, in general, ranging from <u>ab initio</u> VB formalism and <u>ab initio</u> VB computations,[9-17] qualitative VB model and their application,[18-21] semiempirical-intuitive HL calculations,[22,23] and DIM calculations referred to in the previous chapter.

References

1. (a) Borden, W.T.; Davidson, E.R. Ann. Rev. Phys. Chem. 1979, 30, 125.

   (b) Borden, W.T.; Davidson, E.R. Acc. Chem. Res. 1981, 14, 69.

2. Döhnert, D.; Koutecky, J. J. Am. Chem. Soc. 1980, 102, 1789.

3. (a) Platz, M.S.; Burns, J.R. J. Am. Chem. Soc. 1979, 101, 4425.

   (b) Rule, M.; Matlin, A.R.; Hilinski, E.F.; Dougherty, D.A.; Berson, J.A. J. Am. Chem. Soc. 1979, 101, 5098.

   (c) Seeger, D.E.; Hilinski, R.F.; Berson, J.A. J. Am. Chem. Soc. 1981, 103, 720.

4. See Figure 13b on p. 142 of Ref. 1.

5. This can be easily proven by using the theoretical procedures described in this work.

6. Wasserman, E.; Hutton, R.S. Acc. Chem. Res. 1977, 10, 27.

7. Fischer, H.; Murrell, J.N. Theor. Chim. Acta 1963, 1, 464.

8. (a) Heilbronner, E. Tetrahedron Lett. 1964, 1923.

   (b) Zimmerman, H.E. "Quantum Mechanics for Organic Chemists"; Academic Press: New York, 1975.

9. (a) Goddard, III, W.A; Dunning, Jr., T.H.; Hunt, W.J.; Hay, P.J. Acc. Chem. Res. 1973, 6, 368.

   (b) Bobrowicz, F.W.; Goddard, III, W.A. in "Modern Theoretical Chemistry", Schaefer, III, H.F., Ed.; Plenum Press: New York, 1976.

10. Simonetta, M.; Raimondi, M.; Tantardini, G.F. Int. J. Quant. Chem. 1981, 15, 225.

11. Balint-Kurti, G.G.; Karplus, M. in "Orbital Theories of Molecules and Solids", Ed., March, N.H.; Oxford University Press: London, 1974.

12. (a) Simonetta, M.; Gianinetta, E.; Vandoni, I. J. Chem. Phys. 1968, 48, 1579

12. (b) Raimondi, M.; Simonetta, M.; Tantardini, G.F. J. Chem. Phys. 1972, 56, 5091.

13. (a) Gallup, G.A. Int. J. Quant. Chem. 1972, 6, 899

    (b) Gallup, G.A. Advan. Quant. Chem. 1973, 7, 113.

    (c) Norbeck, J.M.: Gallup, G.A. J. Am. Chem. Soc. 1974, 96, 3386.

    (d) Gallup, G.A.; Norbeck, J.M. J. Am. Chem. Sec. 1975, 97, 970.

14. (a) Hiberty, P.C.; Leforestier, C. J. Am. Chem. Soc. 1978, 100, 2012.

    (b) Hiberty, P.C.; Ohanessian, G. J. Am. Chem. Soc. 1982, 104, 66.

15. Pauncz, R. "Spin Eigenfunctions", Plenum Press: New York, 1979.

16. Malrieu, J.-P.; Maynau, D. J. Am. Chem. Soc. 1982, 104, 3021; Maynau, D.; Malrieu, J.-P. ibid. 1982, 104, 3029.

17. McWeeny, R. Proc. Roy. Soc. A 1953, 223, 306.

18. Dauben, W.G.; Salem, L.; Turro, N.J. Acc. Chem. Res. 1975, 8, 41.

19. Yoshioka, Y.; Yamaguchi, K.; Fueno, T. Theoret. Chim. Acta 1977, 45, 1, and previous papers.

20. (a) Shaik, S.S. J. Am. Chem. Soc. 1981, 103, 3692.

    (b) Pross, A.; Shaik, S.S. J. Am. Chem. Soc. 1981, 103, 3702.

    (c) Shaik, S.S.; Pross, A. J. Am. Chem. Soc. 1982, 104, 2708.

21. Harcourt, R.D. "Qualitative Valence-Bond Descriptions of Electron-Rich Molecules" in Lecture Notes in Chemistry, Vol. 30; Springer-Verlag: New York and Berlin, 1982.

22. Herndon, W.C.; Ellzey, Jr., M.L. J. Am. Chem. Soc. 1974, 96, 6631.

23. Randic, M. Mol. Phys. 1977, 34, 849.

Chapter 18. Why a Net Bond exists when it appears to be Nonexistent: The
Electronic Structures of F_2 and Inert Gas Fluorides.

In a previous note[1], I gave an example of why an effective bond can exist
when a real bond is absent within Single Determinant SCF-MO (SD-SCF-MO) theory.
I now consider a much more complex problem, which is soluble only at the level
of SCF-MO-CI theory, in order to further illustrate the conceptual and formal
advantages of VB-type theories.

Consider F_2 at the level of Heitler-London-type theory according to which
this species is represented by a single "resonance structure": $|\underline{F} {\longleftrightarrow} \underline{F}|$. Since
there are three four-electron and two three-electron antibonds and only one two-
electron bond[2a], we conclude that F_2 is unstable relative to two isolated F
atoms. As it is well known, this conclusion is wrong.

Next, consider F_2 at the level of EHMO theory. Because of the integral
approximations that cause overdelocalization[2a], F_2 is now bound[3]. This is due
to inappropriate accentuation of the "aromatic" delocalization of six electrons
in four orbitals ($2p_x, 2p_x', 2s, 2s'$). The conclusion is now correct for the wrong
reasons. This is made evident by transition to the level of SCF-MO theory. Now,
electron-electron interaction is explicitly included in the computation,
"aromatic" overdelocalization is averted, and F_2 ceases to be bound[4]. Again,
this conclusion is wrong.

To obtain the correct behavior of the F_2 dissociation curve, one must rise
to the level of SCF-MO-CI, or, VB theory. I now show that the reason why F_2
is bound at this level of theory is only explicable on the basis of a new
fundamental concept which we developed along the way towards MOVB theory[2] and
that simple monodeterminantal MO ideas, such as bond-counting principles, overlap population considerations, etc., are of no use when it comes to dealing with
weakly "covalently"-bound molecules.

Some years ago, we discovered an organizational principle which remained unrecognized mostly because chemists have been unwilling to explore the potentiality of VB theory as a qualitative theoretical tool.[2a] This principle is aromatic, non-aromatic, and antiaromatic Frontier Configuration Wavefunction (FC) interaction. That is to say, depending on the molecule in question, Frontier CW's may interact in a cyclic, or noncyclic manner with the lowest energy ground eigenstate being produced by a Hückel CW ring, i.e., a cycle of CW's connected by interaction matrix elements so that the product of their signs is positive.[5] Thus, pi cyclopropenyl cation was found to be a Configuration-aromatic (C-aromatic) species because the three Frontier CW's form a Hückel ring while pi cyclopropenyl anion is C-antiaromatic because the three Frontier CW's form a Möbius ring. By contrast, pi cyclobutadiene is a C-nonaromatic molecule, and the same is true of pi benzene, since there are only two Frontier CW's, the Kekule structures.

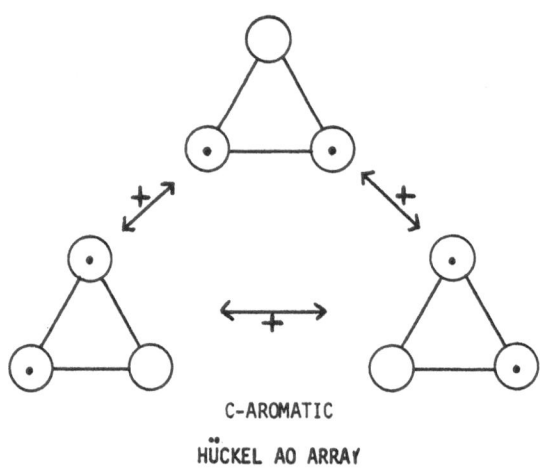

C-AROMATIC

HÜCKEL AO ARRAY

HÜCKEL CW ARRAY

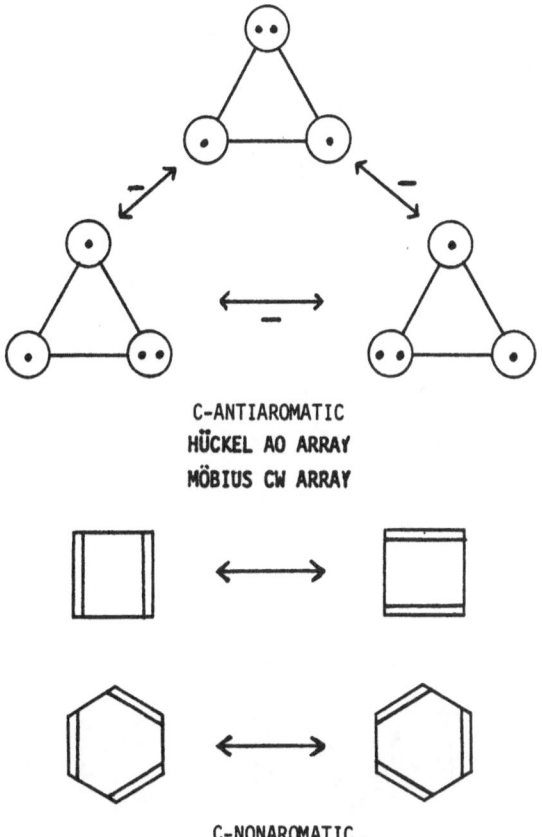

C-ANTIAROMATIC
HÜCKEL AO ARRAY
MÖBIUS CW ARRAY

C-NONAROMATIC

A broader definition of C-aromaticity can be given with respect to the complete set of CW's, rather than the restricted set of Frontier CW's. In a similar fashion, we seek to identify CW rings of the Hückel or Möbius type and predict their chemical consequences. An illustration of the utility of this concept is provided by the molecule F_2. The simplest bond diagrammatic representation of this species (Figure 1a) reveals that there are essentially one six-electron bond and two four-electron antibonds, something which <u>seems</u> to indicate that F_2 will be unbound. The fact

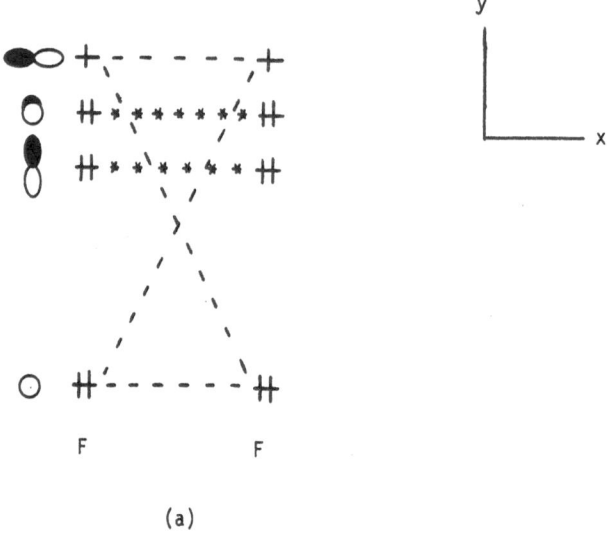

(a)

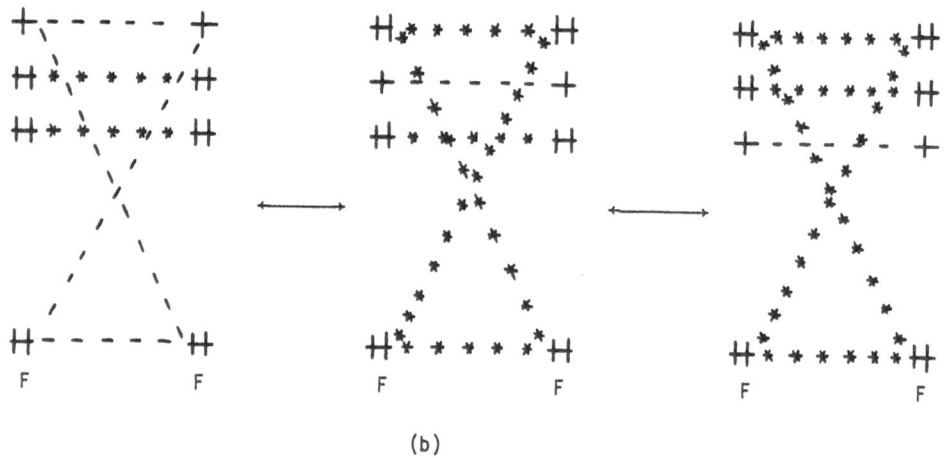

(b)

Figure 1. a. Approximate bond diagrammatic representation of F_2.
          b. The **total** bond diagrammatic representation of F_2.

that F_2 is actually bound and has a dissociation energy of ~ 38.0 kcal/mol implies that there is some bonding effect which is not obvious upon casual perusal of the Lewis electronic dot formula of F_2 and which is primarily responsible for the binding of the two F atoms. We suggest that this bonding effect is C-aromaticity.

Copsey, Murrell, and Stamper performed the following important calculations:[6]

a. Using the complete VB configuration Wavefunction (CW) basis set shown in Figure 2, they computed the optimum dissociation energy of F_2, obtaining a value ~ 16 kcal/mol. At the equilibrium bond distance, the optimal wavefunction is a linear combination of all 16 CW's shown in Figure 2 and it can be represented in bond diagrammatic form as shown in Figure 1b. As pointed out by the authors, the calculations revealed a very noteworthy trend: The coefficients of the "ionic" CW's corresponding to the "covalent" CW's Φ_1, Φ_4, Φ_7, and Φ_{10} as well as the coefficients of the "covalent" CW's themselves were found to be as follows:

Φ_1 : 0.85402 $\Phi_2 + \Phi_3$: 0.32523

Φ_4 : -0.02484 $\Phi_5 + \Phi_6$: -0.03734

Φ_7 : -0.02484 $\Phi_8 + \Phi_9$: -0.03734

Φ_{10} : -0.00400 $\Phi_{11} + \Phi_{12}$: -0.03500

In other words, with only one exception, the "ionic" CW's are more important in stabilizing the system than the corresponding "covalent" ones!

b. By performing computations using truncated basis sets, the authors uncovered the following remarkable trends:

1) F_2 is not bound if all "ionic" CW's are excluded from the basis set.

2) When only the CW's $\Phi_1 - \Phi_3$ and $\Phi_{10} - \Phi_{16}$ were admitted in the basis set, F_2 was found to be only 0.0015 a.u. less bound than predicted by the full calculations. These CW's define a VB wavefunction which is represented by only one bond diagram,

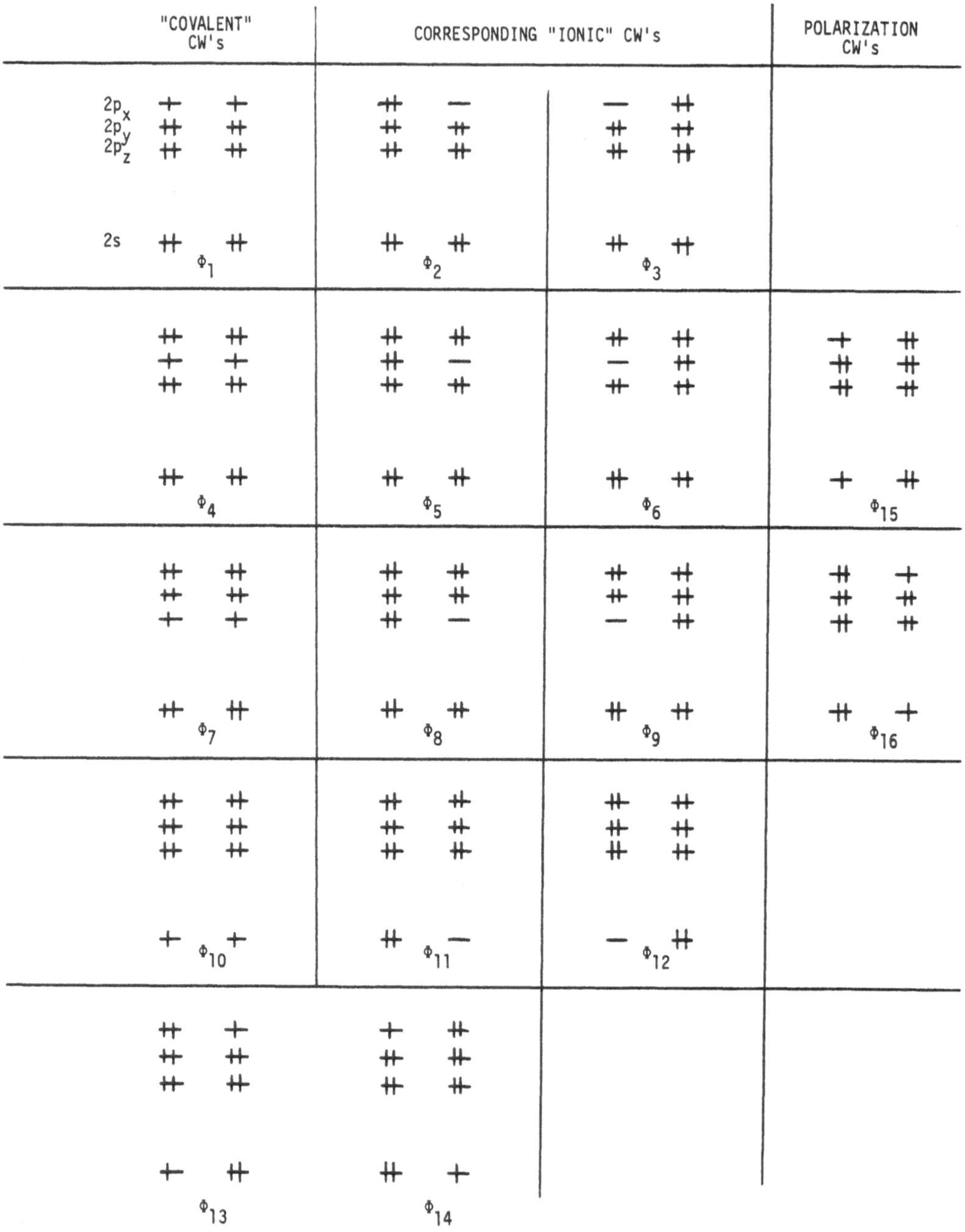

Figure 2. The complete VB CW set for the description of F_2.

i.e., the one shown in Figure 1a. The way in which the various CW's interact is shown diagrammatically in Figure 3.

3) When only the CW's Φ_1 - Φ_9 were admitted to the basis set, F_2 was found again to be only 0.0056 a.u. less bound than predicted by the full calculation. These CW's define a wavefunction which cannot be represented by a single bond diagram. The way in which the various CW's interact is shown diagrammatically in Figure 4.

Finally, it should be restated that <u>monodeterminantal SCF-MO computations predict that F_2 is not bound</u>.[4] These computations are equivalent to performing <u>constrained</u> VB calculations using the Φ_1 - Φ_3 plus Φ_{10} - Φ_{16} CW basis set. Thus, the monodeterminantal MO wavefunction of F_2 is representable by a bond diagram like the one shown in Figure 1a with the stipulation that the weights of the various CW's connoted by the bond diagram are incorrectly assigned.

The Copsey et al. computations are then consistent with the idea of CW "aromaticity". The truncated CW basis set which includes only Φ_1 - Φ_{13} and Φ_{10} - Φ_{16} is capable of reproducing binding because these CW's define a cyclic Hückel array over monoelectronic Charge Transfer (CT) matrix elements, d_{ij}, which produces a low energy ground eigenstate. The importance of cyclic CW interaction is further confirmed by the fact that the truncated basis which includes only Φ_1 - Φ_9 is also capable of reproducing bonding because these CW's again define a cyclic Hückel array with the only difference now being that the array is defined over monoelectronic CT and <u>large</u> bielectronic polarization matrix elements, P_{ij}.

We conclude that the presence of more four-electron antibonds than n-electron bonds in F_2 is prevented from causing repulsion of two fluorine atoms by <u>configuration aromaticity</u> and that <u>configuration aromaticity</u> is responsible for the strong binding of all atoms or fragments which, upon casual inspection, do not seem to be willing to bind to each other.

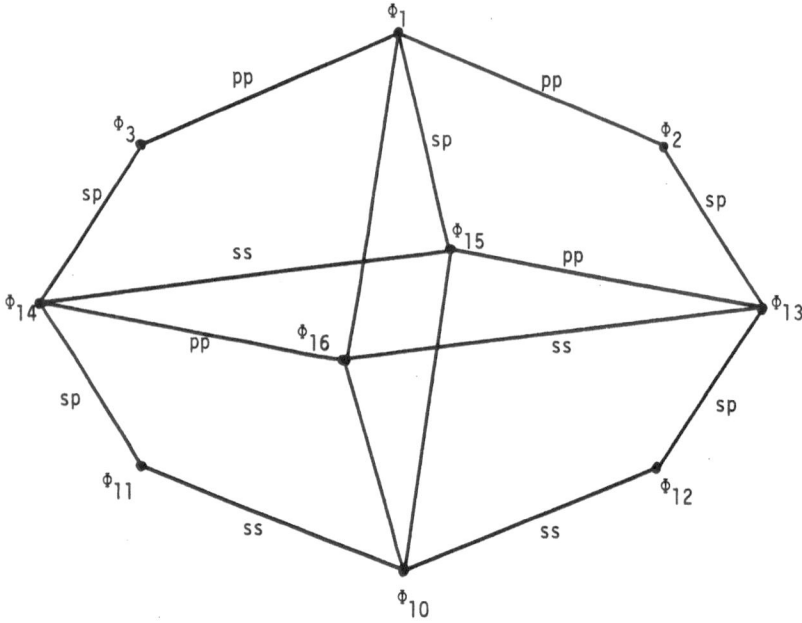

Figure 3. Monoelectronic charge transfer (CT) interaction of Φ_1 - Φ_3 and Φ_{10} - Φ_{16}. Each black circle represents a CW and each line connecting two CW's the corresponding CT interaction matrix element which is proportional to an AO resonance integral of the $<2s|\hat{H}|2s'>$ (ss), $<2s|\hat{H}|2p_x'>$ (sp), or, $<2p_x|\hat{H}|2p_x'>$ (pp) type. Each cycle contains an even number of AO resonance integrals of one type which have the same sign, hence, <u>Hückel</u> CW interaction.

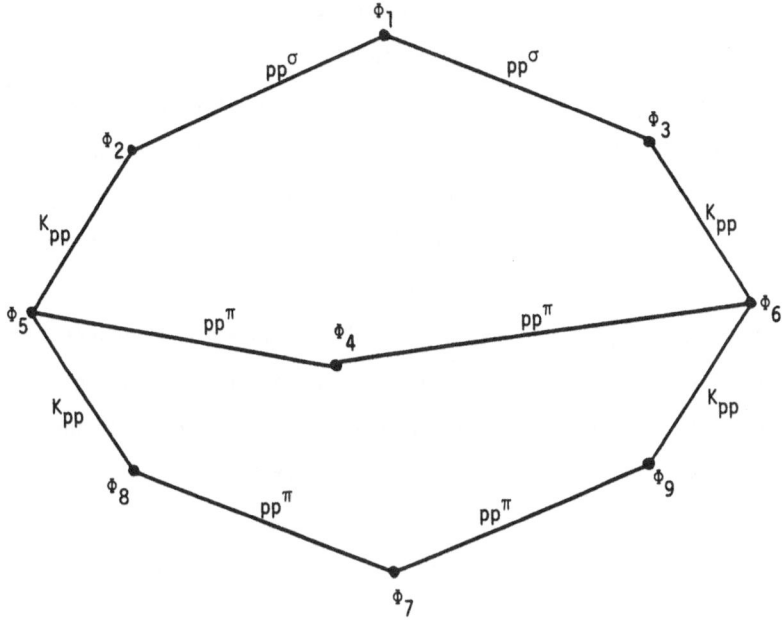

Figure 4. Monoelectronic CT and bielectronic polarization interaction of Φ_1 - Φ_9 which defines a Hückel CW array. There are now sigma and pi AO resonance integrals and, in addition, bielectronic exchange integrals:

$<2p_x|\hat{H}|2p_x'> = pp^\sigma$, $<2p_y|\hat{H}|2p_y'> = <2p_z|\hat{H}|2p_z'> = pp^\pi$, and

$<2p_x(1)2p_y(1)|2p_x(2)2p_y(2)> = <2p_x(1)2p_z(1)|2p_x(2)2p_z(2)> =$

$<2p_y(1)2p_z(1)|2p_y(2)2p_z(2)> = K_{pp}$. Again, each cycle contains an even number of AO resonance and bielectronic exchange integrals of one type which have the same sign, hence Hückel CW interaction.

F_2 is a relatively unstable yet conventional diatomic. KrF_2 is a still less stable unconventional triatomic. Monodeterminantal MO theory cannot properly describe either, and Schaefer[7] points out that more extensive CI is needed for reproducing correctly the binding of KrF_2. Why? Inspection of the approximate bond diagram of KrF_2 reveals that the difference between F_2 and KrF_2 (Figure 5) is simple: The former enjoys monoelectronic and bielectronic C-aromaticity while the latter can only benefit from bielectronic C-aromaticity, where the term "mono-electronic" signifies that a set of CW's defines a Hückel cycle over monoelectronic interaction matrix elements and the term "bielectronic" implies that the cyclic Hückel CW interaction is defined over monoelectronic <u>and</u> large bielectronic inter-action matrix elements. The CW cycles referred to above are constructed in the way indicated below:

    a. F_2 .   Monoelectronic C-aromaticity: Six electrons permuted in

                       $2p_x$, $2p_x'$, $2s$, and $2s'$.[8]

                 Bielectronic C-aromaticity: Ten electrons permuted among

                 $2p_x$, $2p_x'$, $2p_y$, $2p_y'$, $2p_z$, and $2p_z'$.

    b. KrF_2 .   Bielectronic C-aromaticity: Six electrons permuted among

                 $2p_x$, $2s$, σ, and σ^*.

The term "large bielectronic matrix element" denotes a <u>one-center</u> two-electron integral.[9-12]

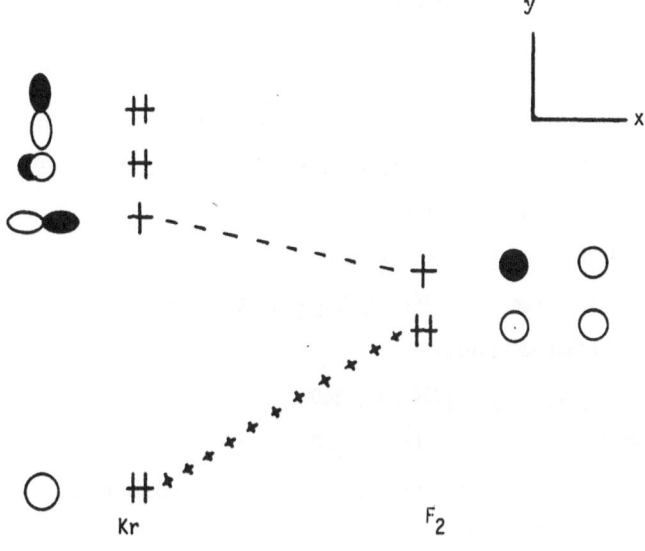

Figure 5. Approximate bond diagrammatic representation of <u>linear</u> KrF$_2$, with the fluorine lone pairs neglected. As in Figure la, a bond is matched by an antibond of different spatial symmetry.

References

1. See Chapter 13.

2. (a) Epiotis, N.D.; Larson, J.R.; Eaton, H. "Unified Valence Bond Theory of Electronic Structure" in Lecture Notes in Chemistry, Vol. 29; Springer-Verlag: New York and Berlin, 1982.

   (b) Epiotis, N.D. Pure Appl. Chem. 1983, 000.

   (c) Epiotis, N.D.; Larson, J.R. Israel J. Chem. 1983, 000.

3. Epiotis, N.D. unpublished results.

4. (a) Wahl, A.C. J. Chem. Phys. 1964, 41, 2600.

   (b) Das, D.; Wahl, A.C. J. Chem. Phys. 1966, 44, 87.

5. The product sign is defined with respect to the signs preceding each negative interaction matrix element.

6. Copsey, D.N.; Murrell, J.N.; Stamper, J.G. Mol. Phys. 1971, 21, 193.

7. Schaefer, III, H.F. "The Electronic Structure of Atoms and Molecules"; Addison-Wesley: Reading, Mass., 1972.

8. According to the Independent Bond Model (IBM) F_2 can be viewed as a composite of three subsystems; A, B, and C with the total wavefunction Ψ being a product of the subsystem wavefunctions Ω_A, Ω_B, and Ω_C.

   $$\Omega_A \; : \; 2p_x, \; 2p_x', \; 2s, \; 2s' \; / \; 6$$

   $$\Omega_B \; : \; 2p_y, \; 2p_y' \; / \; 4$$

   $$\Omega_C \; : \; 2p_z, \; 2p_y' \; / \; 4$$

   $$\Psi = \Omega_A \cdot \Omega_B \cdot \Omega_C$$

   That subsystem A is C-aromatic in a monoelectronic CT sense can be seen without writing down the individual CW's but, rather, by observing that the four orbitals $2p_x$, $2p_x'$, $2s$, and $2s'$ interact in-phase in a cyclic manner and they contain six electrons.

9. The distinction between Hückel and Möbius <u>atomic orbital</u> overlap was first brought to the attention of chemists by Heilbronner and Zimmerman. The chemical implications of this distinction were first discussed by Zimmerman and Woodward and Hoffmann:

    (a) Heilbronner, E. <u>Tetrahedron</u> <u>Lett.</u> <u>1964</u>, 1923.

    (b) Zimmerman, H.E. <u>J</u>. <u>Am</u>. <u>Chem</u>. <u>Soc</u>. <u>1966</u>, 88, 1564, 1566.

    (c) Woodward, R.B.; Hoffmann, R. "The Conservation of Orbital Symmetry"; Academic Press: New York, 1971.

10. The way in which the signs of VB matrix elements determine stereoselection in N-electron/N-orbital systems was discussed in: Mulder, J.J.C.; Oosterhoff, L.J. <u>Chem</u>. <u>Commun</u>. <u>1970</u>, 305, 307.

11. The way in which the signs of VB matrix elements determine spin selection is discussed in: Fischer, H.; Murrell, J.N. <u>Theor</u>. <u>Chim</u>. <u>Acta</u> <u>1963</u>, 1, 464.

12. For recent attempts toward the development of a qualitative VB theory, see, <u>inter</u> <u>alia</u>:

    (a) Hiberty, P.C.; Leforestier, C. <u>J</u>. <u>Am</u>. <u>Chem</u>. <u>Soc</u>. <u>1978</u>, 100, 2012.

    (b) Dauben, W.G.; Salem, L.; Turro, N.J. <u>Acc</u>. <u>Chem</u>. <u>Res</u>. <u>1975</u>, 8, 41.

    (c) Shaik, S.S. <u>J</u>. <u>Am</u>. <u>Chem</u>. <u>Soc</u>. <u>1981</u>, 103, 3692.

    (d) Harcourt, R.D. "Qualitative Valence-Bond Descriptions of Electron-Rich Molecules" in Lecture Notes in Chemistry, Vol. 30; Springer-Verlag: New York and Berlin, 1982.

    (e) Herndon, W.C.; Ellzey, Jr., M.L. <u>J</u>. <u>Am</u>. <u>Chem</u>. <u>Soc</u>. <u>1974</u>, 96, 6631.

    (f) Klein, D.J. <u>Pure</u> <u>Appl</u>. <u>Chem</u>. <u>1983</u>, 000.

Chapter 19. Chemical Anticooperativity and Sigma-Pi Hybridization.

In every scientific discipline, the accumulated knowledge regarding the structure of fundamental entities raises the question of their interaction. Thus, in zoology, we inquire as to how bone and tissue are connected (i.e., interact) so that their combined actions can produce motion. In molecular biology, we inquire as to how an enzyme and an effector molecule combine (i.e., interact) in order to produce a complex which is catalytically active, or, inactive. Finally, reaching further down to the foundation of physical reality, one can inquire as to how bonds interact within molecules in order to produce the lowest possible energetic state. Now, one of the difficulties which thwart studies of the latter type is the mere fact that, in order to investigate interaction, one must be able to define the interacting elements. This is not a problem in, e.g., zoology, where we can unequivocally define bone, tissue, etc., but, it does constitute a problem in quantum chemistry because clear definitions of "non-interacting bonds" and "after-interaction-bonds" are possible within the framework of one but not of another theoretical approach. That is to say, one must appropriately choose among different theoretical vehicles before undertaking a study of interaction at the electronic level since the construct of "bond" is only a model-dependent construct. The purpose of this paper is to exploit the formal and conceptual advantages of Valence Bond (VB) theory[1] in order to provide a blueprint for the study of interaction at the electronic level through formulation and illustrative application of concepts that may ultimately lead to a good understanding of electronic control mechanisms within atoms and molecules.

I. A Broader Look at Hybridization

The advantages of VB theory are formal as well as conceptual.[1] In a formal
sense, we start with a set of VB Configuration Wavefunctions (CW's), constructed
from a set of nonorthogonal Atomic Orbitals (AO's), which ultimately yield the
proper electronic states with no constraints placed on the way in which the CW's are
combined to form these various electronic states. In a conceptual sense, we can
define elementary bonds, "lone" electrons, and vacant orbitals (or holes) so that
their interaction can be conveniently studied.

Next, in order to bypass parity assignments which would render the theory
inapplicable to large systems in a "back-of-the-envelope" sense, we "tailor" VB
theory, thus generating the MOVB theory of chemical bonding, an economical form of
VB theory. The key construct of MOVB theory is the so-called <u>bond diagram</u> which
makes possible the representation of a molecule as a sum of atomic nuclei and interacting n-electron bonds, "lone" electrons, and holes. Using this construct and
the MOVB concepts necessary for interpretation, we were able to provide new explanations for a variety of interesting chemical trends and make predictions which cannot
be arrived at on the basis of conventional qualitative theory. Furthermore, it was
pointed out[1] that, when the AO's of the system at hand do not overlap in a cyclic
manner, one can execute VB theory by using MOVB theoretical concepts. The reason
is that, <u>in the absence of cyclic AO overlap,</u> one can always define two fragments such
that the MO's of at least one fragment are actually AO's. For example, if three
AO's x_1, x_2, and x_3 overlap in the way shown below, the two symmetry MO's of the
fragment A are spatially degenerate and they can be replaced by linear combinations
which are the two AO's themselves.

FRAGMENT DEFINITION

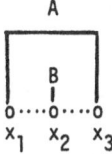

AO OVERLAP INTEGRALS

$$s_{12} \neq 0$$
$$s_{23} \neq 0$$
$$s_{13} = 0$$

FRAGMENT ORBITALS

$N(x_1 - x_3)$ or x_3 ——

$N(x_1 + x_3)$ or x_1 —— —— x_2

Fragment A   Fragment B

For such systems, not only can we define elementary bonds in the usual VB sense, but we can use MOVB-type bond diagrams in order to classify them as U-, H-, or D-bound. Illustrative examples are given below.

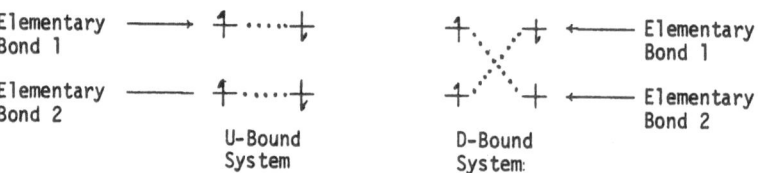

In summary, we can use VB theory in order to study the interactions of elementary n-electron-m-orbital bonds, "lone" pairs, and holes by falling back on MOVB concepts if the system of interest is made up of AO's which do not overlap in a cyclic manner. In this paper we shall assume that the latter condition is always met unless otherwise stated.

Let us now acquaint ourselves with the dimensions of the problem by considering how hybridization enters in MOVB theory and how it is represented in bond diagrammatic fashion. First of all, we recall that a set of orbitals which interact in a one-electron sense and the electrons allocated to them constitute a subsystem in which hybridization is mainly due to the one-electron interaction of all CW's of the subsystem. For example, the MOVB wavefunction of a four-electron-three-orbital subsystem can be represented in either of the two ways indicated in Figure 1 with the hybrid representation being the most chemically informative one. We can say that one-electron hybridization is very clearly projected by MOVB theory and, in fact, it is satisfactorily reproduced for qualitative purposes at all levels of MOVB and corresponding MO theory.

The situation becomes <u>apparently</u> different when we consider the bielectronic interaction of two subsystems which, by symmetry, cannot interact with each other in a one-electron sense. At the level of MOVB theory, we can neglect this bielectronic interaction when comparing, in a qualitative way, two related systems, e.g., two conformational isomers, two geometric isomers, etc., and this approximation is the basis of the Independent Bond Model (IBM). The associated basic notions can be best illustrated by reference to a specific example, namely, the carbon-carbon double bond of ethylene which can be considered as a total system comprised of two subsystems, one being the sigma and the other the pi bond which can only interact in a bielectronic sense. According to the IBM, the total wavefunction can be expressed as a product of two subsystem wavefunctions as long as the Hamiltonian operator is written as a sum of two operators, one operating on one subsystem and the second on the other, with neither containing interelectronic $(1/r_{ij})$ coupling terms. Thus, we can write:

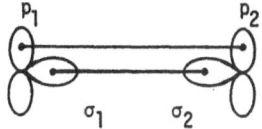

474

DETAILED BOND DIAGRAM

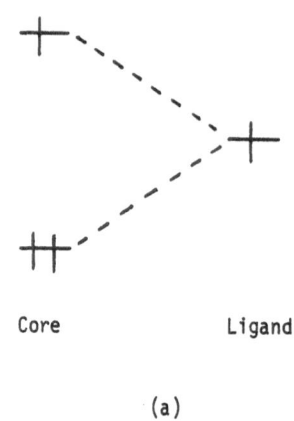

Core   Ligand

(a)

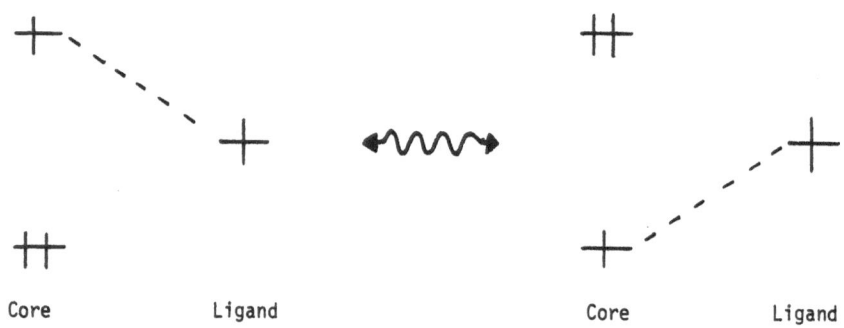

Core   Ligand   Core   Ligand

APPROXIMATE RESONANCE BOND DIAGRAMMATIC REPRESENTATION

(b)

Figure 1. MOVB representations of a four-electron-three-orbital system.

$$\Psi_\Sigma = \lambda_1(\sigma_1{}^1\sigma_2{}^1) + \lambda_2(\sigma_1{}^2 + \sigma_2{}^2) \tag{1}$$

$$\Psi_\Pi = \mu_1(p_1{}^1p_2{}^1) + \mu_2(p_1{}^2 + p_2{}^2) \tag{2}$$

where $\sigma_1{}^1\sigma_2{}^1$ is the "covalent" and $\sigma_1{}^2$ and $\sigma_2{}^2$ the "ionic" CW's necessary for the description of an isolated sigma bond, with $p_1{}^1p_2{}^1$ and $p_1{}^2$ and $p_2{}^2$ having analogous meaning.

$$\Psi_T = \Psi_\Sigma \cdot \Psi_\Pi \tag{3}$$

$$\hat{H} = \hat{H}_\Sigma + \hat{H}_\Pi \tag{4}$$

$$E = E_\Sigma + E_\Pi \tag{5}$$

Notice now that one can use equation (3) to generate total wavefunctions expressed as linear combinations of CW products in the way illustrated in Figure 2 and that these wavefunctions will have form which is dependent on the representation of each bond. At the SD MO level, the constraint approximation "forces" this relationship of the weight factors λ_i and μ_i:

$$\lambda_1 = \lambda_2 = \mu_1 = \mu_2$$

By contrast, individual bonds are correlated at the MOVB-IBM level:

$$\lambda_1 > \lambda_2$$

$$\mu_1 > \mu_2$$

Accordingly, the SD-MO and MOVB-IBM wavefunctions are different whether or not bielectronic subsystem coupling terms are introduced in the Hamiltonian. Furthermore, the MOVB-IBM wavefunction is better than the SD-MO wavefunction for a very simple reason: The SD-MO model makes a grave mistake by assigning the same weight

The Ψ_T Representation in Terms of Subsystem CW Products

	$\lambda_1 \sigma_1^1 \sigma_2^1$	$\lambda_2 \sigma_1^2$	$\lambda_2 \sigma^2$
$\mu_1 p_1^1 p_2^1$	$\mu_1\lambda_1\Phi_1$	$\mu_1\lambda_2\Phi_2$	$\mu_1\lambda_2\Phi_3$
$\mu_2 p_1^2$	$\mu_2\lambda_1\Phi_4$	$\mu_2\lambda_2\Phi_5$	$\mu_2\lambda_2\Phi_6$
$\mu_2 p_2^2$	$\mu_2\lambda_1\Phi_7$	$\mu_2\lambda_2\Phi_8$	$\mu_2\lambda_2\Phi_9$

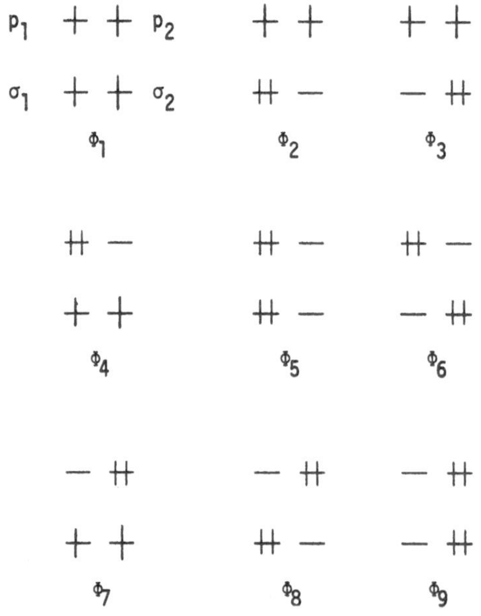

Figure 2. The MOVB representation of two subsystems: a sigma and a pi two-electron bond. The Φ_i's represent the total four-electron-four-orbital system, with each one being the product of two independent subsystem CW's.

to all nonequivalent Φ_i's (Figure 3b) while the MOVB-IBM makes a smaller mistake by assigning equal weights to the nonequivalent Φ_5, Φ_9 and Φ_6, Φ_8 (Figure 3c). However, as implied, the MOVB-IBM wavefunction still does not have "perfect" form. The "perfect" MOVB wavefucntion, shown in Figure 3d, assigns different weights to all nonequivalent Φ_i's and this wavefunction can only be produced by the unconstrained variational treatment of all Φ_i's with respect to a complete Hamiltonian. Thus, it is clear that bielectronic Σ - Π subsystem interaction, or, in plain language, bielectronic sigma-pi hybridization can only be dealt with at the level of non-constrained MOVB theory with SD MO and, of course, HMO, type of theories being incapable of dealing with the problem.

It is then apparent that there are two types of interactions that need to be considered: One-electron hybridization and two-electron hybridization. However, at the level of VB theory the two problems merge into a single one since in both cases we are concerned with the mode of interaction within a manifold of CW's and whether the interaction is brought about by monoelectronic or bielectronic (or both) matrix elements is immaterial. Chemists know intuitively what is "good" or "bad" one-electron hybridization. On the other hand, the notion of bielectronic hybridization, e.g., the way in which pi and sigma bonds interact in planar con-jugated systems, may appear to some readers as strange. We shall see that com-prehending hybridization, in the most general sense of the word, is a very simple matter at the level of VB theory.

CW PRODUCTS

(a)

Φ_1	Φ_2	Φ_3
Φ_4	Φ_5	Φ_6
Φ_7	Φ_8	Φ_9

CORRESPONDING SD-MO COEFFICIENTS

(b)

a	a	a
a	a	a
a	a	a

$(\lambda_1 = \lambda_2 = \mu_1 = \mu_2)$

CORRESPONDING MOVB-IBM COEFFICIENTS

(c)

a	c	c
d	b	b
d	b	b

$(\lambda_1 \neq \lambda_2, \ \mu_1 \neq \mu_2)$

CORRESPONDING "PERFECT" MOVB COEFFICIENTS

(d)

a	c	c
d	b	e
d	e	b

Figure 3. The description of two subsystems , a pi and a sigma two-electron bond, at various levels of theory.

II. High Order Perturbation Theory

There exist two conventional ways for generating the approximate eigenfunctions and eigenvalues of an operator: Through the variation method and through perturbation theory (PT). The latter approach has the conceptual advantage that each order of correction of the unperturbed eigenfunctions and eigenvalues can be associated with one or more physical mechanisms by which elementary bonds, "lone" pairs, holes and systems made up of them interact with each other. This point will be better understood as we develop the theory and apply the emerging concepts to specific chemical problems. We start this process by reviewing the fundamental Rayleigh-Schrödinger (RS) PT equations[2] for low and high order correction of zero order wavefunctions and their associated energies.

The RS - PT equations for the perturbed wavefunction $\psi_m^{(p)}$ and the energy correction $E_m^{(q)}$, expressed in terms of the unperturbed wavefunctions ϕ_m and associated unperturbed energies E_m, respectively, are given below for p = 0,1,2 and q = 0,1,2,3,4, where p and q represent perturbation orders.

$$\psi_m^{(0)} = \phi_m \tag{6}$$

$$\psi_m^{(1)} = \sum_t \frac{\langle\phi_t|\hat{P}|\phi_m\rangle}{(E_m- E_t)} \phi_t + a_{mm}^{(1)} \phi_m \tag{7}$$

$$\psi_m^{(2)} = \sum_{st} \frac{\langle\phi_m|\hat{P}|\phi_t\rangle\langle\phi_t|\hat{P}|\phi_s\rangle}{(E_m-E_s)(E_m-E_t)} \phi_s - \frac{1}{2} \sum_t \frac{\langle\phi_m|\hat{P}|\phi_t\rangle^2}{(E_m-E_t)^2} \phi_t \tag{8}$$

$$E_m^{(0)} = E_m \tag{9}$$

$$E_m^{(1)} = \langle\phi_m|\hat{P}|\phi_m\rangle \tag{10}$$

$$E_m^{(2)} = \sum_t \frac{\langle\phi_m|\hat{P}|\phi_t\rangle^2}{(E_m-E_t)} \tag{11}$$

$$E_m^{(3)} = \sum_{st} \frac{<\phi_m|\hat{P}|\phi_s><\phi_s|\acute{\hat{P}}|\phi_t><\phi_t|\hat{P}|\phi_m>}{(E_m-E_s)(E_m-E_t)} \qquad (12)$$

$$E_m^{(4)} = \sum_{rst} \frac{<\phi_m|\hat{P}|\phi_r><\phi_r|\acute{\hat{P}}|\phi_s><\phi_s|\acute{\hat{P}}|\phi_t><\phi_t|\hat{P}|\phi_m>}{(E_m-E_r)(E_m-E_s)(E_m-E_z)}$$

$$- E_m^{(2)} \sum_t \frac{<\phi_m|\hat{P}|\phi_t>^2}{(E_m-E_t)^2} \qquad (13)$$

The above equations are written subject to the following assumptions and conventions:

a. $<\phi_m|\hat{P}|\phi_t> = <\phi_t|\hat{P}|\phi_m>$ $\qquad (14)$

b. $\acute{\hat{P}} = \hat{P} - <\phi_m|\hat{P}|\phi_m> = \hat{P} - E_m^{(1)}$ $\qquad (15)$

c. In the summation, we exclude r, s, t = m.

Finally, we rewrite equations (8) and (13) in the following way:

$$\psi_m^{(2)} = X_m^{(2)} + Y_m^{(2)} \qquad (16)$$

$$E_m^{(4)} = W_m^{(4)} + Z_m^{(4)} \qquad (17)$$

with

$$X_m^{(2)} = \sum_{st} \frac{<\phi_m|\hat{P}|\phi_t><\phi_t|\acute{\hat{P}}|\phi_s>}{(E_m-E_s)(E_m-E_t)} \phi_s \qquad (18)$$

$$Y_m^{(2)} = -\frac{1}{2} \sum_t \frac{<\phi_m|\hat{P}|\phi_t>^2}{(E_m-E_t)^2} \phi_t \qquad (19)$$

$$W_m^{(4)} = \sum_{rst} \frac{<\phi_m|\hat{P}|\phi_r><\phi_r|\acute{\hat{P}}|\phi_s><\phi_s|\acute{\hat{P}}|\phi_t><\phi_t|\hat{P}|\phi_m>}{(E_m-E_r)(E_m-E_s)(E_m-E_t)} \qquad (20)$$

$$Z_m^{(4)} = - E_m^{(2)} \sum_t \frac{<\phi_m|\hat{P}|\phi_t>^2}{E_m-E_t} \qquad (21)$$

We now focus attention on the RS-PT energy expressions, noting that illustrative computations of $E_m^{(q)}$, for q = 1, 2, and 3, interpretations of their physical significance, and applications to chemical problems can be found in the literature almost always in connection with MO theory[3,4] i.e., when the ϕ_m's are MO's. In a more general sense, the ϕ_m's can be Atomic Orbitals (AO's), MO's, CW's etc. With this in mind, and assuming familiarity of the reader with the computation of $E_m^{(q)}$ with q = 1, 2, and 3, we now enter discussion of the form and implications of $E_m^{(4)}$.

As we have stated before, $E_m^{(4)}$ can be written as a sum of two terms, $W_m^{(4)}$ and $Z_m^{(4)}$ and the characteristic properties of the two terms are the following:

a. $W_m^{(4)}$ vanishes unless there is cyclic interaction of the ϕ's. This means that this term becomes responsible for parity stereoselection and that a geometry can always be found for which $E_m^{(4)}$ is negative or zero.

b. $Z_m^{(4)}$ is nonzero whether or not there is cyclic interaction of the ϕ's. Furthermore, when ϕ_m is a ground state function, it acts as an anticorrection of the second order energy correction, $E_m^{(2)}$, and we all are aware that current qualitative MO theory is fundamentally based on $E_m^{(2)}$ considerations. Since we have already seen how parity stereoselection arises using a VB variational approach[1] i.e., since we already understand how cooperativity ("aromaticity") and anticooperativity ("antiaromaticity") come into play when CW's interact in a cyclic manner, we can now restrict our attention to the role of the $Z_m^{(4)}$ term, which brings about anticooperativity regardless of the nature of the interaction of the ϕ's.

Let us attempt to understand what exactly $Z_m^{(4)}$ does by reference to the hypothetical three function system shown in Figure 4a. If we assume that the only nonzero interaction matrix elements are P_{12} and P_{13}, then the energy of the perturbed ϕ_1 can be written as a sum of the zero order term, $E^{(0)}$, plus the energy corrections ε_{12} and ε_{13} due to the $\phi_1 - \phi_2$ and $\phi_1 - \phi_3$ interactions:

$$E = E^{(0)} + \varepsilon_{12} + \varepsilon_{13} \tag{22}$$

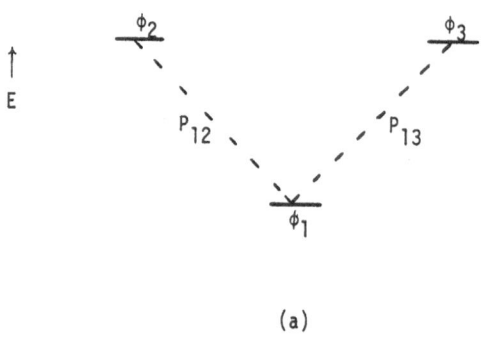

(a)

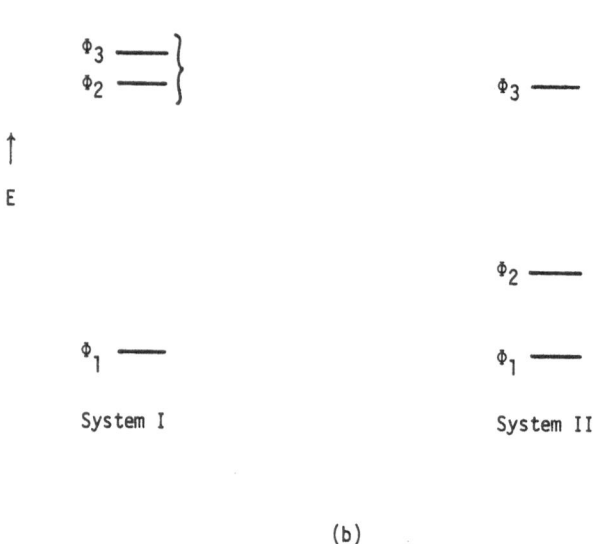

(b)

Figure 4. a. Interaction of prototypical functions. b. Two different manifolds of three interacting CW's differing only in the "energy position" of Φ_2.

Now, let us examine the form of one of the ϵ_{1t}'s (t=2,3), say, ϵ_{12}:

$$\epsilon_{12} = \frac{<\phi_1|\hat{P}|\phi_2>^2}{E_1-E_2} - E_1^{(2)} \frac{<\phi_1|\hat{P}|\phi_2>^2}{(E_1-E_2)^2} \tag{23}$$

where

$$E_1^{(2)} = \sum_t \frac{<\phi_1|\hat{P}|\phi_t>}{E_1-E_t} \tag{24}$$

The first term of equation (23) represents the contribution of the ϕ_1 - ϕ_2 inter-action to the total second order correction and the second term the contribution of the same interaction to the total fourth order correction of the energy of ϕ_1. Now, when ϕ_1 is the lowest energy function, as in the case at hand, the first term of equation (23) and $E_1^{(2)}$ have the same sign. Hence, the second term subtracts from the first due to the negative sign preceding $E_1^{(2)}$ in equation (23). Further-more, note the extremely important fact that this anticorrection depends on the second order interaction of ϕ_1 with each of ϕ_2 and ϕ_3. That is to say, the anticorrection depends on the sum of all second order ϕ_1 - ϕ_t interactions. Since ϵ_{12} is overall negative in the case at hand, the message is that the magnitude of the energy depression of ϕ_1, due to its interaction with a higher lying ϕ_t, is a function of all 1-t interactions. In this way, we recognize that electronic anti-cooperativity is embodied in the $Z_m^{(4)}$ term of $E_m^{(4)}$.

An examination of the RS - PT wavefunction corrections $\psi_m^{(q)}$, reveals that electronic anticooperativity is expressed at the level of $\phi_m^{(2)}$ by the $\gamma_m^{(2)}$ term. This is precisely what is expected since knowledge of wavefunction to order p implies knowledge of energy to order 2p + 1. Electronic anticooperativity is not contained in $\psi_m^{(1)}$ because this takes us only to third order correction of the energy but it is contained in $\psi_m^{(2)}$ as this takes us to fifth order correction of the energy.

With the above in mind, let us now consider the systems shown in Figure 4b in which Φ_i represents a VB CW. Furthermore, let us assume the following interaction matrix elements to be constant throughout:

$$\langle\Phi_1|P|\Phi_2\rangle \neq 0$$

$$\langle\Phi_1|P|\Phi_3\rangle \neq 0$$

$$\langle\Phi_2|P|\Phi_3\rangle = 0$$

This means that the ground CW, Φ_1, can be corrected to first and second order in wavefunction but only second and fourth orders in energy. Now, it is evident that as we sweep from I to II, the Φ_1 - Φ_3 interaction is kept __constant__ while the Φ_1 - Φ_2 interaction becomes __stronger__ at the level of __first order__ RS-PT with respect to the wavefunction. By contrast, at the level of __second order__ PT with respect to wavefunction, the Φ_1 - Φ_3 interaction becomes weaker as the Φ_1 - Φ_2 interaction becomes __stronger__. Now, if Φ_1 and Φ_2 constitute CW's which, in part, define one elementary bond and Φ_1 and Φ_3 constitute CW's which, in part, define a second elementary bond and if we are interested in the ground state properties of the system, we can say that, as we sweep from I to II, __the second bond becomes actually weaker as the first becomes stronger__. Since any two elementary bonds[1] have common CW's (in our example, Φ_1), a rule follows: __A perturbation which strengthens one bond automatically weakens a second bond__. We illustrate this principle by two different examples.

III. Monelectronic (Charge Transfer) Hybridization

Consider the linear pi system formed by the three pi AO's of X, C, and Y which contain a total of four electrons.

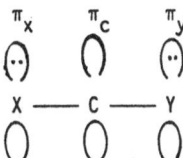

This can be viewed as an H-bound four-electron-three-orbital system and the wavefunction can be represented as shown below:

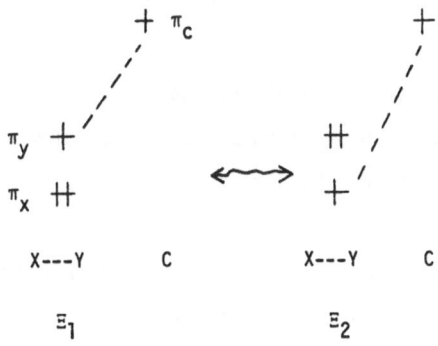

or, in terms of chemical formulae:

$$X = C - \ddot{Y} \longleftrightarrow \ddot{X} - C = Y$$

Since the concepts of MOVB theory have been derived on the basis of the variational treatment of a trial MOVB wavefunction and since a "correct" VB-type theory can only be developed using the variational method or high order PT, the relative weights of the Ξ_1 and Ξ_2 resonance bond diagrams are properly reproduced and the change of the $\lambda_1:\lambda_2$ ratio as a response to a perturbation is immediately predictable, in a qualitative sense. We can say, that the MOVB wavefunction is appropriately normalized and that a perturbation that increases λ_1 must necessarily decrease λ_2. That this "renormalization" is a result of high order interaction can be shown explicityly by comparing the related pi systems XCY_1 and XCY_2 in which the one electron energy of π_y varies relative to the energies of the other orbitals as shown in Figure 5a. The interacting VB CW's of the general XCY system are shown in Figure 5b, where it has been assumed that the CW's are connected only only via monoelectronic Charge Transfer (CT) matrix elements, D_{ij}, which are proportional to resonance integrals over the AO's which define the origin and terminus of the electron hop. The way in which the CW manifolds differ in XCY_1 and XCY_2 and the coupling modes are illustrated in Figure 6a. If attention is restricted to the interaction of the three lowest energy CW's in each case (Figure 6b), the problem becomes exactly analogous to the one dealt with before (see Figure 4b) with Φ_1 and Φ_2 being the main descriptors of the C-Y and Φ_1 and Φ_3 being the main descriptors of the C-X elementary bond and with Φ_1 being the common valve (V) CW which is responsible for anticooperativity. As the energy of Φ_2 relative to Φ_3 decreases, the Φ_1 - Φ_2 interaction turns off the Φ_1 - Φ_3 interaction, or, in different words, the C-Y bond "eliminates" the C-X bond, or, yet in a third

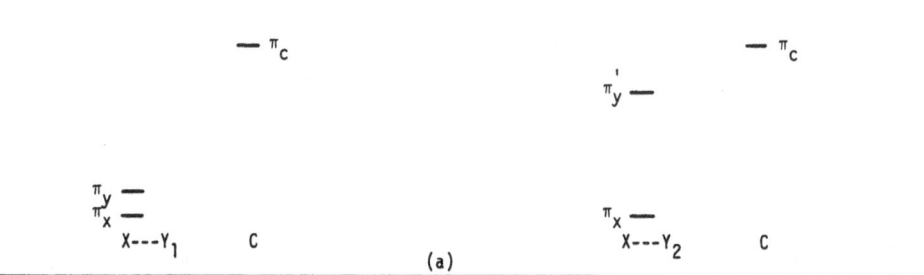

(a)

C - Y ELEMENTARY BOND

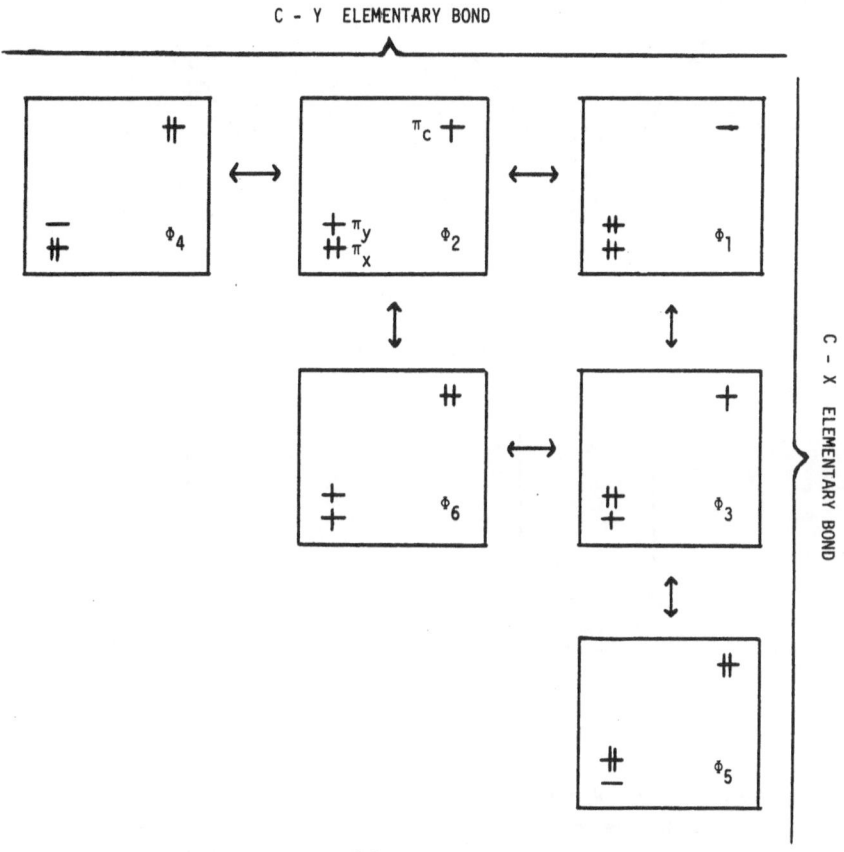

(b)

Figure 5. a. Orbital manifolds of XCY₁ and XCY₂. b. The MOVB description of a four-electron-three-orbital system. Note the definition of C - Y and C - X elementary bonds.

MONOELECTRONIC CHARGE TRANSFER CW COUPLING

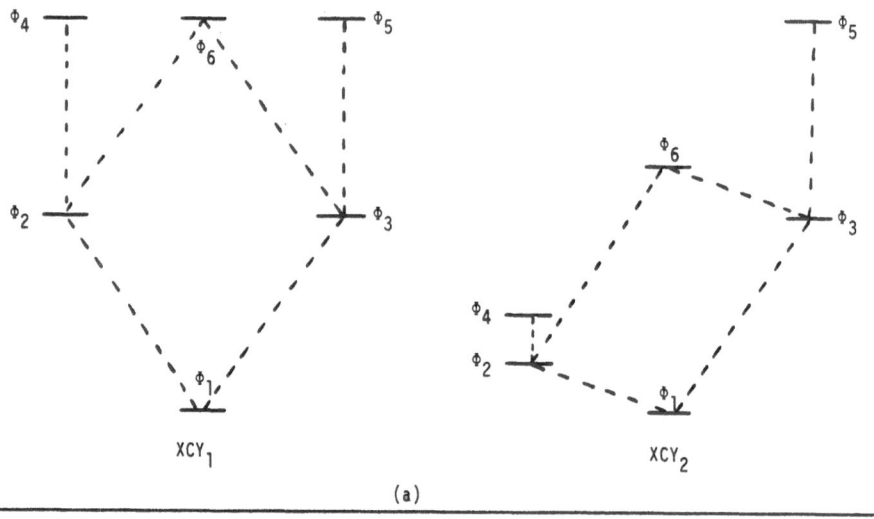

(a)

KEY MONOELECTRONIC CHARGE TRANSFER CW INTERACTION

(b)

Figure 6. a. The CW interaction patterns of XCY_1 and XCY_2 wherein it is assumed that the various CW's are connected only by d_{ij} (Charge Transfer) matrix elements. b. The key Charge Transfer CW interactions within XCY_1 and XCY_2.

language, the relative weights of the two resonance bond diagrams $\equiv_1$ and $\equiv_2$ change so that one of them, $\equiv_1$, becomes an increasingly more accurate representation of the system. In general, we can say that as the energy gap between the C and Y pi AO's decreases and/or the AO resonance integral β_{cy} increases in absolute magnitude, a two-electron pi C-Y bond is formed at the expense of keeping the second pi electron pair localized on X.

The relevance of this analysis to actual mechanistic problems is probably apparent to the reader who has been exposed to carbocation chemistry. For if C is a carbenium ion center and X and Y pi donating substituents, the analysis presented above constitutes the theoretical justification of the so-called Principle of Electron Demand demonstrated marvelously by Gassman and Fentiman[5] through their studies of the solvolysis of norbornane and norbornene derivatives and exploited brilliantly by Brown[6] as a tool for probing anchimeric assistance. This principle is nothing other than the empirical statement of the more general bond anticooperativity phenomenon which is manifested in all types of chemical problems and which can be described as follows: As the pi donating ability of two substituents, $\ddot{X}$ and $\ddot{Y}$, capable of stabilizing a carbenium ion center by pi donation diverges, then the stabilizing effect of the superior donor, $\ddot{Y}$, expressed through C-Y bond formation, will tend to turn off the stabilizing effect of the inferior donor, $\ddot{X}$, resulting in localization of the electron pair on X. For illustrative purposes, experimental data of Gassman and Fentiman are shown in Figure 7. Their kinetic studies demonstrate that a superior donor (OMe) can "turn off" the effect of an inferior one (-CH = CH-). The reader is also referred to the work of H. C. Brown and his associates for further examples.

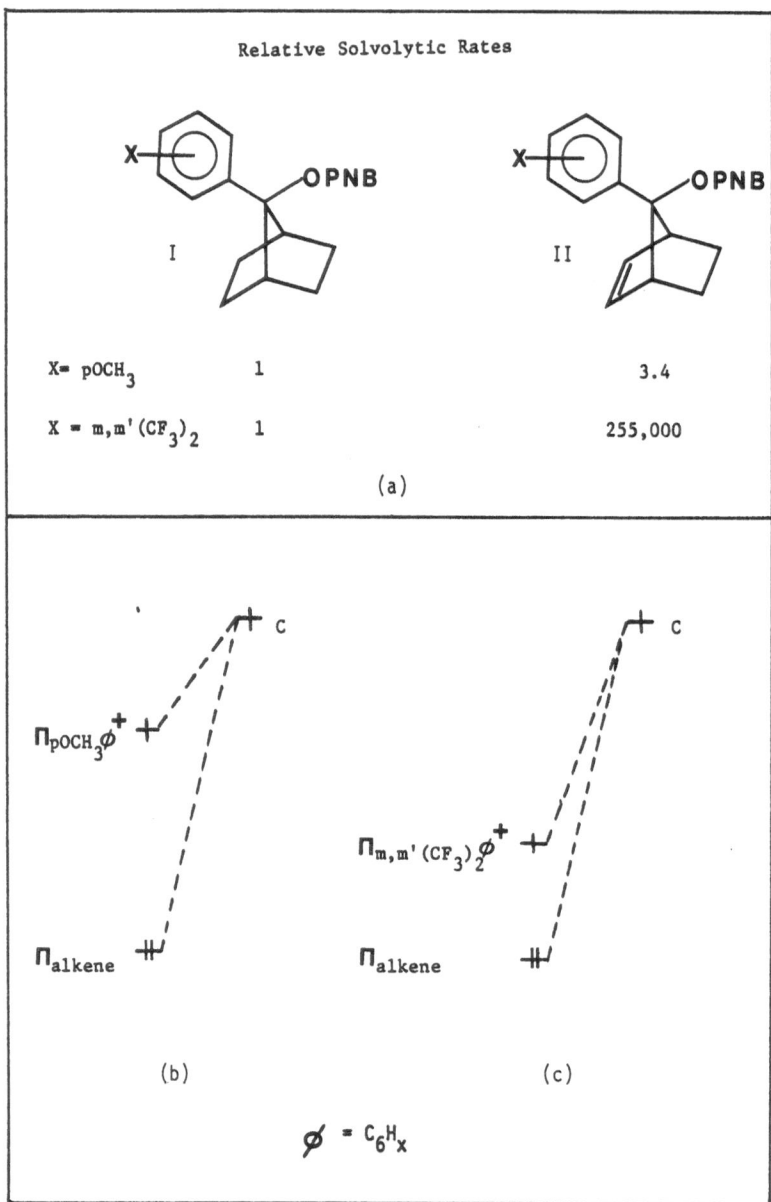

Relative Solvolytic Rates

I II

| X= pOCH$_3$ | 1 | 3.4 |
| X = m,m'(CF$_3$)$_2$ | 1 | 255,000 |

(a)

$\Pi_{pOCH_3\phi}^+$

Π_{alkene}

$\Pi_{m,m'(CF_3)_2\phi}^+$

Π_{alkene}

(b)   (c)

ϕ = C$_6$H$_x$

Figure 7. MOVB diagrammatic representations of II. When X = pOCH$_3\phi$, the aryl group forms a strong bond with the incipient carbonium ion by eliminating the action of the alkene fragment and the rate ratio of systems II and I is near unity. When X = m, m' (CF$_3$)$_2\phi$, the aryl group and the alkene fragment form a <u>hybrid</u> bond with the incipient carbonium ion. Hence, elimination of the alkene moiety is highly unfavorable and the rate ratio becomes very large.

IV. Sigma-Pi Hybridization

The "resonance theory" one is taught in elementary undergraduate courses would be sufficient for predicting bond anticooperativity in the case discussed in the preceding section, which is more than can be said for standard Perturbation MO (PMO) theory of the type often used nowadays for the qualitative analysis of many problems.[7] For, if high order perturbation theory is not implemented, rehybridization due to perturbation cannot be "seen", i.e., Φ_2 and Φ_3 mix independently with Φ_1 and one elementary bond cannot "adjust" to the other. In short, we recognize that "resonance theory" - trained intuition suffices to develop the concept of bond anticooperativity and this is the reason why the Principle of Electron Demand was discovered long before experimentalists became fascinated with MO theory. We now show that the apparently unrelated sigma-pi hybridization concept is yet another consequence of high order interaction which can be revealed by exactly the same analysis as the one presented in the previous section.

Consider, the system shown below in which a pi and a sigma bond are rigorously defined and, thus, cannot interact in a monoelectronic sense. This system can be represented by the bond diagram of Figure 8 which, at the level of the IBM, does not tell us anything as to how the sigma and pi bonds interact. However, if we abandon the IBM and identify the low energy CW's, it is easy to see that we generate a problem identical to the one discussed before. More specifically, we recognize the following types of CW's:

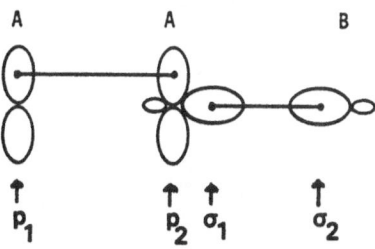

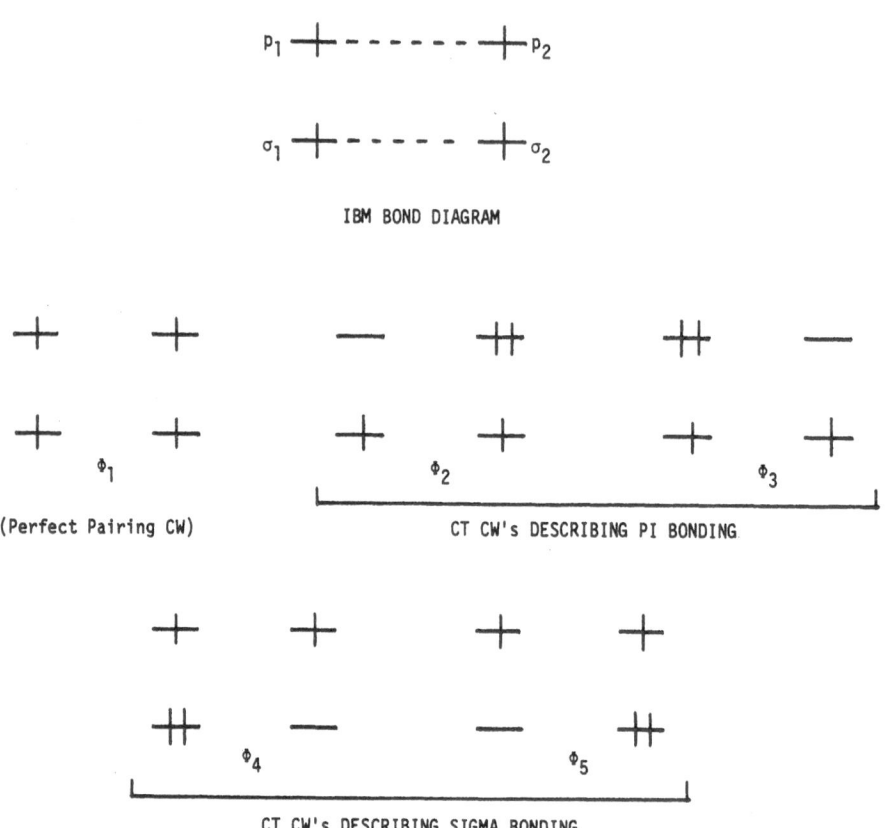

Figure 8. IBM bond diagram and the key CW's describing two interacting bonds (a sigma and a pi) at the level of unconstrained MOVB theory.

a. The "perfect pairing" CW, Φ_1, which is common to both sigma and pi bonds as it describes both sigma and pi bond formation through spin pairing.

b. The low energy CT CW's generated by a one electron shift within the sigma or the pi bond, Φ_2 - Φ_5.

For qualitative purposes, it is now sufficient to consider the interaction of Φ_1 with Φ_2 - Φ_5 as B becomes increasingly electronegative while AO resonance integrals remain constant. Thus, we can envision two orbital manifolds as shown in Figure 9a which are responsible for two different CW manifolds shown in Figure 9b. Since Φ_1, Φ_2, and Φ_3 are the main descriptors of the pi while Φ_1, Φ_4, and Φ_5 are the main descriptors of the sigma bond, we can say that as the electronegativity of B increases the sigma bond will become stronger due to greater Φ_1 - Φ_5 interaction while simultaneously the pi bond will become weaker because the Φ_1 - $(\Phi_2+\Phi_3)$ interaction will be turned off.

Note now an important aspect of this analysis: The mixing of the critical CW's in this and the previous example is accomplished via monoelectronic CT interaction matrix elements while previously we have stated that CW's within a subsystem interact in a monoelectronic sense while different subsystems can interact only in a bielectronic sense. Thus, it <u>appears</u> that there is a contradiction. The reason why "normal" and sigma-pi hybridization are really identical problems at the level of <u>unconstrained</u> VB theory while they appear to be different at the SD MO or MOVB-IBM level, lies in the fact that the latter two theories involve <u>constrained</u> CW products. By contrast, pure VB theory places no constraints on the CW's. Hence, all problems are conceptually identical at the level of VB (or MOVB) theory as they all revolve about the optimal interaction of a set of CW's, <u>in the absence of constraints</u>, to ultimately generate what we subsequently call subsystems.

In conclusion, we can say that the seemingly noninteracting sigma and pi bonds share common CW's, and, as a result, they interact with each other exactly as ele-

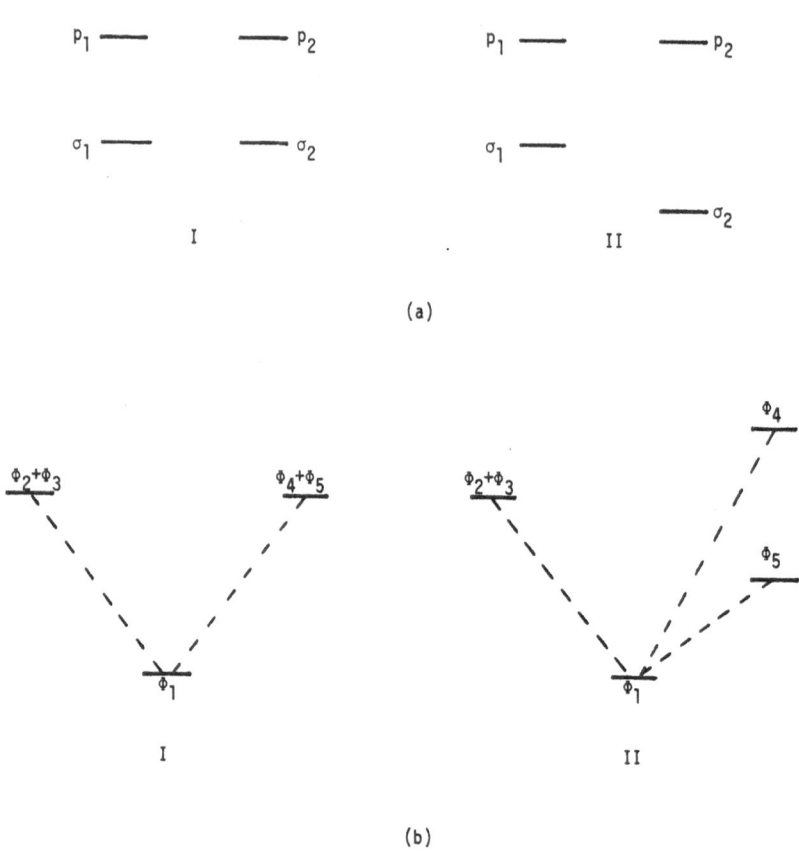

(a)

(b)

Figure 9. a. Two different sigma-pi orbital manifolds. System II differs from I only to the extent that a homopolar has been replaced by a heteropolar sigma bond. b. Principal CW interactions in systems I and II. In II, Φ_5 turns off the action of $\Phi_2 + \Phi_3$.

mentary bonds interact with each other in "normal" hybridization. As a sigma bond becomes stronger due to increased "ionicity", or, because of increased overlap binding ability of a sigma bond atom (B, is our example), the pi bond must get weaker. To put it crudely, <u>the pi bond strength depends on the identities of the surrounding sigma ligands</u>.

The quintessential example of sigma-pi rehybridization due to structural modification is the structural consequences of replacing H by F in ethylene (C_2H_4). As the sigma bonds become stronger, the C-C pi bond becomes weaker as the following pi Bond Dissociation Energies (BDE's) indicate:

pi BDE (kcal/mol):[8]	59.1	52.3
r_{CC}(Å):[9]	1.339	1.311

If one interprets heats of addition to be measures of pi bond strengths, then the trends shown below are also explicable, in part, in terms of the concept of sigma-pi hybridization.[10]

	Heat of Addition for Addend (kcal/mol)		
Molecule	HBr	Br_2	Cl_2
$F_2C = CF_2$	-32.99	-38.48	-57.32
$H_2C = CH_2$	-16.80	-23.80	-41.50

The reader may now legitimately ask: Why do fluorines <u>deionize</u> and, thus, weaken a pi bond in $CF_2 = CF_2$ while no such comparable effect is found in the hydrogenation product $CHF_2 - CHF_2$? The reason is very simple: In $CF_2 = CF_2$, we have a C-F bond of high "ionicity" competing with a pi C-C bond of maximum "covalency". As a result, the more "ionic" bonds deionize the more "covalent" bond. By contrast, in $CHF_2 - CHF_2$, the fluorines cannot weaken the C-H bonds simply because both C-F and C-H bonds have "ionic" character and, thus, neither of the two can deionize the other. This is made evident by the compact bond diagram shown in Figure 10 which shows how the core-ligands bonds would be formed if nonbonded AO overlap were zero.

How does sigma-pi hybridization affect the geometry of a molecule? According to MOVB theory, replacement of all hydrogens by fluorines in $CH_2 = CH_2$ has two distinct effects:

a. It modifies the extent and/or direction of charge transfer in the core-ligand bonds.

b. It causes a weakening of the pi bond.

The bond diagram of $CH_2 = CH_2$ shown in Figure 11 makes now evident that the first pure sigma effect will tend to shorten the C = C bond because replacement of H by F will tend to accentuate the depopulation of the ω_6 core MO which is antibonding with respect to the two carbons. By contrast, sigma-pi hybridization will act in the opposite direction. In fact, the first effect predominates and the central C = C bond of $CF_2 = CF_2$ is actually shorter than that of $CH_2 = CH_2$ despite the fact that the pi bond is found by independent means to be weaker!!

We now reformulate the problem of sigma-pi interaction in the hypothetical linear $L_1-A-A-L_2$ system in the language of MOVB theory, departing from the AO basis shown below, constructing the MO basis by defining L_1---L_2 to be the ligand fragment and A_2 the core fragment, and assuming a linear geometry in which the two A atoms are linked by a sigma and a pi bond and the two ligands are attached to the core fragment by two pure sigma bonds.

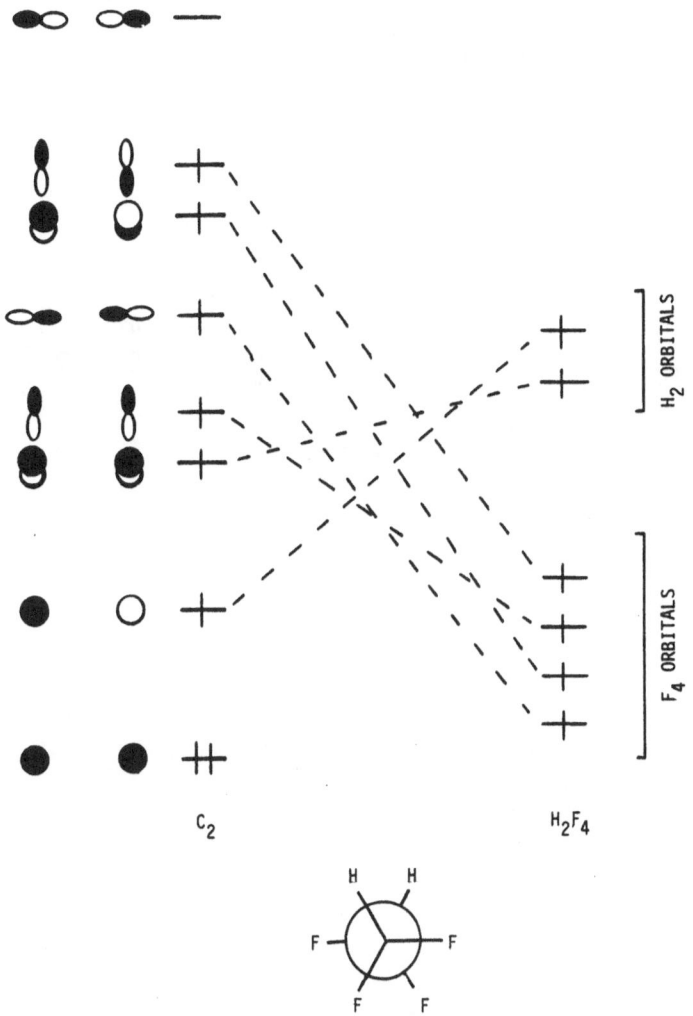

Figure 10. Compact bond diagram of gauche $C_2F_2H_2$ assuming segregation of the F_4 and H_2 ligand MO's. Note how high lying core orbitals are used to make bonds with low lying fluorine orbitals and low lying core orbitals are used to make bonds with higher lying hydrogen orbitals (D-type bonding). The reader may determine that, e.g., trans $C_2F_2H_2$ is not compatible with D-type core-ligand bonding.

D_{2h}

E T H Y L E N E

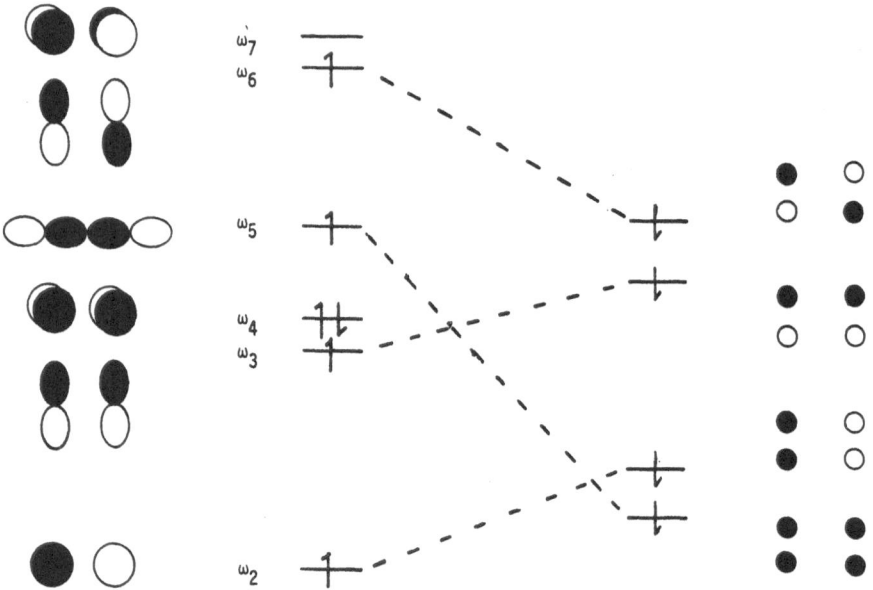

Figure 11. Compact bond diagram of planar C_2H_4.

AO BASIS

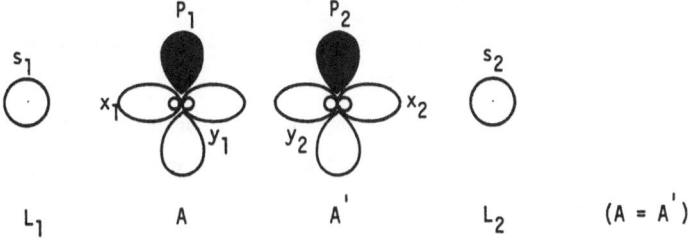

L_1 A A' L_2 $(A = A')$

MO BASIS

$\omega_4 \; \alpha \; y_1 - y_2$ ——

$\pi^* \; \alpha \; p_1 - p_2$ ——

$\pi \; \alpha \; p_1 + p_2$ ——

↑
E

$\omega_3 \; \alpha \; x_1 - x_2$ ——   —— $\sigma^* \; \alpha \; s_1 - s_2$

$\omega_2 \; \alpha \; x_1 + x_2$ ——   —— $\sigma \; \alpha \; s_1 + s_2$

$\omega_1 \; \alpha \; y_1 + y_2$ ——

$A - A'$ $L_1 \; \text{---} \; L_2$

With these conventions we now construct the bond diagrams of L_1AAL_1 for different
combinations of L_1 and L_2:

    a. $L_1 = L_2 = L$. In this case, the correct bond diagrammatic representation
is shown in Figure 12a. The resonance bond diagrams, Ξ_1 and Ξ_2 imply that for each
CW of Ξ_1 having two electrons in π there exists a corresponding CW of Ξ_2 having these
two electrons promoted to π^*. In this example, the weight of Ξ_1, λ_1, far exceeds
that of Ξ_2, λ_2.

    b. $L_1 = L_2 = L'$, where L' is much more electronegative than L. In this case,
we obtain the same bond diagrammatic representation as before with one important
difference: The ratio $\lambda_1:\lambda_2$ has decreased and, as a result, the strength of the pi
bond has been reduced. Why? The reason is very simple: The Ξ_1 - Ξ_2 interaction,
brought about exclusively by bielectronic polarization (P_{ij}) matrix elements,
decreases as L becomes increasingly electronegative and the "ionic" CW's of Ξ_1 play
an increasingly important role because the bielectronic polarization interaction of
two CW's increases as the number of open shell electrons decreases. This is
illustrated below.

$$P_{mn} < P_{pq}$$

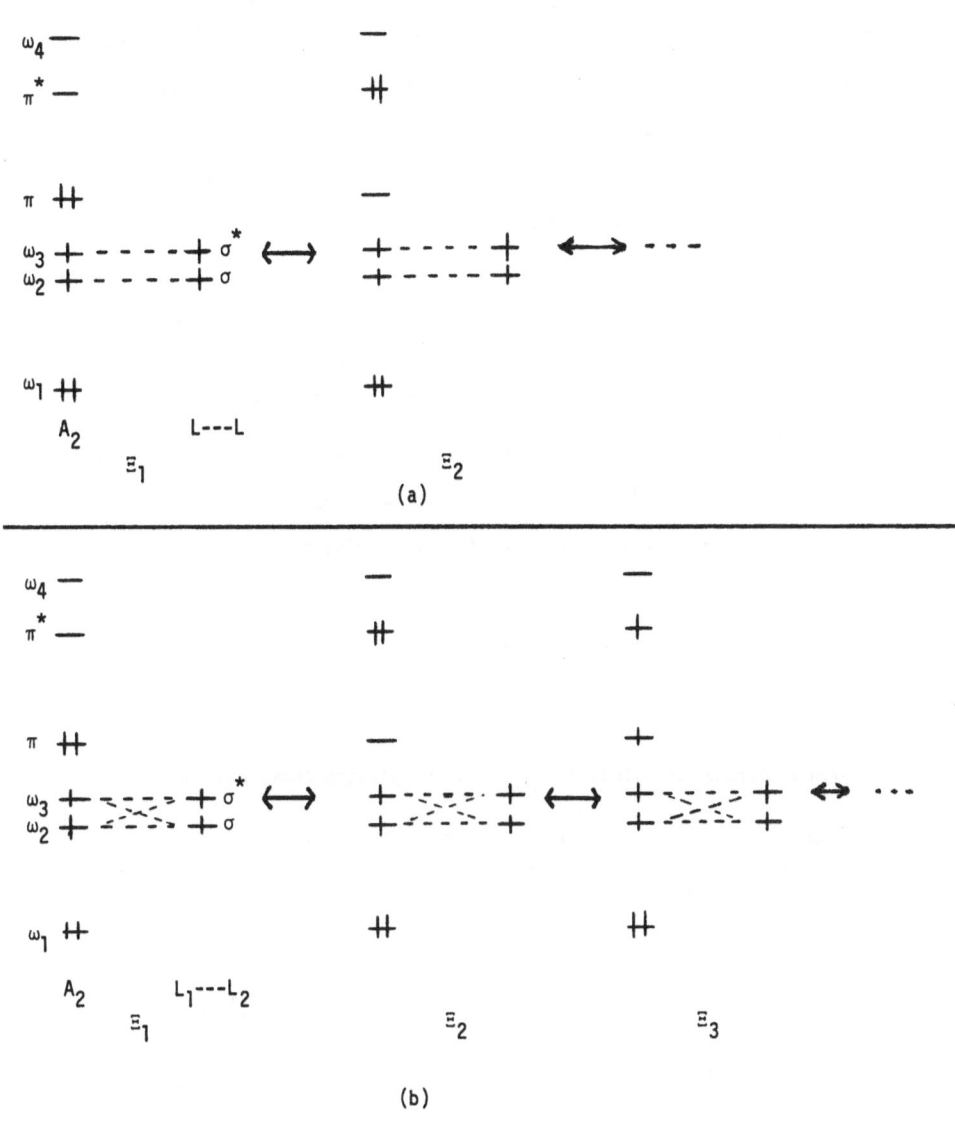

Figure 12. a. The bond diagrammatic representation of L-A-A-L.
b. The bond diagrammatic representation of L_1-A-A-L_2.

The result is that as L becomes more and more electronegative, there is greater two-electron pi excitation and progressive reduction of pi bonding. Thus, we have reached the same conclusions as before.

c. $L_1 \neq L_2$ with L_2 much more electronegative than L_1. In this case, the correct bond diagrammatic representation changes to the one shown in Figure 12b. What now happens is that the weight of the Ξ_2 bond diagram in Figure 12a is reduced because of the contribution of the Ξ_3 bond diagram of Figure 12b. The result is that the pi A - A bond becomes stronger because <u>the presence of single excitation turns off double excitation in the unsymmetrical L_1ALL_2 species.</u> In a quali-tative sense, we can then see that anticooperativity can either weaken or strengthen bonds depending upon the orbital basis which is the departure point for the develop-ment of theory. In this particular case, anticooperativity causes a sigma induced strengthening of a pi bond and this effect is brought into play not any longer by charge transfer (CT) but rather by the less notorious monelectronic polarization (P_{ij}) CW interaction matrix elements. A potential example of how an unsymmetrical sigma frame causes a strengthening of the overlying pi bond relative to the one coexisting with a symmetrical sigma frame is the comparison shown below.

```
    H         H          F         F          F          H
     \       /            \       /            \        /
      C === C              C === C              C === C
     /       \            /       \            /        \
    H         H          F         F          F          H
```

Pi BDE
(kcal/mol):   59.1[8]   52.3[8]   62.1[11]

We have just seen that, in making a transition from ethylene to perfluoro-ethylene, the increased ionicity of the sigma bonds causes a decreased ionicity of the pi bond with accompanying decrease of pi bond strength. In other words, sigma bond ionicity in perfluoroethylene <u>deionizes</u> the C-C pi bond. As a result, <u>this</u> <u>pi bond begins to look like an electron pair bond as described by Heitler-London</u> <u>theory</u>. This difference between ethylene and perfluoroethylene becomes responsible for important reactivity differences between the two compounds.[12] The reasons are explained below.

V. Heitler-London Reactions

Consider the "allowed" and "forbidden" modes of thermal ethylene plus ethylene cycloaddition to cyclobutane treated as a "four-electrons-in-four-pi-AO's" problem. The MOVB CW's, constructed from the π and π^* MO's of the two reactants and necessary for the MOVB description of the reaction, are shown in Figure 13. In the same Figure 13, we also show the VB CW's, constructed from the four pi AO's and necessary for the VB description of the same process. For the time being, we use MOVB theory for the discussion of the problem. At this level of theory, the qualitative energy surfaces for the 2s + 2s and the 2s + 2a cycloaddition modes are shown in Figure 14, with the principal MOVB contributor CW's shown in parenthesis. In both cases, pi bond destruction causes the Φ_1 CW to rise in energy. Sigma bond formation is described mainly by the "perfect pairing" Φ_5 CW, in connection with CW's Φ_6 - Φ_{14} as appropriate. The difference between the "allowed" and "forbidden" path is due to the fact that the low lying "ionic" CW's Φ_{11} and Φ_{12} can only contribute to the 2s + 2a transition state complex on symmetry grounds. This analysis leads to the following conclusions:

a. As the $\pi\pi^*$ triplet excitation energies of the reactants decrease, sigma bond formation is fascilitated because the energy of Φ_5 decreases, the product curve in Figure 14 is translated downwards in energy, and the reaction becomes faster <u>regardless of reaction path</u>.

b. As the triplet excitation energies of the reactants decrease, the energy difference between the "perfect pairing" CW describing sigma bond formation and the low energy CT CW's, Φ_{11} and Φ_{12}, which can only contribute to the 2s + 2a transition state, increases. Hence, the reaction stereoselectivity is diminished and, at the limit of a very large energy difference, the 2s + 2a and 2s + 2s transition state complexes are only weakly differentiable on the basis of the Φ_{11} and Φ_{12} CW's (<u>vide infra</u>).

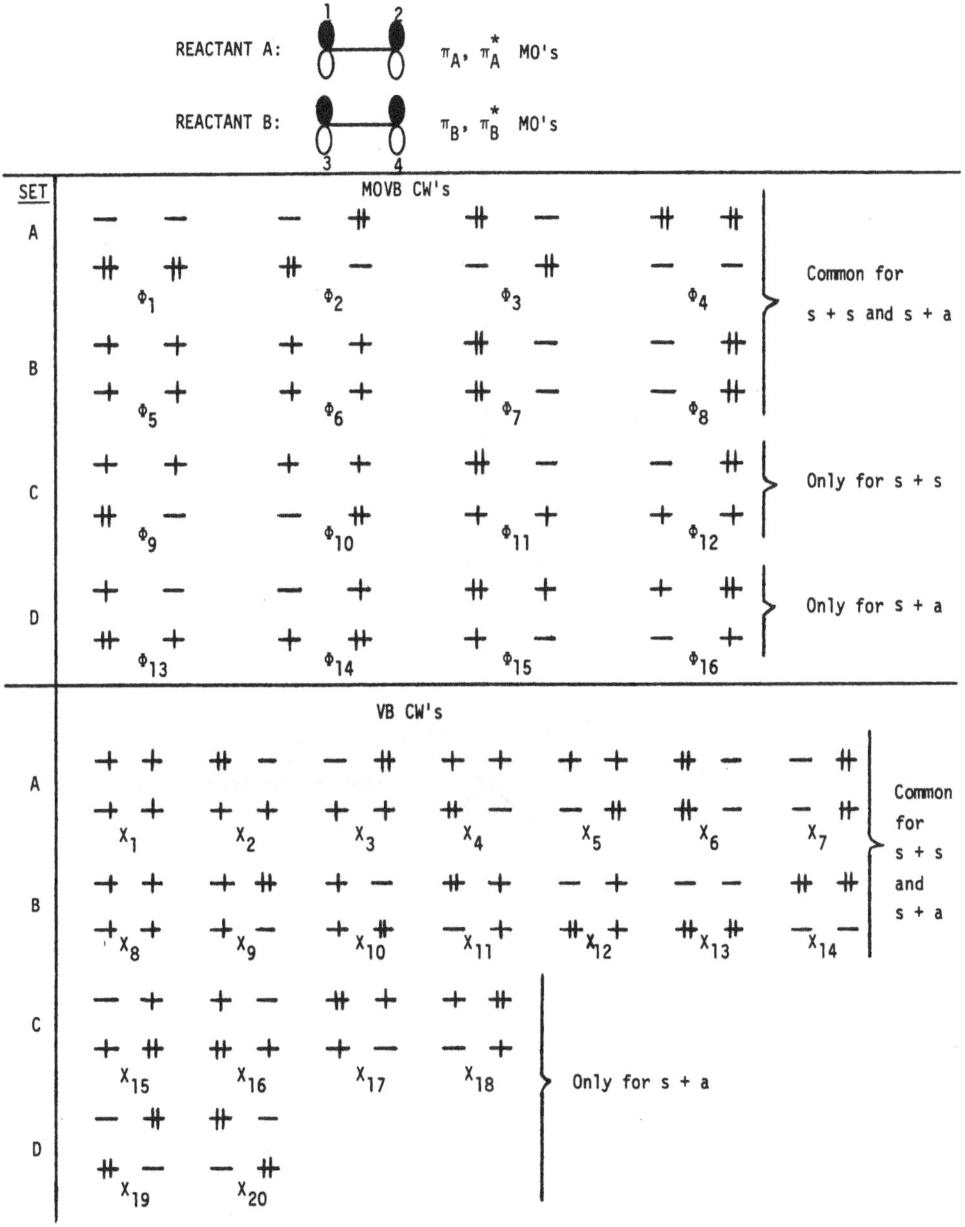

Figure 13. The MOVB and VB CW's required for the treatment of the $2\pi + 2\pi$ cycloaddition of two ethylenes and their contributions to the 2s + 2s and 2s + 2a transition states.

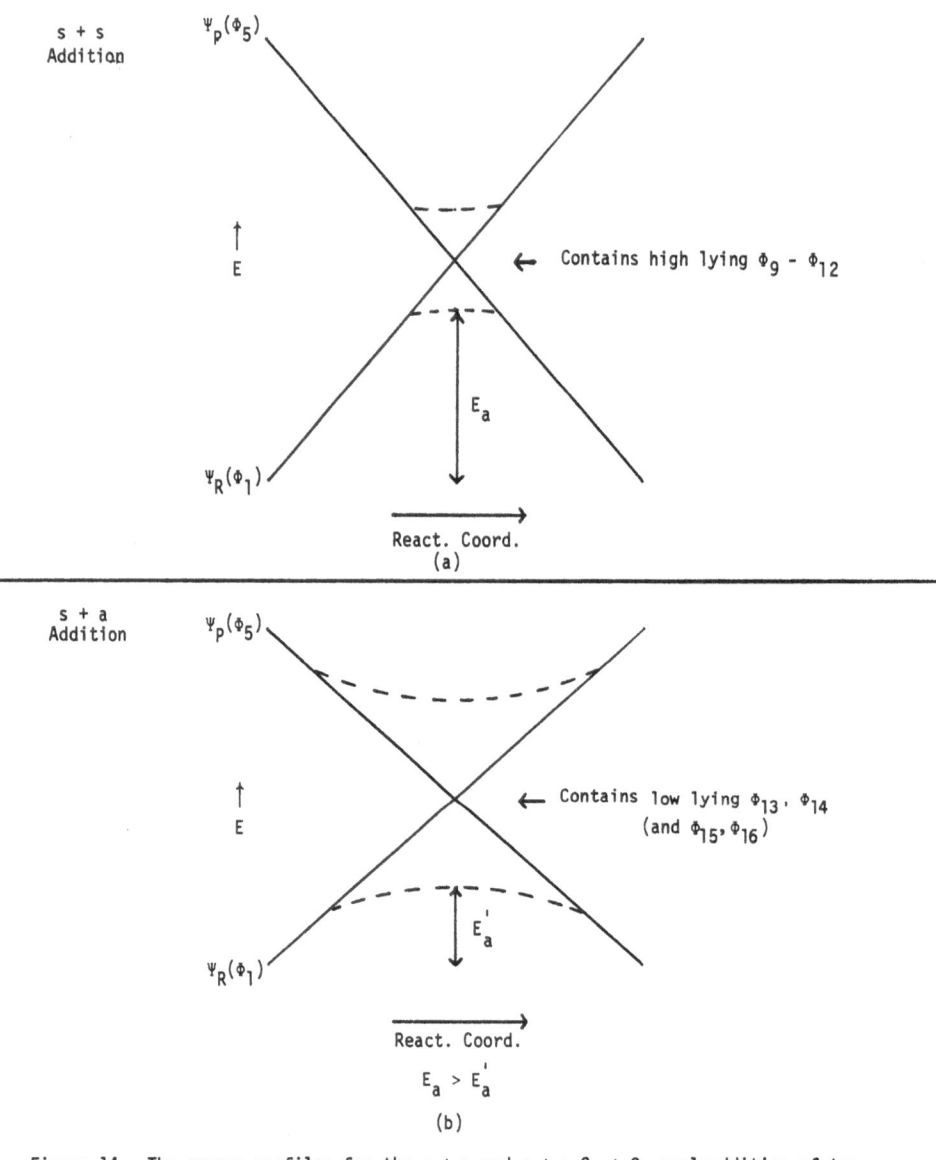

Figure 14. The energy profiles for the s + s and s + a 2π + 2π cycloaddition of two ethylenes. In constructing these surfaces the effect of the sigma framework has implicitly been taken into account. <u>Caution</u>: If only the carbon pi orbitals are considered E_a can be actually negative.

The second point can be better understood by restating the argument in VB terms. In this case, we say that, if only the HL CW's, X_1 and X_8, are considered, a cyclic four-electron-four-orbital Hückel AO (HAO) system (2s + 2s complex) has higher energy than the corresponding Möbius AO (MAO) system (2s + 2a complex) due to high order AO overlap terms.[1] Upon inclusion of all additional VB CW's in the basis set, the MAO system is further significantly stabilized relative to the HAO system mainly because of the exclusive contributions of X_{15} - X_{18} to the MAO wavefunction. This leads to the conclusion that stereoselection at the HL level is small while stereoselection at the VB level is much greater. A change of sigma ligands which weakens the pi bonds of the reactants is effectively tantamount to a transition from a VB to an HL description of the transition state complex and the consequence is reduced stereoselectivity. Thus, for example, the ethylene plus ethylene transition state complex has high "ionic" character and can be described only by VB theory. By contrast, the ethylene plus perfluoroethylene transition state complex has low "ionic" character because of the deionizing action of the sigma ligands and it can be adequately described by HL theory, i.e., in this case HL theory is an excellent approximation of VB theory. Because of the different extents of pi "ionic" delocalization in the two systems, the "allowed"-"forbidden" gap is expected to become narrowed in the case of the perfluoroethylene plus ethylene reaction.

There are two remarkable experimental trends which, at last, can be reasonably rationalized by theory. The first is that cyclobutane formation is kinetically unfavorable when the reactants are ethylenes with the reaction becoming significantly accelerated when one ethylene is replaced by a perfluoroethylene.[13] A second and still more astounding fact is that, when perfluoroethylene is presented with the alternatives of an "allowed" 4s + 2s and a "forbidden" 2s + 2s reaction, it chooses the latter over the former with ethylene making exactly the opposite choice![12]

We say that in the case of reaction (II), we have identified an HL reaction of minimal stereoselectivity and we conclude that deionization of the pi system by the sigma framework is responsible for it.

At this point, we mention that the above discussion is based on the assumption that the 2s + 2a transition state is sterically encumbered and lies well above that for the "2s + 2s" path, where by "2s + 2s" we imply the stepwise path which is the energetically more favorable version of the concerted 2s + 2s path. Finally, we mention that the differential effect of hydrogen and fluorine on thermodynamic equilibria is due to the factor discussed above as well as to rebonding caused by fluorine substitution in the ground state product. The way in which fluorine causes rebonding of the hydrocarbon precursors is exemplified in following sections.

Some years ago, intrigued by the eccentric behavior of fluoroalalkenes and, at the same time, enthralled by the apparent successes of the HMO approach, I suggested that the stereochemistry of nonpolar $2\pi + 2\pi$ cycloadditions of dienes and olefins, first studied by Bartlett and his coworkers, could be explained by assuming concerted 2s + 2a union of the reactants with steric effects determining which of the two cycloaddition components acted as the moiety sustaining bond rotation.[14] This explanation must now be rejected for it is difficult to accept that a sterically encumbered 2s + 2a "allowed" process can compete and often dominate the also "allowed" and sterically unencumbered 4s + 2s process whenever an olefin reacts with a fluoro-diene, or, a fluoro-olefin reacts with a diene. The mystery of why fluorinated olefins and dienes often shun the presumably accessible $4\pi + 2\pi$ cycloaddition route has now been solved because we have finally crossed the threshold of polydeterminantal MO theory and developed concepts using high order perturbation theory! Thus, I now conclude that the original interpretation of the mechanism of nonpolar cycloadditions advocated by Bartlett[15] was correct and my tenuous love-affair with concerted nonpolar cycloadditions a mistake attributable to HMO theory itself. For only if one rises to a high level of conceptual theoret- ical sophistication can he understand why fluoro olefins behave in the way that they do. Of course, this situation is in no way unique since in every single chapter of this work we find that MOVB theory has truly revolutionized our under- standing of chemistry. Indeed, many structure and reactivity trends have a source which is substantially different from the one we used to think they did within the confines of monodeterminantal MO theory!

References

1. Epiotis, N.D.; Larson, J.R.; Eaton, H. "Unified Valence Bond Theory of Electronic Structure" in Lecture Notes in Chemistry, Vol. 29; Springer-Verlag: New York and Berlin, 1982.

2. Dalgarno, A. in "Quantum Theory", Vol. 1, Bates, D.R., Ed.; Academic Press: New York, 1961

3. Dewar, M.J.S. "The Molecular Orbital Theory of Organic Chemistry"; McGraw-Hill: New York, 1969.

4. Libit, L.; Hoffmann, R. J. Am. Chem. Soc. 1974, 96, 1370.

5. Gassman, P.G.; Fentiman, A.F. J. Am. Chem. Soc. 1970, 92, 2549.

6. Brown, H.C. "The Nonclassical Ion Problem"; Plenum Press: New York, 1977.

7. (a) Houk, K. N. Accounts Chem. Res. 1975, 8, 361.
   (b) Fleming, I. "Frontier Orbitals and Organic Chemical Reactions"; John Wiley and Sons, Inc.: New York, 1976.

8. The pi C-C bond dissociation energies have been extracted from the sources listed below:
   C_2H_4: Benson, S.W. J. Chem. Educ. 1965, 42 502.
   C_2F_4: Wu, E-C.; Rodgers, A.S. J. Am. Chem. Soc. 1976, 98, 6112.

9. Callomon, J.E.; Hirota, E.; Kuchitsu, K.; Lafferty, W.; Maki, A.G.; Pote, C.S. in Landolt-Bornstein, "Numerical Data and Function Relationships in Science and Technology", Vol. 7 , New Series, "Structure Data on Free Polyatomic Molecules"; Hellwege, K.H., Ed.; Springer-Verlag: West Berlin, 1976.

10. Data taken from: Sheppard, W.A.; Sharts, C.M. "Organic Fluorine Chemistry"; Benjamin, W.A.: New York, 1969.

11. Pickard, J.M.; Rodgers, A.S. J. Am. Chem. Soc. 1976, 98, 6115.

12. For review of polyfluoroalkene and polyfluorodiene cycloadditions, see: Chambers, R.D. "Fluorine in Organic Chemistry"; John Wiley and Sons: New York, p. 179-189.

13. Roberts, J.D.; Sharts, C.M. Org. React. 1962, 12, 1.

14. Epiotis, N.D. J. Am. Chem. Soc. 1972, 94, 1935.

15. Bartlett, P.D. Quart. Rev. 1970, 24, 473.

Chapter 20. The Stereochemical Consequences of Coulomb Polarization in Ground State Molecules.

In the original monograph,[1] we distinguished five different types of MOVB interaction matrix elements shown in Figure 1. In previous chapters, we have focused primary attention on the chemical implications of CT interaction and we have also seen what role bielectronic polarization plays in problems of molecular electronic structure, with the confines of VB and MOVB theory. Our intention now is to rationally design systems in which monoelectronic polarization and bielectronic correlation becomes as important as CT interaction and probe the stereochemical consequences of CW interaction brought about by the p_{ij} and W_{ij} interaction matrix elements. For reasons that will become apparent, we shall refer to bonding effected by the p_{ij} and W_{ij} matrix elements as coulomb polarization refering to the specific mechanisms as either p or W polarization. For pedagogical reasons, we develop the theory of p and W polarization using elementary systems from which we ultimately generalize to the polyelectronic species. As we shall see, coulomb polarization becomes important when the constituent atoms of a system are weak overlap binders. Thus, this is yet another dimension of the general problem of weak binding first discussed in chapters 1 and 2.

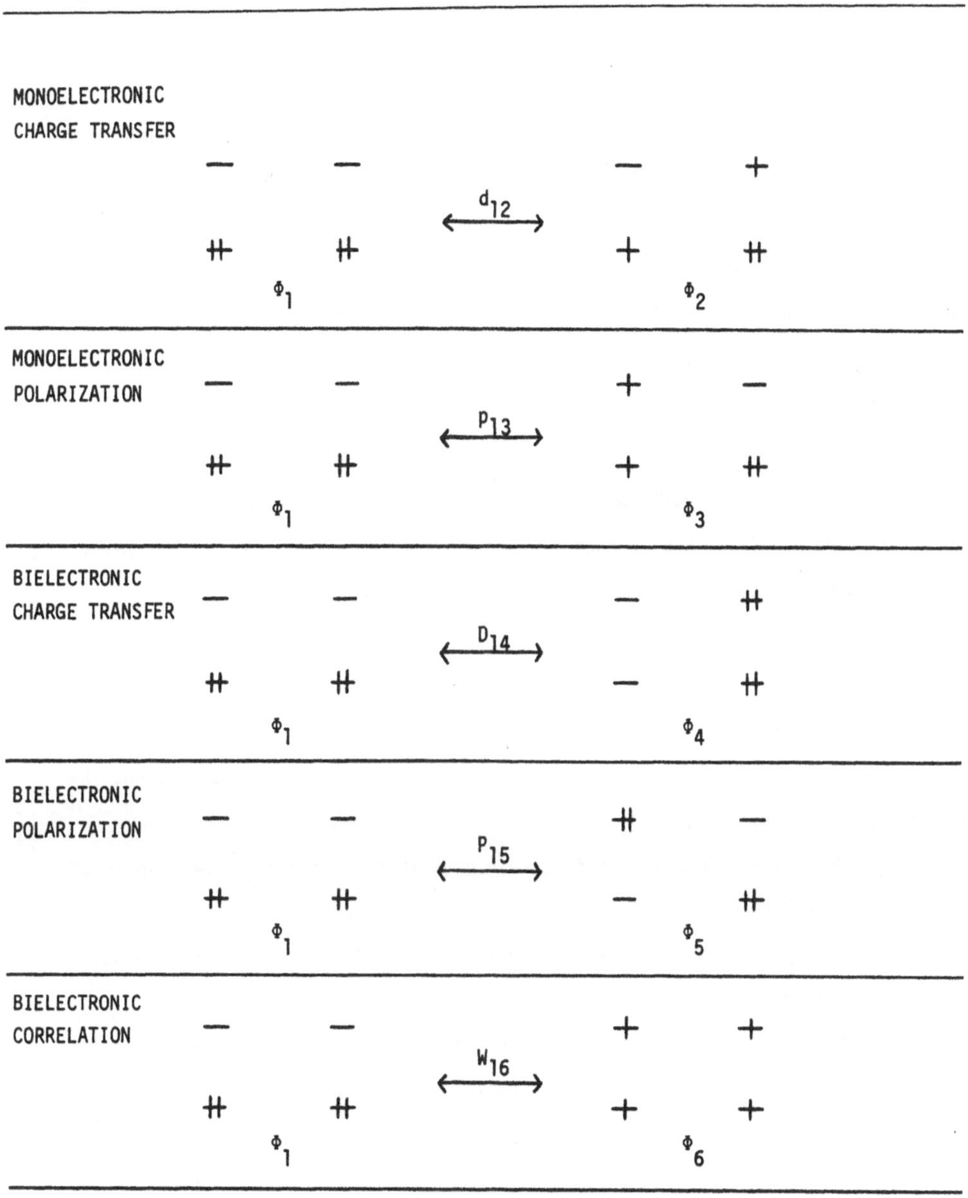
MONOELECTRONIC CHARGE TRANSFER
d_{12}
Φ_1
Φ_2
MONOELECTRONIC POLARIZATION
p_{13}
Φ_1
Φ_3
BIELECTRONIC CHARGE TRANSFER
D_{14}
Φ_1
Φ_4
BIELECTRONIC POLARIZATION
P_{15}
Φ_1
Φ_5
BIELECTRONIC CORRELATION
W_{16}
Φ_1
Φ_6

Figure 1. Key interaction matrix elements of MOVB theory.

I. Theory

The design of a four-electron-three-orbital (4e-3o) system in which p polarization plays a dominant role is immediately obvious by inspection of the wavefunction of this system, shown in Figure 2, and recognition of the following facts:

a. The d_{ij} elements describe the CT interaction of two CW's and this is a function of the overlap binding ability of the atoms which constitute the two fragments. By contrast, the p_{ij}'s describe a "classical" coulomb interaction of the two fragments and they are, to a first approximation, unrelated to the overlap binding ability of the atoms.

b. When Φ_3 is the dominant contributor to the total wavefunction, p polarization is turned off because Φ_3 cannot promote such an interaction. At the other extreme, when Φ_1 becomes the dominant contributor to the total wavefunction, p polarization becomes much more prominent since Φ_1 promotes such an interaction.

c. HMO (HMOVB) and EHMO (EHMOVB) theories contain the CT but not the p polarization interaction.

The strategy for giving prominence to monoelectronic polarization involves now the following two steps.

a. Identify core and ligand fragments such that the direction of primary CT is from core to ligands so that the Φ_1 CW (i.e., the I CW) becomes a major contributor to the total wavefunction. The Φ_1 CW is the lowest energy CW which promotes strong p polarization.

b. Identify core and ligand fragments which, in addition to meeting condition (a), are made up of weak overlap binding atoms. This has two consequences:

1) The d_{ij} matrix elements become small, in absolute magnitude.

2) The energy requirement for local fragment excitation is reduced.

As a result, the key CW interaction responsible for bonding is:

H-BOUND FOUR-ELECTRON-THREE-ORBITAL SYSTEM

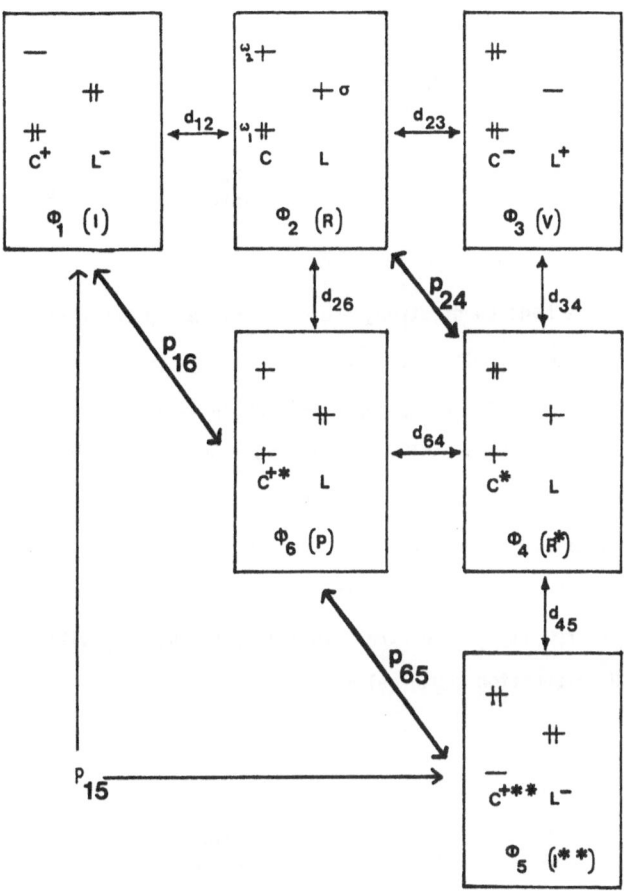

Figure 2. The various types of CW interactions in a four-electron-three-orbital system. Bielectronic charge transfer and "mixed" bielectronic interactions are omitted. It is important to note that monoelectronic polarization can operate only within an H- but not within a D- or U- bound system, by symmetry.

By contrast, if C and L are made up of strong overbinders, then polarization will play a secondary role and bonding will occur primarily through the strong CT interaction of all CW's, and, of course, through spin pairing in Φ_2 and Φ_4.

The analysis given above can be easily repeated for the case of a two-electron-three-orbital $\overset{\circ}{C}$-L system, where the circle on top of C represents a hole (unoccupied orbital), with the strategy for enforcing strong monoelectronic polarization differing from the one outlined above only to the extent that C and L should now be such as to promote CT from ligands to core. In this case, bonding will be primarily due to the CW interaction shown below.

We conclude that a core which has both electron pairs and holes can be polarized by electronegative and electropositive ligands, with the former "opening holes" to which electrons can be promoted and with the latter "creating pairs" which can be uncoupled by single excitations. Next, we consider in some detail the nature of p polarization.

Consider the p polarization interaction of the two CW's shown below:

The total Hamiltonian operator, $\hat{H}$, can be written as a sum of two fragment Hamiltonians, $\hat{H}_C$ and $\hat{H}_L$, plus a coupling term which describes the mutual interaction of the two fragments, $\hat{V}$. At infinite interfragmental distance, we have:

$$<\Phi_1|\hat{H}_C + \hat{H}_L + \hat{V}|\Phi_2> \;=\; <\Phi_1|\hat{H}_C+\hat{H}_L|\Phi_2> = 0 \;,\; \Gamma(\omega_1) \;\neq\; \Gamma(\omega_2)$$

At equilibrium interfragmental distance, we have:

$$<\Phi_1|\hat{H}_C + \hat{H}_L + \hat{V}|\Phi_2> \;=\; <\Phi_1|\hat{V}|\Phi_2> \;\neq\; 0$$

The physical significance of this integral is that it allows the mixing of Φ_1 and Φ_2 so that the resulting electron distribution of the core optimally interacts, in a "classical" sense, with the existing electron distribution of the ligands. In order to find, the core electron distribution, all we have to do is add the wavefunctions

of C and C^*. For example, if C is π^2 ethylene, C^* $\pi\pi^*$ ethylene, and L a doubly
occupied carbon π AO, we have:

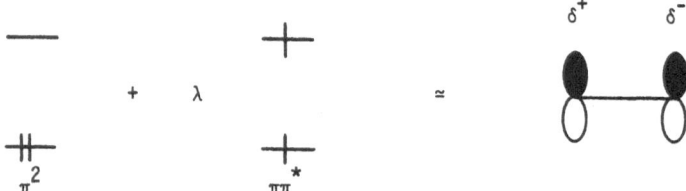

The resulting total wavefunction generated by the polarization interaction of
Φ_1 and Φ_2 can then be described pictorially by placing the ligand fragment in such
a way as to generate core-ligand coulomb attraction.

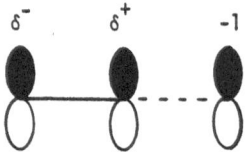

The pictorial representations of charge distributions shown above are obtained by
simply expanding the MO or MOVB CW's into VB CW's as illustrated before[1] (also,
<u>vide infra</u>).

An alternative approach which leads to the same conclusions is to simply write
down explicitly the p_{ij} interaction matrix element and assess its physical significance. If $\hat{V}$ is split into a mono- ($\hat{V}_1$) and bi-electronic ($\hat{V}_2$) part, we have:

$$p_{12} = \langle\Phi_1|\hat{V}_1|\Phi_2\rangle + \langle\Phi_1|\hat{V}_2|\Phi_2\rangle \; \alpha \; \langle\omega_1|\hat{V}_1|\omega_2\rangle + \langle\Phi_1|\hat{V}_2|\Phi_2\rangle$$

This matrix element is large only when ω_1 and ω_2 are localized on the same AO's,
as **in pi ethylene**, and it is only in such an event that there can be significant
charge separation in one fragment which can lead to profitable coulomb interaction
with the other fragment.

Bielectronic correlation cannot occur in the model systems discussed before.
In order to "see" this bonding mechanism, we must go to a larger model system
such as the H-bound four-electron-four-orbital (4e-4o) system in which the various
CW's interact as shown in Figure 3. Because $\underline{W}$ polarization involves a simultaneous
promotion or demotion of two electrons, one in each fragment, such a mechanism
operates regardless of the type of bonding of the 4e-4o system, i.e., irrespective
of whether this is U-, H-, or D-bound. Furthermore, it becomes dominant when, again,
the core and ligand fragments contain weak overlap binders. In such systems,
bonding is due primarily to the CW interaction depicted in Figure 4a. At the
limit of zero overlap interaction between the two fragments, there is the
possibility of Van der Waals bonding, i.e., weak intermolecular bonding due predominantly to coulomb polarization as illustrated in Figure 4b.[2]

Let us now briefly examine the nature of $\underline{W}$ polarization by considering the
interaction of the two CW's shown below:

ω_2 — ω_2 +

— n_2 $\xrightarrow{W_{12}}$ + n_2

+ n_1 + n_1

ω_1 ω_1

C   L   C*   L*

Φ_1 Φ_2

Assuming that ω_1 and ω_2 are the bonding and antibonding orbitals of pi ethylene
and n_1 and n_2 the in- and out-of-phase combination of two 2p AO's placed on either
side of the pi ethylene as illustrated below, we can determine that the wavefunction

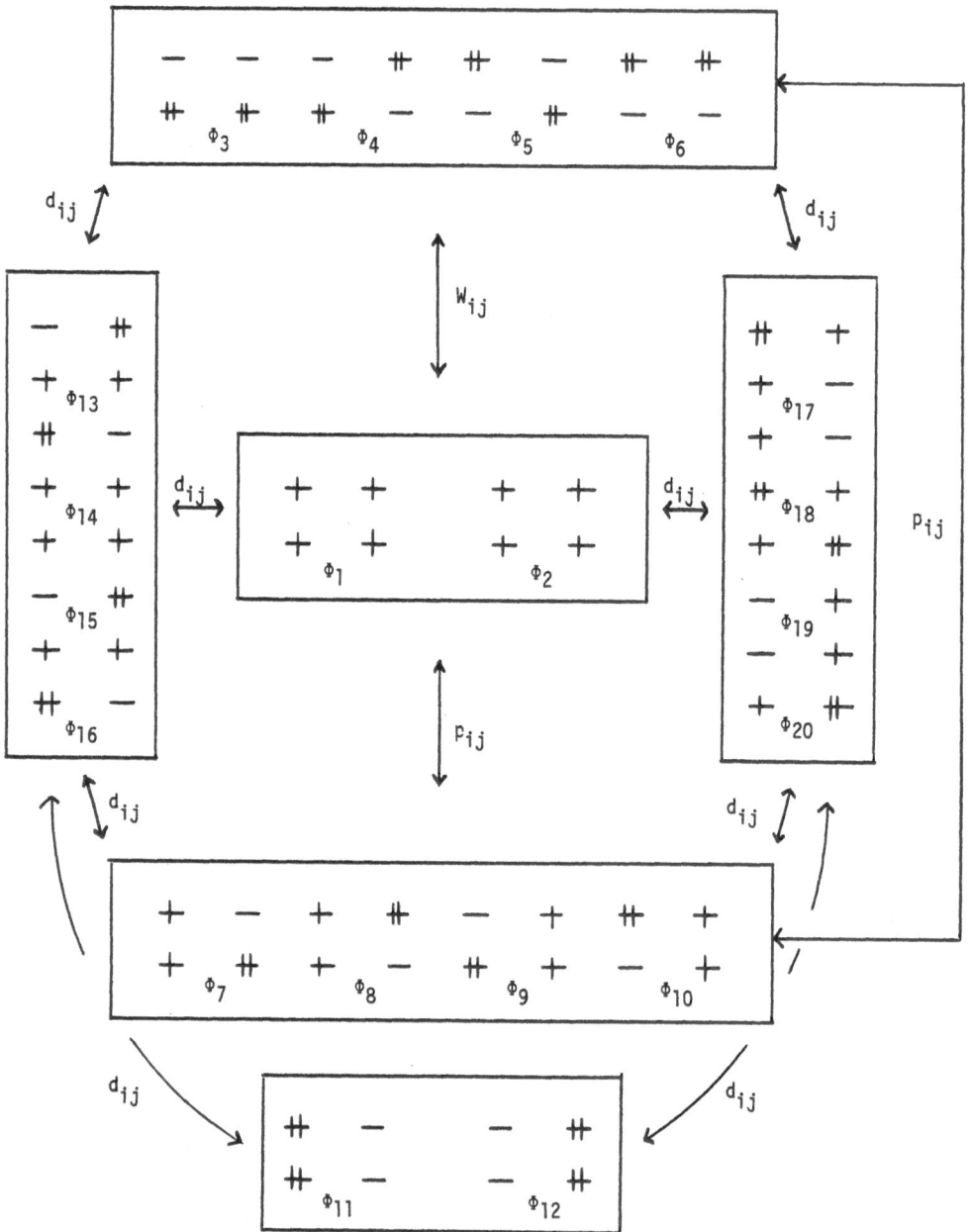

Figure 3. The various types of CW interactions in a four-electron-four-orbital system.

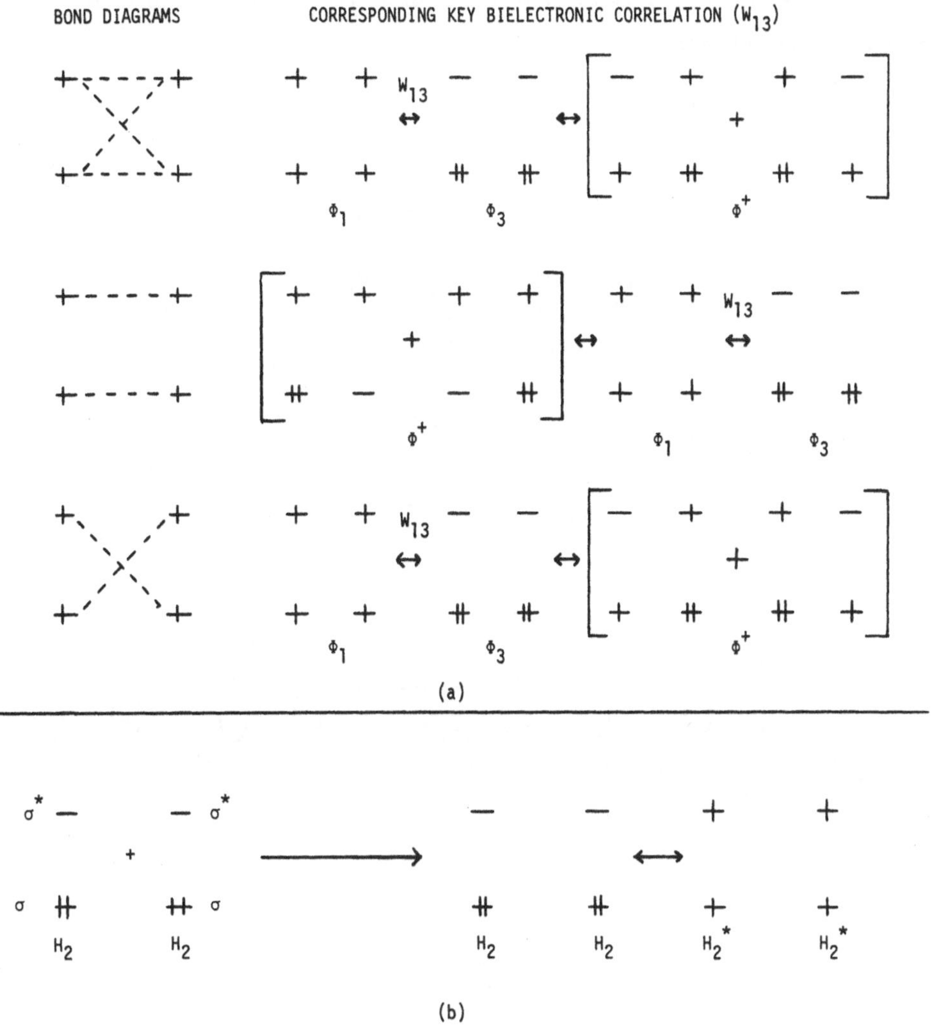

BOND DIAGRAMS   CORRESPONDING KEY BIELECTRONIC CORRELATION (W_{13})

(a)

(b)

Figure 4. a. Bielectronic correlation competing with charge transfer interaction in systems involving weak-core-ligand binding.
   b. Bielectronic correlation in Van der Waals supermolecules when the charge transfer interaction is zero due to long intermolecular distance.

resulting from the $\underline{W}$ polarization interaction of Φ_1 and Φ_2 can be pictorially represented as follows:

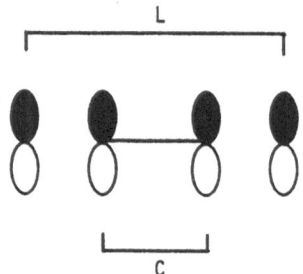

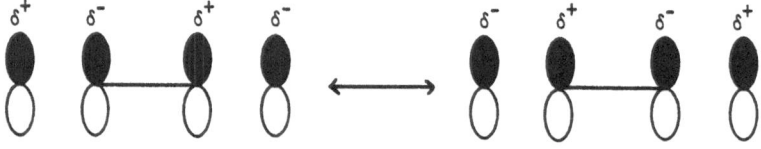

An alternative approach which leads to the same conclusions is to simply write down explicity the W_{12} interaction matrix element.

$$<\Phi_1|\hat{H}|\Phi_2> = <\Phi_1|\hat{V}_2|\Phi_2> \quad \alpha \quad <\omega_1(1)\omega_2(1)|n_1(2)n_2(2)>$$

Again, this matrix element is large only when ω_1 and ω_2 as well as n_1 and n_2 are localized on the same AO's, as in C and L, and only under such circumstances can we obtain interfragmental coulomb attraction through correlated charge alternation within each fragment.

Finally, we note that, much like p polarization, W polarization is not "contained" within EHMO theory. Furthermore, it is not "contained" within SCF-MO theory and it is only reproduced at the level of SCF-MO-CI theory in the case of U- and D-bound 4e-4o systems. For example, the complete description of a U-bound 4e-4o system in terms of MOVB bond diagrams is that shown below.

At the level of SCF-MO theory, the same system is described by Ξ_1', which is Ξ_1 with the coefficients of the various CW's being "screwed up". Hence, the bielectronic correlation interaction shown below is not contained in SCF-MO theory.

CW belonging to Ξ_1

CW belonging to Ξ_2

Furthermore, note that the U-bound 4e-4o system can be thought of as a composite of two subsystems and large polyelectronic molecules are made up of many different subsystems because of operative symmetry constraints. This means, that, in general, a large part of bielectronic correlation in molecules is reproduced only at the SCF-MO-CI (VB or MOVB) level, a situation fundamentally different from the one encountered before in the case of monoelectronic polarization.

Let us now consider an H-bound system in which both p and W polarizations can occur, and ask the question: How do the electronic properties of the core and ligands determine the relative importance of the two coulomb polarization mechanisms in a ground state molecule? By reference to the prototypical system shown in Figure 5, we can easily determine that W polarization will tend to be most important when the core and ligands are made up of atoms of comparable or the same electronegativity so that the dominant contributor is the perfect pairing CW which can promote W polarization. By contrast, p polarization will become dominant when the core and ligands are made up of atoms having different electronegativities and the dominant contributors are CT CW's which can promote only p polarization.

For illustrative purposes, we now investigate the stereochemical consequences of p polarization in systems which fulfill the following two conditions.

a. The atoms making up the core and the ligands have very different electro-negativities.

b. The core and/or the ligands are made up of weak overbinding atoms.
Under these conditions, we can assume that p polarization plays a more important role than W polarization.

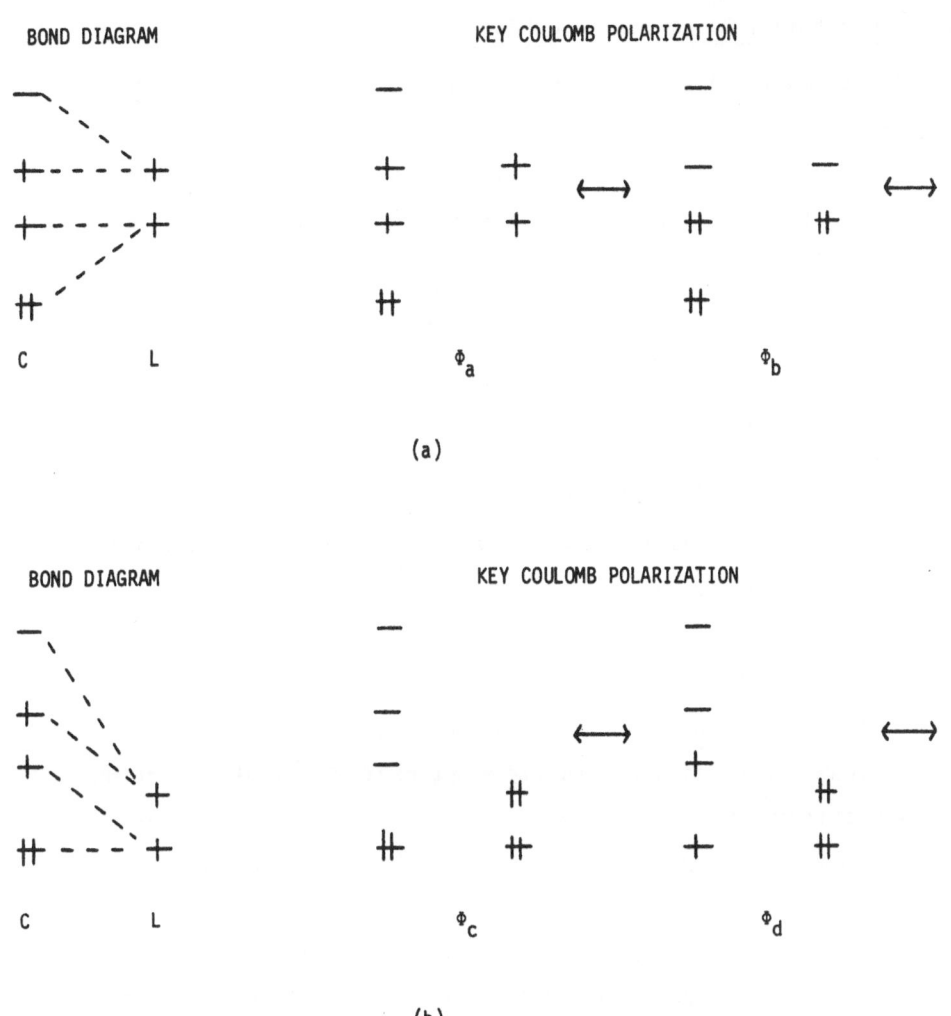

Figure 5. The way in which the difference in electronegativity of the core and ligand atoms determines the predominant type of coulomb polarization. In (a) bielectronic correlation is dominant. In (b) monoelectronic polarization plays the key role. In both cases, it is assumed that the resonance integrals of core and ligand AO's are relatively small.

II. Stereochemical Consequences of p Polarization

Let us take a look at a simple stereochemical problem: Structural isomerism
in hydrogen peroxide, known to exist preferentially in a gauche conformation. We
ask the question: Why is the gauche 1,2 isomer more stable than the hitherto
unobserved 1,1 isomer? A direct answer can be given by constructing the corres-

Gauche 1,2   1,1

ponding bond diagrams, shown in Figure 6, and noting the following:

a. Both isomers involve identical core-ligand bonding to the extent that, in
both, the two ligands are connected to a deexcited core in the principal resonance
contributor bond diagram.

b. The key difference between the two isomers is that the 1,1 isomer suffers
from greater natural Ligand Nonbonded Repulsion (LNP) due to the proximity of the
two hydrogens. Hence, the gauche must be much more stable than the 1,1 isomer,
as found to be the case.

c. Bonding due to p polarization is relatively unimportant because both O and
H are strong overlap binders when in combination.

d. EHMO theory correctly accounts for the greater stability of the gauche
isomer simply because p polarization effects, which are not "contained" in this
brand of theory, are relatively unimportant.

Let us now design a related molecule in which p polarization will be important.
Since we know that the overlap binding ability of an atom decreases down a column
and towards the left of a row of the Periodic Table, we immediately recognize that

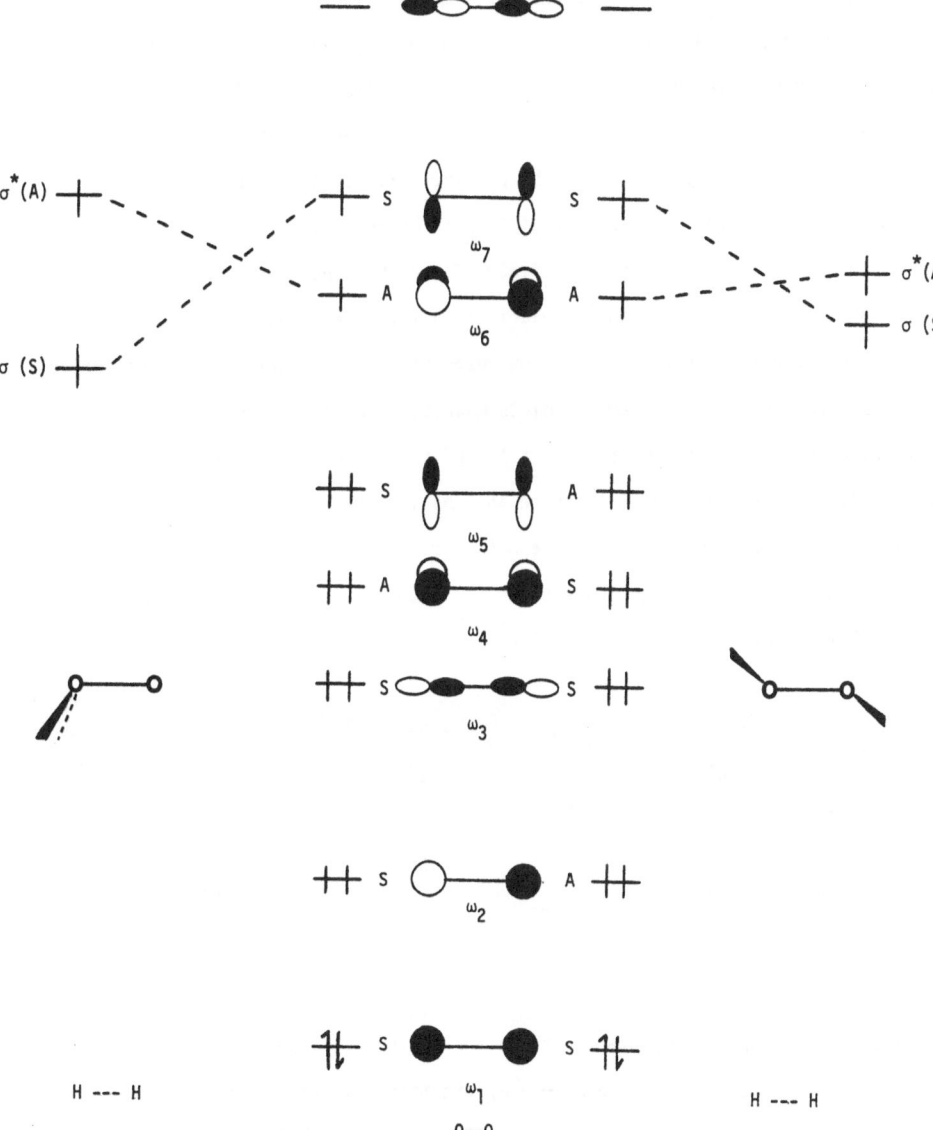

Figure 6. Compact bond diagrams of H_2OO and gauche HOOH.

replacement of the O_2 by an S_2 core and employment of electronegative, or, electro-positive ligands (for the purpose of properly directing primary CT) will produce a system in which $\underline{p}$ polarization will play a much more important role than in hydro-gen peroxide. The question now arises: Which of the two isomers of A_2X_2 (A=O, S, etc., X=monovalent ligand) is favored by $\underline{p}$ polarization? In order to answer this question, let us consider gauche and 1,1 S_2F_2 in which we treat F as a monovalent ligand neglecting its lone pairs. In Figure 7, we show the key CW interaction brought about by the monoelectronic polarization matrix elements. It is immediately apparent that the type of core excitation responsible for polarization is different in the two isomers and so are the absolute magnitudes of the corresponding p_{ij}'s. The question now is: Which of the two types of single core excitations creates stronger (coulombic) bonding of the core and the ligands? In order to answer this question, all we have to do is draw the wavefunctions which result from the $\underline{p}$ polarization interaction of the CW's shown in Figure 7 and determine which one involves better core-ligand coulomb attraction. The "pictures" shown below make evident that the 1,1 isomer is superior.

$$\overset{\delta^-}{F} \text{---} S \text{---} S \text{---} \overset{\delta^-}{F}$$

[actually F-S-S-F]

Let us see eactly why the $\omega_5 \longrightarrow \omega_7$ excitation in the 1,1 isomer produces strong core-ligand interaction, while the $\omega_5 \longrightarrow \omega_6$ excitation in the gauche isomer has no such effect, by expanding the prototype MOVB CW's shown below into VB CW's, recognizing that Φ_1, Φ_2, and Φ_3 simulate X_1, X_2, and X_3 of Figure 7.

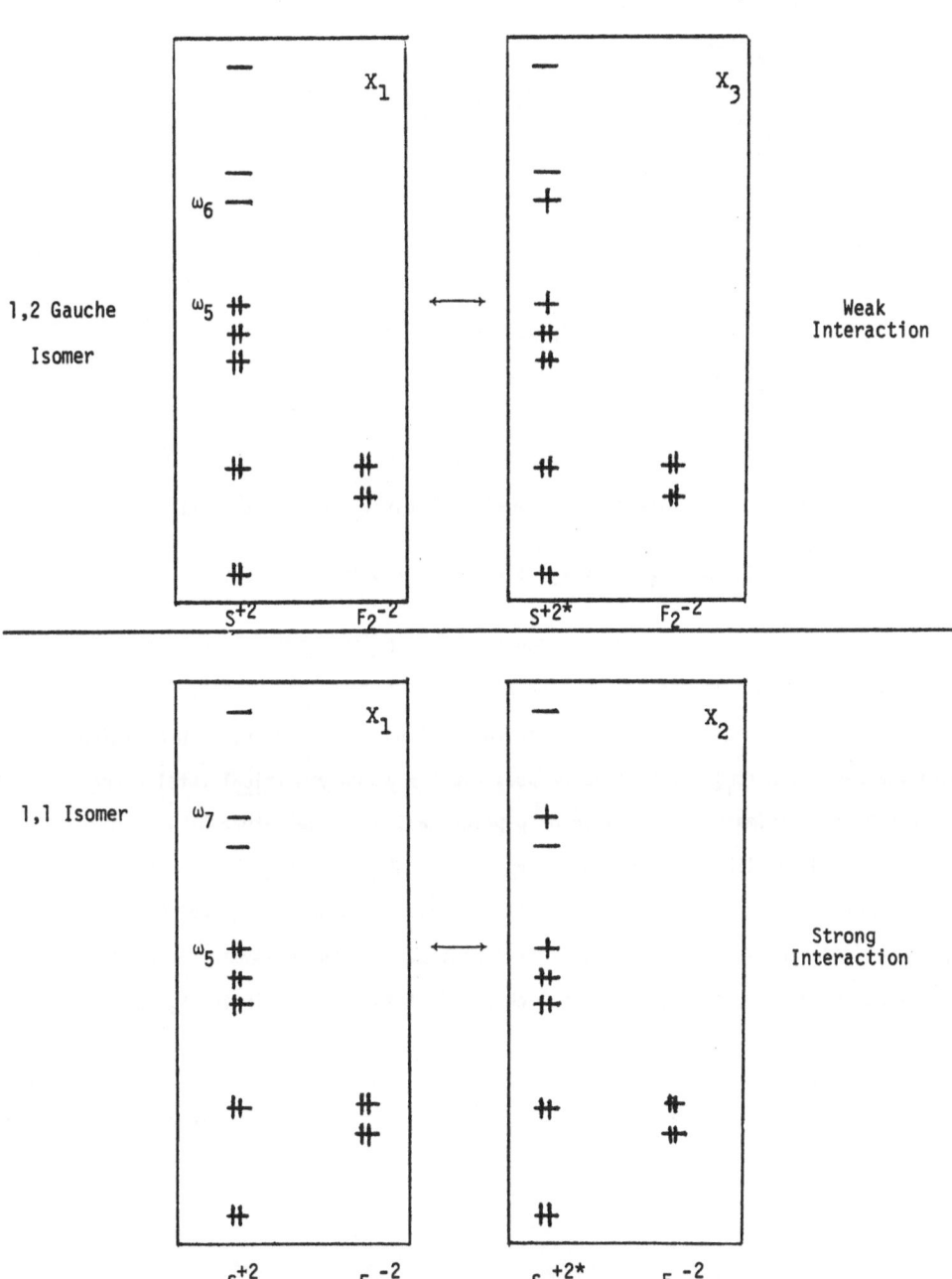

Figure 7. Monoelectronic polarization in 1,1 and gauche 1,2 S_2F_2.

$x_1 - x_2$ +

$$\Phi_2 \quad \alpha \quad (x_1\bar{x}_1 - x_2\bar{x}_2)$$

$x_1 + x_2$ +

Φ_2

$x_1 - x_2$ —

$$\Phi_1 \quad \alpha \quad (x_1\bar{x}_2 + x_2\bar{x}_1) + (x_1\bar{x}_1 + x_2\bar{x}_2)$$

$x_1 + x_2$ #

Φ_1

By adding, Φ_1 and $\lambda \Phi_2$ we obtain an unsymmetrical total change distribution:

$$\Phi_1 + \lambda \Phi_2 \quad \alpha \quad (x_1\bar{x}_2 + x_2\bar{x}_1) + \mu (x_1\bar{x}_1) + \nu (x_2\bar{x}_2) \quad \mu > \nu.$$

The reason why this occurs is that Φ_2 generates an <u>antisymmetric</u> charge distribution, i.e., it reduces to a linear combination of "ionic" structures which have opposite signs, so that, upon addition to a linear combination of "ionic" structures which have the same sign (Φ_1), it becomes responsible for an <u>unsymmetrical</u> total charge distribution. Ultimately, the cause of <u>p</u> polarization is the nature of Φ_2: <u>A singly excited CW with two electrons occupying two MO's spanning the same AO's</u>. By contrast, Φ_3 is a singly excited CW with two electrons occupying two MO's spanning different AO's. As a result, it cannot produce an antisymmetric charge distribution nor an unsymmetrical total charge distribution within the S_2 core.

$y_1 - y_2$ +

$$\Phi_3 \quad = \qquad \alpha \quad (x_1\bar{y}_1 + y_1\bar{x}_1) + (x_1\bar{y}_2 + y_2\bar{x}_1) + (x_2\bar{y}_1 + y_1\bar{x}_2) + (x_2\bar{y}_2 + y_2\bar{x}_2)$$

$x_1 + x_2$ +

In summary, the analysis presented above tells us that polarization is stronger in the 1,1 isomer and it gives rise to the following recipe: In the absence of core-ligand overlap binding differences, an A_2X_2 system isoelectronic to O_2H_2, or any other analogous system, will prefer a geometry which maximizes p polarization provided that A, X, or both are weak overlap binders so that CT bonding does not overwhelm polarization bonding and provided that the direction of primary CT is such as to render the ligands either electron rich or electron poor. Indeed, this is a recipe for charge transfer induced coulomb polarization.

A word of caution: p polarization can be brought about by single, double, etc., charge transfer starting from the perfect pairing CW which defines a neutral core and a neutral ligand fragment. We have arbitrarily chosen the $S_2^{+2}F_2^{-2}$ CW in order to illustrate our approach.

III. Stereochemical Consequences of <u>W</u> Polarization

For illustrative purposes, we now investigate the stereochemical consequences of <u>W</u> polarization in systems which fulfill the following two conditions:

a. The atoms making up the core and the ligands have similar electronegativities.

b. The core and/or the ligands are made up of weak overbinding atoms.

Under these conditions, we can assume that <u>W</u> polarization plays a more important role than <u>p</u> polarization. A molecule which fulfills both of the above conditions is H_2S_2. This molecules can exist in the 1,2 gauche and 1,1 geometries, much like H_2O_2, with the corresponding diagrams being entirely analogous to those shown in Figure 6 for gauche and 1,1 H_2O_2. Which of the two isomers is favored by <u>W</u> polarization?

Recall that the detailed bond diagram of the gauche isomer of H_2S_2 contains all possible R CW's such as the one shown in Figure 8a. The same is true of the detailed bond diagram of the 1,1 isomer (Figure 8b). Now, <u>W</u> polarization introduced by the lowest energy R CW's of the two H_2S_2 isomers can be symbolized as follows:

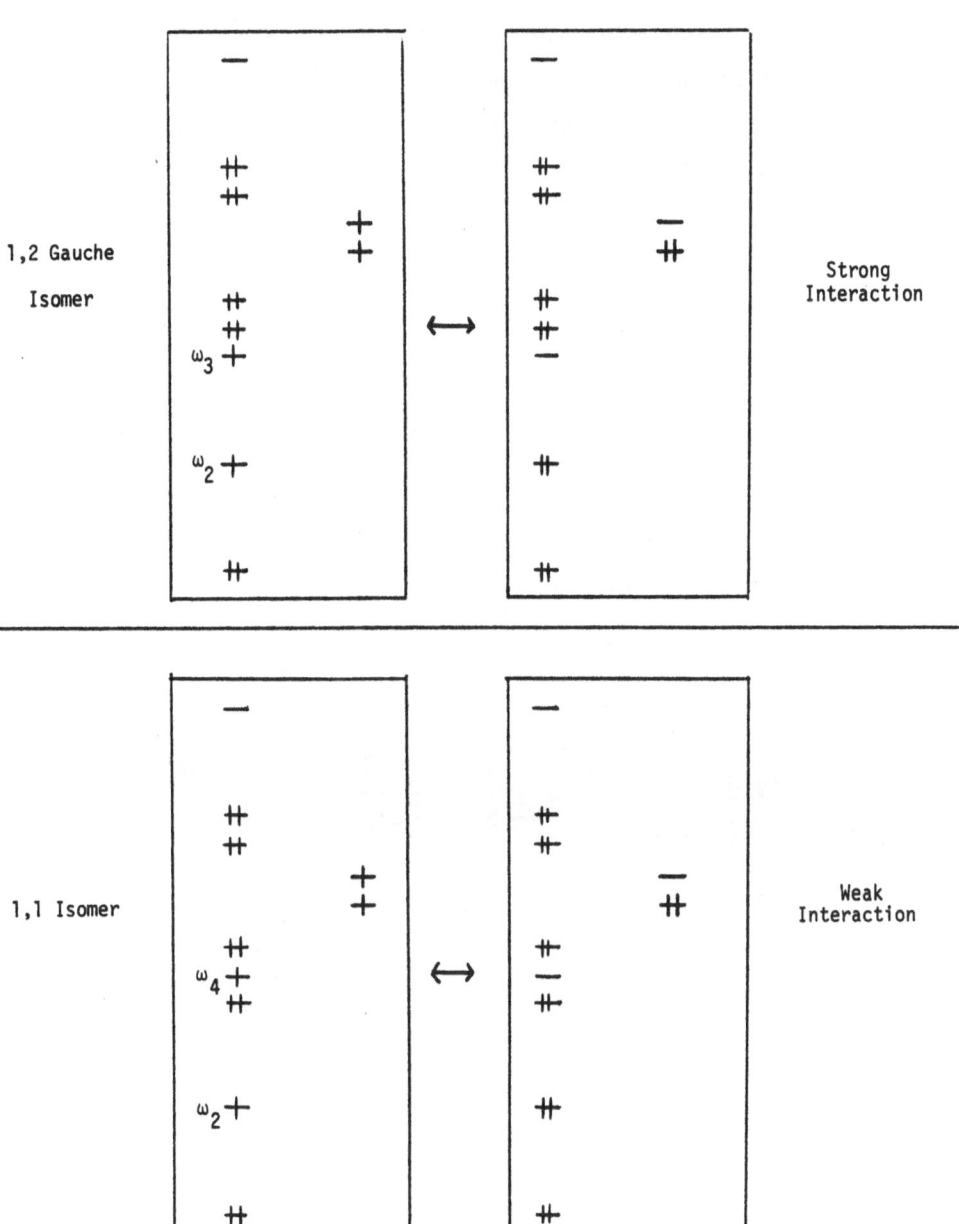

Figure 8. Bielectronic correlation in 1,1 and gauche 1,2 S_2H_2 brought about by higher energy perfect pairing CW's.

Since the core orbitals span different AO's in both isomers $W_a \simeq W'_a \simeq 0$. However, higher lying R CW's give an advantage to the gauche isomer because only in this species can one identify singly occupied core orbitals in these higher lying R CW's spanning the same AO's. A typical example is given below.

1,2
Isomer

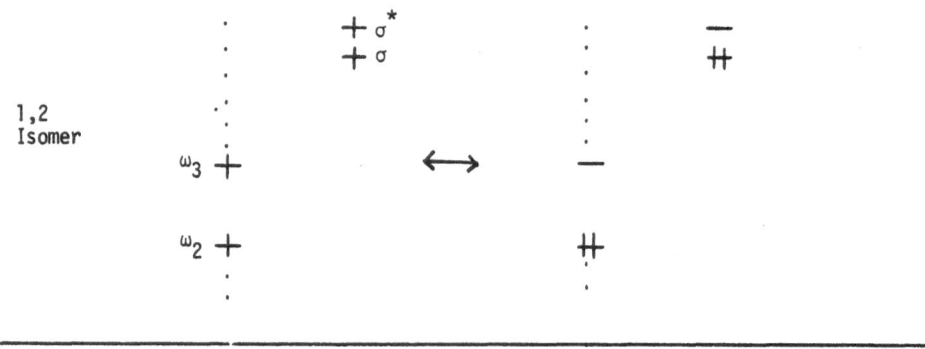

1,1
Isomer

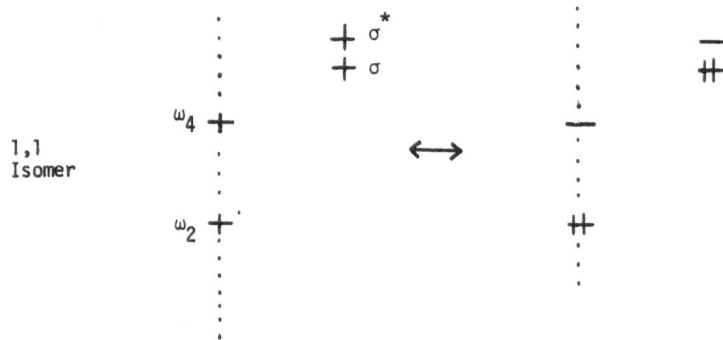

Thus, we conclude that $\underline{W}$ polarization in H_2S_2 will favor the gauche isomer. Further-more, this type of polarization will only be reproduced by polydeterminantal theory since it is not contained in monodeterminantal MO theory. Assuming that the R CW's make a dominant contribution to the total wavefunctions, the result of $\underline{W}$ polari-zation in the two isomers is as follows:

$$\overset{\delta^+}{H} - \overset{\delta^-}{S} - \overset{\delta^+}{S} - \overset{\delta^-}{H}$$

$$\begin{array}{c} \delta^+H \\ \searrow S - S \\ \delta^-H \nearrow \end{array}$$

$$\left[\text{actually} \quad \begin{array}{c} H \searrow \\ S - S \\ H \nearrow \end{array} \right]$$

Let us now see exactly why MO's spanning the same AO's guarantee large bielectronic correlation by reference to the prototypical CW interaction shown below.

$$\begin{array}{cccc} \phi^* - & - \psi^* & \quad\quad & + \quad + \\ \\ \phi \; \text{⧺} & \text{⧺} \; \psi & \quad\quad & + \quad + \\ C \quad L & & C^* \quad L^* \\ \\ \Phi_1 & & \Phi_2 \end{array}$$

First of all, we recognize that there are two independent singlet four open shell electron wavefunctions, ϕ_2' and ϕ_2'', such that ϕ_2' is essentially triplet C^* plus triplet L^* coupled into an overall singlet and ϕ_2'' is essentially singlet C^* plus singlet L^* also coupled into an overall singlet. ϕ_2' is the perfect pairing CW and ϕ_2'' is a higher energy CW which plays an important role in long range interaction, i.e., it can interact strongly with ϕ_1 via a W_{12} matrix element at long inter-fragmental distance. Thus, $\Phi_2 = \phi_2''$. In short, Φ_2 can be thought of as representing two interacting $^1C^*$ and $^1L^*$ fragments. By using the same arguments as before, we can show that charge separation within $^1C^*$ and $^1L^*$ can only occur if ϕ and ϕ^* as well as ψ and ψ^* span the same AO's in which case the $\Phi_1 - \Phi_2$ bielectronic correlation generates strong interfragmental coulomb attraction. If either ϕ and ϕ^* or ψ and ψ^* span different AO's, charge separation cannot be effected and bielectronic correlation due to the $\Phi_1 - \Phi_2$ interaction is very weak.

An alternative explanation which better exposes how antisymmetric charge distribution becomes responsible for bielectronic correlation in a manner exactly analogous to the way it becomes the source of monoelectronic polarization is to simply expand Φ_1 and Φ_2 into VB CW's. Assuming the geometry indicated below, we have:

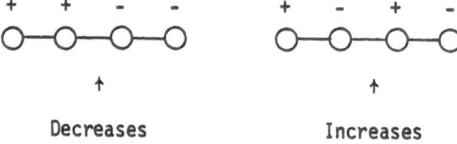

$$\Phi_1 \; \alpha \; - \; - \quad + (x_1\bar{x}_1y_1\bar{y}_1 + x_2\bar{x}_2y_2\bar{y}_2) + (y_1\bar{y}_1x_2\bar{x}_2 + x_1\bar{x}_1y_2\bar{y}_2)$$

$$\Phi_2 \; \alpha \; - \; - \quad - (x_1\bar{x}_1y_1\bar{y}_1 + x_2\bar{x}_2y_2\bar{y}_2) + (y_1\bar{y}_1x_2\bar{x}_2 + x_1\bar{x}_1y_2\bar{y}_2)$$

The result of adding $\lambda\Phi_2$ into Φ_1 is that unfavorable is replaced by favorable coulomb interaction as shown below.

+	+	-	-		+	-	+	-
O—O—O—O					O—O—O—O			

Decreases   Increases

If the MO's of C and L do not span the same AO's, the "ionic" terms vanish and the effect described above either becomes very small or it vanishes completely.

In summary, the analysis presented above tells us that $\underline{W}$ polarization is stronger in the 1,2 gauche isomer, i.e., it operates in a direction which is opposite to the one of $\underline{p}$ polarization. An A_2X_2 system isoelectronic to H_2O_2, or any other analogous system, will adopt a geometry which maximizes $\underline{W}$ polarization provided that A, X, or both are weak overlap binders and that A and X have similar electronegativities. When A and X are strong overbinders, overlap effects take over and cause geometrical preferences which are again compatible with maximal $\underline{W}$ polarization, in most cases. For example, 1,2 is favored over 1,1 H_2O_2 by overlap effects and $\underline{W}$ polarization. $\underline{p}$ and $\underline{W}$ polarization operate in opposite directions because each of the 1,2 and 1,1 isomers of A_2X_2 is made up of two subsystems, one of S and another of A symmetry (Figure 9) which have the following properties:

    a. In the gauche isomer, only fragment orbitals of different subsystems have common AO's.

    b. In the 1,1 isomer, only fragment orbitals of each one and the same subsystems have common AO's.

Since bielectronic correlation couples subsystems while pair or hole monoelectronic polarization operates within each subsystem, it follows immediately that the gauche isomer will be favored by $\underline{W}$ and the 1,1 isomer by $\underline{p}$ polarization since these mechanisms come strongly into play only if fragment orbitals spanning the same AO's are available. This is the first quantum chemical derivation of coulomb polarization selection in molecules.

    Finally, we open a parenthesis in order to point out that the types of interaction matrix elements shown in Figure 1 are the major (not all) determinants of ground state chemistry. Other interaction matrix elements, such as the one shown below, play an important role in excited state chemistry. Thus, Z_{ij} plays the same role in excimer formation that W_{ij} plays in ground state van de Waals complex

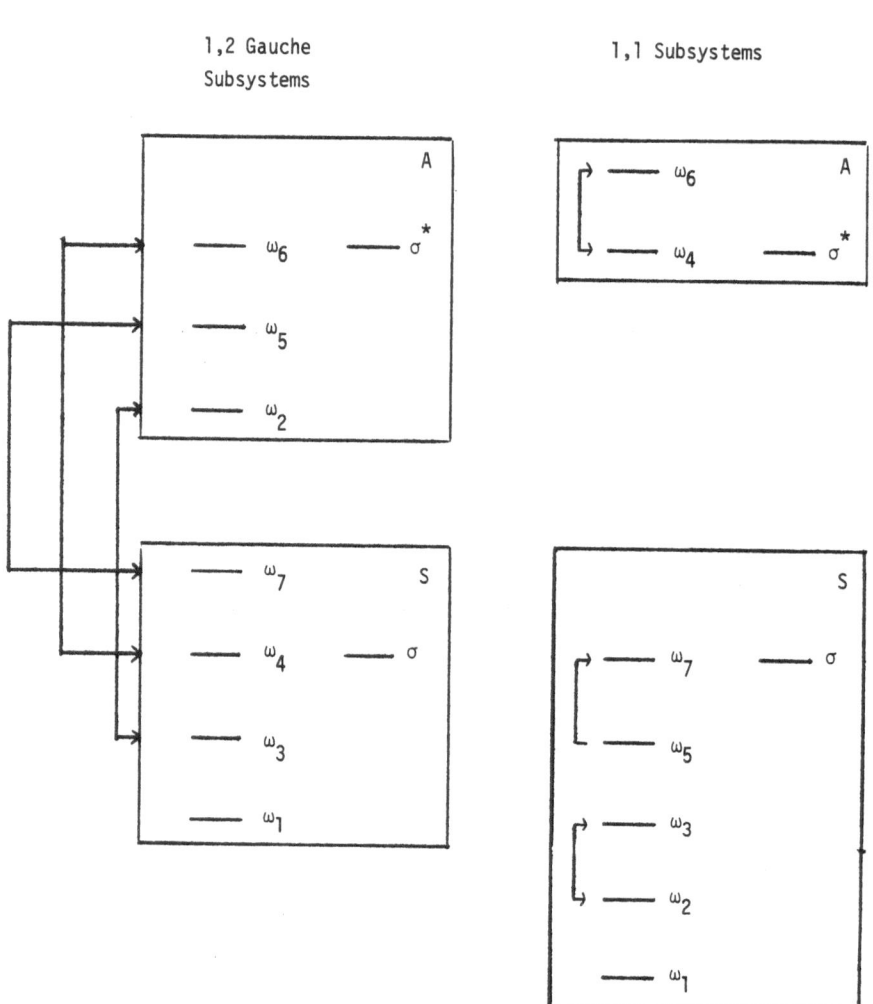

Figure 9. The orbital manifolds and the S and A subsystems of 1,1 and gauche 1,2 H_2O_2. Double arrows indicate MO's spanning the same AO's. This type of spanning occurs in an intersubsystem manner in 1,2 and an intrasubsystem way in 1,1 H_2O_2. As a result, bielectronic correlation is important only in the former and monoelectronic polarization plays a key role only in the latter.

formation, i.e., they both cause bielectronic correlation which results in coulomb attraction of neutral species.

$$
\begin{array}{ccc}
- \ + & & + \ - \\[4pt]
& \xleftarrow{\ \ Z_{12}\ \ } & \\[4pt]
{\text{\tiny ++}} \ + & & + \ {\text{\tiny ++}} \\[4pt]
\Phi_1 & & \Phi_2
\end{array}
$$

IV. The Rational Design of Geminal Molecules

An isomer in which ligands are symmetrically distributed at the two ends of a diatomic core can be called a vicinal isomer; if the ligands are unsymmetrically distributed, it can be called a geminal isomer.

```
                                     H
H - O - O - H                         \
                                       O - O
                                     /
                                    H
```

```
     VICINAL                      GEMINAL
     ISOMER                       ISOMER
```

Most molecules which are constituted of strong overbinders exist in isomeric forms with the vicinal isomers being always more stable. The general strategy for generating exceptions to this rule amounts to replacing strong by weak core overlap binding atoms, A, or strong by weak ligand overlap binding atoms, X, or both, so that significant interfragmental CT is directed from core to ligands, or vice versa. Specific predictions are given in Table 1, where A is a core atom of the same column of the Periodic Table as the corresponding atom on the left side of the Table and X is a ligand which has higher or lower electronegativity than A. Note, however, that A and X should be picked in such a way that their electronegativity difference is not so great that it takes the system near the limit of "pure" ionic binding, in which case the ligands will tend to avoid each other in order to minimize coulomb repulsion and the geminal will become destabilized relative to the vicinal isomer which places the ligands further away from each other. Table 1 defines an area of research which has yet to be explored by the synthetic chemist. Nonetheless, the following facts are known:

    a. H_2O_2 and F_2O_2 exist in gauche conformation (global ground state minimum) and no isomers of the 1,1 type (H_2OO and F_2OO) have yet been reported in the literature. By contrast, Kuczkowski[3] has identified by microwave and mass-spectrometric studies two stable isomers of F_2S_2 : FSSF and F_2SS; and he claimed that

Table 1. Strategies for Generation of Stable Geminal Molecules.

e.g., A = S
X = F, Li

e.g., A = P
X = F, Li

e.g., A = P
X = F, Li

e.g., A = Si
X = F, Li

e.g., A = Si
X = F, Li

this type of isomerism was the first known case in an S_2R_2 molecule. Subsequent SCF MO calculations by Hinchliffe[4] showed that H_2SS is as stable with respect to HSSH as F_2SS is to FSSF. We suggest that this will no longer be the case once these species are computed at the SCF-MO-CI level in which case the 1,2 isomer will be stabilized more than the 1,1 isomer and the effect will be more pronounced in H_2S_2 than F_2S_2. Calculations of the latter type will be useful not only for understanding the electronic structure of H_2O_2 and derivatives but also for testing the qualitative MOVB method.

b. Experimental data suggest that both XSSX and X_2SS (X = F, Cl, Br) exist with the 1,1 being more stable than the 1,2 isomer in F_2S_2 and with the order reversed in Cl_2S_2 and Br_2S_2.[5] This very nicely illustrates that charge transfer induced coulomb polarization, which stabilizes the 1,1 isomer, can occur only when the core and ligand atoms have substantially different electronegativities.

At this point, we interject some cautionary remarks:

a. $F_2C = CH_2$ is more stable than CHF = CHF not because of p polarization but for reasons explained before. $C_2H_2F_2$ is a system made up of strong overlap binding atoms and coulomb polarization effects are hardly expected to be as important as CT effects. This is confirmed by the fact that the greater stability of the 1,1 isomer is reproduced by EHMO computations.

b. $R_2O = O$ has never been isolated while $R_2S = O$ exists, again not because of p polarization. The reason is that the AO coefficient of the π^* and $\pi^{*'}$ MO's of the S - O core (isoelectronic to the O - O core) are such as to now encourage core-ligand binding which is stronger if the two R's are attached on S simply because the $\pi^* - \sigma$ and $\pi^{*'} - \sigma^*$ overlap integrals are larger in this geometry.

c. Organolithium molecules constitute another class of molecules in which one may observe preference for the geminal isomer due to the fact that Li is a very weak overlap binder. For example, it could very well be that the most stable structure of C_2Li_2 is not the rhomboidal but rather the geminal one.

$$\text{Li} - \text{C} \equiv \text{C} - \text{Li}$$

$$\begin{array}{c} \text{Li} \\ \diagdown \\ \text{C} = \text{C} \\ \diagup \\ \text{Li} \end{array}$$

$$\begin{array}{c} \diagup \text{Li} \diagdown \\ \text{C} = \text{C} \\ \diagdown \text{Li} \diagup \end{array}$$

Vicinal	Geminal	Rhomboidal

In fact, this would actually be the case, had we not come very close to the ionic limit at which the dominance of $[(C_2)^{-2}(Li_2)^{+2}]$ CW's causes a destabilization of the geminal isomer.

Coulomb polarization is the underlying cause of many interesting stereochemical trends and what we have already discussed is a mere isolated application of the concept. A related stereochemical problem is the question of whether a reaction complex, e.g., the four-electron-three-orbital (4/3) system XAX, is symmetrical or unsymmetrical assuming that the three centers lie on the same line. If we view the system as a composite of X_2 and A and we assume zero nonbonded AO overlap, the approximate wavefunction of the symmetrical species is as shown below, with Φ_i being defined as in Figure 2.

$$\Psi_S \simeq \lambda_1 \Phi_1 + \lambda_2 \Phi_2 + \lambda_3 \Phi_3$$

We distinguish two limiting cases:

a. X is much more electronegative than A. In this case, Φ_3 will tend to be the dominant CW and this is not connected by a p-type matrix element to any other CW of the complete set (see Figure 2). This means that a distortion of the system, i.e., the symmetrical $\longrightarrow$ unsymmetrical transformation, can have no stabilizing effect. Hence, XAX will be symmetrical so as to maximize core(A)-ligand(X_2) bonding. An example is FHF^-.

b. X is much more electropositive than A. In this case, Φ_1 will tend to be the dominant CW and this is now connected by a p-type matrix element to Φ_6

provided that there is distortion which removes the symmetry constraint that prevents the interaction of these two CW's in the symmetrical form. This means that the symmetrical $\longrightarrow$ unsymmetrical transformation will now have a stabilizing effect and XAX will be unsymmetrical so as to prevent "classical" coulomb repulsion.

Finally, it should be added that coulomb polarization is the reason why thiocarbanions ($-\overset{..}{\underset{|}{C}}-S-$) are more stable than oxocarbanions ($-\overset{..}{\underset{|}{C}}-O-$)[6] and why some transition metal ions form organometallic complexes (e.g., Ag^+ --- olefin).[7]

In closing, we can say that we were able to develop a recipe for predicting the stereochemical consequences of coulomb polarization because we realized three important things:

a. Monoelectronic polarization operates only within one H (not D or U) sub-system while bielectronic correlation couples two H, D, or U subsystems. This is strictly a consequence of operative orbital symmetry constraints.

b. Monoelectronic polarization involves excitation between two core or ligand orbitals of the same spatial symmetry while bielectronic correlation involves excitation between two core (and two ligand) orbitals having different spatial symmetry.

c. The geometry of a system determines whether core and ligand MO's spanning the same AO's have the same or different spatial symmetry.

I end with a prophetic statement, to be explained in a separate paper: A calculational method which artificially magnifies the p_{ij}/d_{ij} ratio will unduly favor geometries characterized by H-bonding, i.e., low symmetry geometries.

References

1. Epiotis, N.D.; Larson, J.R.; Eaton, H. "Unified Valence Bond Theory of Electronic Structure" in Lecture Notes in Chemistry, Vol. 29; Springer-Verlag: New York and Berlin, 1982.

2. For important theoretical studies which make essential use of MOVB theory for the treatment of intermolecular interactions, see:

    (a) Murrell, J.N. "Orbital Theories of Molecules and Solids", March, N.H., Ed.; Clarendon Press: Oxford, 1974.

    (b) Kitaura, K.; Morokuma, K. Int. J. Quantum Chem. 1976, 10, 325.

3. Kuczkowski, R.L. J. Am. Chem. Soc. 1963, 85, 3047; ibid. 1964, 86, 3617.

4. Hinchliffe, A. J. Mol. Struct. 1979, 55, 127.

5. Steele, F. "Lower Sulfur Fluorides" in Advan. Inorg. Chem. Radiochem. 1974, 16, 297.

6. Streitwieser, Jr., A.; Williams, Jr., J.E. J. Am. Chem. Soc. 1975, 97, 190, 191.

7. Basch, H. J. Chem. Phys. 1972, 56, 441.

Chapter 21. The Qualitative Rationalization and Prediction of "Correlation Effects" in "Complex" Ground State Molecules.

One of the great advantages of MOVB theory is that it forms the basis for a logical and coherent interpretation of chemical phenomena. The situation is quite different in the case of MO theory: Within this framework, one typically solves the problem at the SCF-MO level and then corrects the solution by re-solving it at the SCF-MO-CI level. In the process, one generates two apparently distinct conceptual frameworks with the result that the appearance is created that there are "MO effects" and "CI effects". This is quite inappropriate since the "CI effects" are nothing else but consequences of the solution of the SCF-MO equations, unless by "CI effects" one implies the chemical consequences of nonvalence orbitals which are not included in the monodeterminantal calculation. Perusal of the vast theoretical literature reveals that this point has not been properly appreciated. We believe that this is due to the fact that there has been no con-ceptual tool capable of revealing the nature of error involved at the Single Determinant (SD) MO level and how it is linked to fundamental electronic mechanisms which are grossly reproduced by SD MO theory. With MOVB theory as our weapon, we now attempt to answer the following question: What is the meaning of the term, "valence correlation effect", or, more briefly, "correlation effect"?

Consider the four popular and commonplace brands of MO theory and their MOVB counterparts stated below. We now have the luxury to ascend the ladder of MOVB

MO Theory	Equivalent MOVB Theory
Hückel MO (HMO)	HMOVB
Extended Hückel MO (EHMO)	EHMOVB
SCF-MO	Constrained MOVB ("MOVB")
SCF-MO-CI	MOVB

theory, from HMOVB to MOVB, and in every transition (e.g., from HMOVB to EHMOVB) describe exactly how the removal of an approximation changes the nature of bonding in a target system. Using the equivalence relationships, we can then state in MOVB language the deficiencies of different brands of MO theory in a way that is operationally meaningful.

What is the "error" committed at the level of SCF-MO theory, or, how does an MOVB-type wavefunction change as we make a transition from "MOVB" to MOVB theory? We can immediately answer this question by saying that the constraint approxima- tion made at the level of SCF-MO, or, equivalently, "MOVB" theory, causes the wavefunction to be unreasonably "ionic". This is rather well known and needs no further discussion.[1-3] For our purposes, we need to know that this error becomes increasingly important under the following two circumstances:

a. As the "covalent" character of a bond increases, e.g., as we go from NaCl to Na_2, or, Cl_2.

b. As a "covalent" bond becomes progressively weaker, e.g., as the H - H bond of H_2 is "stretched" and, ultimately, breaks.

Next, we recognize two important features of the MOVB bond diagram:

a. It projects the degree of "ionic" character of each core-ligand multi- center bond.

b. It projects the type of hybridization of core and ligand MO's involved in a given geometry. Hybridization is intimately connected with bond strengths. Hence, a bond diagram reveals in a direct way the strength of core-ligand bonds. Thus, it follows that, by revealing the types of core-ligand bonds of a molecule ("ionic" versus "covalent" and "weak covalent" versus "strong covalent"), its correct MOVB bond diagrammatic representation sets up the stage for identifying the mistake made at the "MOVB" or SCF-MO level of theory.

Let us consider, as an example, the structure of NH_3. The MOVB bond diagrams for planar and pyramidal ammonia are shown in Figure 1. A comparison of these two

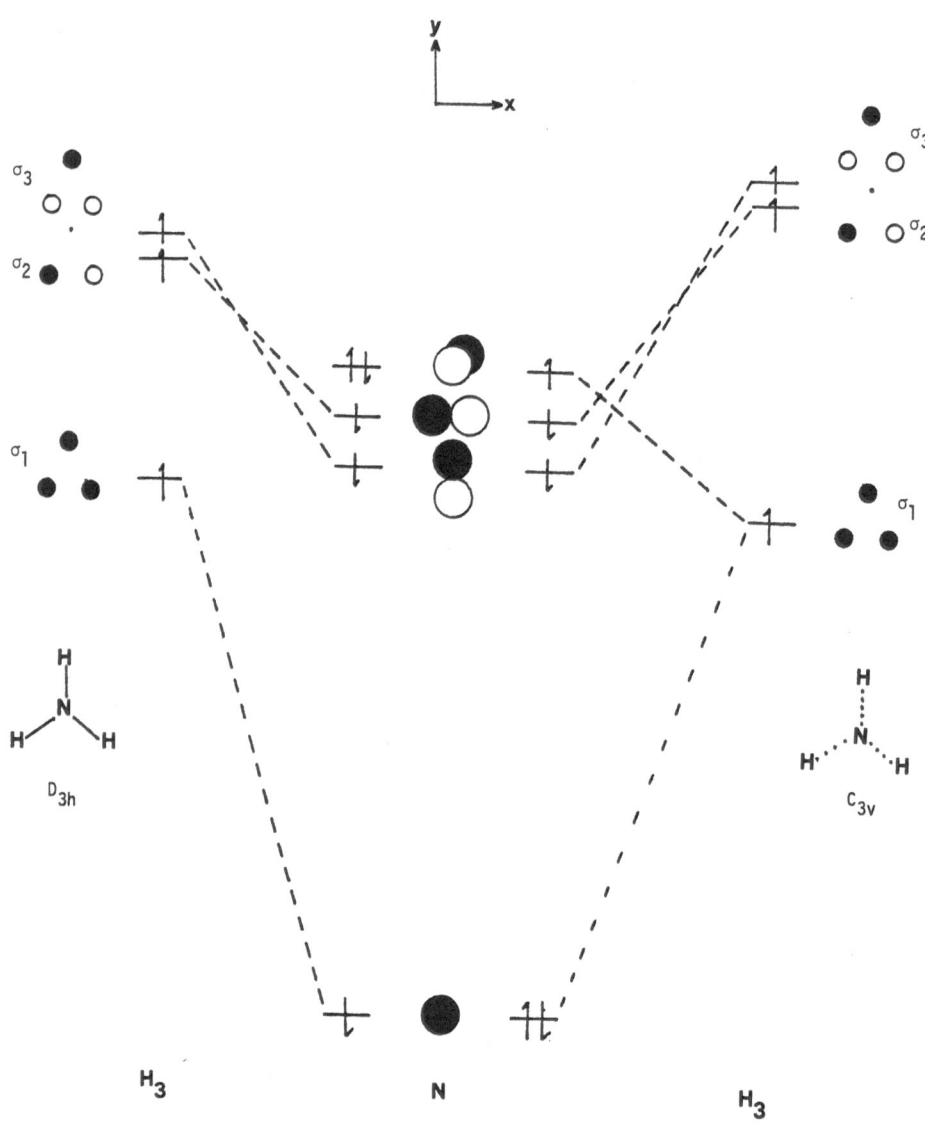

Figure 1. Detailed bond diagrams of planar (D_{3h}) and pyramidal (C_{3v}) NH_3.

bond diagrams using MOVB concepts is tantamount to a solution of the stereo-chemical problem of NH_3 and this has been discussed elsewhere. The same compari-son also revelas how removal of the error made at the SCF-MO level by recalcula-tion at the SCF-MO-CI level may favor one or the other of the two forms of NH_3. To put it crudely, these two bond diagrams can be used to answer the question: Do "correlation effects" favor the planar or the pyramidal form? The answer to this question is simple: All that one has to do is ascertain which of the two forms has more "covalent" and weaker bonds. Figure 1 provides the following information:

a. The $2p_x$ - σ_2 "covalent" bonds have comparable strengths, <u>to a first approximation</u>, in the two isomers.

b. The $2p_y$ - σ_3 "covalent" bond is <u>stronger</u> in the planar form due to greater spatial overlap.

c. The $2s$ - σ_1 bond of the planar species ought to be compared with a bond in the pyramidal species which is a hybrid of $2p_z$ - σ_1 and $2s$ - σ_1 bonds having principal $2p_z$ - σ_1 character. Because the $2s$ AO is a stronger overlap binder than the $2p$ AO of a first row atom, the $2s$ - σ_1 bond of D_{3h} NH_3 is more "ionic" as well as stronger than the (principal) $2p_z$ - σ_1 bond of C_{3v} NH_3.

We conclude that pyramidal NH_3 has more "covalent" and weaker bonds than planar NH_3 and, thus, it will be favored by "correlation effects". In accordance with this analysis, it has been found that "correlation effects" may account for as much as half the inversion barrier.[4] The reader is reminded that SCF-MO theory is successful in dealing with this problem only if a "good" AO basis set is employed.[5] The proper rationalization of the geometry of NH_3 can now be stated as follows: Planar NH_3 is favored by spatial overlap while pyramidal NH_3 by core deexcitation with the latter tending to dominate the former factor because of the very large 2s - 2p energy gap of N. Because there exist two opposing factors, only a "perfect" SCF-MO-CI, MOVB, or, VB, computation has quantitative meaning and it can inform us as to which side the balance of opposing effects has been tilted, although it must be said that oftentimes SD MO theory gives answers which are not qualitatively (or even quantitatively) different from those obtained by polydeterminantal MO theory. Thus, the qualitative explanation of why NH_3 is pyramidal is founded on electronic mechanisms which are reproduced, albeit somewhat incorrectly, by SD-MO theory.

We now turn our attention to a more subtle and interesting problem which is impossible to approach intellectually using current qualitative theory and which becomes trivial when formulated in the language of MOVB theory. The riddle can be phrased as follows: Is the cis or the trans isomer of 1,2 difluoroethylene more stabilized by "correlation effects"? In a previous chapter, we analyzed the electronic structures of the two isomers (each formulated as C_2 core plus H_2F_2 ligands) and uncovered the following trends:

a. The cis isomer is sigma D-bound whereas the trans isomer is sigma U-bound. This means that the core-ligand sigma bonds are more "ionic" in the cis isomer. It follows that, if the lone pairs of the fluorines are excluded in a computation, the trans form should be lowered in energy relative to the cis form as a transition is made from SCF-MO to SCF-MO-CI theory. Indeed, this is precisely what happens as the data of Table 1 clearly reveal! Note that H is more positive and F more negative in the cis isomer exactly as predicted by the MOVB analysis.

b. The cis isomer enjoys greater pi fluorine lone pair delocalization, i.e., the cis isomer has a more "covalent" core-ligand pi bond. It follows that, if the sigma framework is excluded in a computation, the cis form will be lowered in energy relative to the trans form as a transition is made from SCF-MO to SCF-MO-CI theory.

Table 1. STO-3G SCF-MO-CI Computations of Cis and Trans CH_2H_2'.[a]

Isomer [b]	Rel. Energies (kcal/mol)	H' Atomic Charge	H Atomic Charge
cis $C_2H_2H_2'$	0.000	-0.250	+0.196
trans $C_2H_2H_2'$	1.356	-0.242	+0.188
cis $C_2H_2H_2'$	0.000 SCF-MO		
trans $C_2H_2H_2'$	1.531		

[a] H' is a pseudohydrogen with Z = 1.2 simulating F without lone pairs. CI includes all single and double excitations of the SCF-MO wavefunctions.

[b] Standard geometries were used throughout.

We conclude that the energy difference between the cis and trans forms of 1,2 difluoroethylene can increase or decrease as one shifts from monodeterminantal to polydeterminantal MO theory depending on whether the sigma or the pi mechanism of stabilization of the cis form is less or more accurately described at the SCF-MO level of theory. Recent studies by Cremer in Cologne[6] suggest that "correlation effects" favor the cis isomer. This finding, interpreted by MOVB theory, implies that monodeterminantal MO theory fails to "appreciate" the pi delocalization factor, _partly_ responsible for the greater stability of the cis isomer, as much as it does "appreciate" the sigma bonding factor, also responsible for the same trend. In any event, both factors are qualitatively reproduced at the monodeterminantal level, albeit incorrectly in a quantitative sense. Again the message is clear: "Correlation effects" are corrections of fundamental electronic mechanisms which are qualitatively describable by monodeterminantal MO theory.

Our conclusion then is: In most ground stereochemical problems, monodeterminantal SCF-MO theory provides reasonable quantitative estimates of geometrical preferences. In addition, an analysis of the SCF-MO (or EHMO) wavefunction can reveal the reasons for stereoselection, i.e., the operative electronic mechanisms responsible for the preference for one geometry over another. These same mechanisms are responsible for differential CI corrections of the energies of isomers so that "CI effects", or, "correlation effects" are reflections of the electronic structures of molecules as grossly reproduced by monodeterminantal MO theory. To put it crudely, a _proper_ analysis of the electronic structure of molecules at the SD MO level automatically predicts the way in which "correlation effects" will come into play!

In order to further illustrate how bond "ionicity" controls the CI correction, we now consider the transition states of two different $2\pi + 2\pi$ cyclodimerizations of ethylene to form cyclobutane: One involving s + s and the other s + a union of the two reactants. In bond diagrammatic terms these two transition states can be symbolized as follows:[7]

a. s + s

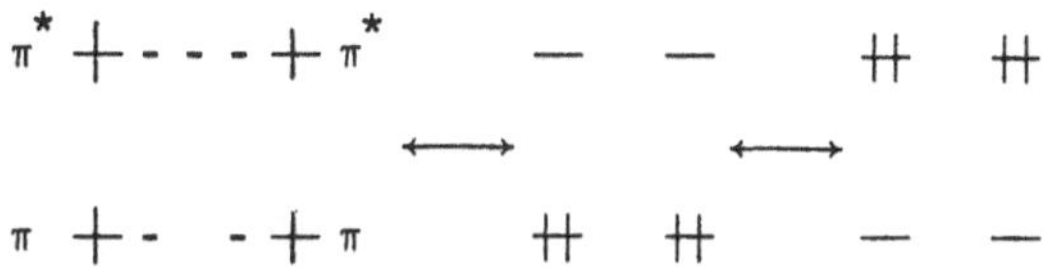

b. s + a

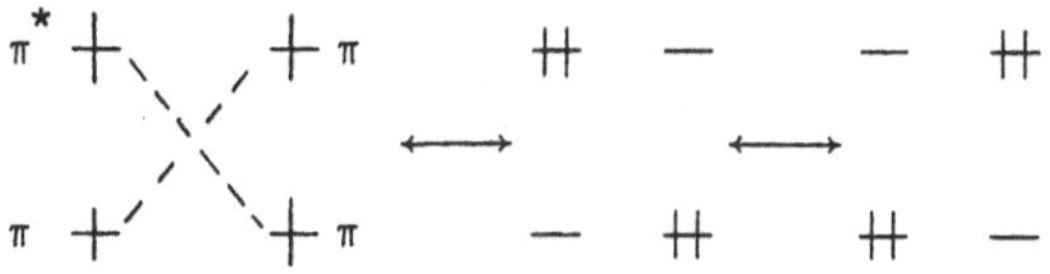

The first involves formation of two "covalent" while the second formation of two "ionic" bonds linking the two reactants. It follows that the former should be favored over the latter by "correlation effects" and this should be true with regards to all "antiaromatic" — "aromatic" transition state comparisons. The literature of computational quantum chemistry contains many suggestions that this is indeed the case.

The same argument can be restated in VB theoretical terms as follows: "Correlation effects" favor the molecular complex with the least favorable electron delocalization, i.e., the structure which involves inefficient mixing of "covalent" and "ionic" VB CW's. This can be exemplified by the lowest energy singlet and triplet states of a Hückel AO (HAO) and a Möbius AO (MAO) four-electron-four-orbital system, the NDO-VB wavefunctions of which are shown in Table 2.[7] Note that the lowest singlet HAO system involves poor delocatization as the four CW's shown below cannot, by symmetry, contribute to its wavefunction. By contrast, the lowest singlet MAO and lowest triplet HAO systems involve better delocalization, with X_1 - X_4 contributing to the wavefunctions of both species. We conclude

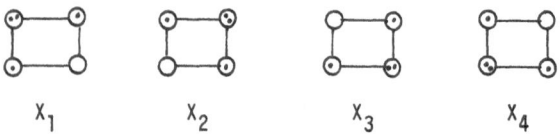

X_1 X_2 X_3 X_4

that "correlation effects" will favor singlet HAO relative to singlet MAO and triplet HAO due to greater "covalency".

The analyses presented above are based on the reasonable assumption that, since the isomers which are being compared do not exhibit vicinal-geminal dichotomy, bielectronic correlation effects (i.e., intrafragment "correlation effects") remain relatively constant. This is no longer true in comparing, e.g., gauche and 1,1 H_2S_2. In this case, "correlation effects", in the form of bielectronic correlation (a form of <u>coulomb polarization</u>), will favor the vicinal (gauche) isomer for reasons which we have explained before. The same thing will be true in the following comparisons:

$$\overset{H}{\underset{H}{\diagdown}}C \equiv C: \qquad \text{versus} \qquad \dot{C} \equiv C \overset{/H}{\underset{H}{\diagup}}$$

$$\diagdown C - \ddot{C} \diagdown \qquad \text{versus} \qquad \dot{C} = C.$$

Table 2

NDO VB WAVEFUNCTIONS OF FOUR-ELECTRON-FOUR-ORBITAL HÜCKEL (HAO) AND MÖBIUS (MAO) AO SYSTEMS

		E (eV)	$\uparrow\uparrow/\uparrow\uparrow$	(2)	(3)	(4)	(5)	(6)	(7)	(8)	(9)	(10)	(11)	(12)	(13)	(14)	(15)	(16)	(17)	(18)	(19)	$\cdot\!\Updownarrow/\Updownarrow\!\cdot$	$\Updownarrow\!\cdot/\cdot\!\Updownarrow$
HAO Singlet	Ψ_1	−13.7	.65	.38	⧄	−.23	.23	−.23	.23	.23	−.23	.23	−.23	0	0	0	0	0	0	0	0	0	0
HAO Singlet	Ψ_2	−10.8	−.29	.51	⧄	.24	.24	.24	.24	.24	.24	.24	.24	0	0	0	0	.10	.10	.10	.10	.28	.28
HAO Singlet	Ψ_3	−9.3	0	0	⧄	−.26	.26	.26	.26	−.26	.26	−.26	−.26	.25	−.25	.25	−.25	0	0	0	0	.33	−.33
HAO Triplet	Ω	−12.9	.47	.27	−.38	−.23	.23	−.23	.23	.23	−.23	.23	−.23	.19	.19	−.19	−.19	⧄	⧄	⧄	⧄	⧄	⧄
MAO Singlet	χ	−19.3	.50	.29	⧄	.24	.24	−.24	.24	.24	−.24	.24	−.24	.15	.15	−.15	−.15	0	0	0	0	.21	.21
MAO Triplet	Φ	−9.9	.57	.33	−.46	−.21	−.21	.21	−.21	−.21	.21	−.21	−.21	0	0	0	0	⧄	⧄	⧄	⧄	⧄	⧄

In fact it is tempting to speculate that even the more strongly bound linear C_2H_2
and planar C_2H_4 systems will be favored by bielectronic correlation relative to
their carbenoid analogues.

Again, as an illustration of the utility of qualitative theory in appraising
the reliability of quantum chemical computations, let us consider the structure of
$CH_2 = CH_2$. As we have discussed before, linear ($D_{\alpha h}$) C_2H_2 and planar (D_{2h}) C_2H_4
are U-bound molecules with trans-bent (C_{2h}) C_2H_2 and trans-bent (C_{2h}) C_2H_4 being
their deexcited $H^{\ddagger}$-bound analogues. It is only because core deexcitation is not
large enough to counteract loss of spatial overlap that C_2H_2 is $D_{\alpha h}$ and C_2H_4 is D_{2h}.
However, when C is replaced by a weaker overlap binder, the deexcitation factor
becomes relatively more important and MOVB theory predicts that, e.g., Sn_2H_4 will
tend to adopt a trans-bent rather than planar geometry. Indeed, experiment shows
that Sn_2H_4 <u>does</u> have a C_{2h} structure.[8]   Now, an excellent opportunity for
arriving at a <u>wrong</u> prediction is afforded by monodeterminantal SCF-MO theory
which will tend to project $H_3Sn - \overset{\cdot\cdot}{S}nH$ as the most stable structure simply because
bielectronic polarization which favors the $H_2Sn \equiv SnH_2$ isomers is not "contained"
in the theory. This prediction, which was made long before we became aware of
the studies of Poirier and Goddard and those of Trinquier, et al. referred to in
Table 3, has been confirmed by the aforementioned studies, the results of which
are shown in Table 3.

In summary, we have predicted "correlation effects" (in MO theory) by
focusing attention on the principal complete bond diagrams (e.g., Ψ_1 in Scheme 1
on p. 4) of two isomers whenever these describe fundamentally different core-
ligand binding (e.g., D_{3h} versus C_{3v} NH_3, cis versus trans $C_2H_2F_2$, and HAO versus
MAO A_4), or, by considering higher energy complete bond diagrams (e.g., Ψ_2 in

Table 3. $E(HX-XH_3)-E(H_2X=XH_2)$ at Different Levels of Theory (in kcal/mol).

X	Basis Set	SCF-MO	SCF-MO-CI	Ref.
Si	3-21 G	- 0.1	+ 10.1	a
	4-31 G	- 8.2		b
	66-31 G	- 2.1		a
Ge	Double ʃ + d	- 8.4	+ 4.6	c

a. Poirier, R.A.; Goddard, J.D. Chem. Phys. Lett. 1981, 80, 37.

b. Snyder, L.C.; Wasserman, Z.R. J. Am. Chem. Soc. 1979, 101, 5222.

c. Trinquier, G.; Malrieu, J.-P.; Riviere, P. J. Am. Chem. Soc. 1982, 104, 4529.

Scheme 1 on p. 4) whenever the principal complete bond diagrams describe similar core-ligand binding (e.g., trans-bent $CH_2=CH_2$ versus $CH_3-\ddot{C}H$). It goes without saying that much will have to be done before we know whether the rationale presented in this chapter amounts to a general procedure for qualitatively predicting "correlation effects".

References

1. Monodeterminantal SCF-MO theory is unsuccessful in treating weak homopolar
    bonds but relatively successful in dealing with "ionic" bonds which can be
    qualitatively described by one "ionic" resonance structure. For review of
    pertinent computational data, see: Schaefer, III, H.F. "The Electronic
    Structure of Atoms and Molecules"; Addison-Wesley: Reading, MA, 1972, p.
    153-160.

2. The manner in which CI corrects the deficiency of the monodeterminantal MO
    wavefunction is often illustrated by reference to the simple example of a two
    electron-two orbital system, e.g., H_2, pi ethylene, etc. Elementary discus-
    sions of this type can be found in a number of elementary texts. See, inter
    alia: Borden, W.T. "Modern Molecular Orbital Theory for Organic Chemists";
    Prentice-Hall: Englewood Cliffs, NJ, 1975. The following are some of the
    methodologies used for obtaining correlated wavefunctions:

    (a) Many Body Perturbation Theory: Paldus, J.: Cizek, J. Advan. Quant.
         Chem. 1975, 9, 105 and references therein. Pople, J.A.; Binkley, J.S.;
         Seeger, R. Int. J. Quant. Chem. 1976, 510, 1.

    (b) Cluster Expansions: Sinanoglou, O. J. Chem. Phys. 1962, 36, 706.

    (c) Second Order Bethe-Goldstone Method: Nesbet, R.K. Adv. Chem. Phys. 1969,
         14, 1.

    (d) Independent Electron Pair Approximation: Ahlrichs, R.; Lischka, H.;
         Staemmler, V.; Kutzelnigg, W. J. Chem. Phys. 1975, 12, 1225 and references
         therein [Note: This method is related to those of (b) and (c) above].

    (e) Coupled Electron Pair Approximation: Meyer, W. Int. J. Quant. Chem.
         1971, 55, 341.

3. The near cancellation of "correlation effects" in some chemical processes has been noted early in the following works:

(a) Nesbet, R.K. J. Chem, Phys. 1962, 36, 1518.

(b) McLean, A.D. J. Chem. Phys. 1963, 39, 2653.

(c) Nesbet, R.K. Advan. Chem. Phys. 1965, 9, 321.

In this work, we have outlined a methodology for identifying the molecular isomer which is differentially **favored** by "correlation effects". This type of information is useful particularly when the SCF-MO energy differences are smaller.

4. (a) Pipano, A.; Gilman, R.D.; Bender, C.F.; Shavitt, I. Chem. Phys. Letters 1970, 4, 583.

(b) Pipano, A.; Gilman, R.D.; Shavitt, I. Chem. Phys. Letters 1970, 5, 285.

5. (a) Body, R.G.; McClure, D.S.; Clementi, E. J. Chem. Phys. 1968, 49, 4916.

(b) Rauk, A.; Allen, L.C.; Clementi, E. J. Chem. Phys. 1970, 52, 4133.

(c) Stevens, R.M. J. Chem. Phys. 1971, 55, 1725.

6. Cremer, D. Chem. Phys. Letters 1981, 81, 481.

7. Epiotis, N.D.: Larson, J.R.; Eaton, H. "Unified Valence Bond Theory of Electronic Structure" in Lecture Notes in Chemistry, Vol. 29; Springer-Verlag: New York and Berlin, 1982.

8. (a) Lappert, M.F. in "Inorganic Compounds with Unusual Properties"; King, R.B., Ed.; Advances in Chemistry Series, No. 150; American Chemical Society: Washington, DC, 1976.

(b) Davidson, P.J.; Harris, D.H.; Lappert, M.F. J. Chem. Soc. Dalton 1976 2268.

EPILOGUE

As the amount of experimental research reported in the literature continually increases, as the number os scientific journals grows unabated, and as theory becomes more and more sophisticated because of continual forward strides in computer technology, experimentalists can no longer afford to use primitive concepts in their daily planning of syntheses and their quests of new mechanisms as well as in communications with collegues and educating students. For the same reasons, theoreticians can no longer pretend to interpret the results of highly complex computations through usage of elementary qualitative MO concepts. This two part work attempts to establish a common language for experimentalists and theoreticians alike for discussing, analyzing, and resolving chemical problems and for rationally designing "new chemistry". This new language makes principal use of the MOVB bond diagrammatic representation of the electronic structure of molecules. Our goal has been to teach this new formalism so that the reader can come to the point that, by a few strokes of the pen, i.e., by the construction of one or more bond diagrams, he can resolve controversial issues, rectify mistaken impressions, and predict new phenomena, without the need of explicit computations and without the necessity of lengthy discussions and clarifications. As a final illustration of "how bond diagrams speak for themselves", consider the following problems and their MOVB resolution:

a. NCO^- and NCS^- are ambident species which can find Z^+ by either of the two terminal atoms. Using many-fragment MOVB theory, the bond diagrammatic representation of NCO^- is shown in Figure 1. A major contributor CW, Φ, is the one which is common to Ξ_1 and Ξ_2 and which describes ground triplet N^- and C and low lying singlet O (Figure 2a). Recognizing that N and O have roughly equal

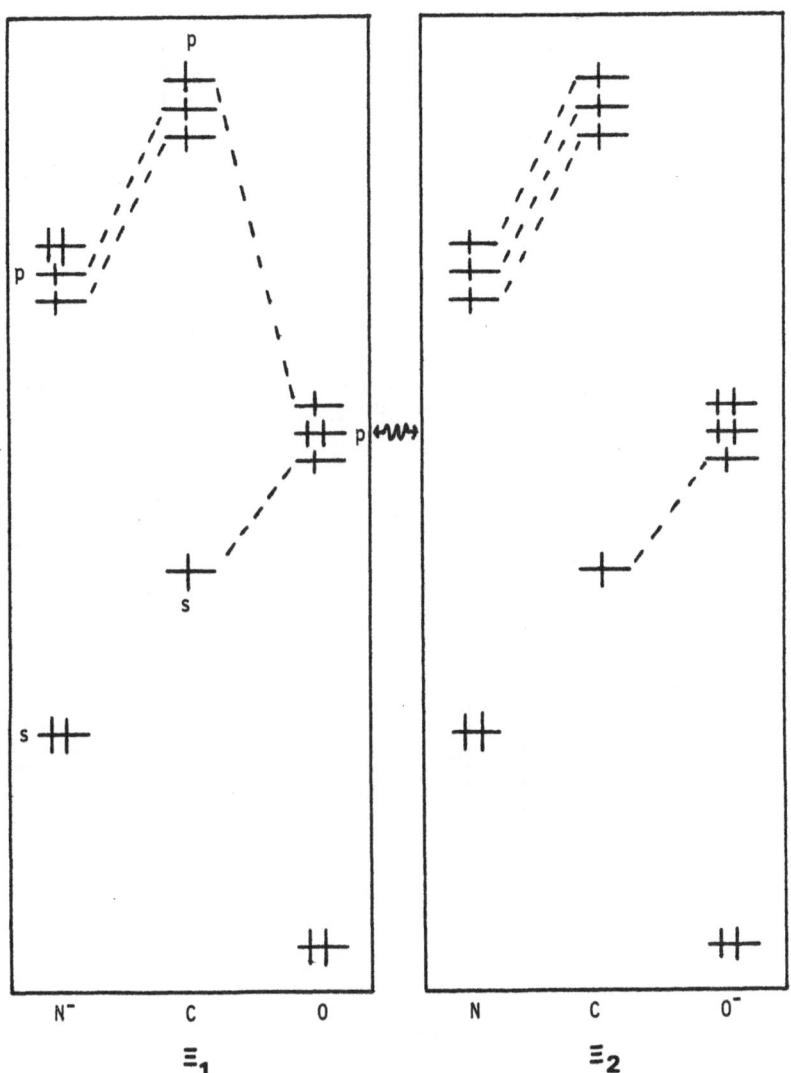

Figure 1

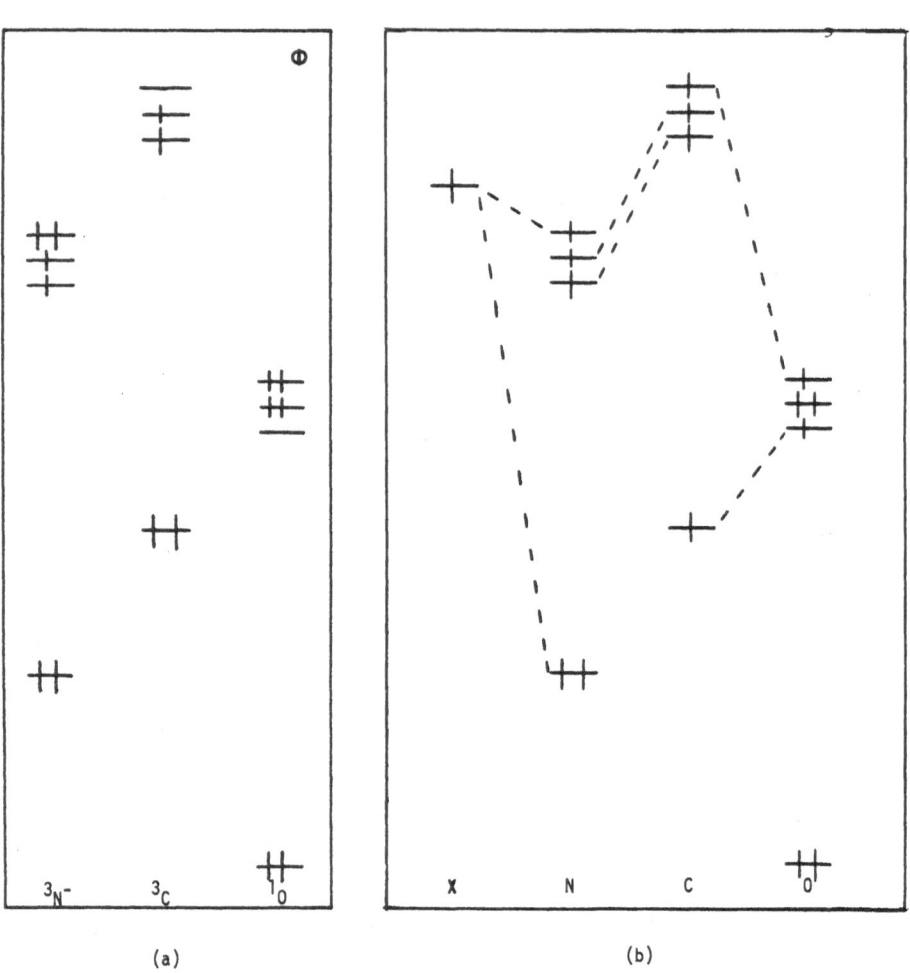

(a)   (b)

Figure 2

overlap binding ability and noting that, in φ, the 2p and 2s electron pairs of the less electronegative N⁻ have higher energy than the 2p and 2s electron pairs of O, we conclude that NCO⁻ will bind X⁺ via the two higher energy lone pairs of N⁻ so that the remaining lone pairs are permitted to occupy the lower energy orbitals of O. The appropriate bond diagram is shown in Figure 2b. We say that X-NCO will be more stable than NCO-X because the former is D- and the latter U-bound. By contrast, we predict that although N is more electronegative than S, U-bound X-NCS will be more stable than D-bound NCS-X because N is a much stronger overlap binder than S (primarily because of the greater overlap binding ability of N2s relative to S3s AO) provided that X is a strong overlap binder (e.g., X = H, alkyl, etc.). Both predictions are in accord with thermochemical data.[1] The novel insight now provided by MOVB theory is that the relative stability of the two X(NCS) isomers will critically depend on the overlap binding ability of X with NCS-X becoming the more stable isomer when X becomes a weak overlap binder since, once formation of strong core-ligand bonds is not possible, core (or, ligand) excitation is "unjustified". With this background, one can now understand the trend defined by the three molecules shown below.[2]

$$H_3C \diagdown \atop N = C = S \qquad {H_3N \diagdown \atop H_3N \diagup} Pt {\diagup SCN \atop \diagdown SCN} \qquad {R_3P \diagdown \atop R_3P \diagup} Pt {\diagup NCS \atop \diagdown NCS}$$

$$\underline{1} \qquad\qquad\qquad \underset{\sim}{2} \qquad\qquad\qquad \underset{\sim}{3}$$

In 1 , the N-isomer is more stable because excitation (i.e., utilization of the lower lying electron pairs of N⁻ in NCS⁻ in making an R-N bond) is more than "paid back" by strong R-N bond formation. In 2, the weak overlap binding ability of Pd (due to utilization of weak overlap binding d orbitals) is the cause

564

for the S-isomer becoming more stable. Finally, replacement of NH_3 by PH_3 in 3,
i.e., replacement of NH_3 by a ligand having low lying unoccupied orbitals, causes
depletion of d orbital electron density from Pt, increases the metal positive
charge and promotes ionic bonding between Pt^{n+} and the negative end of the
$(SCN)^-$ group. This is the simplest illustration of "molecular recognition",
i.e., the way in which the electronic properties of one molecule determine the
position of attachment on a second polydentate molecule.[3]

b. Recently, Schleyer and his collaborators[4] "discovered calculationally a
large number of AX_n molecules (e.g., OLi_4) comprised of first-row elements whose
unusual stoichiometries suggest violations of the octet rule". This rule,
translated into MOVB language, says that AX_n molecules (viewed as core-ligand
composites) are stable when there are as many multicenter bonds as ligand atoms
(or fragments). The compact bond diagrams of D_{4h} and T_d OLi_4 (Figure 3) show
that by utilization of the Li 2s and 2p AO's, one can generate four multicenter
bonds connecting O and Li_4 in both geometries. Hence, there has been no
violation of the octet rule! In a pedological sense this can be conveyed by
the Lewis structure shown below. The unusual thing about it is that it involves

$$\begin{array}{c} Li \\ | \\ Li - O \quad :Li \\ || \\ Li \end{array}$$

violations of the "duet rule", i.e., the prediction that H within a molecule
will tend to look like He through covalent bond formation (e.g., H-H) and that
the same thing will be true of Li through ionic bond formation (e.g., $Li^+ F^-$).
Furthermore, many formalisms have been advanced to account for "hypervalency".
However, most of these models are impotent in making explicit stereochemical

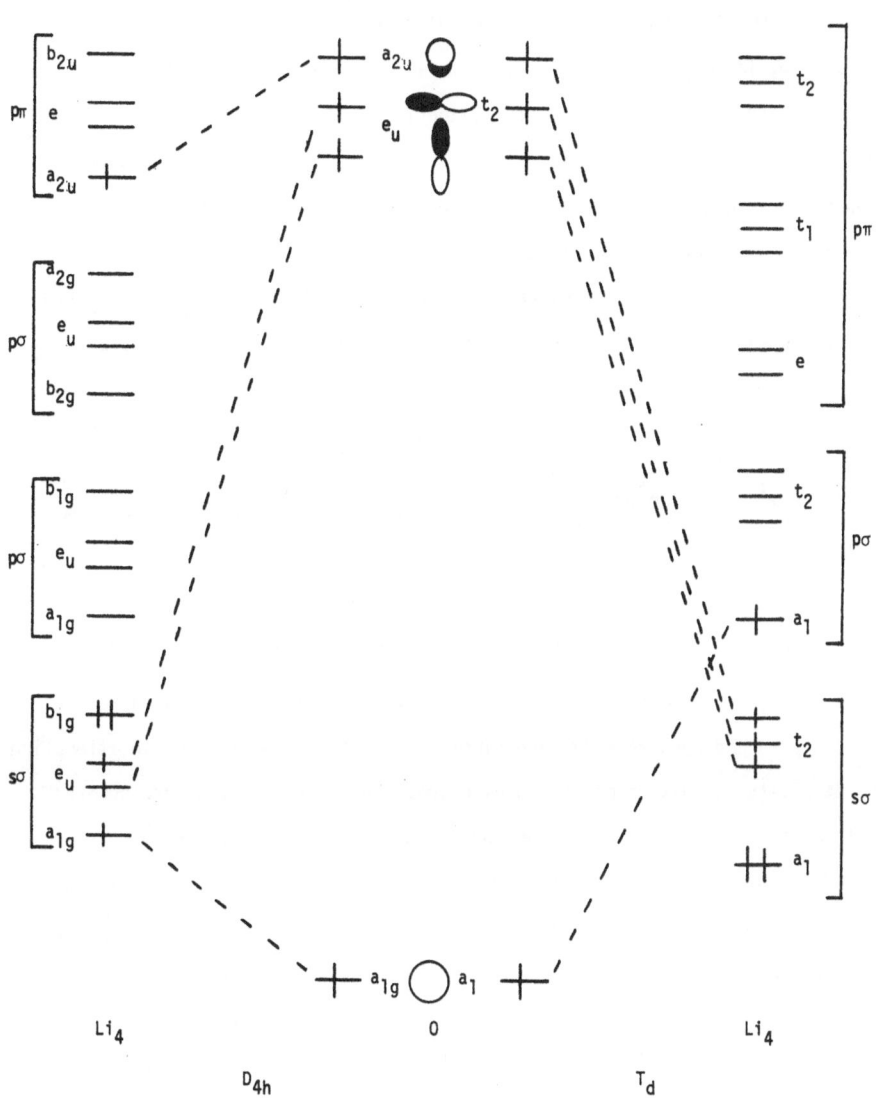

Figure 3

predictions for reasons that the reader can fully understand after reading
Chapter 9. Now, in the case of OLi_4, T_d is more favorable than D_{4h} not only
because the ligand lone pair is more stabilized in this geometry but also
because T_d OLi_4 has <u>four</u> sigma while D_{4h} OLi_4 has <u>three</u> sigma and <u>one</u> pi multicenter bonds. By further construction of bond diagrams, one can easily show that
T_d is the best choice of OLi_4, as found by computations.[4]

The compact bond diagrams of planar and perpendicular "hypervalent" I_2Cl_6 and
those of staggered and bridged "electron deficient" B_2H_6 are shown in Figures 4
and 5.

D_{2h}	D_{2h}	D_{3d}	D_{2h}
Planar	Perpendicular	Staggered	Bridged

One can now forget the imperfect rules of chemistry popular today, disregard the
notions of hyper- and hypo-valency, and simply "read" the bond diagrams in attempting
to infer stability and determine stereoselection. In our example, a mere count of
two-electron bonds and four-electron antibonds tells us that planar is more stable
than perpendicular I_2Cl_6. On the other hand, bridged may be more or less stable
than staggered B_2H_6 depending on the cost of core (B_2) excitation and the payback
of (B_2) - (H_6) multicenter bond formation. Furthermore, MOVB theory teaches us
design: B_2H_6 opts for a bridged structure, but an A_2X_6 isoelectronic species
where x is more electropositive than A and the $2s\sigma$ and $2p\pi_g$ gap of A_2 large,
may well choose the staggered geometry.

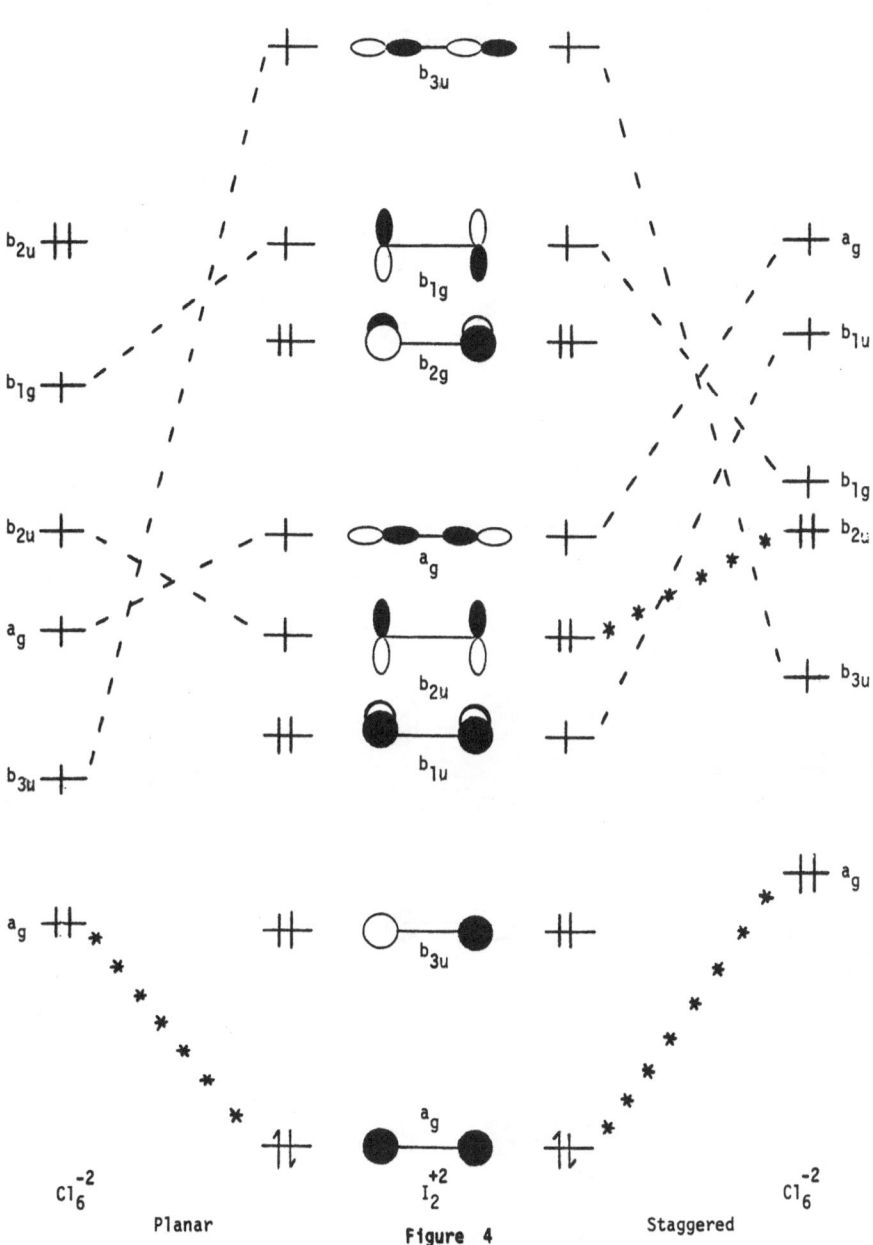

Figure 4

a2u
b3u
eg
b1g
eg
b1g
eu
b2g
b2u
a1g
ag
ag
b1u
b2u
e
b3u
b1u
a2u
a2u
b3u
ag
a1g
a1g
ag
H6+
B2-
B2
H6
D3d
D2h

Figure 5

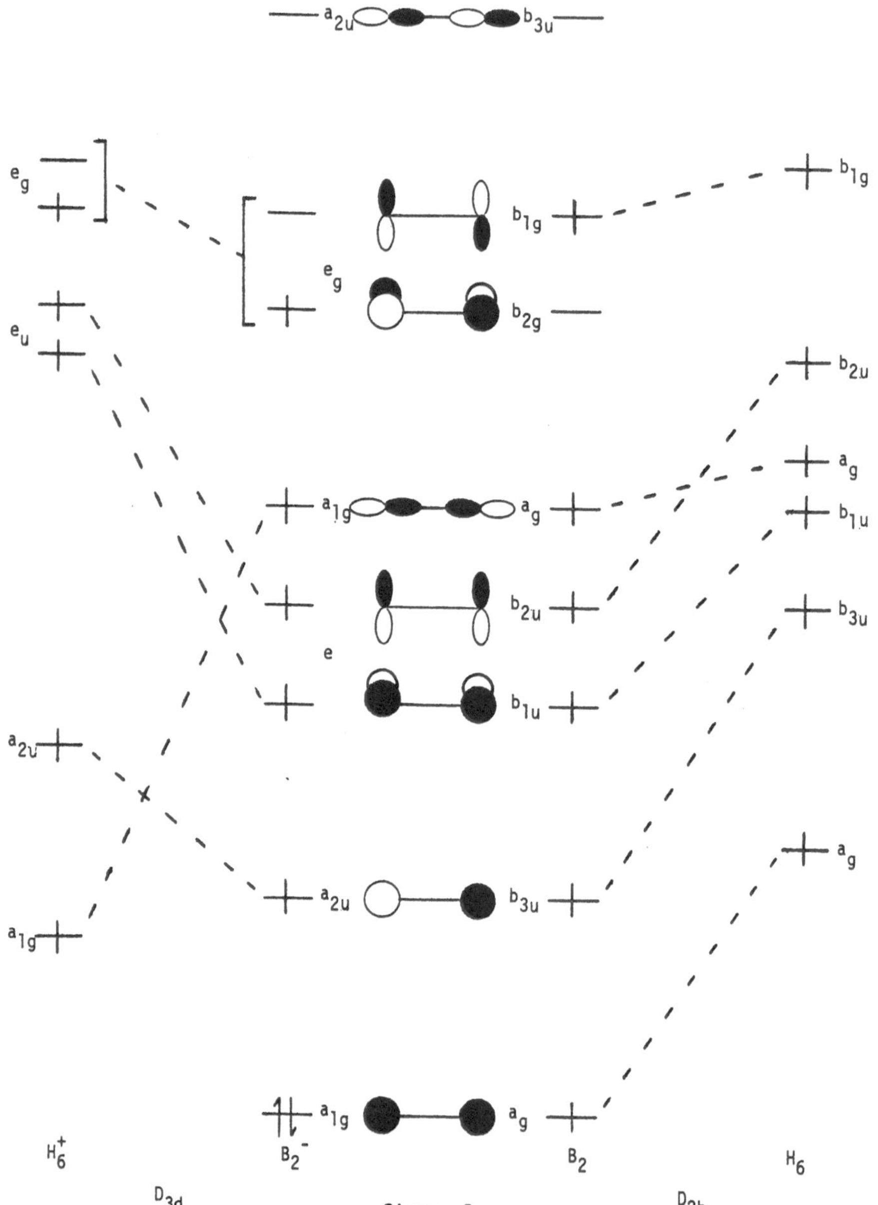

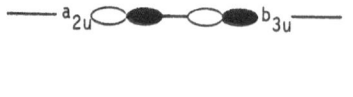

c. In our willingness to apply theory to chemistry, we often forget that
the results of "gas-phase theory" are applicable to "gas-phase chemistry" and not
necessarily to "solution-phase chemistry". This has often created heated contro-
versies, which, when finally resolved, provided good examples of the importance
of "solvent effects".[5] Recently, it has been found that, while S_N2 at carbon
occurs always by inversion of configuration, S_N2 at silicon occurs either by
retention or inversion of configuration depending on the nature of the nucleophile,
leaving group, and solvent.[6] If we assume that D_{3h} AH_3F_2 (A=C, Si) is a good
model of the inversion transition state (or intermediate) and C_s AH_3F_2 a good
model of the retention transition state (or intermediate), we may ask the
question: Is there any reason for C_s to attain lower energy than D_{3h} AH_3F_2, in
the absence of any external perturbations ("solvent effects", etc.)? The answer
can be immediately obtained by writing and interpeting the bond diagrammatic representation for D_3h (Figure 6) and C_s (Figure 7) CH_3F_2 (viewed as C core plus H_3F_2
ligand). In both cases, the large geminal nonbonded overlap of H and F AO's tends
to make $\equiv_1$ the dominant contributor so that C_s has two strikes againt it: It
involves impaired core-ligand spation overlap (e.g., the $2p_y$ - σ_4 overlap integral
in C_s is smaller than in D_{3h}) and it features a more destabilized ligand "lone
pair" (σ_5 has higher energy in C_s than in D_{3h}). Replacement of C by Si will
tend to eliminate only the second effect through abolition of geminal nonbonded
interaction (due to the fact that Si-R are longer than C-R bonds). Hence, we
predict that, while the energetic advantage of D_{3h} over C_s will diminish as C is
replaced by Si[7], it will never cease to exist. That is to say we predict that S_N2
on C or Si in the gas phase should occur with predominant inversion of configuration
but less so in the case of Si. Hence, the predominant retention of configuration
occasionally observed in S_N2 at carbon is dictated by some other environmental
factor such as solvation, counterion participation, etc.

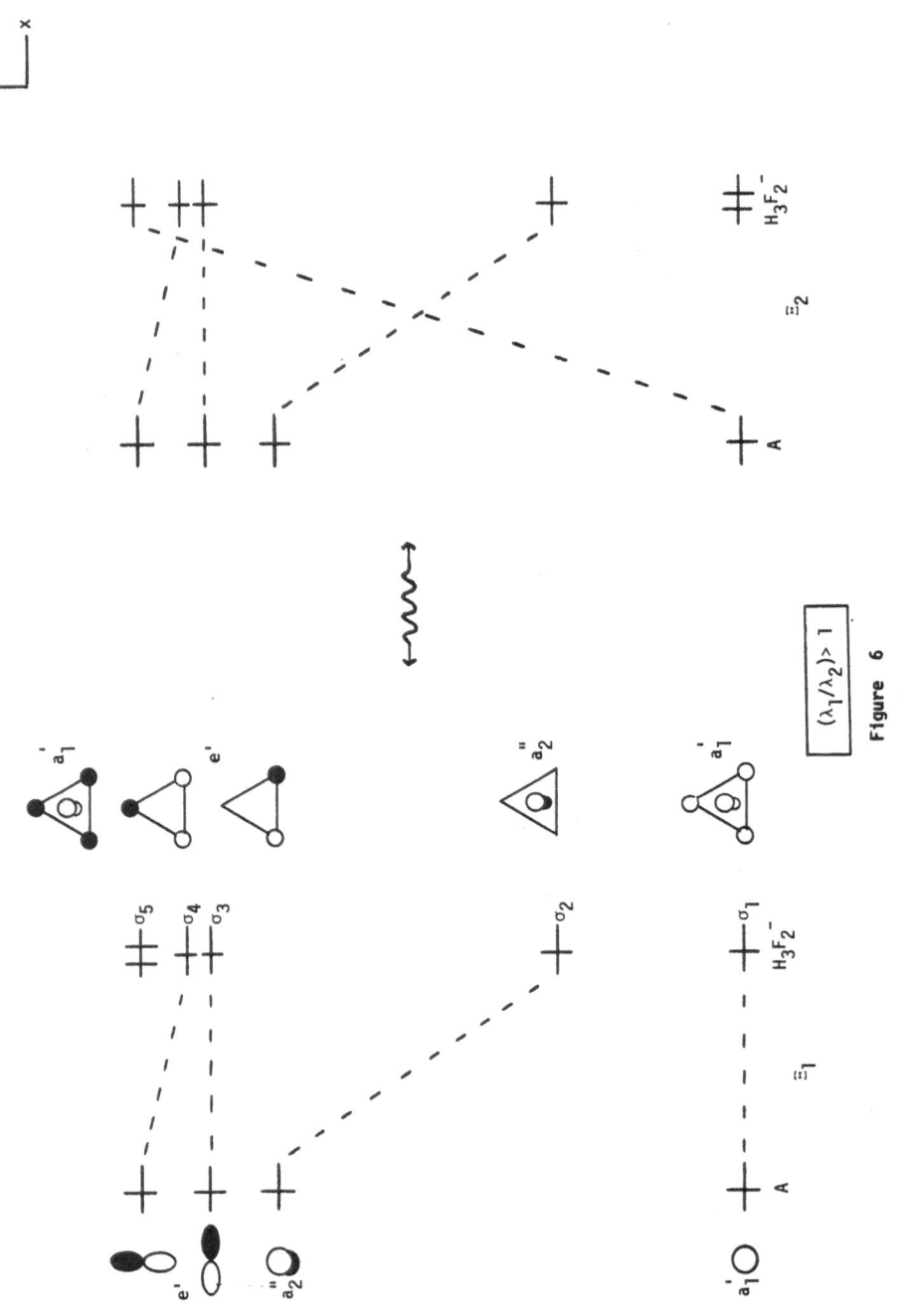

Figure 6

571

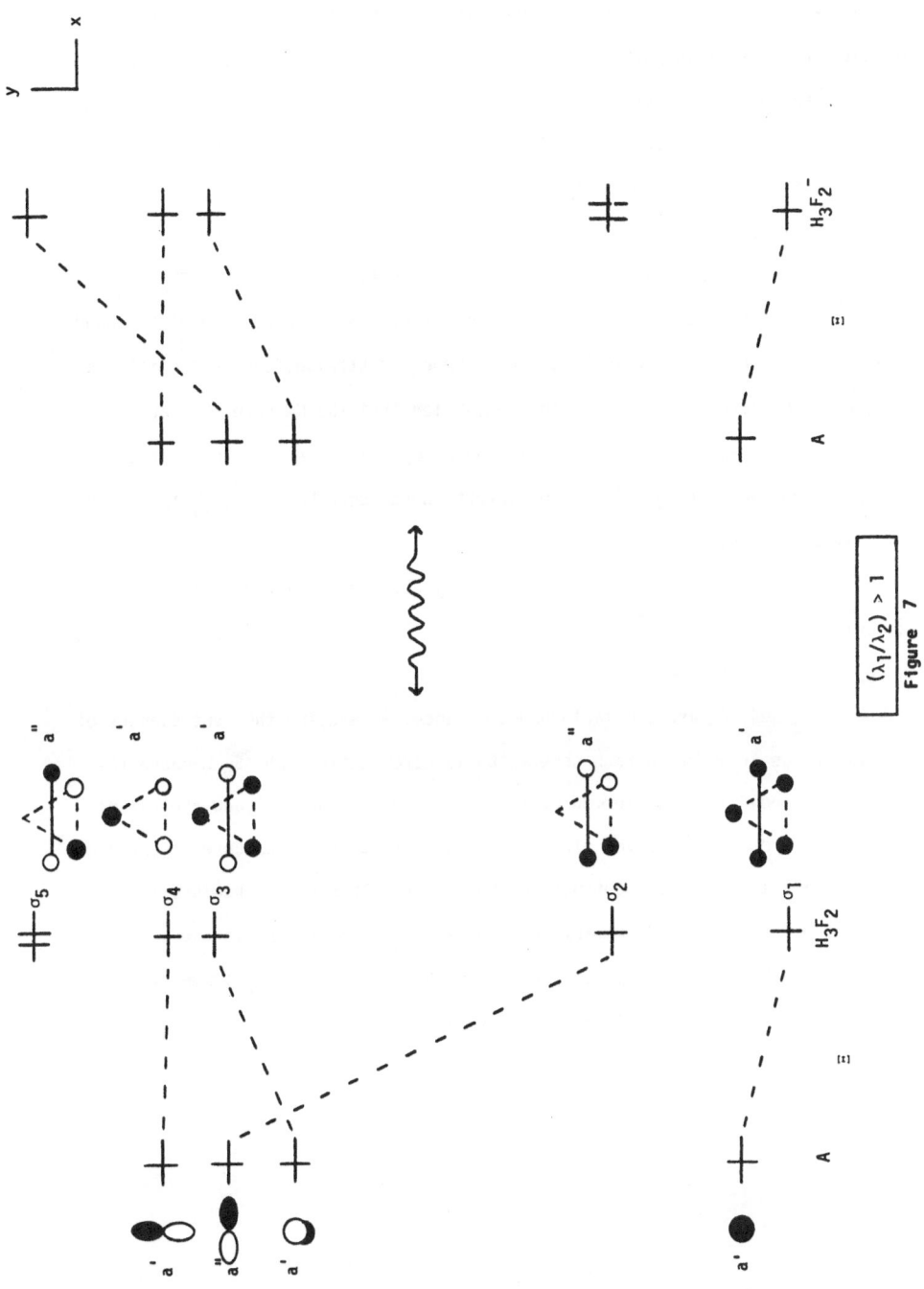

Figure 7

d. While it is true that some concepts developed from studies of ground state molecules are useful in photochemistry,[8] it is well recognized that the electronic structure and properties of excited molecules can only be accounted well by high level SCF-MO-CI type theory. Thus it is interesting to investigate the stereochemistry of such species by MOVB theory. As an illustration, we compare excited $^3\Sigma_u^+$ and excited ^3A and ^3B cis and trans acetylene by routine construction of the corresponding detailed bond diagrams. For simplicity, we substitute $^3\Sigma_u^+$ by a single $^3\pi\pi^*$ configuration of linear acetylene (^3L). Mere inspection of the bond diagrams shown in Figures 8-10 leads to a number of straightforward predictions. These are enunciated below, using the convention that the C_2 core MO's are symbolized by ω_n and those of the H_2 ligand by σ_n with n varying in order of increasing MO energy (e.g., ω_1 is the lowest energy core MO, etc., σ_1 is the in-phase ligand MO, etc.):

1) ^3L is unstable relative to $^3B_2$ simply because bending allows an electron occupying an antibonding π^* MO (ω_7) to find its way (via delocalization to a low energy core (ω_2) or ligand (σ_2) orbital.

2) $^3B_2$ and ^3B_u are similarly bound. Hence, we examine the consequences of primary charge transfer in each case. $^3B_2$ is more stable than ^3B_u because the most "ionic" bond, the one formed by the overlap of ω_7 and a ligand orbital of appropriate symmetry, involves electron transfer that generates core-ligand overlap repulsion in the latter but not in the former. This occurs because the doubly occupied ω_1 has a_g symmetry in trans and a_1 symmetry in cis and the $\omega_7 - \sigma_4$ bond in trans has a_g symmetry but the $\omega_7 - \sigma_2$ bond in cis has b_2 symmetry. The CW's responsible for the differentation are shown below.

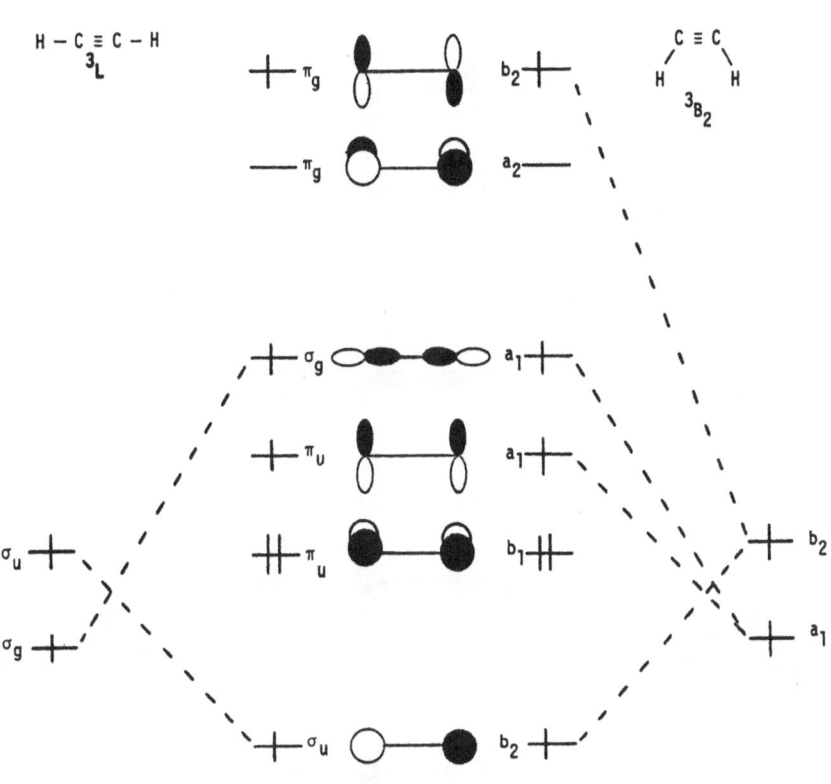

Figure 8

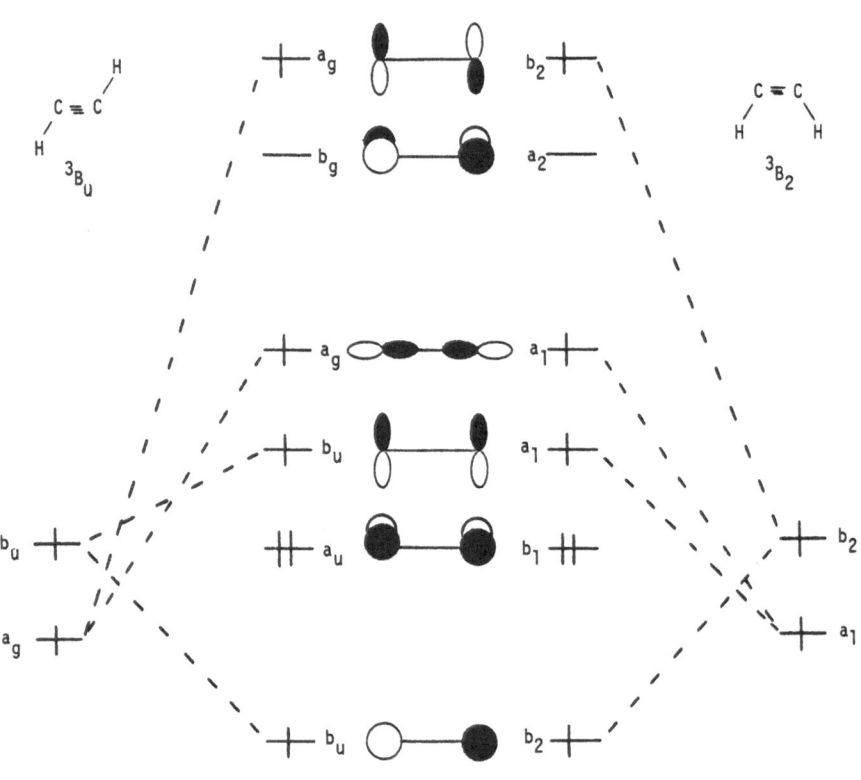

Figure 9

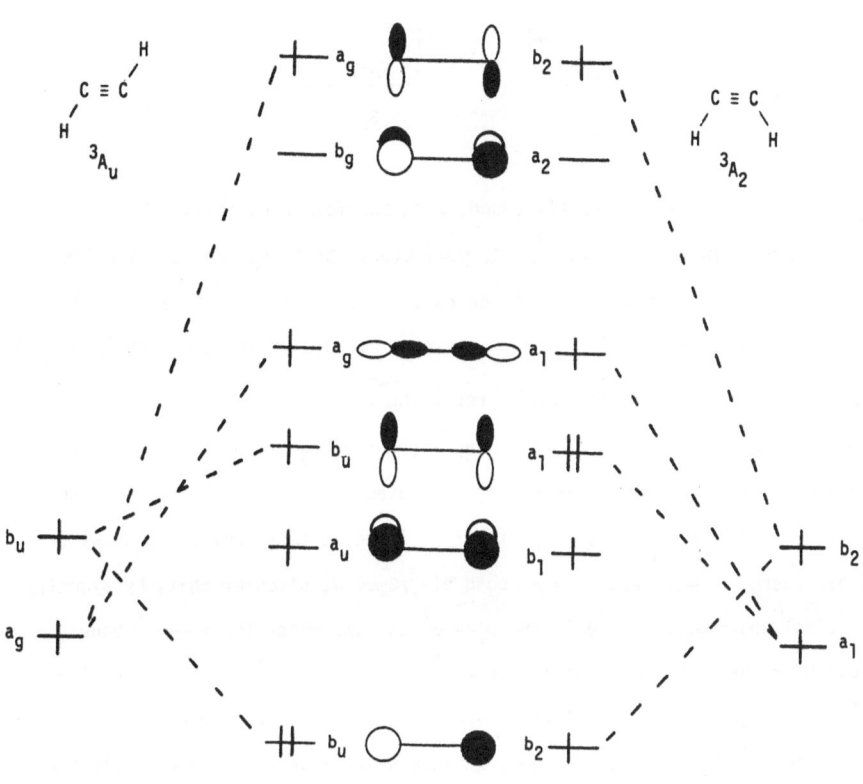

Figure 10

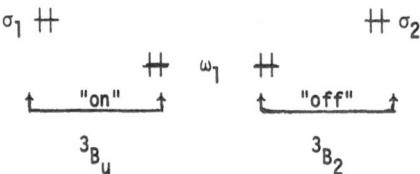

$$- \omega_7 -$$

σ_1 ⧺   ⧺ σ_2

⧺ ω_1 ⧺

↑___"on"___↑   ↑___"off"___↑

^3B_u $^3B_2$

3) ^3A_u and $^3A_2$ are differently bound, with the former exhibiting D- and the latter U-bonding. In other words, ^3A_u is more stable than $^3A_2$ because it allows for core deexcitation, i.e., in the former an electron pair may occupy ω_2 while in the latter it must reside in ω_4 or ω_5, insofar as core orbitals are concerned.

We can make further predictions by recalling that an electronic state is a hybrid of one (or more) principal bond diagram(s) and higher energy bond diagrams. Using MO-theory-inspired terminology, the bond diagrams of Figures 9 and 10 can be labeled $\sigma \rightarrow \sigma^*$ and $\pi \rightarrow \sigma^*$, respectively. If we further proceed to construct the higher energy $\pi \rightarrow \pi^*$ and $\sigma \rightarrow \pi^*$ bond diagrams, we discover that, by symmetry, the principal contributor to ^3B is the $\sigma \rightarrow \sigma^*$ and the minor the $\pi \rightarrow \pi^*$ bond diagram, while the principal contributor to ^3A is the $\pi \rightarrow \sigma^*$ and the minor the $\sigma \rightarrow \pi^*$ bond diagram. Note that the minor contributors "force" electrons into the high lying π^* MO of the core. Now, we have already seen that n even-electron bonds are "better" than n odd-electron bonds at the strong overlap limit. Furthermore, it is safe to assume that optimization of the stronger sigma core-ligand bonds is energetically more important than optimization of the weaker pi core bond. On this basis the ^3B must lie below the ^3A states, as only the former receive (significant) contributions from bond diagrams ($\pi \rightarrow \pi^*$) which permit core-ligand sigma binding exclusively with even-electron bonds.

Finally, recall that two <u>different</u> MO's which span the same AO's cause severe coulomb repulsion when occupied by two singlet coupled electrons. No such effect operates when the two electrons are triplet coupled. For example, $^1\pi\pi^*$ ethylene has 100% "zwitterionic" character while $^3\pi\pi^*$ ethylene has 100% "diradical" character. In the triplet manifold, overlap effects render the ^3B more stable than the ^3A state because coulomb repulsion plays a secondary role in differentiating the two types of states. In the singlet manifold, the situation changes dramatically: The π_z (π_y) and π_z^* (π_y^*) MO's span the same AO's and the ^1B and ^1A states allocate electrons as follows:

1) The ^1B states have one electron in π_y and another in π_y^* in the $\sigma \to \sigma^*$ bond diagram and one electron in π_z and a second to π_z^* in the $\pi \to \pi^*$ bond diagram.

2) The ^1A states have on electron in π_z and another in π_y^* in the $\pi \to \sigma^*$ bond diagram and one electron in π_y and a second in π_z^* in the $\sigma \to \pi^*$ bond diagram. We conclude that interelectronic coulomb repulsion will tend to cause the ^1A states to attain lower energy than the ^1B states.

The analysis presented above is consistent with the results of good quality <u>ab initio</u> SCF-MO-CI calculations.[9] A consistent picture cannot be generated by falling back on concepts founded on low-level theory (e.g., Walsh diagrams, etc.).[10]

e. Myoglobin, hemoglobin, and synthetic models thereof feature a "gathering" of Fe(II), porphine dianion (P), an axial base (B:), and O_2 in the geometry shown below and with O_2 attached to the metal in an "angular" (I) rather than "triangular" (II) manner.[11]

B   I   II

Why is this so? Construction of compact bond diagrams for the staggered and eclipsed forms of I and II reveals that, in staggered I, one may form two metal-oxygen bonds by coupling ground triplet O_2 to ground triplet Fe(II)PB into an overall singlet and allowing for the appropriate delocalization as indicated by the bond diagram of Figure 11. By contrast, formation of two metal-oxygen bonds with concommitant minimization of the number of four-electron antibonds in staggered II can only be achieved by coupling a $\pi \longrightarrow \pi^*$ excited triplet O_2 to ground triplet Fe(II)PB into an overall singlet. The proper bond diagram is shown in Figure 11.[12] Because of the large energy gap separating π and π^* in O_2, I is predicted to be more stable than II. It is important to note a probable cancellation effect: Because of multicenter orbital overlap, the two metal oxygen bonds in II are stronger than in I. On the other hand, I has two very weak four-electron antibonds (due to the overlap of π_z with $d_{x^2-y^2}$ and d_{xz}) while II has one very strong four-electron antibond (due to the overlap of π_y and d_{yz}).

I now use MOVB theory to "build bridges" between fundamental "small molecule" chemistry and the problem at hand. We have seen that linear H_2O involves excited O making strong bonds with H_2 while, by contrast, bent H_2O involves ground O making weaker bonds with H_2. Similarly, cis or trans planar HOOH involves excited O_2 making strong bonds with H_2 while, by contrast, gauche HOOH or 1,1 H_2OO involve ground O_2 making weaker bonds with H_2. "Triangular" II is then analogous to linear H_2O and cis HOOH and "angular" I is analogous to bent H_2O and 1,1 H_2OO (which suffers from no nonbonded ligand repulsion). Hence, C_{2v} H_2O, C_2 H_2O_2, and "angular" I are equilibrium structures simply because core excitation is too prohibitive. How can we bias all these systems towards the alternative geometry? We have already seen that as X in X_2O and H_2O_2 becomes more electropositive rehybridization occurs favoring the isomer involving core excitation. Thus, $(CH_3)_2O$ has a

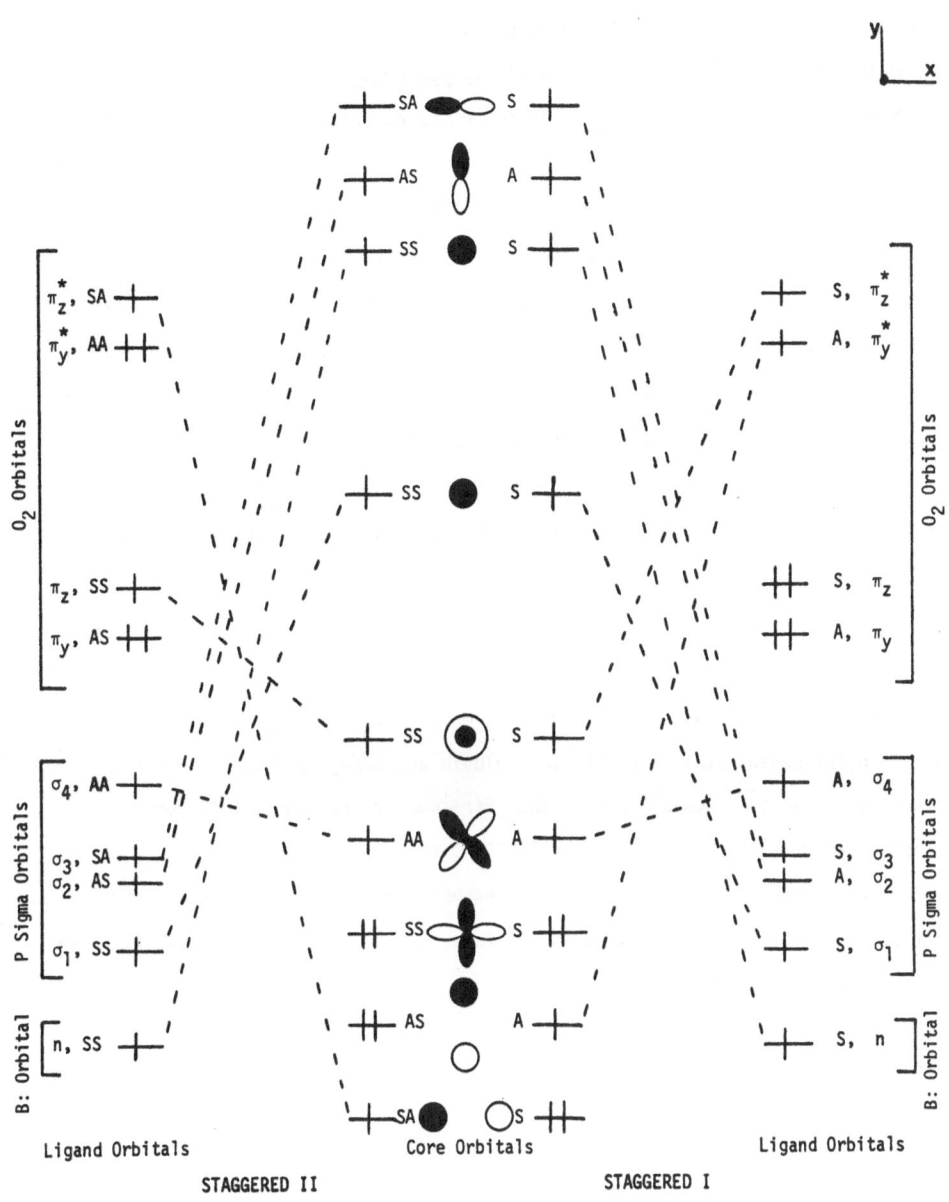

Figure 11

larger angle than F_2O and H_2O_2 has a larger dihedral angle than F_2O_2. Accordingly, we predict that as d^6 M(II) electronegativity <u>decreases</u>, "triangular" coordination will become increasingly favorable. There are no experimental results yet available to check this predicition.

The situation changes dramatically when the central metal atom has fewer d electron pairs and more d holes which can combine with doubly occupied O_2 MO's to define additional metal-oxygen bonds. Indeed, it is found that the maximum number of metal-oxygen bonds in Ti(II)PBO$_2$ are now generated in the <u>eclipsed</u> "angular" and "triangular" forms. Since there are no longer four-electron antibonds, since the two forms have the same number of metal-oxygen bonds, and since multicenter orbital overlap renders the bonds in the "triangular" stronger than those in the "angular" form, we predict a reversal of the stereochemical preference encountered in the case of Fe(II)PBO$_2$. Now, the "triangular" is expected to be more stable than the "angular" eclipsed form. The corresponding bond diagrams are shown in Figure 12. This prediction is consistent with the available experimental results.[13]

This discussion constitutes a complete answer to the oft-posed question: Why is O_2 singlet-coordinated in Fe(II) in myglobin and hemoglobin and why does it attach itself in an "angular" rather than "triangular" fashion? This question cannot be answered by monodeterminantal SCF-MO theory which predicts that O_2 in I is triplet rather than singlet.[14] The reason: Such a brand of theory cannot properly describe weak two-electron bonds (singlet I) while it does well with one- and three-electrons bonds (triplet I).[15]

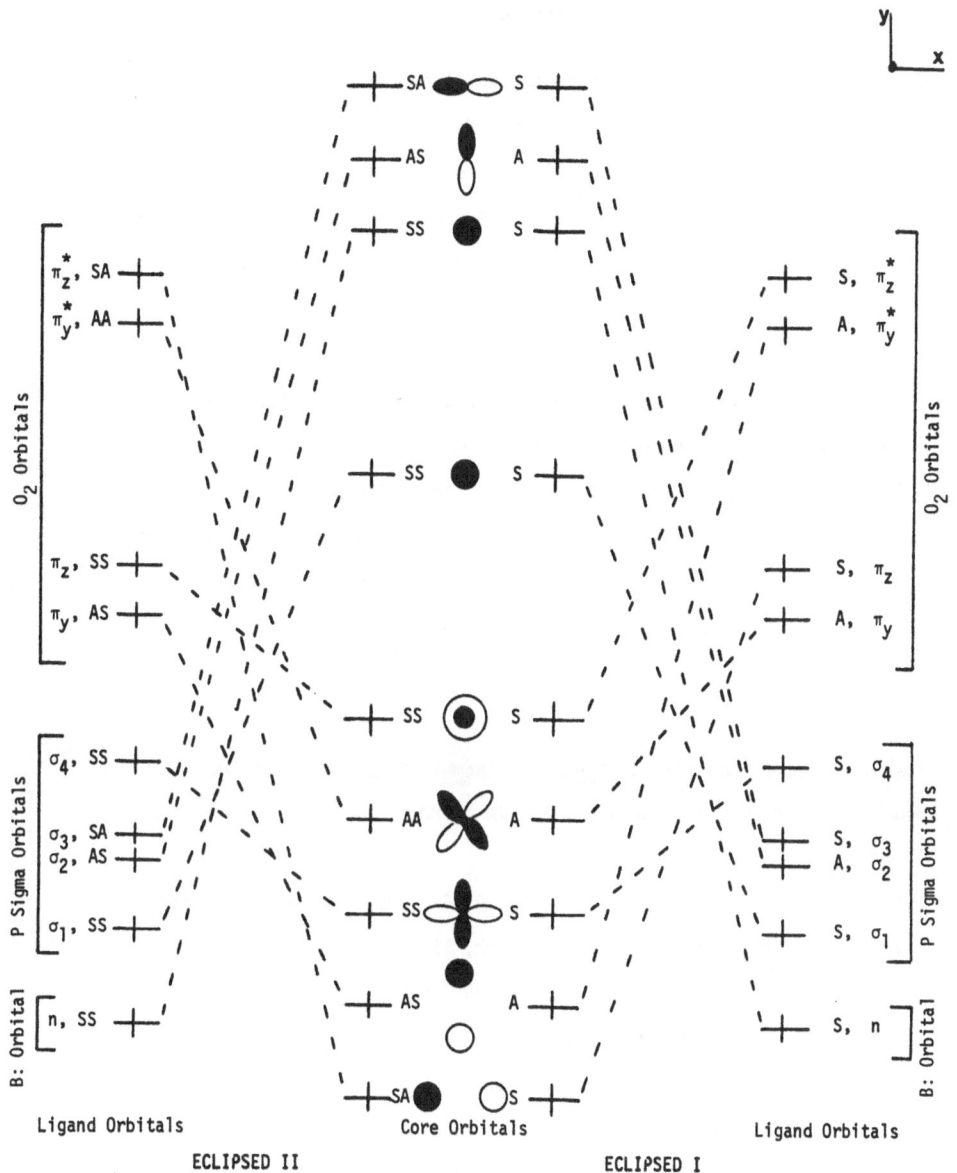

Figure 12

References

1. Shaw, R. in "The Chemistry of Cyanates and their Thio Derivatives", Part 1, Patai, S., Ed.; John Wiley and Sons: New York, 1977.

2. The trend defined by 2 and 3 was first noted by: Turco, A.; Pecile, C *Nature* 1961, 191, 66. For review, see: Burmeister, J. L. *Coord. Chem. Rev.* 1968, 3, 225. The "pi bonding competition" hypothesis advanced to explain this trend is not valid. This can be easily shown by construction and comparison of appropriate bond diagrams.

3. Ambident selectivity in the weak binding regime has been correctly analyzed in an MO frame by: Klopman, G. *J. Am. Chem. Soc.* 1968, 90, 223. See also: Hudson, R. F. *Angew. Chem., Int. Ed. Engl.* 1973, 12, 36.

4. Schleyer, P. von R.; Würthwein, E-U.; Pople, J. A. *J. Am. Chem. Soc.* 1982, 104, 5839.

5. For a discussion of one such controversy, see: Epiotis, N. D.; Cherry, W. R.; Shaik, S.; Yates, R. L.; Bernardi, F. *Topics Curr. Chem.* 1977, 70, 1 (and especially p. 160).

6. Corriu, R. J. P.; Guerin, C. *Advan. Organomet. Chem.* 1982, 20, 265.

7. This conclusion has also been reached in an MO frame: Anh, N. T.; Minot, C. *J. Am. Chem. Soc.* 1980, 102, 103.

8. For example, see Chapter 6 in Pearson, R. G. "Symmetry Rules for Chemical Reactions"; John Wiley and Sons: New York, 1976.

9. (a) Demoulin, D. *Chem. Phys.* 1975, 11, 329. This paper first reported calculations, which show that ^3B lie below ^3A but ^1A lie below ^1B states!

    (b) Schaefer, III, H. F.; Wetmore, R. W. *J. Chem. Phys.* 1978, 69, 1648.

    (c) Janoschek, R.; Winkelhofer, G.; Fratev, F. presented in part at the IUPAC Symposium on Theoretical Organic Chemistry, Dubrovnik, August 1982. I thank Professor Janoschek for a copy of the manuscript prior to publication.

10. The (expected) impotence of Walsh diagrams (see, inter alia: Walsh, A. D. J. Chem. Soc. 1953, 2288) to deal with many facets of the problems discussed in this section has been pointed out in ref. 9b and 9c.

11. Collman, J. P. Accounts Chem. Res. 1977, 10, 265.

12. Construction of bond diagrams reveal that the following species are disfavored, insofar as bonds are concerned, relative to the staggered II because:

    1) Eclipsed II with excited triplet O_2 has one pi and one sigma Fe(II) - O_2 bonds and one pi and one weak delta Fe(II) - O_2 antibonds.

    2) Eclipsed II with ground triplet O_2 has one pi and one weak delta Fe(II) - O_2 bonds and a sigma and a pi Fe(II) - O_2 antibonds.

    3) Staggered II with ground triplet O_2 has only one pi Fe(II) - O_2 bond (if all Fe(II) - P sigma bonds are to be preserved) and three Fe(II) - O_2 antibonds (two pi and one sigma).

    The reader will find checking these results an excellent practice in MOVB theory.

13. (a) Guilard, R.; Fontesse, M; Fournari, P.; Lecomte, C.; Protas, J. Chem. Commun. 1976, 161.

    (b) Guilard, R.; Latour, J.-M.; Lecomte, C.; Marchon, J.-C.; Protas, J.; Ripoll, D.; Inorg. Chem. 1978, 17, 1228.

14. Dedieu, A.; Rohmer, M.-M.; Veillard, H.; Veillard, A. Nouveau J. Chem. 1979, 3, 653. This paper contains also an excellent review of the experimental facts and the theoretical computations targeted to them.

15. Triplet II is obtained by relocating an electron from π_z^* to d_{yz}. Thus, two 2-electron are replaced by one 1-electron and one 3-electron bonds.

These are corrections to be applied to the original monograph "Unified Valence Bond Theory of Electronic Structure":

a. Table 8 on page 128 is followed by the page 128A shown below which is followed by page 129. Table 8 on page 130 should be deleted.

128A

The results of the HVB computations which have been carried out in the manner prescribed above are shown in Figure 11. As the HL CW's do not interact, no differentiation between a Möbius and a Hückel AO system can be made on this basis. Thus, in the first stage, we compute the eigenstates by including the HL and all unique singly "ionic" CW's. In the second stage, we compute the eigenstates by interacting all elementary CW's. Finally, the correct HVB eigenstates are computed by diagonalizing the energy matrix over the entire CW basis set. The final wavefunctions of the ground states of the Möbius and Hückel AO systems are given in Table 8. It is immediately apparent that the lower energy of the ground Möbius AO system relative to the ground Hückel AO system is due to the fact that the doubly ionic and the extrinsic CW's can couple efficiently the elementary structures only in the former case.

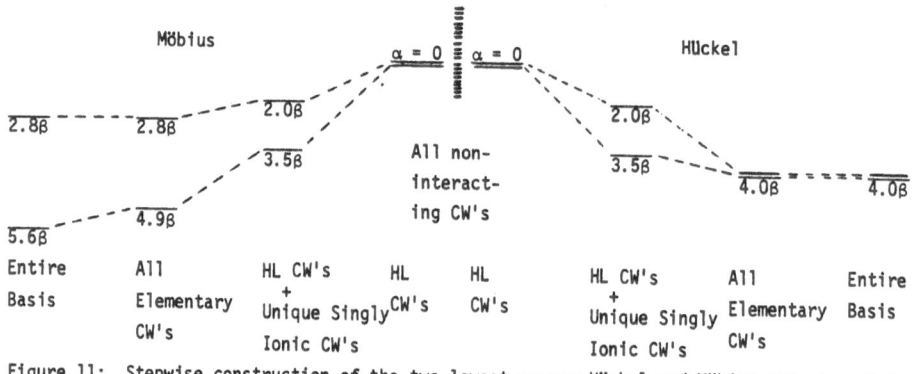

Figure 11: Stepwise construction of the two lowest energy Hückel and Möbius HVB eigenstates.

b. Equation (54) on p. 249 should read: $\Delta\varepsilon = \varepsilon(\omega_1) - \varepsilon(\omega_2)$.

c. The top line on p. 251 should end: ... "D" and "U", $\Delta\varepsilon'$ is such so that,

....

d. On p. 266, item 2, "This ... multiplity" should be replaced by: "This is the open shell CW which places core and ligand electron pairs in the lowest energy orbitals subject to the requirement that it generates the maximum number of core-ligand bonds through spin pairing".

e. Dashed lines connecting σ_2' with ω_7 and σ_1'' with ω_7 should be added to Figure 28 on p. 287.

f. On p. 293, item c "the same" should be changed to "a different" and "different labels" to "the same label.".

g. The HL CW's used in the computations are linear combinations of the Kekule structures.

N. D. Epiotis

Unified Valence Bond Theory of Electronic Structure

With collaboration of J. R. Larson, H. L. Eaton

1982. VIII, 305 pages
(Lecture Notes in Chemistry, Volume 29)
ISBN 3-540-11491-2

Contents: Qualitative Valence Bond Theory of Model Systems. – Qualitative Molecular Orbital-Valence Bond Theory.

Qualitative monodeterminantal MO theory, though undeniably stimulating, suffers from lack of conceptual clarity. After some years of preoccupation with this theory, the authors abandoned it. In its place, they have sought to develop a new way of thinking about molecular electronic structure which exploits the positive aspects of MO and VB theories and which eliminates conceptual interdisciplinary barriers. The new approach is described in this work, starting with elementary VB notions and culminating with the presentation of the compact MOVB method a "back of the envelope" theory which operates near the same level as "state of the art" quantum chemical computations. Though much of the space is devoted to the presentation of the theory, a number of chemically significant illustrative applications are provided with the aim of making the reader fully qualified to use and extend the approach in any field of personal interest. The ultimate goal is to establish a new language of chemistry which can eventually lead to a higher level of understanding of why and how molecules exist and react and pave the way to predictions which lie outside the intuitive range of chemists and physicists.

Springer-Verlag
Berlin
Heidelberg
New York
Tokyo

N. D. Epiotis

Theory of Organic Reactions

1978. 69 figures, 47 tables. XIV, 290 pages
(Reactivity and Structure, Volume 5)
ISBN 3-540-08551-3

Contents: One-determinental theory of chemical reactivity. - Configuration interaction overview of chemical reactivity. - The dynamic linear combination of fragment configurations method. - Even-even intermolecular multicentric reactions. - The problem of correlation imposed barriers. - Reactivity trends of thermal cycloadditions. - Reactivity trends of singlet photochemical cycloadditions. - Miscellaneous intermolecular multicentric reactions. - $\pi + \sigma$ addition reactions. - Even-odd multicentric intermolecular reactions. - Potential energy surfaces for odd-odd multicentric intermolecular reactions. - Even-even intermolecular bicentric reactions. - Even-odd intermolecular bicentric reactions. - Odd-Odd intermolecular bicentric reactions. Potential energy surfaces for geometric isomerization and radical combination. - Odd-odd intramolecular multicentric reactions. - Even-even intramolecular multicentric reactions. - Mechanisms of electrocyclic reactions. - Triplet reactivity. - Photophysical processes. - The importance of low lying non-valence orbitals. - Divertissements. - A contrast of "accepted" concepts of organic reactivity and the present work.

Structural Theory of Organic Chemistry

By **N. D. Epiotis, W. R. Cherry, S. Shaik, R. L. Yates, F. Bernardi**
1977. 60 figures, 58 tables. VIII, 250 pages
(Topics in Current Chemistry, Volume 70)
ISBN 3-540-08099-6

Contents: Theory. - Nonbonded Interactions. - Geminal Interactions. - Conjugative Interactions. - Bond Ionicity Effects.

This work constitutes the first attempt to develop a new overview of structural chemistry based on quantum theory. A combination of qualitative molecular orbital theory and explicit quantum mechanical computations leads to the identification of the key factors controlling the shape of a molecule. On this basis, the experimentalist can anticipate structural trends for simple as well as complex molecules. In all cases, the theoretical principles are illustrated by reference to available experimental data. (420 references)

Springer-Verlag
Berlin
Heidelberg
New York
Tokyo

Lecture Notes in Chemistry